Developments in Mathematics

Volume 54

More information about this series at http://www.springer.com/series/5834

Patrik Eklund · Javier Gutiérrez García
Ulrich Höhle · Jari Kortelainen

Semigroups in Complete Lattices

Quantales, Modules and Related Topics

Springer

Patrik Eklund
Department of Computing Science
Umeå University
Umeå
Sweden

Ulrich Höhle
School of Mathematics
and Natural Sciences
University of Wuppertal
Wuppertal
Germany

Javier Gutiérrez García
Department of Mathematics
University of the Basque Country UPV/EHU
Bilbao
Spain

Jari Kortelainen
Department of Information Technology
South-Eastern Finland University
of Applied Sciences
Mikkeli
Finland

ISSN 1389-2177 ISSN 2197-795X (electronic)
Developments in Mathematics
ISBN 978-3-030-07687-0 ISBN 978-3-319-78948-4 (eBook)
https://doi.org/10.1007/978-3-319-78948-4

Mathematics Subject Classification (2010): 06F07, 18C20, 18D15, 46L05

Printed on acid-free paper

This Springer imprint is published by the registered company Springer International Publishing AG part of Springer Nature
The registered company address is: Gewerbestrasse 11, 6330 Cham, Switzerland

To Aija, Ana, Felix, Heidi-Tuulia and Kesia

Foreword

This book is about quantales, lattice-ordered semigroups structured in such a way that they generalize the locales of point-free topology and various multiplicative lattices of ideals from the theories of rings and C^*-algebras. The formal notion of a quantale—a complete lattice equipped with a semigroup multiplication which distributes over arbitrary joins—arose in the 1980s, and the subject has flourished and diversified in content and range of mathematical connections since that beginning, with developments appearing in a diversity of journals.

The most notable and distinctive feature of this monograph is a shift from regarding quantales as the principal objects of study to a categorical emphasis that encompasses the category Sup (morphisms preserve all suprema) of complete lattices, the category of quantales with Sup-semigroup morphisms, other related categories naturally connected with quantales, and a variety of categorical constructions that enhance quantale theory. Categorical properties of Sup such as existence of tensor products and the fact it is symmetric and monoidal closed led the authors to a whole new categorical perspective and approach of applying enriched category to the study of quantales. The authors have pioneered work along these lines in recent years, and now for the first time seek to bring it together in one organized, comprehensive, and mostly self-contained source.

An attractive additional feature of the text is the effort to make some connections and applications to other areas. One is the introduction and study of the spectrum of a C^*-algebra as a quotient of the tensor product of the quantales of closed left and right ideals and the use of quantale modules of this quantale to study algebra representations. Another rather vast direction of study arises by replacing standard objects such as $\{0, 1\}$ as the range for characteristic functions for subsets of a set or the positive reals for metric spaces with quantales, so that one gets many-valued versions of sets, or metric spaces, or other standard objects. Thus, the authors seek to show how the foundations of a theory of many-valuedness can be based on quantale theory. The authors are to be commended for collecting and putting together in an attractive and accessible manner the necessary categorical background, the systematic, organized presentation from their perspective of quantale theory, and for the interesting applications to C^*-algebra theory and to many-valued structures.

This book can serve a wide audience from being an introduction to quantale theory to providing a valuable source of reference.

Baton Rouge, Louisiana
September 2017

Jimmie Lawson

Preface

Since the mid-1980s, quantales have found an increasing interest in various areas of mathematics. This development has its origin in ideas coming from non-commutative geometry and is driven by the desire to create a non-commutative and non-idempotent set theory known as the theory of quantale sets. Since that time, an extensive literature has grown and is scattered over a diversity of journals. Furthermore, and as algebra at the same time became connected with developments using category theory, dedicated workshops and discussion fora emerged to focus on enriched category theory in further support of studying quantales. The 33rd Linz Seminar 2012 devoted to enriched category theory, and organized by U. Höhle and E.P. Klement, was typical in that respect. In that particular workshop, the general feeling emerged that there exists a necessity to have a new comprehensive and coherent representation of the field. This insight became the starting point of the cooperation between the authors leading to the present book.

The foundations of many-valuedness, as enabling non-commutativity for its related operators, are another thematic and applicative aspect overlaying the formal algebraic treatment in this book. Indeed, the book aims to provide a treatment of semigroups in complete lattices with a special focus on quantales and modules and offers a new way to approach many-valuedness in mathematics—in particular as it is related to algebraic, logical, and topological methodology.

We give an introduction to the theory of quantales with applications to C^*-algebras and many-valued order theory from the perspective of the category of complete lattices and join-preserving maps, which also includes a brief sketch on automata in this context. The book aims to be self-contained. However, some basic knowledge in category theory is desirable, and similarly knowledge in functional analysis is helpful with respect to applications.

This book does not exclude the possibility of being used as a textbook. Therefore, every section is accompanied by a subsection of exercises ranging in difficulty from a trivial to a rather advanced level. The more advanced exercises are supplied with various hints which can be considered as an extension of the previous results. We hope that the book forms a useful basis for further research in enriched category theory, non-classical logic, many-valued and non-commutative topology, and computer science.

During the preparation of the book, we received support from various people. Explicitly we would like to mention the passionate commitment of T. Kubiak, who collected a large variety of papers on quantales, was involved in the discussion on how to organize such a book, and made most valuable suggestions during its preparation. We also acknowledge the stimulating advice we received from M. Barr and J. Lawson at advanced stages of its development. In addition, and in fact over the past few years, from idea generation to finalizing the book, several colleagues and collaborators have been supportive and provided encouragement on various occasions.

This book would not have seen daylight without the support of our working environments and people in our vicinity. Our working environments were diverse but always supportive and flexible, enabling all of us to use required time for the production of the book. Some directed financial support was enabled by the Spanish Government provided by MTM2015-63608-P (MINECO/FEDER), and further indirect financial support was available through various parallel projects. We also express our gratitude to Springer-Verlag, especially to Dr. Remi Lodh for his stimulating advice and efficient assistance during the preparation of the final manuscript.

Last but certainly not least, we acknowledge the invaluable support we received from our families and friends.

Umeå, Sweden — Patrik Eklund
Bilbao, Spain — Javier Gutiérrez García
Wuppertal, Germany — Ulrich Höhle
Mikkeli, Finland — Jari Kortelainen
December 2017

Contents

Symbols

Categories

Introduction

Quantales appear in various areas of mathematics—in quantaloid-enriched category theory, in non-classical logics as completions of the Lindenbaum algebra, and in different representations of the spectrum of C^*-algebras as many-valued and non-commutative topologies.

The theory of quantales has a long history, and its beginning dates back to the 1920s. Even though the terminology goes back to C. J. Mulvey when he proposed a quantization of the theory of locales (cf. [58]) at the Oberwolfach Category Meeting, 1983 (cf. [82]), it was W. Krull in 1924 and 1928 who formulated the axioms of ideal lattices as those axioms of a two-sided quantale (cf. [68, 69]). In this context, we have of course anticipated contemporary terminology. Moreover, it is interesting to see that W. Krull also formulated the axioms of prime elements of two-sided quantales and noticed the existence of left residuals and right residuals. The first simple properties of residuals had already been collected in [69], whose origin can even be traced back to some formulas derived by R. Dedekind for modules in the commutative setting (cf. [22]).

The continuation of this development took place in the 1930s. In 1939, M. Ward and R. P. Dilworth gave a detailed account of the residuation theory associated with quantales (cf. [24, 109]). A first overview of this theory appeared in G. Birkhoff's famous book on lattice theory, which ran through three editions in 1940, 1948, 1967 (cf. [14]). Here, G. Birkhoff used the language of completely lattice-ordered (groupoids) semigroups with zero for the description of (pre)quantales. For a long time, this book was the standard reference in this area until 1990 when K. I. Rosenthal's book on quantales appeared (cf. [101]).

These developments are characterized by an object-based presentation—this means that the question of which type of morphisms is appropriate for the theory of quantales did not find a satisfactory solution. It is therefore not surprising that colimit constructions of quantales are not treated in this type of literature, because the standard approach in this context would first consist in the construction of free objects and subsequently in a quotient construction in order to fulfil "certain equations". This problem can be brought to a solution, if we exhaust all possibilities provided by the tensor product of complete lattices.

The tensor product goes back to the Ph.D. thesis by D. G. Mowat (1968) and is accessible to a larger audience in the papers by Z. Shmuely (1974) and B. Banaschewski and E. Nelson (1976) (cf. [6, 106]). In this context, it is important to see how elements of the tensor product appear as Galois connections between complete lattices—a concept which had already been covered in the first edition of Birkhoff's book on lattice theory.

The existence of the tensor product together with the important fact that the category of complete lattices and join-preserving maps is symmetric and monoidal closed opens a *completely different view* on the *theory of quantales* than the previous perspective from the point of view of residuation. The category `Sup` of complete lattices and join-preserving maps is star-autonomous (cf. [8]) and is the Eilenberg–Moore category of the monad of downclosed sets on the category of preordered sets. Secondly, (pre)quantales are (magmas) semigroups in `Sup`, and the formation of right- (left-)residuals determines a right (left) action on the dual lattice of the underlying quantale. Hence, module theory on unital quantales turned out to be a natural field of applications for quantales.

This second approach to the theory of quantales was initiated by A. Joyal and M. Tierney in their booklet *An Extension of the Galois Theory of Grothendieck* 1984 (cf. [59]). Here, they observed the interesting fact that frames (which occur as complete dual Brouwerian lattices in [14]) are special commutative monoids in `Sup` (cf. Proposition 1 on p. 21 in [59]).

Furthermore, the existence of free quantales is an immediate corollary of the general theorem that in every cocomplete, symmetric, and monoidal closed category free semigroups exist. Based on this observation, the cocompleteness of the category of semigroups in `Sup` is inherited by `Sup`.

The aim of this book is to tie up these branches of different developments and to harmonize and reshape the theory of quantales as a significant algebraic structure in complete lattices. We describe a coherent theory which also includes the monadic basis of quantales expressed by the composition of the term monad with the monad of downclosed sets. Since we have tried to keep the book more or less self-contained, we begin with basic, categorical properties of preordered sets. For this purpose, we also recall some aspects of universal algebra in monoidal categories and the construction of composite monads. Further, this approach means that in the theory of quantales we have to fill some gaps which, unfortunately, are still present—e.g. a definition of prime elements satisfying an appropriate level of generality. W. Krull only formulated these axioms in the case of two-sided quantales.

In this context, we present a new concept of prime elements which permits a non-commutative generalization of the well-known adjunction between locales and topological spaces (cf. [58]) and leads to a satisfactory representation of semiunital quantales by six-valued topological spaces. Moreover, prequantales occur as algebras of the composite monad of the term monad with the monad of downclosed sets on the category of preordered sets—a property which also explains the rôle of the Minkowski multiplication in the theory of quantales. Finally, we introduce the spectrum of a non-commutative C^*-algebra as a quotient of the tensor product of the quantales of closed left and closed right ideals. It is interesting to see that this construction collapses to the standard concept of the spectrum in the case of commutative C^*-algebras.

The last chapter, as already mentioned above, deals with applications of unital quantales to module theory. In comparison with the traditional theory of modules over rings, this approach means the replacement of the category of abelian groups by the category of complete lattices and join-preserving maps. Here, involutive and unital quantales play a special rôle at certain places. On the one hand, the existence of an involution guarantees the self-duality of the category of left modules, but on the other hand involutions on the underlying quantale are also necessary for the concept of an involutive module. The latter concept plays an important rôle in the representation theory of C^*-algebras. As a fundamental result, we refer to the fact that irreducible representations of a C^*-algebra A and irreducible involutive left modules on the unital quantale of all closed linear subspaces of A are equivalent concepts (cf. [70]). In this sense, module theory on unital quantales has non-trivial applications to the theory of operator algebras.

Another direction of applications is the well-known fact that the free right module on a unital quantale $\mathfrak{Q}$ generated by the power set of X is isomorphic to the right $\mathfrak{Q}$-module of all "$\mathfrak{Q}$-fuzzy subsets" of X (cf. p. 10 in [59]). This observation opens the door towards many-valuedness inside mathematics as well as to applications outside of mathematics, and it is not surprising that right modules on a unital quantale $\mathfrak{Q}$ and join-complete $\mathfrak{Q}$-valued lattices are equivalent concepts — an observation which goes back to I. Stubbe (2006) (cf. [107]) in a more general context given by quantaloid-enriched categories. Since the chain $\mathbb{1} = \{0,1\}$ of two elements is the unit object of $\mathtt{Sup}$, we can summarize this situation from the point of view of many-valued structures as follows: The category of right modules on a unital quantale plays the same rôle as $\mathtt{Sup}$ has done in the binary setting given by $\mathbb{1} = \{0,1\}$.

If we restrict our interest to commutative and unital quantales $\mathfrak{Q}$ as A. Joyal and M. Tierney did in their booklet 1984, then the category $\mathtt{Mod}(\mathfrak{Q})$ of modules has a tensor product which is a quotient of the tensor product in $\mathtt{Sup}$. In this context, $\mathtt{Mod}(\mathfrak{Q})$ turns into a star-autonomous category, and we can obviously repeat all those algebraic constructions with $\mathtt{Mod}(\mathfrak{Q})$ which we have previously done with $\mathtt{Sup}$. However, it is an open question whether such an approach is desirable.

The answer might depend on the perspective of possible applications. A typical context of applications is computer science, where modules are sometimes related to program transformation techniques for imperative languages (cf. [23]).

Chapter 1
Foundations

Categorical aspects of preordered sets seem to be the natural point of departure for a more fundamental understanding of quantales and their related topics. The following basic properties of the category of preordered sets are explained:

- completeness and cocompleteness,
- cartesian closedness,
- free preordered groupoids,
- the term monad (of a single binary operator symbol),
- the monad of downclosed sets.

In this context we emphasize the important fact that the composition of the term monad on preordered sets with the monad of downclosed sets exists in the sense of some distributive law. This insight opens the door for a comprehensive study of prequantales and quantales.

Evidently this approach requires some knowledge of universal algebra in certain categorical settings. For this purpose we recall in Sect. 1.1 the construction of free magmas and free monoids which is based on an application of the free algebra algorithm due to J. Adámek 1974 (cf. [1, 3]). We also describe the tensor product of semigroups in symmetric monoidal categories, which will later play a significant rôle in the construction of coproducts of certain quantales. Further, in Sect. 1.2 we add a detailed analysis of the concept of distributive laws and their applications which goes back to J. Beck 1969 (cf. [10]). This concept might be understood as the way to circumvent the symmetry axiom of symmetric monoidal categories. A reader who is familiar with all these topics might skip both Sects. 1.1 and 1.2 and move directly to Sect. 1.3 on categorical properties of preordered sets.

P. Eklund et al., *Semigroups in Complete Lattices*, Developments in Mathematics 54, https://doi.org/10.1007/978-3-319-78948-4_1

1.1 Some Properties of Universal Algebra in Monoidal Categories

In this section it is assumed that the reader is familiar with the fundamental concepts of category theory as they have been developed in [73]. We will follow the notation in [73]; only at those places where we deviate from [73] we will give special explanations. First we recall the concept of monoidal categories.

Definition 1.1.1. A sextuple $\mathsf{C} = (\mathsf{C}_0, \otimes, \mathbb{1}, a, \ell, r)$ is called a *monoidal category* if C_0 is a category, $\otimes\colon \mathsf{C}_0 \times \mathsf{C}_0 \to \mathsf{C}_0$ is a bifunctor, $\mathbb{1}$ is an object of C_0 and the transformations $a\colon\ \otimes \circ(\otimes \times \mathrm{id}_{\mathsf{C}_0}) \to \otimes \circ (\mathrm{id}_{\mathsf{C}_0} \times \otimes)$, $\ell\colon \mathbb{1} \otimes \_ \to \mathrm{id}_{\mathsf{C}_0}$ and $r\colon \_ \otimes \mathbb{1} \to \mathrm{id}_{\mathsf{C}_0}$ are natural isomorphisms with the components

$$(X \otimes Y) \otimes Z \xrightarrow{a_{XYZ}} X \otimes (Y \otimes Z), \quad \mathbb{1} \otimes Y \xrightarrow{\ell_Y} Y \ \text{ and } \ X \otimes \mathbb{1} \xrightarrow{r_X} X$$

subjected to the following coherence axioms:

(1) The bifunctor $\otimes$ is associative — i.e. the following pentagonal diagram is commutative:

$$\begin{array}{ccccc} ((X \otimes Y) \otimes Z) \otimes U & \xrightarrow{a_{(X\otimes Y)ZU}} & (X \otimes Y) \otimes (Z \otimes U) & \xrightarrow{a_{XY(Z\otimes U)}} & X \otimes (Y \otimes (Z \otimes U)) \\ \downarrow{\scriptstyle a_{XYZ}\otimes 1_U} & & & & \uparrow{\scriptstyle 1_X \otimes a_{YZU}} \\ (X \otimes (Y \otimes Z)) \otimes U & & \xrightarrow{a_{X(Y\otimes Z)U}} & & X \otimes ((Y \otimes Z) \otimes U) \end{array}$$

(2) $\ell_{\mathbb{1}} = r_{\mathbb{1}}$ and the following triangle is commutative:

$$\begin{array}{ccc} (X \otimes \mathbb{1}) \otimes Y & \xrightarrow{a_{X\mathbb{1}Y}} & X \otimes (\mathbb{1} \otimes Y) \\ & {\scriptstyle r_X\otimes 1_Y} \searrow \quad \swarrow {\scriptstyle 1_X\otimes \ell_Y} & \\ & X \otimes Y & \end{array}$$

If C is a monoidal category, then the bifunctor $\otimes$ is called the *tensor product* of C and $\mathbb{1}$ the *unit object* of C.

The next example goes back to J. Bénabou 1963 (cf. [12]).

Example 1.1.2. In order to avoid foundational problems we consider a small category C. Then the category $\mathtt{End}(\mathsf{C})$ of endofunctors of C consists of the following data:

- objects of $\mathtt{End}(\mathsf{C})$ are all endofunctors $\mathsf{C} \xrightarrow{\mathsf{F}} \mathsf{C}$,
- morphisms of $\mathtt{End}(\mathsf{C})$ are all natural transformations — i.e. with every pair (F, G) of endofunctors of C we associate the set $\hom(\mathsf{F}, \mathsf{G})$ of all natural transformations $\tau\colon \mathsf{F} \to \mathsf{G}$.

The composition of natural transformations is given by the (vertical) composition and the identity of F is the identity transformation 1_{F} of F.

Since the Interchange Law is valid, we are in a position to introduce a monoidal structure on $\mathtt{End(C)}$ in the following way. The tensor product — i.e. the bifunctor

$$\mathtt{End(C)} \times \mathtt{End(C)} \xrightarrow{\otimes} \mathtt{End(C)}$$

acts as follows:

(i) On each pair of objects (F, G) the bifunctor $\otimes$ is given by the composition of functors — i.e. $\mathsf{F} \otimes \mathsf{G} = \mathsf{GF}$.
(ii) On each pair of morphisms (τ, τ'), where $\tau\colon \mathsf{F} \to \mathsf{G}$ and $\tau'\colon \mathsf{F}' \to \mathsf{G}'$, the bifunctor $\otimes$ is given by the star composition of natural transformations — i.e. $\tau \otimes \tau' = \tau' \star \tau\colon \mathsf{F'F} \to \mathsf{G'G}$.

The unit object is the identity functor id_C of C. Since the composition of functors and the star composition of natural transformations are associative and the identity transformation $1_{\mathrm{id}_\mathsf{C}}$ of id_C is the unit w.r.t. the star composition, the natural isomorphisms a, ℓ and r coincide with the respective identity transformations Id — i.e.

$$\mathsf{HGF} \xrightarrow{1_{\mathsf{HGF}}} \mathsf{HGF}, \quad \mathrm{id}_\mathsf{C}\mathsf{F} \xrightarrow{1_\mathsf{F}} \mathsf{F}, \quad \mathsf{G}\mathrm{id}_\mathsf{C} \xrightarrow{1_\mathsf{G}} \mathsf{G}.$$

Hence the sextuple $(\mathtt{End(C)}, \otimes, \mathrm{id}_\mathsf{C}, \mathrm{Id}, \mathrm{Id}, \mathrm{Id})$ is a monoidal category.

A pair (X, m) is called a *magma* in C if $X \in |\mathsf{C}_0|$ and m is a binary operation on X — i.e. a morphism $X \otimes X \xrightarrow{m} X$ of C_0.
A magma (X, m) is a *semigroup* if m is associative — this means the commutativity of the following diagram:

$$\begin{array}{ccccc} (X \otimes X) \otimes X & & \xrightarrow{a_{XXX}} & & X \otimes (X \otimes X) \\ {\scriptstyle m\otimes 1_X}\downarrow & & & & \downarrow{\scriptstyle 1_X\otimes m} \\ X \otimes X & \xrightarrow{m} & X & \xleftarrow{m} & X \otimes X \end{array} \qquad \text{(Associativity Axiom)}$$

A magma (X, m) is called *unital* if there exists a morphism $\mathbb{1} \xrightarrow{e} X$ such that the diagram

$$\begin{array}{ccccc} \mathbb{1} \otimes X & \xrightarrow{e\otimes 1_X} & X \otimes X & \xleftarrow{1_X\otimes e} & X \otimes \mathbb{1} \\ & {\scriptstyle \ell_X}\searrow & \downarrow{\scriptstyle m} & \swarrow{\scriptstyle r_X} & \\ & & X & & \end{array} \qquad \text{(Unit Axiom)}$$

commutes. A unit e of a magma (X, m) is *uniquely* determined. Hence e is *the* unit. In fact, if e' is a further unit, then we obtain $e' \circ \ell_\mathbb{1} = m \circ (e \otimes e') = e \circ r_\mathbb{1}$. Since the isomorphisms $\ell_\mathbb{1}$ and $r_\mathbb{1}$ coincide, the relation $e' = e$ follows.

A *monoid* in C is a unital semigroup in C.

Let (X, m) and (X', m') be magmas in C. A morphism $X \xrightarrow{h} X'$ in C_0 is called a *homomorphism* if the following diagram is commutative:

$$\begin{array}{ccc} X \otimes X & \xrightarrow{\;h \otimes h\;} & X' \otimes X' \\ {\scriptstyle m}\downarrow & & \downarrow{\scriptstyle m'} \\ X & \xrightarrow{\quad h \quad} & X' \end{array}$$

If (X, m) and (X', m') are unital and e and e' are the respective units, then a homomorphism h is *unital* if the following additional diagram is commutative:

$$\begin{array}{ccc} & \mathbb{1} & \\ {\scriptstyle e}\swarrow & & \searrow{\scriptstyle e'} \\ X & \xrightarrow{\quad h \quad} & X' \end{array}$$

Finally, if we assume that C_0 has finite coproducts and the tensor product distributes over finite coproducts, then the unitalization of magmas exists. For this purpose we fix some further notation.

The coproduct of two objects X and Y is denoted by $X \sqcup Y$. If $X \xrightarrow{j_X} X \sqcup Y$ and $Y \xrightarrow{j_Y} X \sqcup Y$ are the coproduct injections, then for every pair of morphisms $X \xrightarrow{f} Z$ and $Y \xrightarrow{g} Z$ the unique morphism $X \sqcup Y \xrightarrow{h} Z$ with $h \circ j_X = f$ and $h \circ j_Y = g$ is denoted by $h = f \sqcup g$.

If (X, m) is now any magma, then on the coproduct $\widehat{X} = X \sqcup \mathbb{1}$ we introduce the following binary operation $\widehat{X} \otimes \widehat{X} \xrightarrow{\widehat{m}} \widehat{X}$ and the unit $\mathbb{1} \xrightarrow{\widehat{e}} \widehat{X}$ by:

- $\widehat{m} = (j_X \circ m) \sqcup (j_X \circ \ell_X) \sqcup (j_X \circ r_X) \sqcup (j_{\mathbb{1}} \circ \ell_{\mathbb{1}})$, where j_X and $j_{\mathbb{1}}$ are the respective coproduct injections,
- $\widehat{e} = j_{\mathbb{1}}$, where $j_{\mathbb{1}}$ is the coproduct injection of $\mathbb{1}$.

Then $(\widehat{X}, \widehat{m}, \widehat{e})$ is a unital magma and is called the *unitalization* of (X, m). It is not difficult to verify that the unitalization preserves the associativity axiom — i.e. the unitalization of a semigroup is a monoid.

The aim of the following considerations is to show that under certain sufficient but rather general conditions, free magmas and free monoids exist. For this purpose we recall some further terminology of monoidal categories.

Definition 1.1.3. (a) A monoidal category $\mathsf{C} = (\mathsf{C}_0, \otimes, \mathbb{1}, a, \ell, r)$ is *symmetric* if C is provided with a *symmetry* c — i.e. a natural isomorphism $c\colon \otimes \to \otimes$ whose components

$$X \otimes Y \xrightarrow{c_{XY}} Y \otimes X$$

satisfy the subsequent coherence axioms expressed by the commutativity of the following diagrams:

(3)

$$\begin{array}{ccc} X \otimes Y & \xrightarrow{c_{XY}} & Y \otimes X \\ & \searrow^{1_{X\otimes Y}} & \downarrow c_{YX} \\ & & X \otimes Y \end{array} \qquad \begin{array}{ccc} X \otimes \mathbb{1} & \xrightarrow{c_{X\mathbb{1}}} & \mathbb{1} \otimes X \\ & \searrow^{r_X} & \downarrow \ell_X \\ & & X \end{array}$$

(4)

$$\begin{array}{ccccc} (X \otimes Y) \otimes Z & \xrightarrow{a_{XYZ}} & X \otimes (Y \otimes Z) & \xrightarrow{c_{X(Y\otimes Z)}} & (Y \otimes Z) \otimes X \\ \downarrow c_{XY}\otimes 1_Z & & & & \downarrow a_{YZX} \\ (Y \otimes X) \otimes Z & \xrightarrow{a_{YXZ}} & Y \otimes (X \otimes Z) & \xrightarrow{1_Y \otimes c_{XZ}} & Y \otimes (Z \otimes X) \end{array}$$

(b) A monoidal category C is *closed* if for each object $X \in |\mathsf{C}_0|$ the endofunctor $\_ \otimes X : \mathsf{C}_0 \to \mathsf{C}_0$ has a right adjoint $[X, \_] : \mathsf{C}_0 \to \mathsf{C}_0$. The respective components of the unit and counit (the latter is also called the *evaluation arrow*) of this adjoint situation $\_ \otimes X \dashv [X, \_]$ are denoted by $Z \xrightarrow{\eta_Z} [X, Z \otimes X]$ and $[X, Z] \otimes X \xrightarrow{\mathrm{ev}_Z} Z$.
(c) A monoidal category C is *biclosed* if for each object $X \in |\mathsf{C}_0|$ the endofunctors $\_ \otimes X : \mathsf{C}_0 \to \mathsf{C}_0$ and $X \otimes \_ : \mathsf{C}_0 \to \mathsf{C}_0$ have right adjoint functors.

Evidently, every symmetric and monoidal closed category is biclosed, and the right adjoint functor of $X \otimes \_$ coincides with $[X, \_]$, where the corresponding counit $\varepsilon = (\varepsilon_Y)_{Y \in |\mathsf{C}_0|}$ has the form $\varepsilon_Y = \mathrm{ev}_Y \circ c_{X[X,Y]}$.

1.1.1 Free Magmas

The next theorem recalls the important fact that in monoidal biclosed categories the tensor product always preserves direct limits.

Theorem 1.1.4. *Let* $\mathsf{C} = (\mathsf{C}_0, \otimes, \mathbb{1}, a, \ell, r)$ *be a monoidal biclosed category and* $(I, \leq)$ *be a directed preordered set. Further, let* $(X_i, f_{ji})_{i\in I}$ *and* $(Y_i, g_{ji})_{i\in I}$ *be direct systems. If* $(X_0, (f_i)_{i\in I})$ *and* $(Y_0, (g_i)_{i\in I})$ *are the respective direct limits of* $(X_i, f_{ji})_{i\in I}$ *and* $(Y_i, g_{ji})_{i\in I}$*, then*

$$(X_0 \otimes Y_0, (f_i \otimes g_i)_{i\in I})$$

is the direct limit of $(X_i \otimes Y_i, f_{ji} \otimes g_{ji})_{i\in I}$.

Proof. Since for $i \le j$ the relation $f_i \otimes g_i = (f_j \otimes g_j) \circ (f_{j\,i} \otimes g_{j\,i})$ is obvious, we only verify the universal property of direct limits. For this purpose we choose $D \in \mathsf{C}_0$ and consider a system $(h_i)_{i \in I}$ of morphisms $X_i \otimes Y_i \xrightarrow{h_i} D$ with $h_j \circ (f_{j\,i} \otimes g_{j\,i}) = h_i$ for $i \le j$. Since the index set I is directed, for each pair $(i, k) \in I \times I$ there exists a morphism $X_i \otimes Y_k \xrightarrow{\sigma_{i\,k}} D$ such that the diagram

$$\begin{array}{ccc} X_i \otimes Y_k & \xrightarrow{f_{m\,i} \otimes g_{m\,k}} & X_m \otimes Y_m \\ & \searrow^{\sigma_{i\,k}} & \downarrow h_m \\ & & D \end{array}$$

is commutative for all $m \in I$ with $i \le m$ and $k \le m$.

(a) We verify the existence of $X_0 \otimes Y_0 \xrightarrow{h_0} D$ satisfying $h_0 \circ (f_k \otimes g_k) = h_k$ for all $k \in I$. In a first step we fix $k \in I$. Since C is closed, the endofunctor $\_ \otimes Y_k$ has a right adjoint and hence preserves direct limits. Since $\sigma_{i\,k} = \sigma_{j\,k} \circ (f_{j\,i} \otimes 1_{Y_k})$ $(i \le j)$, there exists a unique morphism $X_0 \otimes Y_k \xrightarrow{\sigma_k} D$ such that $\sigma_k \circ (f_i \otimes 1_{Y_k}) = \sigma_{i\,k}$ holds for all $i \in I$. Now we apply $\sigma_{i\,k} = \sigma_{i\,l} \circ (1_{X_i} \otimes g_{l\,k})$ and observe:

$$\sigma_k = \sigma_l \circ (1_{X_0} \otimes g_{l\,k}), \qquad k \le l.$$

Since C is biclosed, the endofunctor $X_0 \otimes \_$ also preserves direct limits. Hence there exists a unique morphism $X_0 \otimes Y_0 \xrightarrow{h_0} D$ satisfying $h_0 \circ (1_{X_0} \otimes g_k) = \sigma_k$ for all $k \in I$. Obviously, the construction of h_0 implies:

$$h_0 \circ (f_k \otimes g_k) = \sigma_k \circ (f_k \otimes 1_{Y_k}) = \sigma_{k\,k} = h_k.$$

Thus the existence of h_0 is verified.

(b) In order to prove the uniqueness of h_0 we fix some morphism $X_0 \otimes Y_0 \xrightarrow{h} D$ with the property $h \circ (f_k \otimes g_k) = h_k$ for all $k \in I$. Then we observe:

$$\begin{aligned} h_0 \circ (f_i \otimes g_k) &= \sigma_k \circ (f_i \otimes 1_{Y_k}) \\ &= \sigma_{i\,k} \\ &= h_m \circ (f_{m\,i} \otimes g_{m\,k}) \\ &= h \circ (f_m \otimes g_m) \circ (f_{m\,i} \otimes g_{m\,k}) \\ &= h \circ (f_i \otimes g_k). \end{aligned}$$

Hence h_0 coincides with h. □

If we now combine Theorem 1.1.4 with Theorem A.1.1 in Appendix A.1, then we obtain the following important result.

Corollary 1.1.5. *Let C be a monoidal biclosed category with an underlying cocomplete category C_0. Then for each object $X \in |\mathsf{C}_0|$ there exists a magma $(X^\sharp, m_{X^\sharp})$ and a morphism $X \xrightarrow{\eta_X} X^\sharp$ such that for every magma (Y, m_Y) and for every morphism $X \xrightarrow{h} Y$ there exists a unique homomorphism $X^\sharp \xrightarrow{h^\sharp} Y$ making the following diagram commutative*:

$$\begin{array}{ccc} X & \xrightarrow{\eta_X} & X^\sharp \\ & {\scriptstyle h}\searrow & \vdots\, {\scriptstyle h^\sharp} \\ & & Y \end{array}$$

Proof. Let us consider the composition of the diagonal functor $\Delta\colon \mathsf{C}_0 \to \mathsf{C}_0 \times \mathsf{C}_0$ with the bifunctor $\otimes$. Then the resulting endofunctor $\mathsf{F}_\otimes$ of C_0 has the following explicit form:

$$\mathsf{F}_\otimes(X) = X \otimes X, \quad X \xrightarrow{h} Y, \quad \mathsf{F}_\otimes(h) = h \otimes h.$$

Obviously $\mathsf{F}_\otimes$-algebras (cf. Appendix A.1) and magmas are the same things. Since $\mathsf{F}_\otimes$ preserves direct limits (cf. Theorem 1.1.4), the assertion follows from Theorem A.1.1. □

We can also reformulate the assertion of the previous corollary as the statement that the forgetful functor from the category of magmas to the underlying category C_0 has a left adjoint functor. Hence $(X^\sharp, m_{X^\sharp})$ with the morphism $X \xrightarrow{\eta_X} X^\sharp$ is also called the *free magma generated by* X.

Finally, we do not fail to mention that in the case of monoidal biclosed categories with an underlying complete and cocomplete category the category of magmas is always complete and cocomplete.

1.1.2 Free Monoids

In this subsection we assume that the monoidal category $\mathsf{C} = (\mathsf{C}_0, \otimes, \mathbb{1}, a, \ell, r)$ is biclosed and its underlying category C_0 is cocomplete.

We fix an object X of C_0. In order to construct the free monoid generated by X we deviate here from the standard approach (cf. [73]) and base our arguments on the endofunctor $\mathsf{F}_X = \_ \otimes X$. It is interesting to see that F_X also plays a significant rôle in the theory of automata (cf. Sect. 3.4). We will show that the free F_X-algebra generated by the unit object of C_0 is the free monoid generated by X.

For this purpose we first have to make some preparations. We maintain here the notation from Theorem A.1.1 in Appendix A.1 and add some further notation and comments.

Notation and Comment. Since C is biclosed, the tensor product is distributive over coproducts. Hence, if $B \xrightarrow{j_B} B \sqcup C$ and $C \xrightarrow{j_C} B \sqcup C$ are the coproduct injections of $B \sqcup C$, then $1_A \otimes j_B$ and $1_A \otimes j_C$ are the coproduct injections of $A \otimes (B \sqcup C)$. Further, the coproduct $B_1 \sqcup B_2 \xrightarrow{h_1 \oplus h_2} C_1 \sqcup C_2$ of $B_1 \xrightarrow{h_1} C_1$ and $B_2 \xrightarrow{h_2} C_2$ is given by $h_1 \oplus h_2 = (j_{C_1} \circ h_1) \sqcup (j_{C_2} \circ h_2)$.

In a first step we apply the natural isomorphism a occurring in the associativity axiom of the tensor product. We fix the endofunctor $\mathsf{F}_X = \_ \otimes X \colon \mathsf{C}_0 \to \mathsf{C}_0$ and recall the following notation from the proof of Theorem A.1.1. For every $A \in |\mathsf{C}_0|$ we put

$$\begin{aligned} Z_1(A) &= A \otimes X, & W_1(A) &= Z_1(A) \sqcup A, & \\ Z_n(A) &= W_{n-1}(A) \otimes X, & W_n(A) &= Z_n(A) \sqcup A, & n \geq 2. \end{aligned}$$

Further, let $A \xrightarrow{j_A} W_1(A) = Z_1(A) \sqcup A$ be the coproduct injection. We state the following definitions of morphisms (see again the proof of Theorem A.1.1):

$$\begin{aligned} &Z_1(A) \xrightarrow{e^A_{21} = \mathsf{F}_X(j_A)} Z_2(A), && W_1(A) \xrightarrow{\ell^A_{21} = \mathsf{F}_X(j_A) \oplus 1_A} W_2(A), \\ &Z_n(A) \xrightarrow{e^A_{n+1\,n} = \mathsf{F}_X(\ell^A_{n\,n-1})} Z_{n+1}(A), && W_n(A) \xrightarrow{\ell^X_{n+1\,n} = e^A_{n+1\,n} \oplus 1_A} W_{n+1}(A), \quad n \geq 2. \end{aligned}$$

Then for every morphism $A \xrightarrow{\xi} B \otimes C$ we define recursively a sequence $(\varphi^\xi_n)_{n \in \mathbb{N}}$ of morphisms $Z_n(A) \xrightarrow{\varphi^\xi_n} B \otimes Z_n(C)$ by

$$\varphi^\xi_1 = a_{BCX} \circ \mathsf{F}_X(\xi) \quad \text{and} \quad \varphi^\xi_n = a_{BW_{n-1}(C)X} \circ \mathsf{F}_X(\varphi^\xi_{n-1} \oplus \xi), \quad n \geq 2,$$

and we will use the notation $W_n(A) \xrightarrow{\psi^\xi_n = \varphi^\xi_n \oplus \xi} B \otimes W_n(C)$ in the following considerations. First we notice that $(1_B \otimes e^C_{21}) \circ \varphi^\xi_1 = \varphi^\xi_2 \circ e^A_{21}$. Hence the relation

$$(1_B \otimes e^C_{n+1\,n}) \circ \varphi^\xi_n = \varphi^\xi_{n+1} \circ e^A_{n+1\,n}, \qquad n \in \mathbb{N}$$

follows by induction. Now we refer to the construction of direct limits and obtain a unique morphism $Z_\infty(A) \xrightarrow{\varphi^\xi_\infty} B \otimes Z_\infty(C)$ satisfying the conditions

$$\varphi^\xi_\infty \circ e^A_n = (1_B \otimes e^C_n) \circ \varphi^\xi_n, \qquad n \in \mathbb{N}. \tag{1.1}$$

In this context we introduce a morphism $A^\sharp \xrightarrow{\psi^\xi_\infty} B \otimes C^\sharp$ by $\psi^\xi_\infty = \varphi^\xi_\infty \oplus \xi$.

Further, we refer to the formula (A.2) in the proof of Theorem A.1.1 and recall the morphism $A^\sharp \otimes X \xrightarrow{\vartheta_A} Z_\infty(A)$. Then a repeated application of (1.1) leads to the following relation:

$$\begin{aligned}
\varphi_\infty^\xi \circ \vartheta_A \circ \mathsf{F}_X(e_n^A \oplus 1_A) &= \varphi_\infty^\xi \circ e_{n+1}^A = (1_B \otimes e_{n+1}^C) \circ \varphi_{n+1}^\xi \\
&= (1_B \otimes \vartheta_C) \circ \big(1_B \otimes \mathsf{F}_X(e_n^C \oplus 1_C)\big) \circ a_{BW_n(C)X} \circ \mathsf{F}_X(\psi_n^\xi) \\
&= (1_B \otimes \vartheta_C) \circ a_{BC^\sharp X} \circ \mathsf{F}_X\big(1_B \otimes (e_n^C \oplus 1_C)\big) \circ \mathsf{F}_X(\psi_n^\xi) \\
&= (1_B \otimes \vartheta_C) \circ a_{BC^\sharp X} \circ \mathsf{F}_X\big(\big(1_B \otimes (e_n^C \oplus 1_C)\big) \circ \psi_n^\xi\big) \\
&= (1_B \otimes \vartheta_C) \circ a_{BC^\sharp X} \circ \mathsf{F}_X(\psi_\infty^\xi) \circ \mathsf{F}_X(e_n^A \oplus 1_A).
\end{aligned}$$

Thus

$$(1_B \otimes \vartheta_C) \circ a_{BC^\sharp X} \circ \mathsf{F}_X(\psi_\infty^\xi) = \varphi_\infty^\xi \circ \vartheta_A \tag{1.2}$$

follows from the universal property of direct limits.

After these preparations we consider the situation

$$A \xrightarrow{r_A^{-1}} A \otimes \mathbb{1}, \quad A^\sharp \xrightarrow{\psi_\infty^{r_A^{-1}}} A \otimes \mathbb{1}^\sharp, \quad A \otimes \mathbb{1}^\sharp \xrightarrow{r_{A\otimes\mathbb{1}^\sharp}^{-1}} (A \otimes \mathbb{1}^\sharp) \otimes \mathbb{1}, \quad \mathbb{1}^\sharp \xrightarrow{r_{\mathbb{1}^\sharp}^{-1}} \mathbb{1}^\sharp \otimes \mathbb{1}$$

and abbreviate these morphisms as follows:

$$\xi = \psi_\infty^{r_A^{-1}}, \qquad \pi = r_{A\otimes\mathbb{1}^\sharp}^{-1}, \qquad \rho = r_{\mathbb{1}^\sharp}^{-1}.$$

As an immediate corollary of the coherence axioms of monoidal categories the property

$$a_{A\mathbb{1}^\sharp\mathbb{1}} \circ \pi = 1_A \otimes \rho \tag{1.3}$$

is obviously valid (cf. Exercise 1.1.1). Consequently the following diagram is commutative:

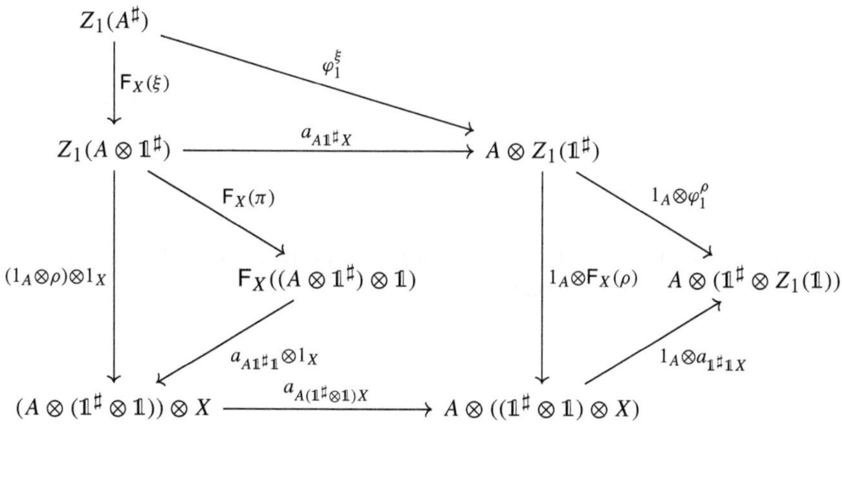

Now we insert the pentagonal diagram and obtain the relation

$$(1_A \otimes \varphi_1^{\rho}) \circ \varphi_1^{\xi} = a_{A\mathbb{1}^\sharp Z_1(\mathbb{1})} \circ \varphi_1^{\pi} \circ \mathsf{F}_X(\xi). \tag{1.4}$$

In a next step we apply the F_X-term functor T and prove the following technical, but important property:

$$a_{A\mathbb{1}^\sharp\mathbb{1}^\sharp} \circ \psi_\infty^{\pi} \circ \mathsf{T}(\xi) = (1_A \otimes \psi_\infty^{\rho}) \circ \psi_\infty^{\xi}. \tag{1.5}$$

Because of (1.1) and (1.4) we first notice:

$$\begin{aligned}(1_A \otimes \varphi_\infty^{\rho}) \circ \varphi_\infty^{\xi} \circ e_1^{A^\sharp} &= (1_A \otimes \varphi_\infty^{\rho}) \circ \left(1_A \otimes e_1^{\mathbb{1}^\sharp}\right) \circ \varphi_1^{\xi}\\ &= \left(1_A \otimes \left(\varphi_\infty^{\rho} \circ e_1^{\mathbb{1}^\sharp}\right)\right) \circ \varphi_1^{\xi}\\ &= \left(1_A \otimes \left((1_{\mathbb{1}^\sharp} \otimes e_1^{\mathbb{1}}) \circ \varphi_1^{\rho}\right)\right) \circ \varphi_1^{\xi}\\ &= \left(1_A \otimes (1_{\mathbb{1}^\sharp} \otimes e_1^{\mathbb{1}})\right) \circ (1_A \otimes \varphi_1^{\rho}) \circ \varphi_1^{\xi}\\ &= \left(1_A \otimes (1_{\mathbb{1}^\sharp} \otimes e_1^{\mathbb{1}})\right) \circ a_{A\mathbb{1}^\sharp Z_1(\mathbb{1})} \circ \varphi_1^{\pi} \circ \mathsf{F}_X(\xi)\\ &= a_{A\mathbb{1}^\sharp Z_\infty(\mathbb{1})} \circ \left((1_A \otimes 1_{\mathbb{1}^\sharp}) \otimes e_1^{\mathbb{1}}\right) \circ \varphi_1^{\pi} \circ \mathsf{F}_X(\xi)\\ &= a_{A\mathbb{1}^\sharp Z_\infty(\mathbb{1})} \circ \varphi_\infty^{\pi} \circ e_1^{A\otimes\mathbb{1}^\sharp} \circ \mathsf{F}_X(\xi).\end{aligned}$$

Now we are going to construct $\mathsf{T}(\xi)$ (see Appendix A.1) and put

$$\widehat{\xi}_1 = e_1^{A\otimes\mathbb{1}^\sharp} \circ \mathsf{F}_X(\xi).$$

Hence the previous relation can be rewritten as follows:

$$(1_A \otimes \varphi_\infty^{\rho}) \circ \varphi_\infty^{\xi} \circ e_1^{A^\sharp} = a_{A\mathbb{1}^\sharp Z_\infty(\mathbb{1})} \circ \varphi_\infty^{\pi} \circ \widehat{\xi}_1. \tag{1.6}$$

We assume that (1.6) holds for some $n \in \mathbb{N}$, and not only for $n = 1$ as we have explained. Since $a_{A\mathbb{1}^\sharp\mathbb{1}} \circ \pi = 1_A \otimes \rho$ and $a_{A\mathbb{1}^\sharp Z_\infty(\mathbb{1})} \oplus a_{A\mathbb{1}^\sharp\mathbb{1}} = a_{A\mathbb{1}^\sharp(Z_\infty(\mathbb{1})\sqcup\mathbb{1})}$, we obtain:

$$(1_A \otimes \psi_\infty^{\rho}) \circ \psi_\infty^{\xi} \circ (e_n^{A^\sharp} \oplus 1_{A^\sharp}) = a_{A\mathbb{1}^\sharp\mathbb{1}^\sharp} \circ \psi_\infty^{\pi} \circ \left(\widehat{\xi}_n \oplus \xi\right). \tag{1.7}$$

If we apply the endofunctor F_X to (1.7) and use repeatedly (1.2), then we arrive at the following commutative diagram[1]

[1] To simplify the notation we have dropped the subscript X of F_X in this diagram.

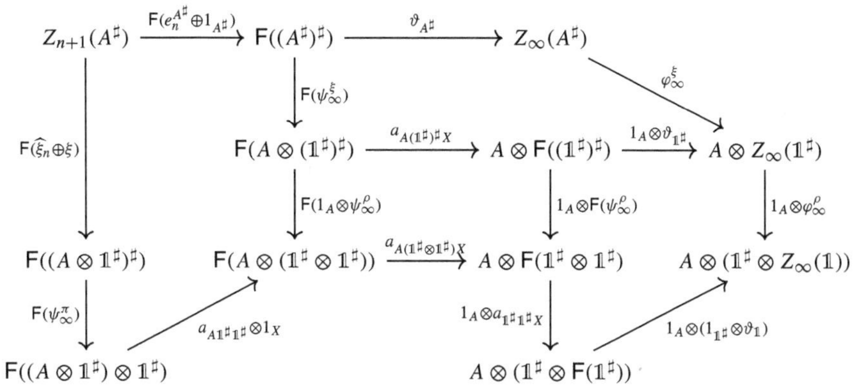

If we now intend to insert again the pentagonal diagram, then we first make use of (1.2) and notice:

$$\begin{aligned}
&(1_A \otimes (1_{\mathbb{1}^\sharp} \otimes \vartheta_{\mathbb{1}})) \circ a_{A\mathbb{1}^\sharp \mathsf{F}_X(\mathbb{1}^\sharp)} \circ a_{(A\otimes\mathbb{1}^\sharp)\mathbb{1}^\sharp X} \circ \mathsf{F}_X(\psi_\infty^\pi) \\
&\qquad = a_{A\mathbb{1}^\sharp Z_\infty(\mathbb{1})} \circ (1_{A\otimes\mathbb{1}^\sharp} \otimes \vartheta_{\mathbb{1}}) \circ a_{(A\otimes\mathbb{1}^\sharp)\mathbb{1}^\sharp X} \circ \mathsf{F}_X(\psi_\infty^\pi) \\
&\qquad = a_{A\mathbb{1}^\sharp Z_\infty(\mathbb{1})} \circ \varphi_\infty^\pi \circ \vartheta_{A\otimes\mathbb{1}^\sharp}.
\end{aligned}$$

Because of this observation the previous commutative diagram entails the following relation (see also formula (A.2) in the proof of Theorem A.1.1):

$$\begin{aligned}
(1_A \otimes \varphi_\infty^\rho) \circ \varphi_\infty^\xi \circ e_{n+1}^{A^\sharp} &= a_{A\mathbb{1}^\sharp Z_\infty(\mathbb{1})} \circ \varphi_\infty^\pi \circ \vartheta_{A\otimes\mathbb{1}^\sharp} \circ \mathsf{F}_X(\widehat{\xi}_n \oplus \xi) \\
&= a_{A\mathbb{1}^\sharp Z_\infty(\mathbb{1})} \circ \varphi_\infty^\pi \circ \widehat{\xi}_{n+1}.
\end{aligned}$$

Hence (1.6) also holds for $n+1$.

Finally we pass to the direct limit of $Z_n(A^\sharp)$ and conclude from the previous results that the relation

$$(1_A \otimes \varphi_\infty^\rho) \circ \varphi_\infty^\xi = a_{A\mathbb{1}^\sharp Z_\infty(\mathbb{1})} \circ \varphi_\infty^\pi \circ \widehat{\xi}_\infty$$

holds. Hence (1.5) follows.

Let $\mathbf{T} = (\mathsf{T}, \eta, \mu)$ be the F_X-term monad (cf. Appendix A.1). In this context we now turn to the monoidal structure on the free $\mathbf{T}$-algebra $\mathbb{1}^\sharp$ generated by the unit object $\mathbb{1}$ of C. Therefore in the previous considerations we consider the special case $A = \mathbb{1}^\sharp$ and recall for the convenience of the reader the explicit construction of the $\mathbb{1}$-component $(\mathbb{1}^\sharp)^\sharp \xrightarrow{\mu_{\mathbb{1}}} \mathbb{1}^\sharp$ of the multiplication μ (see Appendix A.1):

$$\begin{aligned}
&\mu_{\mathbb{1}} = \mu_\infty^{\mathbb{1}} \sqcup 1_{\mathbb{1}^\sharp}, \quad \mu_\infty^{\mathbb{1}} \circ e_n^{\mathbb{1}^\sharp} = \mu_n^{\mathbb{1}}, \qquad n \in \mathbb{N}, \\
&\mu_1^{\mathbb{1}} = \delta_{\mathbb{1}^\sharp} = j_\infty^{\mathbb{1}} \circ \vartheta_{\mathbb{1}}, \quad \mu_{n+1}^{\mathbb{1}} = \delta_{\mathbb{1}^\sharp} \circ \mathsf{F}_X(\mu_n^{\mathbb{1}} \sqcup 1_{\mathbb{1}^\sharp}), \qquad n \in \mathbb{N}.
\end{aligned}$$

Lemma 1.1.6. *If $\rho = r_{\mathbb{1}^\sharp}^{-1}$ and $\xi = \psi_\infty^\rho$, then the following diagram is commutative:*

$$\begin{array}{ccc} ((\mathbb{1}^\sharp)^\sharp)^\sharp & \xrightarrow{\mu_{\mathbb{1}^\sharp}} & (\mathbb{1}^\sharp)^\sharp \\ {\scriptstyle \psi_\infty^\xi}\downarrow & & \downarrow{\scriptstyle \psi_\infty^\rho} \\ \mathbb{1}^\sharp \otimes (\mathbb{1}^\sharp)^\sharp & \xrightarrow{1_{\mathbb{1}^\sharp} \otimes \mu_{\mathbb{1}}} & \mathbb{1}^\sharp \otimes \mathbb{1}^\sharp \end{array}$$

Proof. By the construction of direct limits it is sufficient to show that for all $n \in \mathbb{N}$ the diagram

$$\begin{array}{ccc} Z_n((\mathbb{1}^\sharp)^\sharp) & \xrightarrow{\mu_n^{\mathbb{1}^\sharp}} & (\mathbb{1}^\sharp)^\sharp \\ {\scriptstyle \varphi_n^\xi}\downarrow & & \downarrow{\scriptstyle \psi_\infty^\rho} \\ \mathbb{1}^\sharp \otimes Z_n(\mathbb{1}^\sharp) & \xrightarrow{1_{\mathbb{1}^\sharp} \otimes \mu_n^{\mathbb{1}}} & \mathbb{1}^\sharp \otimes \mathbb{1}^\sharp \end{array} \tag{1.8}$$

is commutative. We begin with the observation:

$$\begin{aligned} \psi_\infty^\rho \circ \delta_{(\mathbb{1}^\sharp)^\sharp} &= (1_{\mathbb{1}^\sharp} \otimes j_\infty^{\mathbb{1}}) \circ \varphi_\infty^\rho \circ \vartheta_{\mathbb{1}^\sharp} \\ &= (1_{\mathbb{1}^\sharp} \otimes j_\infty^{\mathbb{1}}) \otimes (1_{\mathbb{1}^\sharp} \otimes \vartheta_{\mathbb{1}}) \circ a_{\mathbb{1}^\sharp \mathbb{1}^\sharp X} \circ \mathsf{F}_X(\psi_\infty^\rho) \\ &= (1_{\mathbb{1}^\sharp} \otimes \delta_{\mathbb{1}^\sharp}) \circ \varphi_1^\xi . \end{aligned}$$

Hence (1.8) is commutative for $n = 1$. Now we assume that (1.8) holds for n. Since $\xi = \psi_\infty^\rho$, the relation $\psi_\infty^\rho \circ (\mu_n^{\mathbb{1}^\sharp} \sqcup 1_{(\mathbb{1}^\sharp)^\sharp}) = \big(1_{\mathbb{1}^\sharp} \otimes (\mu_n^{\mathbb{1}} \sqcup 1_{\mathbb{1}^\sharp})\big) \circ (\varphi_n^\xi \oplus \xi)$ is obvious. Hence we obtain (where we apply again (1.2)):

$$\begin{aligned} \psi_\infty^\rho \circ \mu_{n+1}^{\mathbb{1}^\sharp} &= \psi_\infty^\rho \circ \delta_{(\mathbb{1}^\sharp)^\sharp} \circ \mathsf{F}_X(\mu_n^{\mathbb{1}^\sharp} \sqcup 1_{(\mathbb{1}^\sharp)^\sharp}) \\ &= (1_{\mathbb{1}^\sharp} \otimes j_\infty^{\mathbb{1}}) \circ \varphi_\infty^\rho \circ \vartheta_{\mathbb{1}^\sharp} \circ \mathsf{F}_X(\mu_n^{\mathbb{1}^\sharp} \sqcup 1_{(\mathbb{1}^\sharp)^\sharp}) \\ &= (1_{\mathbb{1}^\sharp} \otimes \delta_{\mathbb{1}^\sharp}) \circ a_{\mathbb{1}^\sharp \mathbb{1}^\sharp X} \circ \mathsf{F}_X(\psi_\infty^\rho) \circ \mathsf{F}_X(\mu_n^{\mathbb{1}^\sharp} \sqcup 1_{(\mathbb{1}^\sharp)^\sharp}) \\ &= (1_{\mathbb{1}^\sharp} \otimes \delta_{\mathbb{1}^\sharp}) \circ a_{\mathbb{1}^\sharp \mathbb{1}^\sharp X} \circ \mathsf{F}_X\big(1_{\mathbb{1}^\sharp} \otimes (\mu_n^{\mathbb{1}} \sqcup 1_{\mathbb{1}^\sharp})\big) \circ \mathsf{F}_X(\psi_n^\xi) \\ &= (1_{\mathbb{1}^\sharp} \otimes \delta_{\mathbb{1}^\sharp}) \circ \big(1_{\mathbb{1}^\sharp} \otimes \mathsf{F}_X(\mu_n^{\mathbb{1}} \sqcup 1_{\mathbb{1}^\sharp})\big) \circ a_{\mathbb{1}^\sharp W_n(\mathbb{1}^\sharp) X} \circ \mathsf{F}_X(\psi_n^\xi) \\ &= (1_{\mathbb{1}^\sharp} \otimes \mu_{n+1}^{\mathbb{1}}) \circ \varphi_{n+1}^\xi . \end{aligned}$$

□

If $A \xrightarrow{h} B$ is any morphism, then it follows from the free algebra algorithm (see Appendix A.1) in connection with (1.1) and (1.2) that it is not difficult to verify the commutativity of the following diagram

$$\begin{array}{ccc} A^\sharp & \xrightarrow{\mathsf{T}(h)} & B^\sharp \\ {\scriptstyle \psi_\infty^{r_A^{-1}}}\downarrow & & \downarrow{\scriptstyle \psi_\infty^{r_B^{-1}}} \\ A \otimes \mathbb{1}^\sharp & \xrightarrow{h \otimes 1_{\mathbb{1}^\sharp}} & B \otimes \mathbb{1}^\sharp \end{array}$$

Hence $\psi_\infty = (\psi_\infty^{r_C^{-1}})_{C \in \mathsf{C}_0}$ is a natural isomorphism from the F_X-term functor T to $\_ \otimes \mathbb{1}^\sharp$.

As a last step of the preparations we introduce a binary operation $m_{\mathbb{1}^\sharp}$ on $\mathbb{1}^\sharp$ by the diagram

$$\begin{array}{ccc} (\mathbb{1}^\sharp)^\sharp & \xrightarrow{\mu_{\mathbb{1}}} & \mathbb{1}^\sharp \\ {\scriptstyle \psi_\infty^\rho} \downarrow & \nearrow_{m_{\mathbb{1}^\sharp}} & \\ \mathbb{1}^\sharp \otimes \mathbb{1}^\sharp & & \end{array}$$

where $\rho = r_{\mathbb{1}^\sharp}^{-1}$. We will show that $(\mathbb{1}^\sharp, m_{\mathbb{1}^\sharp}, \eta_{\mathbb{1}})$ is the free monoid generated by X. Hence from an intuitive point of view the binary operation $m_{\mathbb{1}^\sharp}$ plays the rôle of "concatenation".

Theorem 1.1.7. *The triple $(\mathbb{1}^\sharp, m_{\mathbb{1}^\sharp}, \eta_{\mathbb{1}})$ is a monoid in C.*

Proof. First we recall $\xi = \psi_\infty^{r_{\mathbb{1}^\sharp}^{-1}}$ and $\pi = r_{\mathbb{1}^\sharp \otimes \mathbb{1}^\sharp}^{-1}$. Since $\psi_\infty : \mathsf{T} \to \_ \otimes \mathbb{1}^\sharp$ is a natural isomorphism, the relation

$$\mu_{\mathbb{1}} \circ \mathsf{T}(\mu_{\mathbb{1}}) = m_{\mathbb{1}^\sharp} \circ (m_{\mathbb{1}^\sharp} \otimes 1_{\mathbb{1}^\sharp}) \circ \psi_\infty^\pi \circ \mathsf{T}(\xi)$$

follows. On the other hand, referring to Lemma 1.1.6 and formula (1.5) we obtain:

$$\begin{aligned} \mu_{\mathbb{1}} \circ \mu_{\mathbb{1}^\sharp} &= m_{\mathbb{1}^\sharp} \circ (1_{\mathbb{1}^\sharp} \otimes m_{\mathbb{1}^\sharp}) \circ (1_{\mathbb{1}^\sharp} \otimes \psi_\infty^\rho) \circ \psi_\infty^\xi \\ &= m_{\mathbb{1}^\sharp} \circ (1_{\mathbb{1}^\sharp} \otimes m_{\mathbb{1}^\sharp}) \circ a_{\mathbb{1}^\sharp \mathbb{1}^\sharp \mathbb{1}^\sharp} \circ \psi_\infty^\pi \circ \mathsf{T}(\xi). \end{aligned}$$

Hence the associativity of $m_{\mathbb{1}^\sharp}$ follows from the associativity axiom of the free T-algebra $(\mathbb{1}^\sharp, \mu_{\mathbb{1}})$ — i.e. $\mu_{\mathbb{1}} \circ \mu_{\mathbb{1}^\sharp} = \mu_{\mathbb{1}} \circ \mathsf{T}(\mu_{\mathbb{1}})$ (cf. Appendix A.1). In order to verify the unit axiom for $\eta_{\mathbb{1}}$ we first observe $\psi_\infty^{r_{\mathbb{1}}^{-1}} = \ell_{\mathbb{1}^\sharp}^{-1}$. Then we obtain:

$$\mu_{\mathbb{1}} \circ \mathsf{T}(\eta_{\mathbb{1}}) = m_{\mathbb{1}^\sharp} \circ (\eta_{\mathbb{1}} \otimes 1_{\mathbb{1}^\sharp}) \circ \ell_{\mathbb{1}^\sharp}^{-1}.$$

Since $\mathbb{1} \xrightarrow{\eta_{\mathbb{1}}} \mathbb{1}^\sharp = Z_\infty(\mathbb{1}) \sqcup \mathbb{1}$ is the coproduct injection of $\mathbb{1}$ and ρ coincides with $r_{\mathbb{1}^\sharp}^{-1}$, the following relation holds:

$$\mu_{\mathbb{1}} \circ \eta_{\mathbb{1}^\sharp} = m_{\mathbb{1}^\sharp} \circ \psi_\infty^\rho \circ \eta_{\mathbb{1}^\sharp} = m_{\mathbb{1}^\sharp} \circ (1_{\mathbb{1}^\sharp} \otimes \eta_{\mathbb{1}}) \circ r_{\mathbb{1}^\sharp}^{-1}.$$

Hence the unit axiom for $\eta_{\mathbb{1}}$ follows from the property $\mu_{\mathbb{1}} \circ \mathsf{T}(\eta_{\mathbb{1}}) = 1_{\mathbb{1}^\sharp} = \mu_{\mathbb{1}} \circ \eta_{\mathbb{1}^\sharp}$ (cf. Appendix A.1) which is related to the unit axiom of the F_X-term monad (see also Sect. 1.2). □

In order to establish the universal property of the monoid $(\mathbb{1}^\sharp, m_{\mathbb{1}^\sharp}, \eta_{\mathbb{1}})$ we introduce a morphism $X \xrightarrow{\eta_X} \mathbb{1}^\sharp$ of C_0 by

$$\eta_X = j_\infty^{\mathbb{1}} \circ e_1^{\mathbb{1}} \circ \ell_X^{-1}.$$

Warning. In this subsection the morphism η_X defined above is *not* the X-component of the unit of the F_X-term monad.

Theorem 1.1.8. *Let* (Y, m_Y, e_Y) *be a monoid in* C *and* $X \xrightarrow{h} Y$ *be a morphism of* C_0. *There exists a unique unital homomorphism* $(\mathbb{1}^\sharp, m_{\mathbb{1}^\sharp}, \eta_{\mathbb{1}}) \xrightarrow{\widehat{h}} (Y, m_Y, e_Y)$ *making the following diagram commutative*:

$$\begin{array}{ccc} X & \xrightarrow{\eta_X} & \mathbb{1}^\sharp \\ & \searrow^{h} & \downarrow{\scriptstyle \widehat{h}} \\ & & Y \end{array}$$

Proof. (a) (Uniqueness) Since $\widehat{h}$ is unital and $\mathbb{1}^\sharp$ is the coproduct of $Z_\infty(\mathbb{1})$ and $\mathbb{1}$, $\widehat{h}$ is uniquely determined by $h_\infty = \widehat{h} \circ j_\infty^{\mathbb{1}}$. Obviously, h_∞ induces a sequence $(h_n)_{n \in \mathbb{N}}$ of morphisms $Z_n(\mathbb{1}) \xrightarrow{h_n} Y$ by $h_n = h_\infty \circ e_n^{\mathbb{1}}$, and $h_1 = h \circ \ell_X$ follows from the commutativity of the previous diagram. If we can prove

$$h_{n+1} = m_Y \circ \big((h_n \sqcup e_Y) \otimes h\big), \qquad n \in \mathbb{N}, \tag{1.9}$$

then $\widehat{h}$ is uniquely determined. By the coherence axiom (2) we first observe that $\varphi_1^\rho = 1_{\mathbb{1}^\sharp} \otimes \ell_X^{-1}$. Then we obtain the following relation:

$$\begin{aligned} h_{n+1} &= \widehat{h} \circ j_\infty^{\mathbb{1}} \circ e_{n+1}^{\mathbb{1}} = \widehat{h} \circ \mu_1^{\mathbb{1}} \circ \mathsf{F}_X(e_n^{\mathbb{1}} \oplus 1_{\mathbb{1}}) = \widehat{h} \circ \mu_{\mathbb{1}} \circ j_\infty^{\mathbb{1}^\sharp} \circ e_1^{\mathbb{1}^\sharp} \circ \mathsf{F}_X(e_n^{\mathbb{1}} \oplus 1_{\mathbb{1}}) \\ &= \widehat{h} \circ m_{\mathbb{1}^\sharp} \circ \psi_\infty^\rho \circ j_\infty^{\mathbb{1}^\sharp} \circ e_1^{\mathbb{1}^\sharp} \circ \mathsf{F}_X(e_n^{\mathbb{1}} \oplus 1_{\mathbb{1}}) \\ &= m_Y \circ (\widehat{h} \otimes \widehat{h}) \circ (1_{\mathbb{1}^\sharp} \otimes j_\infty^{\mathbb{1}}) \circ \varphi_\infty^\rho \circ e_1^{\mathbb{1}^\sharp} \circ \mathsf{F}_X(e_n^{\mathbb{1}} \oplus 1_{\mathbb{1}}) \\ &= m_Y \circ (\widehat{h} \otimes \widehat{h}) \circ (1_{\mathbb{1}^\sharp} \otimes j_\infty^{\mathbb{1}}) \circ (1_{\mathbb{1}^\sharp} \otimes e_1^{\mathbb{1}}) \circ \varphi_1^\rho \circ \big((e_n^{\mathbb{1}} \oplus 1_{\mathbb{1}}) \otimes 1_X\big) \\ &= m_Y \circ \big((h_n \sqcup e_Y) \otimes (h_1 \circ \ell_X^{-1})\big) \end{aligned}$$

where we have used (1.1). Hence (1.9) follows.
(b) (Existence) Since e_Y is the unit of (Y, m_Y), we first observe that for any morphism $A \xrightarrow{f} Y$ the relations

$$m_Y \circ (e_Y \otimes f) = f \circ \ell_A \quad \text{and} \quad m_Y \circ (f \otimes e_Y) = f \circ r_A \tag{1.10}$$

hold. Now we put $h_1 = h \circ \ell_X$ and recursively define a sequence $(h_n)_{n \in \mathbb{N}}$ of morphisms $Z_n(\mathbb{1}) \xrightarrow{h_n} Y$ by (1.9). We apply (1.10) and obtain:

$$h_2 \circ e_{21}^{\mathbb{1}} = m_Y \circ ((h_1 \sqcup e_Y) \otimes h) \circ (j_{\mathbb{1}} \otimes 1_X) = m_Y \circ (e_Y \otimes h) = h \circ \ell_X = h_1.$$

Then $h_{n+1} \circ e_{n+1\,n}^{\mathbb{1}} = h_n$ follows by induction. Hence there exists a unique morphism $Z_\infty(\mathbb{1}) \xrightarrow{h_\infty} Y$ such that $h_\infty \circ e_n^{\mathbb{1}} = h_n$ for all $n \in \mathbb{N}$. We put $\widehat{h} = h_\infty \sqcup e_Y$ and observe that $\widehat{h} \circ \eta_X = h_1 \circ \ell_X^{-1} = h$ and $\widehat{h} \circ \eta_{\mathbb{1}} = e_Y$.

Finally, we show that $\widehat{h}$ is a homomorphism. For this purpose we first prove the following relation for all $k \in \mathbb{N}$:

$$\widehat{h} \circ \mu_k^{\mathbb{1}} = m_Y \circ (\widehat{h} \otimes h_k) \circ \varphi_k^{\rho}. \tag{1.11}$$

We choose $k = 1$ and notice that the relation

$$\begin{aligned}\widehat{h} \circ \mu_1^{\mathbb{1}} \circ \mathsf{F}_X(e_n^{\mathbb{1}} \oplus 1_{\mathbb{1}}) = h_\infty \circ \vartheta_{\mathbb{1}} \circ \mathsf{F}_X(e_n^{\mathbb{1}} \oplus 1_{\mathbb{1}}) = h_{n+1} = m_Y \circ \big((h_n \sqcup e_Y) \otimes h\big)\\ = m_Y \circ (\widehat{h} \otimes h_1) \circ \varphi_1^{\rho} \circ \mathsf{F}_X(e_n^{\mathbb{1}} \oplus 1_{\mathbb{1}})\end{aligned}$$

holds for all $n \in \mathbb{N}$. Hence the universal property of the direct limit of $\mathsf{F}_X(W_n(\mathbb{1}))$ implies the validity of (1.11) when $k = 1$.

Let us now assume that (1.11) holds for some $k \in \mathbb{N}$. Then the following diagram is commutative:

$$\begin{array}{ccccccc}
Z_{k+1}(\mathbb{1}^\sharp) & \xrightarrow{\mathsf{F}_X(\mu_k^{\mathbb{1}} \sqcup 1_{\mathbb{1}^\sharp})} \mathsf{F}_X(\mathbb{1}^\sharp) \xrightarrow{\widehat{h} \otimes 1_X} & Y \otimes X & \xrightarrow{1_Y \otimes h} & Y \otimes Y \\
\big\downarrow{\scriptstyle \mathsf{F}_X(\psi_k^{\rho})} & & \big\uparrow{\scriptstyle m_Y \otimes 1_X} & & \big\uparrow{\scriptstyle m_Y \otimes 1_Y} \\
\mathsf{F}_X(\mathbb{1}^\sharp \otimes W_k(\mathbb{1})) & \xrightarrow{\mathsf{F}_X(\widehat{h} \otimes (h_k \sqcup e_Y))} & (Y \otimes Y) \otimes X & \xrightarrow{1_{Y \otimes Y} \otimes h} & (Y \otimes Y) \otimes Y \\
\big\downarrow{\scriptstyle a_{\mathbb{1}^\sharp W_k(\mathbb{1}) X}} & & \big\downarrow{\scriptstyle a_{YYX}} & & \big\downarrow{\scriptstyle a_{YYY}} \\
\mathbb{1}^\sharp \otimes Z_{k+1}(\mathbb{1}) & \xrightarrow{\widehat{h} \otimes ((h_k \sqcup e_Y) \otimes 1_X)} & Y \otimes (Y \otimes X) & \xrightarrow{1_Y \otimes (1_Y \otimes h)} & Y \otimes (Y \otimes Y)
\end{array}$$

where we have used again (1.10). Further, we recall $\mu_1^{\mathbb{1}} = \delta_{\mathbb{1}^\sharp}$ and apply the associativity of m_Y. Then we obtain:

$$\begin{aligned}
\widehat{h} \circ \mu_{k+1}^{\mathbb{1}} &= \widehat{h} \circ \mu_1^{\mathbb{1}} \circ F_X(\mu_n^{\mathbb{1}} \sqcup 1_{\mathbb{1}^\sharp}) \\
&= m_Y \circ (\widehat{h} \otimes h) \circ \mathsf{F}_X(\mu_n^{\mathbb{1}} \sqcup 1_{\mathbb{1}^\sharp}) \\
&= m_Y \circ (m_Y \otimes 1_Y) \circ a_{YYY}^{-1} \circ \big(\widehat{h} \otimes ((h_k \sqcup e_Y) \otimes h)\big) \circ \varphi_{k+1}^{\rho} \\
&= m_Y \circ (1_Y \otimes m_Y) \circ \big(\widehat{h} \otimes ((h_k \sqcup e_Y) \otimes h)\big) \circ \varphi_{k+1}^{\rho} \\
&= m_Y \circ (\widehat{h} \otimes h_{k+1}) \circ \varphi_{k+1}^{\rho},
\end{aligned}$$

Thus (1.11) also holds for $k + 1$. Finally, we pass to the respective direct limits and again make use of (1.10). Then the diagram

$$\begin{array}{ccccc}
Z_\infty(\mathbb{1}^\sharp) \sqcup \mathbb{1}^\sharp & \xrightarrow{\mu_{\mathbb{1}}} & \mathbb{1}^\sharp & \xrightarrow{\widehat{h}} & Y \\
\big\downarrow{\scriptstyle \psi_\infty^{\rho}} & \nearrow{\scriptstyle m_{\mathbb{1}^\sharp}} & & & \big\uparrow{\scriptstyle m_Y} \\
\mathbb{1}^\sharp \otimes \big(Z_\infty(\mathbb{1}) \sqcup \mathbb{1}\big) & & \xrightarrow{\widehat{h} \otimes (h_\infty \sqcup e_Y)} & & Y \otimes Y
\end{array}$$

is commutative, and consequently $\widehat{h}$ is a homomorphism. □

We record the following:

FACT. *The triple* $(\mathbb{1}^\sharp, m_{\mathbb{1}^\sharp}, \eta_{\mathbb{1}})$ *is the free monoid generated by* X *and the morphism* $\eta_X = j^\sharp_\infty \circ e^{\mathbb{1}}_1 \circ \ell_X^{-1}$ *is the corresponding "embedding"* $X \xrightarrow{\eta_X} \mathbb{1}^\sharp$.

1.1.3 Tensor Products of Semigroups

In contrast to Sect. 1.1.1, the biclosedness of the underlying monoidal category is now replaced by the symmetry axiom. Hence in this subsection we assume a symmetry c on C (see Definition 1.1.3 (a)).

Let (X, m_X) and (Y, m_Y) be semigroups in C. In order to define a semigroup structure on the tensor product $X \otimes Y$ we need a construction interchanging the second and third factor of $(X \otimes Y) \otimes (X \otimes Y)$. For this purpose we need some notation and choose therefore four objects X, Y, U, V of C_0. First we introduce a morphism

$$(X \otimes Y) \otimes (U \otimes V) \xrightarrow{\Phi_{XYUV}} (X \otimes (Y \otimes U)) \otimes V$$

by the diagram:

$$\begin{array}{ccc} (X \otimes Y) \otimes (U \otimes V) & & \\ {\scriptstyle a_{X\otimes YUV}}\big\uparrow & \searrow^{\Phi_{XYUV}} & \\ ((X \otimes Y) \otimes U) \otimes V & \xrightarrow{a_{XYU}\otimes 1_V} & (X \otimes (Y \otimes U)) \otimes V \end{array}$$

Then we define a morphism $(X \otimes Y) \otimes (U \otimes V) \xrightarrow{\Theta_{XYUV}} (X \otimes U) \otimes (Y \otimes V)$ interchanging the second and third factor as follows:

$$\Theta_{XYUV} = (\Phi_{XUYV})^{-1} \circ \big((1_X \otimes c_{YU}) \otimes 1_V\big) \circ \Phi_{XYUV}. \tag{1.12}$$

Since a and c are natural transformations, the diagram

$$\begin{array}{ccc} (X_1 \otimes Y_1) \otimes (U_1 \otimes V_1) & \xrightarrow{(f\otimes g)\otimes(h\otimes k)} & (X_2 \otimes Y_2) \otimes (U_2 \otimes V_2)) \\ {\scriptstyle \Theta_{X_1Y_1U_1V_1}}\big\downarrow & & \big\downarrow{\scriptstyle \Theta_{X_2Y_2U_2V_2}} \\ (X_1 \otimes U_1) \otimes (Y_1 \otimes V_1) & \xrightarrow{(f\otimes h)\otimes(g\otimes k)} & (X_2 \otimes U_2) \otimes (Y_2 \otimes V_2) \end{array} \tag{1.13}$$

is commutative — i.e. Θ is also a natural transformation.

Theorem 1.1.9. *Let* C *be a symmetric monoidal category. If* (X, m_X) *and* (Y, m_Y) *are semigroups in* C, *then* $(X \otimes Y, m)$ *with* $m = (m_X \otimes m_Y) \circ \Theta_{XYXY}$ *is again a semigroup in* C.

Proof. If (X, m_X) and (Y, m_Y) are semigroups in C, then we define a binary operation m on $X \otimes Y$ by $m = (m_X \otimes m_Y) \circ \Theta_{XYXY}$. In order to verify the associativity axiom for m we first observe the commutativity of the diagram

$$\begin{array}{ccc}
\big((X \otimes Y) \otimes (X \otimes Y)\big) \otimes (X \otimes Y) & \xrightarrow{a_{(X\otimes Y)(X\otimes Y)(X\otimes Y)}} & (X \otimes Y) \otimes \big((X \otimes Y) \otimes (X \otimes Y)\big) \\
{\scriptstyle \Theta_{XYXY}\otimes 1_{X\otimes Y}}\downarrow & & \downarrow{\scriptstyle 1_{X\otimes Y}\otimes\Theta_{XYXY}} \\
\big((X \otimes X) \otimes (Y \otimes Y)\big) \otimes (X \otimes Y) & & (X \otimes Y) \otimes \big((X \otimes X) \otimes (Y \otimes Y)\big) \\
{\scriptstyle \Theta_{(X\otimes X)(Y\otimes Y)XY}}\downarrow & & \downarrow{\scriptstyle \Theta_{XY(X\otimes X)(Y\otimes Y)}} \\
((X \otimes X) \otimes X) \otimes ((Y \otimes Y) \otimes Y) & \xrightarrow{a_{XXX}\otimes a_{YYY}} & (X \otimes (X \otimes X)) \otimes (Y \otimes (Y \otimes Y))
\end{array}$$

which is a direct consequence of the pentagonal diagram (1) and the coherence axiom (4) (for more details see Proposition A.2.1 in Appendix A.2). Further, we derive the following relations from (1.13)

$$\begin{aligned}
\big((m_X \otimes 1_X) \otimes (m_Y \otimes 1_Y)\big) \circ \Theta_{X\otimes XY\otimes YXY} &= \Theta_{XYXY} \circ \big((m_X \otimes m_Y) \otimes 1_{X\otimes Y}\big), \\
\big(1_X \otimes m_X) \otimes (1_Y \otimes m_Y)\big) \circ \Theta_{XYX\otimes XY\otimes Y} &= \Theta_{XYXY} \circ \big(1_{X\otimes Y} \otimes (m_X \otimes m_Y)\big).
\end{aligned}$$

Then we obtain:

$$\begin{aligned}
& m \circ (m \otimes 1_{X\otimes Y}) \\
&= (m_X \otimes m_Y) \circ \Theta_{XYXY} \circ \big((m_X \otimes m_Y) \otimes 1_{X\otimes Y}\big) \circ (\Theta_{XYXY} \otimes 1_{X\otimes Y}) \\
&= (m_X \otimes m_Y) \circ \big((m_X \otimes 1_X) \otimes (m_Y \otimes 1_Y)\big) \circ \Theta_{X\otimes XY\otimes YXY} \circ (\Theta_{XYXY} \otimes 1_{X\otimes Y}) \\
&= (m_X \otimes m_Y) \circ \big((1_X \otimes m_X) \otimes (1_Y \otimes m_Y)\big) \circ \\
&\qquad\qquad \circ (a_{XXX} \otimes a_{YYY}) \circ \Theta_{X\otimes XY\otimes YXY} \circ (\Theta_{XYXY} \otimes 1_{X\otimes Y}) \\
&= m \circ (1_{X\otimes Y} \otimes m) \circ a_{X\otimes YX\otimes YX\otimes Y}.
\end{aligned}$$

□

The semigroup $(X \otimes Y, m)$ constructed in Theorem 1.1.9 is called the *tensor product* of the semigroups (X, m_X) and (Y, m_Y). If (X, m_X, e_X) and (Y, m_Y, e_Y) are monoids in C, then $e = (e_X \otimes e_Y) \circ \ell_{\mathbb{1}}^{-1}$ is obviously the unit of $(X \otimes Y, m)$ (see also Exercise 1.1.2). Hence the tensor product of monoids in C is again a monoid in C.

Exercises

1.1.1. Show that in any monoidal category the following diagrams are commutative:

$$\begin{array}{ccccc}
(\mathbb{1} \otimes X) \otimes Y & & \xrightarrow{a_{\mathbb{1}XY}} & & \mathbb{1} \otimes (X \otimes Y) \\
& {\scriptstyle \ell_X\otimes 1_Y}\searrow & & \swarrow{\scriptstyle \ell_{X\otimes Y}} & \\
& & X \otimes Y & &
\end{array}
\qquad
\begin{array}{ccccc}
(X \otimes Y) \otimes \mathbb{1} & & \xrightarrow{a_{XY\mathbb{1}}} & & X \otimes (Y \otimes \mathbb{1}) \\
& {\scriptstyle r_{X\otimes Y}}\searrow & & \swarrow{\scriptstyle 1_X\otimes r_Y} & \\
& & X \otimes Y & &
\end{array}$$

1.1.2. Let $\mathsf{C} = (\mathsf{C}_0, \otimes, \mathbb{1}, a, c, \ell, r)$ be a symmetric and monoidal category. Further, let $(\mathbb{1} \otimes \mathbb{1}) \otimes (X \otimes Y) \xrightarrow{\Theta_{\mathbb{1}\mathbb{1}XY}} (\mathbb{1} \otimes X) \otimes (\mathbb{1} \otimes Y)$ be the morphism introduced in (1.12). Show that the following relation holds:

$$(\ell_X \otimes \ell_Y) \circ \Theta_{\mathbb{1}\mathbb{1}XY} = \ell_{X \otimes Y} \circ (\ell_{\mathbb{1}} \otimes 1_{X \otimes Y}).$$

1.1.3. Let $\mathsf{C} = (\mathsf{C}_0, \otimes, \mathbb{1}, a, c, \ell, r)$ be a monoidal biclosed category with underlying cocomplete category C_0. Further, let $X \in |\mathsf{C}_0|$ and $(X^\sharp, m_{X^\sharp})$ be the free magma in C generated by X. Show that the unitalization of $(X^\sharp, m_{X^\sharp})$ is the free unital magma in C.

1.2 Monads and Distributive Laws

Even though we expect that the reader is familiar with the basic definitions and properties of monads, we begin by recalling some of them. The reason for this approach is to fix our notation for later purposes.

Let C be a category. A triple $\mathbf{T} = (\mathsf{T}, \eta, \mu)$ is a *monad* on C if T is an endofunctor of C (i.e. $\mathsf{T}\colon \mathsf{C} \to \mathsf{C}$) and $\eta\colon \mathrm{id}_{\mathsf{C}} \to \mathsf{T}$ and $\mu\colon \mathsf{TT} \to \mathsf{T}$ are natural transformations such that for all $X \in |\mathsf{C}|$ the following diagrams are commutative:

$$\begin{array}{ccccc} \mathsf{T}(X) & \xrightarrow{\eta_{\mathsf{T}(X)}} & \mathsf{TT}(X) & \xleftarrow{\mathsf{T}(\eta_X)} & \mathsf{T}(X) \\ & {\scriptstyle 1_{\mathsf{T}(X)}}\searrow & \downarrow{\scriptstyle \mu_X} & \swarrow{\scriptstyle 1_{\mathsf{T}(X)}} & \\ & & \mathsf{T}(X) & & \end{array} \qquad \text{(Unit Axiom)}$$

$$\begin{array}{ccc} \mathsf{TTT}(X) & \xrightarrow{\mathsf{T}(\mu_X)} & \mathsf{TT}(X) \\ {\scriptstyle \mu_{\mathsf{T}(X)}}\downarrow & & \downarrow{\scriptstyle \mu_X} \\ \mathsf{TT}(X) & \xrightarrow{\mu_X} & \mathsf{T}(X) \end{array} \qquad \text{(Associativity Axiom)}$$

Using the (vertical) composition and the star composition of natural transformations, the previous axioms can be rewritten as follows:

$$\mu \circ (\eta \star 1_{\mathsf{T}}) = 1_{\mathsf{T}} = \mu \circ (1_{\mathsf{T}} \star \eta) \quad \text{and} \quad \mu \circ (\mu \star 1_{\mathsf{T}}) = \mu \circ (1_{\mathsf{T}} \star \mu).$$

Hence, if C is small, then monads on C are monoids in the monoidal category $\mathtt{End}(\mathsf{C})$ (cf. Example 1.1.2). This observation justifies the previous notation as well as the following terminology. If $\mathbf{T}$ is a monad, then η is called the *unit* of $\mathbf{T}$, and μ is called the *multiplication* of $\mathbf{T}$.

Every adjoint situation $(\eta, \mathsf{C} \xrightarrow{\mathsf{F}} \mathsf{D} \xrightarrow{\mathsf{G}} \mathsf{C})$ — i.e. F is a left adjoint functor to G, and η is the corresponding unit of $\mathsf{F} \dashv \mathsf{G}$ — induces a monad $\mathbf{T} = (\mathsf{T}, \eta, \mu)$ on C as follows:

$$\mathsf{T} = \mathsf{GF}, \quad \eta \text{ is the unit of } \mathsf{F} \dashv \mathsf{G}, \quad \mu = (\mu_X)_{X \in |\mathsf{C}|}, \quad \mu_X = \mathsf{G}(\ulcorner 1_{\mathsf{T}(X)} \urcorner)$$

where $\mathsf{FGF}(X) \xrightarrow{\ulcorner 1_{\mathsf{T}(X)} \urcorner} \mathsf{F}(X)$ is the unique morphism making the following diagram commutative

$$\begin{array}{ccc} \mathsf{GF}(X) & \xrightarrow{\eta_{\mathsf{T}(X)}} & \mathsf{GFGF}(X) \\ {\scriptstyle 1_{\mathsf{T}(X)}}\downarrow & \swarrow {\scriptstyle \mathsf{G}(\ulcorner 1_{\mathsf{T}(X)} \urcorner)} & \\ \mathsf{GF}(X) & & \end{array}$$

Comment. If the right adjoint functor G of F is the *forgetful functor*, then the X-component μ_X of the multiplication of the induced monad "can be read as" $\ulcorner 1_{\mathsf{T}(X)} \urcorner$. In this situation μ_X is the *extension* of the identity of $\mathsf{T}(X)$ (cf. e.g. the construction of the multiplication of the F-term monad in Appendix A.1).

The aim of the following considerations is to show that every monad is induced by an adjoint situation. For this purpose we can construct either the Kleisli category or the Eilenberg–Moore category. Here we prefer the latter.

Definition 1.2.1. Let $\mathbf{T} = (\mathsf{T}, \eta, \mu)$ be a monad on a category C. A $\mathbf{T}$-*algebra* is a pair (X, ξ) where $X \in |\mathsf{C}|$ and $\mathsf{T}(X) \xrightarrow{\xi} X$ is a morphism (so-called *structure map*) making the following diagrams commutative:

$$\begin{array}{ccc} X & \xrightarrow{\eta_X} & \mathsf{T}(X) \\ & {\scriptstyle 1_X}\searrow & \downarrow{\scriptstyle \xi} \\ & & X \end{array} \qquad\qquad \begin{array}{ccc} \mathsf{TT}(X) & \xrightarrow{\mathsf{T}(\xi)} & \mathsf{T}(X) \\ {\scriptstyle \mu_X}\downarrow & & \downarrow{\scriptstyle \xi} \\ \mathsf{T}(X) & \xrightarrow{\xi} & X \end{array}$$

(Unit Axiom) (Associativity Axiom)

Let (X, ξ) and (Y, ζ) be $\mathbf{T}$-algebras. A $\mathbf{T}$-*homomorphism* is a morphism $X \xrightarrow{h} Y$ in C such that the following diagram is commutative:

$$\begin{array}{ccc} \mathsf{T}(X) & \xrightarrow{\mathsf{T}(h)} & \mathsf{T}(Y) \\ {\scriptstyle \xi}\downarrow & & \downarrow{\scriptstyle \zeta} \\ X & \xrightarrow{h} & Y \end{array}$$

Obviously, $\mathbf{T}$-algebras and $\mathbf{T}$-homomorphisms form a category denoted by $\mathsf{C}^{\mathbf{T}}$ — the so-called *Eilenberg–Moore category*.

Theorem 1.2.2. *Let* $\mathbf{T} = (\mathsf{T}, \eta, \mu)$ *be a monad on* C. *Then the forgetful functor* $\mathsf{U}\colon \mathsf{C}^{\mathbf{T}} \to \mathsf{C}$ *sending each* $\mathbf{T}$*-algebra to its underlying* C*-object has a left adjoint* $\mathsf{F}\colon \mathsf{C} \to \mathsf{C}^{\mathbf{T}}$. *Moreover, the monad induced by* $\mathsf{F} \dashv \mathsf{U}$ *coincides with* $\mathbf{T}$.

The assertion of the previous theorem follows from the universal property of free $\mathbf{T}$-algebras. Since we will frequently refer to this property, we include here its details (see [74, p. 36]).

Lemma 1.2.3. (Universal Property of Free **T**-Algebras) *Let* **T** *be a monad on* C *and* $X \in |\mathsf{C}|$. *Then* $(\mathsf{T}(X), \mu_X)$ *is a* **T***-algebra, and for every further* **T***-algebra* (Y, ζ) *and morphism* $X \xrightarrow{h} Y$ *in* C *there exists a unique* **T***-homomorphism* $\mathsf{T}(X) \xrightarrow{h^\sharp} Y$ *making the following diagram commutative*:

$$\begin{array}{ccc} X & \xrightarrow{\eta_X} & \mathsf{T}(X) \\ & {\scriptstyle h}\searrow & \downarrow{\scriptstyle h^\sharp} \\ & & Y \end{array} \tag{1.14}$$

Proof. It follows immediately from the unit and associativity axiom of a monad that $(\mathsf{T}(X), \mu_X)$ is a **T**-algebra.

(Uniqueness) Let (Y, ζ) be a **T**-algebra and $\mathsf{T}(X) \xrightarrow{h^\sharp} Y$ a **T**-homomorphism making the diagram in Lemma 1.2.3 commutative. Then we obtain:

$$h^\sharp = h^\sharp \circ \mu_X \circ \mathsf{T}(\eta_X) = \zeta \circ \mathsf{T}(h^\sharp) \circ \mathsf{T}(\eta_X) = \zeta \circ \mathsf{T}(h).$$

(Existence) Motivated by the uniqueness proof we define $\mathsf{T}(X) \xrightarrow{h^\sharp} Y$ by:

$$h^\sharp = \zeta \circ \mathsf{T}(h). \tag{1.15}$$

Since $\eta\colon \mathrm{id}_{\mathsf{C}} \to \mathsf{T}$ is a natural transformation, we obtain:

$$h^\sharp \circ \eta_X = \zeta \circ \mathsf{T}(h) \circ \eta_X = \zeta \circ \eta_Y \circ h = h.$$

Hence $h^\sharp$ makes the diagram (1.14) commutative. Therefore it remains to show that $h^\sharp$ is a **T**-homomorphism. In fact, since $\mu\colon \mathsf{TT} \to \mathsf{T}$ is a natural transformation, we conclude from the associativity axiom of ζ that the following relation holds:

$$\zeta \circ \mathsf{T}(h^\sharp) = \zeta \circ \mathsf{T}(\zeta) \circ \mathsf{TT}(h) = \zeta \circ \mu_Y \circ \mathsf{TT}(h) = \zeta \circ \mathsf{T}(h) \circ \mu_X = h^\sharp \circ \mu_X. \quad \square$$

Referring to [73, 74] the following permanence property is well-known. If C is a complete category, then the Eilenberg–Moore category of a monad **T** on C is also complete.

Since monads are intuitively monoids, the question arises whether a "tensor product of monads" exists in some sense. However in this context, we regret to find out that in the case of small categories C the monoidal category of endofunctors of C is *not* symmetric. Therefore with regard to Sect. 1.1.3 the previous question asks us to find a substitute for the concept of symmetry in monoidal categories. This notion leads to the idea of "swapper maps" or to distributive laws of endofunctors over monads.

In the following we adopt the terminology introduced by E.G. Manes 1976 (see [74, Definition 3.6 on p. 311]).

Definition 1.2.4. Let C be a category. Further, let $\mathsf{P}\colon \mathsf{C} \to \mathsf{C}$ be an endofunctor and $\mathbf{T} = (\mathsf{T}, \eta^{\mathsf{T}}, \mu^{\mathsf{T}})$ be a monad on C. A *distributive law of* P *over* $\mathbf{T}$ is a natural transformation $\sigma\colon \mathsf{TP} \to \mathsf{PT}$ such that the following diagrams are commutative for each $X \in |\mathsf{C}|$

(A)
$$\begin{array}{ccc} \mathsf{TP}(X) & & \\ {\scriptstyle \eta^{\mathsf{T}}_{\mathsf{P}(X)}}\uparrow & \overset{\sigma_X}{\searrow} & \\ \mathsf{P}(X) & \xrightarrow{\mathsf{P}(\eta^{\mathsf{T}}_X)} & \mathsf{PT}(X) \end{array}$$

(B)
$$\begin{array}{ccccc} \mathsf{TTP}(X) & \xrightarrow{\mathsf{T}(\sigma_X)} & \mathsf{TPT}(X) & \xrightarrow{\sigma_{\mathsf{T}(X)}} & \mathsf{PTT}(X) \\ {\scriptstyle \mu^{\mathsf{T}}_{\mathsf{P}(X)}}\downarrow & & & & \downarrow{\scriptstyle \mathsf{P}(\mu^{\mathsf{T}}_X)} \\ \mathsf{TP}(X) & & \xrightarrow{\qquad\sigma_X\qquad} & & \mathsf{PT}(X) \end{array}$$

Given a distributive law of an endofunctor P over a monad $\mathbf{T}$, we first show that for every $\mathbf{T}$-algebra (X, ξ_X) there exists an intrinsic $\mathbf{T}$-algebra structure on $\mathsf{P}(X)$.

Lemma 1.2.5. *Let* $\sigma\colon \mathsf{TP} \to \mathsf{PT}$ *be a distributive law of an endofunctor* P *over a monad* $\mathbf{T}$.

(a) *If* (X, ξ_X) *is a* $\mathbf{T}$*-algebra, then the morphism* $\mathsf{TP}(X) \xrightarrow{\zeta_X} \mathsf{P}(X)$ *defined by*

$$\zeta_X = \mathsf{P}(\xi_X) \circ \sigma_X \tag{1.16}$$

is again a $\mathbf{T}$*-structure map — i.e.* $(\mathsf{P}(X), \zeta_X)$ *is a* $\mathbf{T}$*-algebra.*

(b) *If* $(X, \xi_X) \xrightarrow{h} (Y, \xi_Y)$ *is a* $\mathbf{T}$*-homomorphism, then* $(\mathsf{P}(X), \zeta_X) \xrightarrow{\mathsf{P}(h)} (\mathsf{P}(Y), \zeta_Y)$ *is again a* $\mathbf{T}$*-homomorphism.*

Proof. (a) Let us define ζ_X by (1.16). Then the commutativity of (A) implies the unit axiom. In fact, the relation

$$\zeta_X \circ \eta^{\mathsf{T}}_{\mathsf{P}(X)} = \mathsf{P}(\xi_X) \circ \sigma_X \circ \eta^{\mathsf{T}}_{\mathsf{P}(X)} = \mathsf{P}(\xi_X) \circ \mathsf{P}(\eta^{\mathsf{T}}_X) = 1_{\mathsf{P}(X)}$$

holds. Since σ is a natural transformation, the commutativity of (B) guarantees the commutativity of the diagram:

$$\begin{array}{ccccc} \mathsf{TTP}(X) & \xrightarrow{\mathsf{T}(\sigma_X)} & \mathsf{TPT}(X) & \xrightarrow{\mathsf{TP}(\xi_X)} & \mathsf{TP}(X) \\ \Big| & & {\scriptstyle \sigma_{\mathsf{T}(X)}}\downarrow & & \downarrow{\scriptstyle \sigma_X} \\ {\scriptstyle \mu^{\mathsf{T}}_{\mathsf{P}(X)}}\Big| & & \mathsf{PTT}(X) & \xrightarrow{\mathsf{PT}(\xi_X)} & \mathsf{PT}(X) \\ \downarrow & & {\scriptstyle \mathsf{P}(\mu^{\mathsf{T}}_X)}\downarrow & & \downarrow{\scriptstyle \mathsf{P}(\xi_X)} \\ \mathsf{TP}(X) & \xrightarrow{\sigma_X} & \mathsf{PT}(X) & \xrightarrow{\mathsf{P}(\xi_X)} & \mathsf{P}(X) \end{array}$$

Hence the associativity axiom $\zeta_X \circ \mu^{\mathsf{T}}_{\mathsf{P}(X)} = \zeta_X \circ \mathsf{T}(\zeta_X)$ is also verified.

(b) Let $(X, \xi_X) \xrightarrow{h} (Y, \xi_Y)$ be a **T**-homomorphism. Referring to (1.16) and again to the property that σ is a natural transformation, it is easily seen that the diagram

$$\begin{array}{ccccccc}
\mathsf{TP}(X) & & & \xrightarrow{\mathsf{TP}(h)} & & & \mathsf{TP}(Y) \\
 & \searrow^{\sigma_X} & & & & \swarrow^{\sigma_Y} & \\
 & & \mathsf{PT}(X) & \xrightarrow{\mathsf{PT}(h)} & \mathsf{PT}(Y) & & \\
\zeta_X & & \downarrow{\scriptstyle \mathsf{P}(\xi_X)} & & {\scriptstyle \mathsf{P}(\xi_Y)}\downarrow & & \zeta_Y \\
 & \searrow & \mathsf{P}(X) & \xrightarrow{\mathsf{P}(h)} & \mathsf{P}(Y) & \swarrow &
\end{array}$$

is commutative. Hence $\mathsf{P}(h)$ is a **T**-homomorphism. □

By the previous lemma we introduce the following terminology and notation. If $\sigma : \mathsf{TP} \to \mathsf{PT}$ is a distributive law, then the **T**-algebra $(\mathsf{P}(X), \zeta_X)$ constructed from a given **T**-algebra (X, ξ_X) according to (1.16) is called the *canonical* **T**-*algebra induced by* ξ_X and is denoted by $\widetilde{\mathsf{P}}(X, \xi_X)$. Later we will see that the symbol $\widetilde{\mathsf{P}}$ plays the rôle of a special functor which can be regarded as a lifting of the given functor P.

Corollary 1.2.6. *Let $\sigma : \mathsf{TP} \to \mathsf{PT}$ be a distributive law of an endofunctor P over a monad $\mathbf{T} = (\mathsf{T}, \eta^{\mathsf{T}}, \mu^{\mathsf{T}})$ on C. Then for every C-object X the C-morphism $\mathsf{TP}(X) \xrightarrow{\sigma_X} \mathsf{PT}(X)$ is a* **T**-*homomorphism from $(\mathsf{TP}(X), \mu^{\mathsf{T}}_{\mathsf{P}(X)})$ to $\widetilde{\mathsf{P}}(\mathsf{T}(X), \mu^{\mathsf{T}}_X)$.*

Proof. Referring to (1.16) the **T**-structure map $\zeta_{\mathsf{T}(X)}$ of $\widetilde{\mathsf{P}}(\mathsf{T}(X), \mu^{\mathsf{T}}_X)$ is given by

$$\zeta_{\mathsf{T}(X)} = \mathsf{P}(\mu^{\mathsf{T}}_X) \circ \sigma_{\mathsf{T}(X)}.$$

Hence the commutativity of diagram (B) is equivalent to the statement that σ_X is a **T**-homomorphism. □

In the next step we strengthen Lemma 1.2.5 and give a characterization of the existence of distributive laws.

Theorem 1.2.7. *Let C be a category. Further, let $\mathsf{P} : \mathsf{C} \to \mathsf{C}$ be an endofunctor and let $\mathbf{T} = (\mathsf{T}, \eta^{\mathsf{T}}, \mu^{\mathsf{T}})$ be a monad on C. Then the following assertions are equivalent:*

(i) *There exists a distributive law $\sigma : \mathsf{TP} \to \mathsf{PT}$ of P over* **T**.

(ii) *For every* **T**-*algebra (X, ξ_X) there exists a morphism $\mathsf{TP}(X) \xrightarrow{\alpha_X} \mathsf{P}(X)$ such that $(\mathsf{P}(X), \alpha_X)$ is a* **T**-*algebra, and for every* **T**-*homomorphism $X \xrightarrow{h} Y$ the diagram*

$$\begin{array}{ccc}
\mathsf{TP}(X) & \xrightarrow{\mathsf{TP}(h)} & \mathsf{TP}(Y) \\
{\scriptstyle \alpha_X}\downarrow & & \downarrow{\scriptstyle \alpha_Y} \\
\mathsf{P}(X) & \xrightarrow{\mathsf{P}(h)} & \mathsf{P}(Y)
\end{array}$$

is commutative — i.e. $\mathsf{P}(X) \xrightarrow{\mathsf{P}(h)} \mathsf{P}(Y)$ is again a **T**-*homomorphism.*

Before we prove Theorem 1.2.7, we comment on the meaning of assertion (ii). In fact, if $\mathsf{C}^{\mathbf{T}}$ is the Eilenberg–Moore category of $\mathbf{T}$ and $\mathsf{U}\colon \mathsf{C}^{\mathbf{T}} \to \mathsf{C}$ is the forgetful functor, then assertion (ii) in Theorem 1.2.7 is equivalent to the statement that PU factors through U — i.e. the existence of a lifting $\mathsf{C}^{\mathbf{T}} \xrightarrow{\widetilde{\mathsf{P}}} \mathsf{C}^{\mathbf{T}}$ of P such that the following diagram commutes:

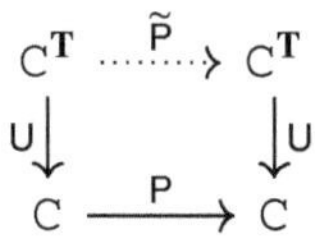

Proof of Theorem 1.2.7. The necessity of assertion (ii) for assertion (i) follows immediately from Lemma 1.2.5. In order to show that assertion (ii) is also sufficient for assertion (i) we proceed as follows.
Let $(\mathsf{T}(X), \mu^{\mathsf{T}}_X)$ be the free $\mathbf{T}$-algebra generated by the C-object X. Then we conclude from assertion (ii) that there exists a $\mathbf{T}$-structure map $\mathsf{TPT}(X) \xrightarrow{\alpha_{\mathsf{T}(X)}} \mathsf{PT}(X)$ such that $(\mathsf{PT}(X), \alpha_{\mathsf{T}(X)})$ is a $\mathbf{T}$-algebra. Motivated by Corollary 1.2.6, for every $X \in |\mathsf{C}|$ we consider the unique extension of the C-morphism $\mathsf{P}(X) \xrightarrow{\mathsf{P}(\eta^{\mathsf{T}}_X)} \mathsf{PT}(X)$ to a $\mathbf{T}$-homomorphism $\mathsf{TP}(X) \xrightarrow{\mathsf{P}(\eta^{\mathsf{T}}_X)^{\sharp}} \mathsf{PT}(X)$ (cf. Lemma 1.2.3), which we will denote by σ_X. Explicitly σ_X is given by:

$$\sigma_X = \alpha_{\mathsf{T}(X)} \circ \mathsf{TP}\big(\eta^{\mathsf{T}}_X\big). \tag{1.17}$$

We put $\sigma = (\sigma_X)_{X \in |\mathsf{C}|}$. Since the assertion (ii) implies that for every morphism $X \xrightarrow{f} Y$ the morphism $\mathsf{PT}(f)$ is also a $\mathbf{T}$-homomorphism, the following diagram is commutative:

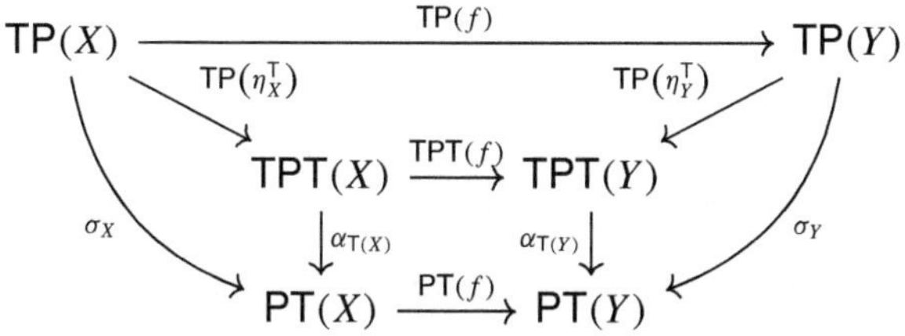

Hence $\sigma\colon \mathsf{TP} \to \mathsf{PT}$ is a natural transformation. By definition, σ_X fills in diagram (A). In order to verify (B) we first use (1.17) and obtain:

$$\begin{aligned}
\mathsf{P}\big(\mu^{\mathsf{T}}_X\big) \circ \sigma_{\mathsf{T}(X)} \circ \mathsf{T}(\sigma_X) \circ \eta^{\mathsf{T}}_{\mathsf{TP}(X)} &= \mathsf{P}\big(\mu^{\mathsf{T}}_X\big) \circ \sigma_{\mathsf{T}(X)} \circ \eta^{\mathsf{T}}_{\mathsf{PT}(X)} \circ \sigma_X \\
&= \mathsf{P}\big(\mu^{\mathsf{T}}_X\big) \circ \mathsf{P}\big(\eta^{\mathsf{T}}_{\mathsf{T}(X)}\big) \circ \sigma_X = \sigma_X \\
&= \sigma_X \circ \mu^{\mathsf{T}}_{\mathsf{P}(X)} \circ \eta^{\mathsf{T}}_{\mathsf{TP}(X)}.
\end{aligned}$$

Since μ_X^{T} is a **T**-homomorphism (see the associativity axiom of μ), we conclude from assertion (ii) that $\mathsf{P}(\mu_X^{\mathsf{T}})$ is also a T-homomorphism. Hence the commutativity of diagram (B) follows from the previous relation and the universal property of free **T**-algebras. □

As a preparation for Beck's Theorem we now characterize the commutativity of two further diagrams. For this purpose we make the following

STANDING ASSUMPTION. Let P be an endofunctor of C and $\mathbf{T} = (\mathsf{T}, \eta^{\mathsf{T}}, \mu^{\mathsf{T}})$ be a monad on C. Then $\sigma : \mathsf{TP} \to \mathsf{PT}$ is always a distributive law of P over $\mathbf{T}$.

Lemma 1.2.8. *Let* $\eta^{\mathsf{P}} : \mathrm{id}_{\mathsf{C}} \to \mathsf{P}$ *be a natural transformation. For every* $X \in |\mathsf{C}|$ *the following assertions are equivalent*:

(i) *The* C*-morphism* $\mathsf{T}(X) \xrightarrow{\eta^{\mathsf{P}}_{\mathsf{T}(X)}} \mathsf{PT}(X)$ *is a* **T***-homomorphism from* $(\mathsf{T}(X), \mu_X)$ *to* $\widetilde{\mathsf{P}}(\mathsf{T}(X), \mu_X)$.

(ii) *The following diagram is commutative*:

(C)

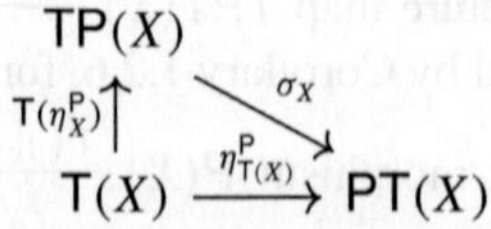

Proof. Since σ_X is always a **T**-homomorphism (cf. Corollary 1.2.6) the implication (ii) $\Longrightarrow$ (i) is obvious. On the other hand, since η^{P} and η^{T} are natural transformations and the diagram (A) is commutative, we obtain:

$$\eta^{\mathsf{P}}_{\mathsf{T}(X)} \circ \eta^{\mathsf{T}}_X = \mathsf{P}(\eta^{\mathsf{T}}_X) \circ \eta^{\mathsf{P}}_X = \sigma_X \circ \eta^{\mathsf{T}}_{\mathsf{P}(X)} \circ \eta^{\mathsf{P}}_X = \sigma_X \circ \mathsf{T}(\eta^{\mathsf{P}}_X) \circ \eta^{\mathsf{T}}_X.$$

If we now assume (i) and use again the fact that σ_X is a **T**-homomorphism, then the commutativity of diagram (C) follows from the universal property of free **T**-algebras. □

Corollary 1.2.9. *Let* $\eta^{\mathsf{P}} : \mathrm{id}_{\mathsf{C}} \to \mathsf{P}$ *be a natural transformation. If the diagram* (C) *is commutative, then for every* **T***-algebra* (X, ξ_X) *the* C*-morphism* $X \xrightarrow{\eta^{\mathsf{P}}_X} \mathsf{P}(X)$ *is a* **T***-homomorphism from* (X, ξ_X) *to* $\widetilde{\mathsf{P}}(X, \xi_X)$.

Proof. Since the diagram (C) is commutative, the assertion follows from the commutativity of the diagram:

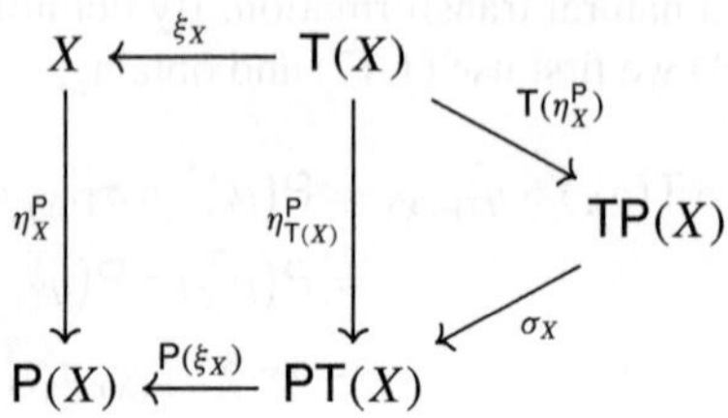

□

Lemma 1.2.10. *Let $\mu^{\mathsf{P}}\colon \mathsf{PP} \to \mathsf{P}$ be a natural transformation and $X \in |\mathsf{C}|$. Then $\mathsf{PPT}(X) \xrightarrow{\mu^{\mathsf{P}}_{\mathsf{T}(X)}} \mathsf{PT}(X)$ is a* **T***-homomorphism from $\widetilde{\mathsf{P}}(\widetilde{\mathsf{P}}(\mathsf{T}(X), \mu^{\mathsf{T}}_X))$ to $\widetilde{\mathsf{P}}(\mathsf{T}(X), \mu^{\mathsf{T}}_X)$ if and only if the following diagram is commutative:*

$$
\text{(D)} \qquad
\begin{array}{ccccc}
\mathsf{TPP}(X) & \xrightarrow{\sigma_{\mathsf{P}(X)}} & \mathsf{PTP}(X) & \xrightarrow{\mathsf{P}(\sigma_X)} & \mathsf{PPT}(X) \\
{\scriptstyle \mathsf{T}(\mu^{\mathsf{P}}_X)}\big\downarrow & & & & \big\downarrow{\scriptstyle \mu^{\mathsf{P}}_{\mathsf{T}(X)}} \\
\mathsf{TP}(X) & & \xrightarrow{\qquad\sigma_X\qquad} & & \mathsf{PT}(X)
\end{array}
$$

Proof. By Corollary 1.2.6 and Theorem 1.2.7 the C-morphism $\mathsf{P}(\sigma_X)$ is a **T**-homomorphism. If $\mu^{\mathsf{P}}_{\mathsf{T}(X)}$ is a **T**-homomorphism, then it is easily seen that the composed morphisms

$$\mu^{\mathsf{P}}_{\mathsf{T}(X)} \circ \mathsf{P}(\sigma_X) \circ \sigma_{\mathsf{P}(X)} \quad \text{and} \quad \sigma_X \circ \mathsf{T}(\mu^{\mathsf{P}}_X)$$

are again **T**-homomorphisms. Since $\mu^{\mathsf{P}}\colon \mathsf{PP} \to \mathsf{P}$ is a natural transformation, we obtain the following relation from the commutativity of (A):

$$
\begin{aligned}
\mu^{\mathsf{P}}_{\mathsf{T}(X)} \circ \mathsf{P}(\sigma_X) \circ \sigma_{\mathsf{P}(X)} \circ \eta^{\mathsf{T}}_{\mathsf{PP}(X)} &= \mu^{\mathsf{P}}_{\mathsf{T}(X)} \circ \mathsf{P}(\sigma_X) \circ \mathsf{P}\big(\eta^{\mathsf{T}}_{\mathsf{P}(X)}\big) \\
&= \mu^{\mathsf{P}}_{\mathsf{T}(X)} \circ \mathsf{PP}\big(\eta^{\mathsf{T}}_X\big) = \mathsf{P}\big(\eta^{\mathsf{T}}_X\big) \circ \mu^{\mathsf{P}}_X \\
&= \sigma_X \circ \eta^{\mathsf{T}}_{\mathsf{P}(X)} \circ \mu^{\mathsf{P}}_X = \sigma_X \circ \mathsf{T}\big(\mu^{\mathsf{P}}_X\big) \circ \eta^{\mathsf{T}}_{\mathsf{PP}(X)}.
\end{aligned}
$$

Hence the necessity of the commutativity of (D) follows from the universal property of free **T**-algebras.
In order to show that the commutativity of (D) is sufficient we proceed as follows. Referring to Lemma 1.2.5 (a) we recall that the structure maps $\mathsf{TPT}(X) \xrightarrow{\alpha} \mathsf{PT}(X)$ of $\widetilde{\mathsf{P}}\big(\mathsf{T}(X), \mu^{\mathsf{T}}_X\big)$ and $\mathsf{TPPT}(X) \xrightarrow{\beta} \mathsf{PPT}(X)$ of $\widetilde{\mathsf{P}}\big(\widetilde{P}\big(\mathsf{T}(X), \mu^{\mathsf{T}}_X\big)\big)$ have the following form (cf. (1.16)):

$$\alpha = \mathsf{P}(\mu^{\mathsf{T}}_X) \circ \sigma_{\mathsf{T}(X)} \quad \text{and} \quad \beta = \mathsf{PP}(\mu^{\mathsf{T}}_X) \circ \mathsf{P}(\sigma_{\mathsf{T}(X)}) \circ \sigma_{\mathsf{PT}(X)}.$$

Now we use again the fact that μ^{P} is a natural transformation and derive the commutative diagram

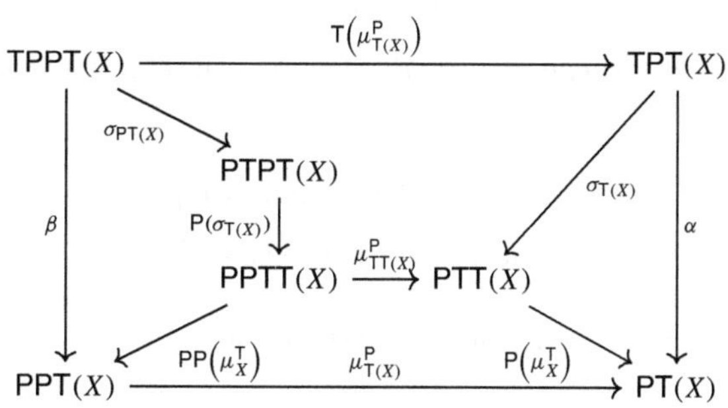

from the commutativity of diagram (D). Hence $\mu^{\mathsf{P}}_{\mathsf{T}(X)}$ is a **T**-homomorphism. □

Corollary 1.2.11. *Let $\mu^{\mathsf{P}} : \mathsf{PP} \to \mathsf{P}$ be a natural transformation. If diagram* (D) *is commutative, then for every* **T**-*algebra* (X, ξ_X) *the* C-*morphism* $\mathsf{PP}(X) \xrightarrow{\mu_X^{\mathsf{P}}} \mathsf{P}(X)$ *is a* **T**-*homomorphism from* $\widetilde{\mathsf{P}}(\widetilde{\mathsf{P}}(X, \xi_X))$ *to* $\widetilde{\mathsf{P}}(X, \xi_X)$.

Proof. The assertion follows immediately from the commutativity of the diagram:

$$\begin{array}{ccccccc}
\mathsf{TPP}(X) & \xrightarrow{\sigma_{\mathsf{P}(X)}} & \mathsf{PTP}(X) & \xrightarrow{\mathsf{P}(\sigma_X)} & \mathsf{PPT}(X) & \xrightarrow{\mathsf{PP}(\xi_X)} & \mathsf{PP}(X) \\
{\scriptstyle \mathsf{T}(\mu_X^{\mathsf{P}})}\downarrow & & & & \downarrow{\scriptstyle \mu_{\mathsf{T}(X)}^{\mathsf{P}}} & & \downarrow{\scriptstyle \mu_X^{\mathsf{P}}} \\
\mathsf{TP}(X) & & \xrightarrow{\qquad\qquad \sigma_X \qquad\qquad} & & \mathsf{PT}(X) & \xrightarrow{\mathsf{P}(\xi_X)} & \mathsf{P}(X)
\end{array}$$
□

Let $\mathbf{T} = (\mathsf{T}, \eta^{\mathsf{T}}, \mu^{\mathsf{T}})$ and $\mathbf{P} = (\mathsf{P}, \eta^{\mathsf{P}}, \mu^{\mathsf{P}})$ be monads on C. As a first step we recall the construction of the star composition of the respective natural transformations:

$$\eta := \eta^{\mathsf{P}} \star \eta^{\mathsf{T}} : \mathrm{id}_{\mathsf{C}} \to \mathsf{PT}, \eta_X = \eta_{\mathsf{T}(X)}^{\mathsf{P}} \circ \eta_X^{\mathsf{T}} = \mathsf{P}(\eta_X^{\mathsf{T}}) \circ \eta_X^{\mathsf{P}}. \tag{1.18}$$

$$\mu := \mu^{\mathsf{P}} \star \mu^{\mathsf{T}} : \mathsf{PPTT} \to \mathsf{PT}, \mu_X = \mu_{\mathsf{T}(X)}^{\mathsf{P}} \circ \mathsf{PP}(\mu_X^{\mathsf{T}}) = \mathsf{P}(\mu_X^{\mathsf{T}}) \circ \mu_{\mathsf{TT}(X)}^{\mathsf{P}}. \tag{1.19}$$

After these preparations we are ready to prove Beck's Theorem.

Theorem 1.2.12. ([10]) *Let* $\mathbf{T} = (\mathsf{T}, \eta^{\mathsf{T}}, \mu^{\mathsf{T}})$ *and* $\mathbf{P} = (\mathsf{P}, \eta^{\mathsf{P}}, \mu^{\mathsf{P}})$ *be monads on* C, *and let* $\sigma : \mathsf{TP} \to \mathsf{PT}$ *be a distributive law of* P *over* **T** *such that the diagrams* (C) *and* (D) *are commutative. Then the triple* $\mathbf{PT} = (\mathsf{PT}, \eta, \mu)$ *is a monad on* C *where the unit* η *and the multiplication* μ *are determined as follows*:

$$\begin{cases} \eta_X = \eta_{\mathsf{T}(X)}^{\mathsf{P}} \circ \eta_X^{\mathsf{T}} = \mathsf{P}(\eta_X^{\mathsf{T}}) \circ \eta_X^{\mathsf{P}} = (\eta^{\mathsf{P}} \star \eta^{\mathsf{T}})_X \quad \textit{and} \\ \mu_X = \mu_{\mathsf{T}(X)}^{\mathsf{P}} \circ \mathsf{PP}(\mu_X^{\mathsf{T}}) \circ \mathsf{P}(\sigma_{\mathsf{T}(X)}) \\ \quad\;\; = \mathsf{P}(\mu_X^{\mathsf{T}}) \circ \mu_{\mathsf{TT}(X)}^{\mathsf{P}} \circ \mathsf{P}(\sigma_{\mathsf{T}(X)}) = ((\mu^{\mathsf{P}} \star \mu^{\mathsf{T}}) \circ (1_{\mathsf{P}} \star \sigma \star 1_{\mathsf{T}}))_X. \end{cases} \tag{1.20}$$

The monad **PT** constructed in the previous theorem is called the *composite monad* of **T** and **P** in the sense of the *distributive law* σ.

For the proof of Theorem 1.2.12 the next lemma is useful.

Lemma 1.2.13. *The commutativity of the diagrams* (B) *and* (D) *implies the following equation*:

$$\begin{aligned} &(\mu^{\mathsf{P}} \star 1_{\mathsf{T}}) \circ (1_{\mathsf{P}} \star \sigma) \circ (1_{\mathsf{P}} \star \mu^{\mathsf{T}} \star 1_{\mathsf{P}}) \circ (\sigma \star 1_{\mathsf{TP}}) \\ &\qquad = (1_{\mathsf{P}} \star \mu^{\mathsf{T}}) \circ (\sigma \star 1_{\mathsf{T}}) \circ (1_{\mathsf{T}} \star \mu^{\mathsf{P}} \star 1_{\mathsf{T}}) \circ (1_{\mathsf{TP}} \star \sigma). \end{aligned}$$

Proof. Since the commutativity of (B) and (D) means

$$\begin{aligned} \sigma \circ (\mu^{\mathsf{T}} \star 1_{\mathsf{P}}) &= (1_{\mathsf{P}} \star \mu^{\mathsf{T}}) \circ (\sigma \star 1_{\mathsf{T}}) \circ (1_{\mathsf{T}} \star \sigma), \\ \sigma \circ (1_{\mathsf{T}} \star \mu^{\mathsf{P}}) &= (\mu^{\mathsf{P}} \star 1_{\mathsf{T}}) \circ (1_{\mathsf{P}} \star \sigma) \circ (\sigma \star 1_{\mathsf{P}}), \end{aligned}$$

we obtain the following relation by a repeated application of the interchange law (cf. [47, 73]):

$$
\begin{aligned}
&(\mu^{\mathsf{P}} \star 1_{\mathsf{T}}) \circ (1_{\mathsf{P}} \star \sigma) \circ (1_{\mathsf{P}} \star \mu^{\mathsf{T}} \star 1_{\mathsf{P}}) \circ (\sigma \star 1_{\mathsf{TP}}) \\
&\qquad = (\mu^{\mathsf{P}} \star 1_{\mathsf{T}}) \circ \big(1_{\mathsf{P}} \star \big((1_{\mathsf{P}} \star \mu^{\mathsf{T}}) \circ (\sigma \star 1_{\mathsf{T}}) \circ (1_{\mathsf{T}} \star \sigma)\big)\big) \circ (\sigma \star 1_{\mathsf{TP}}) \\
&\qquad = (\mu^{\mathsf{P}} \star \mu^{\mathsf{T}}) \circ (1_{\mathsf{P}} \star \sigma \star 1_{\mathsf{T}}) \circ (\sigma \star \sigma) \\
&\qquad = (1_{\mathsf{P}} \star \mu^{\mathsf{T}}) \circ (\mu^{\mathsf{P}} \star 1_{\mathsf{TT}}) \circ (1_{\mathsf{P}} \star \sigma \star 1_{\mathsf{T}}) \circ (\sigma \star 1_{\mathsf{PT}}) \circ (1_{\mathsf{TP}} \star \sigma) \\
&\qquad = (1_{\mathsf{P}} \star \mu^{\mathsf{T}}) \circ (\sigma \star 1_{\mathsf{T}}) \circ (1_{\mathsf{T}} \star \mu^{\mathsf{P}} \star 1_{\mathsf{T}}) \circ (1_{\mathsf{TP}} \star \sigma). \qquad \square
\end{aligned}
$$

Proof of Theorem 1.2.12.
(a) (Unit axiom) Since the commutativity of the diagrams (A) and (C) can be rewritten as follows:

$$
\sigma \circ \big(\eta^{\mathsf{T}} \star 1_{\mathsf{P}}\big) = 1_{\mathsf{P}} \star \eta^{\mathsf{T}} \quad \text{and} \quad \sigma \circ \big(1_{\mathsf{T}} \star \eta^{\mathsf{P}}\big) = \eta^{\mathsf{P}} \star 1_{\mathsf{T}}. \tag{1.21}
$$

we again obtain by a repeated application of the interchange law:

$$
\begin{aligned}
\mu \circ (\eta \star 1_{\mathsf{PT}}) &= (\mu^{\mathsf{P}} \star \mu^{\mathsf{T}}) \circ (1_{\mathsf{P}} \star \sigma \star 1_{\mathsf{T}}) \circ (\eta^{\mathsf{P}} \star \eta^{\mathsf{T}} \star 1_{\mathsf{P}} \star 1_{\mathsf{T}}) \\
&= (\mu^{\mathsf{P}} \star \mu^{\mathsf{T}}) \circ \big(\eta^{\mathsf{P}} \star ((\sigma \star 1_{\mathsf{T}}) \circ (\eta^{\mathsf{T}} \star 1_{\mathsf{P}} \star 1_{\mathsf{T}}))\big) \\
&= (\mu^{\mathsf{P}} \star \mu^{\mathsf{T}}) \circ \big(\eta^{\mathsf{P}} \star (\sigma \circ (\eta^{\mathsf{T}} \star 1_{\mathsf{P}})) \star 1_{\mathsf{T}}\big) \\
&= (\mu^{\mathsf{P}} \star \mu^{\mathsf{T}}) \circ (\eta^{\mathsf{P}} \star 1_{\mathsf{P}} \star \eta^{\mathsf{T}} \star 1_{\mathsf{T}}) \\
&= \big(\mu^{\mathsf{P}} \circ (\eta^{\mathsf{P}} \star 1_{\mathsf{P}})\big) \star \big(\mu^{\mathsf{T}} \circ (\eta^{\mathsf{T}} \star 1_{\mathsf{T}})\big) = 1_{\mathsf{PT}}.
\end{aligned}
$$

Analogously, $\sigma \circ \big(1_{\mathsf{T}} \star \eta^{\mathsf{P}}\big) = \eta^{\mathsf{P}} \star 1_{\mathsf{T}}$ implies $\mu \circ (1_{\mathsf{PT}} \star \eta) = 1_{\mathsf{PT}}$.
(b) (Associativity axiom) First we use the associativity of μ^{T} and apply again repeatedly the interchange law:

$$
\begin{aligned}
&\mu \circ (1_{\mathsf{PT}} \star \mu) \\
&= (\mu^{\mathsf{P}} \star \mu^{\mathsf{T}}) \circ (1_{\mathsf{P}} \star \sigma \star 1_{\mathsf{T}}) \circ \big(1_{\mathsf{PT}} \star \big((\mu^{\mathsf{P}} \star \mu^{\mathsf{T}}) \circ (1_{\mathsf{P}} \star \sigma \star 1_{\mathsf{T}})\big)\big) \\
&= (\mu^{\mathsf{P}} \star \mu^{\mathsf{T}}) \circ (1_{\mathsf{P}} \star \sigma \star 1_{\mathsf{T}}) \circ \big(1_{\mathsf{PT}} \star \big((1_{\mathsf{P}} \star \mu^{\mathsf{T}}) \circ (\mu^{\mathsf{P}} \star 1_{\mathsf{TT}}) \circ (1_{\mathsf{P}} \star \sigma \star 1_{\mathsf{T}})\big)\big) \\
&= (\mu^{\mathsf{P}} \star \mu^{\mathsf{T}}) \circ (1_{\mathsf{P}} \star \sigma \star 1_{\mathsf{T}}) \circ (1_{\mathsf{PTP}} \star \mu^{\mathsf{T}}) \circ (1_{\mathsf{PT}} \star \mu^{\mathsf{P}} \star 1_{\mathsf{TT}}) \circ (1_{\mathsf{PTP}} \star \sigma \star 1_{\mathsf{T}}) \\
&= (\mu^{\mathsf{P}} \star \mu^{\mathsf{T}}) \circ (1_{\mathsf{P}} \star \sigma \star \mu^{\mathsf{T}}) \circ (1_{\mathsf{PT}} \star \mu^{\mathsf{P}} \star 1_{\mathsf{TT}}) \circ (1_{\mathsf{PTP}} \star \sigma \star 1_{\mathsf{T}}) \\
&= (\mu^{\mathsf{P}} \star \mu^{\mathsf{T}}) \circ (1_{\mathsf{PPT}} \star \mu^{\mathsf{T}}) \circ (1_{\mathsf{P}} \star \sigma \star 1_{\mathsf{TT}}) \circ (1_{\mathsf{PT}} \star \mu^{\mathsf{P}} \star 1_{\mathsf{TT}}) \circ (1_{\mathsf{PTP}} \star \sigma \star 1_{\mathsf{T}}) \\
&= (\mu^{\mathsf{P}} \star \mu^{\mathsf{T}}) \circ (1_{\mathsf{PP}} \star \mu^{\mathsf{T}} \star 1_{\mathsf{T}}) \circ (1_{\mathsf{P}} \star \sigma \star 1_{\mathsf{TT}}) \circ (1_{\mathsf{PT}} \star \mu^{\mathsf{P}} \star 1_{\mathsf{TT}}) \circ (1_{\mathsf{PTP}} \star \sigma \star 1_{\mathsf{T}}).
\end{aligned}
$$

By means of the star composition we now multiply the equation in Lemma 1.2.13 from the left by 1_{P} and from the right by 1_{T}. Then we obtain:

$$\mu \circ (1_{\mathsf{PT}} \star \mu)$$
$$= (\mu^{\mathsf{P}} \star \mu^{\mathsf{T}}) \circ (1_{\mathsf{P}} \star \mu^{\mathsf{P}} \star 1_{\mathsf{TT}}) \circ (1_{\mathsf{PP}} \star \sigma \star 1_{\mathsf{T}}) \circ (1_{\mathsf{PP}} \star \mu^{\mathsf{T}} \star 1_{\mathsf{PT}}) \circ (1_{\mathsf{P}} \star \sigma \star 1_{\mathsf{TPT}}).$$

Now we apply the associativity of μ^{P}

$$\mu \circ (1_{\mathsf{PT}} \star \mu)$$
$$= (\mu^{\mathsf{P}} \star \mu^{\mathsf{T}}) \circ (\mu^{\mathsf{P}} \star 1_{\mathsf{PTT}}) \circ (1_{\mathsf{PP}} \star \sigma \star 1_{\mathsf{T}}) \circ (1_{\mathsf{PP}} \star \mu^{\mathsf{T}} \star 1_{\mathsf{PT}}) \circ (1_{\mathsf{P}} \star \sigma \star 1_{\mathsf{TPT}})$$

and reverse the previous "direction" of reasoning. Hence the relation

$$\mu \circ (1_{\mathsf{PT}} \star \mu) = \mu \circ (\mu \star 1_{\mathsf{PT}})$$

follows. □

After having established Beck's Theorem the aim is now to characterize the Eilenberg–Moore category of composite monads. Therefore we fix two monads $\mathbf{T} = (\mathsf{T}, \eta^{\mathsf{T}}, \mu^{\mathsf{T}})$ and $\mathbf{P} = (\mathsf{P}, \eta^{\mathsf{P}}, \mu^{\mathsf{P}})$ on C and assume that there is a distributive law $\sigma : \mathsf{TP} \to \mathsf{PT}$ of P over $\mathbf{T}$ such that the diagrams (C) and (D) are commutative.

Theorem 1.2.14. *If (X, ξ) is a* **PT**-*algebra, then (X, ξ_1) with $\xi_1 = \xi \circ \eta^{\mathsf{P}}_{\mathsf{T}(X)}$ is a* **T**-*algebra, and (X, ξ_2) with $\xi_2 = \xi \circ \mathsf{P}(\eta^{\mathsf{T}}_X)$ is a* **P**-*algebra. Moreover, the following relations hold*:

(i) $\xi = \xi_2 \circ \mathsf{P}(\xi_1)$,
(ii) $\xi \circ \sigma_X = \xi_1 \circ \mathsf{T}(\xi_2)$.

Proof. The respective unit algebra axiom is evident. In order to verify the associativity axiom we proceed as follows.

(a) Referring to the commutativity of (C) (cf. (1.21)) we first obtain:

$$\begin{aligned}
\mu \circ (1_{\mathsf{PT}} \star \eta^{\mathsf{P}} \star 1_{\mathsf{T}}) &= (\mu^{\mathsf{P}} \star \mu^{\mathsf{T}}) \circ (1_{\mathsf{P}} \star \sigma \star 1_{T}) \circ (1_{\mathsf{PT}} \star \eta^{\mathsf{P}} \star 1_{\mathsf{T}}) \\
&= (\mu^{\mathsf{P}} \star \mu^{\mathsf{T}}) \circ \big(1_{\mathsf{P}} \star (\sigma \circ (1_{\mathsf{T}} \star \eta^{\mathsf{P}})) \star 1_{\mathsf{T}}\big) \\
&= (\mu^{\mathsf{P}} \star \mu^{\mathsf{T}}) \circ (1_{\mathsf{P}} \star \eta^{\mathsf{P}} \star 1_{\mathsf{TT}}) \\
&= \big((\mu^{\mathsf{P}} \circ (1_{\mathsf{P}} \star \eta^{\mathsf{P}})\big) \star \mu^{\mathsf{T}} \\
&= 1_{\mathsf{P}} \star \mu^{\mathsf{T}}.
\end{aligned}$$

Now we conclude from the associativity axiom of (X, ξ) that the diagram

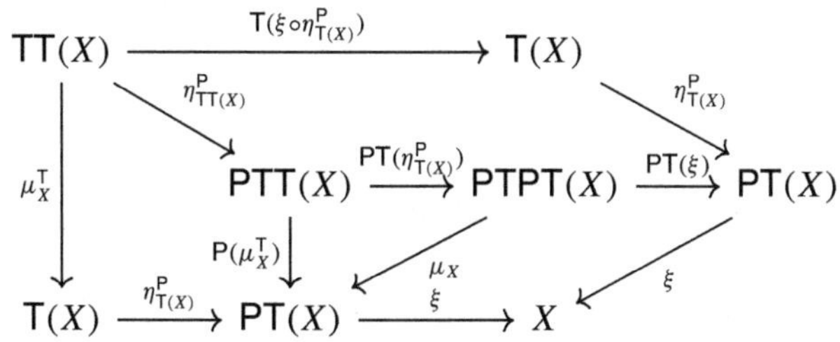

is commutative. Hence $\xi_1 \circ \mu_X^{\mathsf{T}} = \xi_1 \circ \mathsf{T}(\xi_1)$ follows.
(b) First we use the commutativity of (A) (cf. (1.21)) and observe:

$$
\begin{aligned}
(\sigma \star 1_{\mathsf{T}}) \circ (1_{\mathsf{TP}} \star \eta^{\mathsf{T}}) \circ (\eta^{\mathsf{T}} \star 1_{\mathsf{P}}) &= (\sigma \star \eta^{\mathsf{T}}) \circ (\eta^{\mathsf{T}} \star 1_{\mathsf{P}} \star 1_{\mathrm{id}_{\mathbb{C}}}) \\
&= \big(\sigma \circ (\eta^{\mathsf{T}} \star 1_{\mathsf{P}})\big) \star (\eta^{\mathsf{T}} \circ 1_{\mathrm{id}_{\mathbb{C}}}) \\
&= 1_{\mathsf{P}} \star \eta^{\mathsf{T}} \star \eta^{\mathsf{T}} \\
&= 1_{\mathsf{P}} \star \big((\eta^{\mathsf{T}} \star 1_{\mathsf{T}}) \circ (1_{\mathrm{id}_{\mathbb{C}}} \star \eta^{\mathsf{T}})\big) \\
&= 1_{\mathsf{P}} \star \big((\eta^{\mathsf{T}} \star 1_{\mathsf{T}}) \circ \eta^{\mathsf{T}}\big).
\end{aligned}
$$

Now we use the unit axiom of μ^{T} and obtain:

$$
\begin{aligned}
\mu \circ (1_{\mathsf{PTP}} \star \eta^{\mathsf{T}}) \circ (1_{\mathsf{P}} \star \eta^{\mathsf{T}} \star 1_{\mathsf{P}}) &= (\mu^{\mathsf{P}} \star \mu^{\mathsf{T}}) \circ \big(1_{\mathsf{PP}} \star ((\eta^{\mathsf{T}} \star 1_{\mathsf{T}}) \circ \eta^{\mathsf{T}})\big) \\
&= \mu^{\mathsf{P}} \star \eta^{\mathsf{T}} \\
&= (1_{\mathsf{P}} \star \eta^{\mathsf{T}}) \circ \mu^{\mathsf{P}}.
\end{aligned}
$$

Finally, we apply the associativity axiom of (X, ξ)

$$
\begin{aligned}
\xi \circ \mathsf{P}(\eta_X^{\mathsf{T}}) \circ \mu_X^{\mathsf{P}} &= \xi \circ \mu_X \circ \mathsf{PTP}(\eta_X^{\mathsf{T}}) \circ \mathsf{P}(\eta_{\mathsf{P}(X)}^{\mathsf{T}}) \\
&= \xi \circ \mathsf{PT}(\xi) \circ \mathsf{PTP}(\eta_X^{\mathsf{T}}) \circ \mathsf{P}(\eta_{\mathsf{P}(X)}^{\mathsf{T}}) \\
&= \xi \circ \mathsf{P}\big(\mathsf{T}(\xi \circ \mathsf{P}(\eta_X^{\mathsf{T}})) \circ \eta_{\mathsf{P}(X)}^{\mathsf{T}}\big) \\
&= \xi \circ \mathsf{P}\big(\eta_X^{\mathsf{T}} \circ (\xi \circ \mathsf{P}(\eta_X^{\mathsf{T}}))\big) \\
&= (\xi \circ \mathsf{P}(\eta_X^{\mathsf{T}})) \circ \mathsf{P}(\xi \circ \mathsf{P}(\eta_X^{\mathsf{T}}))
\end{aligned}
$$

and obtain $\xi_2 \circ \mu_X^{\mathsf{P}} = \xi_2 \circ \mathsf{P}(\xi_2)$.
(c) In order to verify the validity of property (i) we refer again to (1.21)) and observe:

$$
\begin{aligned}
\mu \circ (1_{\mathsf{P}} \star \eta^{\mathsf{T}} \star 1_{\mathsf{PT}}) \circ (1_{\mathsf{P}} \star \eta^{\mathsf{P}} \star 1_{\mathsf{T}}) &= (\mu^{\mathsf{P}} \star \mu^{\mathsf{T}}) \circ \left(1_{\mathsf{P}} \star \left(\sigma \circ (\eta^{\mathsf{T}} \star 1_{\mathsf{P}}) \circ \eta^{\mathsf{P}}\right) \star 1_{\mathsf{T}}\right) \\
&= (\mu^{\mathsf{P}} \star \mu^{\mathsf{T}}) \circ \left(1_{\mathsf{P}} \star \left((1_{\mathsf{P}} \star \eta^{\mathsf{T}}) \circ \eta^{\mathsf{P}}\right) \star 1_{\mathsf{T}}\right) \\
&= (\mu^{\mathsf{P}} \star \mu^{\mathsf{T}}) \circ (1_{\mathsf{P}} \star \eta^{\mathsf{P}} \star \eta^{\mathsf{T}} \star 1_{\mathsf{T}}) \\
&= 1_{\mathsf{PT}}.
\end{aligned}
$$

Hence (i) follows from the following relation:

$$
\begin{aligned}
\xi_2 \circ \mathsf{P}(\xi_1) &= \xi \circ \mathsf{P}(\eta_X^{\mathsf{T}} \circ \xi \circ \eta_{\mathsf{T}(X)}^{\mathsf{P}}) \\
&= \xi \circ \mathsf{P}(\mathsf{T}(\xi) \circ \eta_{\mathsf{PT}(X)}^{\mathsf{T}} \circ \eta_{\mathsf{T}(X)}^{\mathsf{P}}) \\
&= \xi \circ \mathsf{PT}(\xi) \circ \mathsf{P}(\eta_{\mathsf{PT}(X)}^{\mathsf{T}} \circ \eta_{\mathsf{T}(X)}^{\mathsf{P}}) \\
&= \xi \circ \mu_X \circ \mathsf{P}(\eta_{\mathsf{PT}(X)}^{\mathsf{T}} \circ \eta_{\mathsf{T}(X)}^{\mathsf{P}}) \\
&= \xi.
\end{aligned}
$$

Further, we conclude from the unit axioms of μ^{P} and μ^{T} that the relation

$$
\begin{aligned}
&\mu \circ (\eta^{\mathsf{P}} \star 1_{\mathsf{TPT}}) \circ (1_{\mathsf{TP}} \star \eta^{\mathsf{T}}) \\
&= (\mu^{\mathsf{P}} \star \mu^{\mathsf{T}}) \circ (1_{\mathsf{P}} \star \sigma \star 1_{\mathsf{T}}) \circ (\eta^{\mathsf{P}} \star 1_{\mathsf{TPT}}) \circ (1_{\mathsf{TP}} \star \eta^{\mathsf{T}}) \\
&= (\mu^{\mathsf{P}} \star \mu^{\mathsf{T}}) \circ (\eta^{\mathsf{P}} \star \sigma \star 1_{\mathsf{T}}) \circ (1_{\mathsf{TP}} \star \eta^{\mathsf{T}}) \\
&= (\mu^{\mathsf{P}} \star \mu^{\mathsf{T}}) \circ (\eta^{\mathsf{P}} \star 1_{\mathsf{PTT}}) \circ (1_{\mathrm{id}_{\mathsf{C}}} \star \sigma \star 1_{\mathsf{T}}) \circ (1_{\mathsf{TP}} \star \eta^{\mathsf{T}}) \\
&= (1_{\mathsf{P}} \star \mu^{\mathsf{T}}) \circ (\sigma \star \eta^{\mathsf{T}}) \\
&= (1_{\mathsf{P}} \star \mu^{\mathsf{T}}) \circ (1_{\mathsf{PT}} \star \eta^{\mathsf{T}}) \circ (\sigma \star 1_{\mathrm{id}_{\mathsf{C}}}) \\
&= 1_{\mathsf{PT}} \circ \sigma = \sigma
\end{aligned}
$$

holds. Now we obtain:

$$
\begin{aligned}
\xi_1 \circ \mathsf{T}(\xi_2) = \xi \circ \eta_{\mathsf{T}(X)}^{\mathsf{P}} \circ \mathsf{T}(\xi) \circ \mathsf{TP}(\eta_X^{\mathsf{T}}) &= \xi \circ \mathsf{PT}(\xi) \circ \eta_{\mathsf{TPT}(X)}^{\mathsf{P}} \circ \mathsf{TP}(\eta_X^{\mathsf{T}}) \\
= \xi \circ \mu_X \circ \eta_{\mathsf{TPT}(X)}^{\mathsf{P}} \circ \mathsf{TP}(\eta_X^{\mathsf{T}}) &= \xi \circ \sigma_X.
\end{aligned}
$$

Thus property (ii) is verified. □

Proposition 1.2.15. *If* (X, ξ_1) *is a* **T***-algebra and* (X, ξ_2) *is a* **P***-algebra satisfying the following compatibility condition:*

$$
\xi_1 \circ \mathsf{T}(\xi_2) = \xi_2 \circ \mathsf{P}(\xi_1) \circ \sigma_X, \tag{1.22}
$$

then $(X, \xi_2 \circ \mathsf{P}(\xi_1))$ *is a* **PT***-algebra.*

Proof. The unit axiom is always trivial. From (1.22) it is easily seen that the following diagram is commutative:

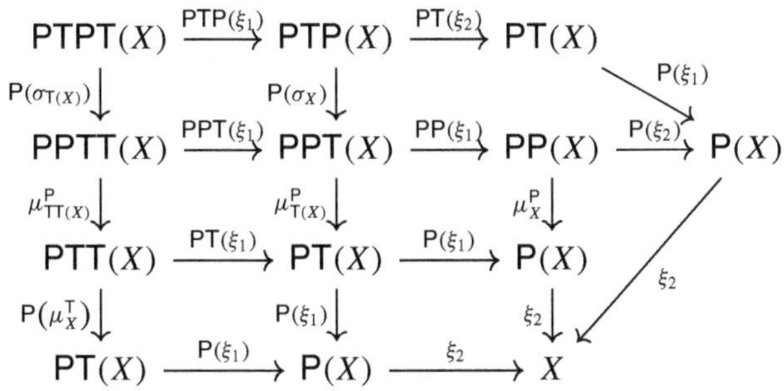

Hence the associativity axiom follows. □

Comment. We conclude from Lemma 1.2.5 (a) that in the light of formula (1.16) the compatibility condition (1.22) is equivalent to the requirement that $\mathsf{P}(X) \xrightarrow{\xi_2} X$ is a $\mathbf{T}$-homomorphism.

The previous results can be summarized as follows.

Theorem 1.2.16. *Let* $\mathbf{T} = (\mathsf{T}, \eta^{\mathsf{T}}, \mu^{\mathsf{T}})$ *and* $\mathbf{P} = (\mathsf{P}, \eta^{\mathsf{P}}, \mu^{\mathsf{P}})$ *be monads on* C *provided with a distributive law* $\sigma : \mathsf{TP} \to \mathsf{PT}$ *such that additionally the diagrams* (C) *and* (D) *are commutative. Then the composite monad* $\mathbf{PT} = (\mathsf{PT}, \eta, \mu)$ *exists, and the Eilenberg–Moore category* $\mathsf{C}^{\mathbf{PT}}$ *is isomorphic to the subcategory of* $\mathsf{C}^{\mathbf{P}} \times \mathsf{C}^{\mathbf{T}}$ *consisting of all*

- *pairs* $((X, \xi_2), (X, \xi_1))$ *of algebras satisfying the compatibility condition* (1.22) *and*
- $\mathsf{C}-morphisms$ $X \xrightarrow{h} Y$ *which are simultaneously* $\mathbf{T}$*-homomorphisms and* $\mathbf{P}$*-homomorphisms.*

In particular, each $\mathbf{PT}$*-algebra* (X, ξ) *has the form* $\xi = \xi_2 \circ \mathsf{P}(\xi_1)$ *where* (X, ξ_2) *is a* $\mathbf{P}$*-algebra and* (X, ξ_1) *is a* $\mathbf{T}$*-algebra.*

Proof. Let $((X, \xi_1), (X, \xi_2))$ and $((Y, \zeta_1), (Y, \zeta_2))$ be pairs of algebras satisfying (1.22). We only have to show that a C-morphism is a $\mathbf{PT}$-homomorphism if and only if h is a $\mathbf{T}$-homomorphism and a $\mathbf{P}$-homomorphism. If h is a $\mathbf{PT}$-homomorphism, then the following diagrams are commutative:

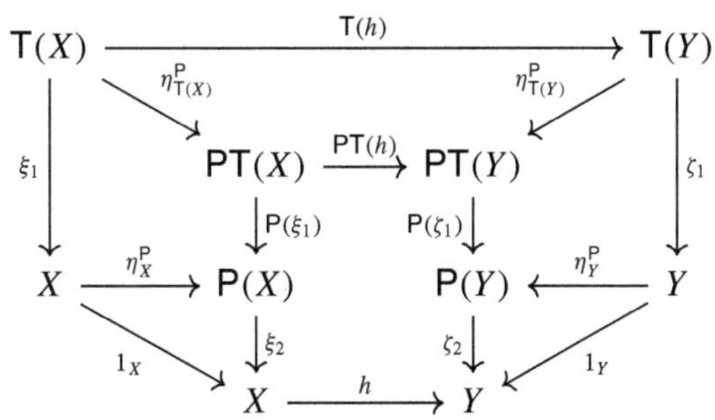

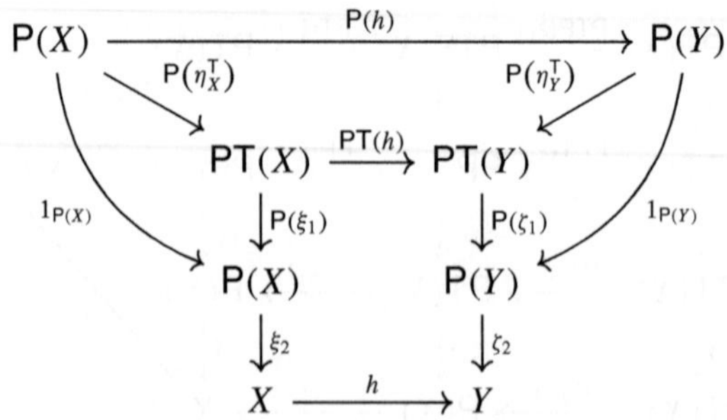

Hence h is a **T**-homomorphism and a **P**-homomorphism. On the other hand, if h is a **T**-homomorphism and a **P**-homomorphism, then h is obviously a **PT**-homomorphism. □

Example 1.2.17. Let $\mathbf{T} = (\mathsf{T}, \eta^{\mathsf{T}}, \mu^{\mathsf{T}})$ and $\mathbf{P} = (\mathsf{P}, \eta^{\mathsf{P}}, \mu^{\mathsf{P}})$ be monads on C provided with a distributive law $\sigma : \mathsf{TP} \to \mathsf{PT}$ such that additionally the diagrams (C) and (D) are commutative. Hence the composite monad $\mathbf{PT} = (\mathsf{PT}, \eta, \mu)$ exists. Further, let (X, ξ_X) be a **T**-algebra and $\mathsf{PTP}(X) \xrightarrow{\vartheta} \mathsf{P}(X)$ be a morphism determined by the following commutative diagram:

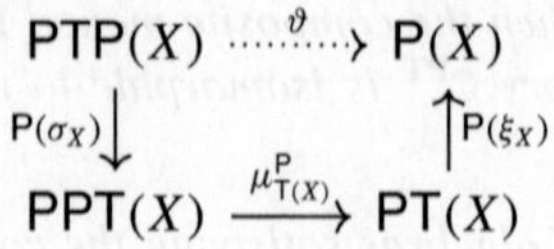

Then the pair $(\mathsf{P}(X), \vartheta)$ is a **PT**-algebra. In order to verify this statement we will apply Proposition 1.2.15. For this purpose we first notice that $(\mathsf{P}(X), \zeta_X)$ with

$$\zeta_X = \mathsf{P}(\xi_X) \circ \sigma_X$$

is a **T**-algebra (cf. Lemma 1.2.5 (a)). As a second algebra structure on $\mathsf{P}(X)$ we can consider the free **P**-algebra $(\mathsf{P}(X), \mu^{\mathsf{P}}_X)$. We verify the compatibility condition (1.22). In fact, because of diagram (D) the following relation holds:

$$\begin{aligned}\mu^{\mathsf{P}}_X \circ \mathsf{P}(\zeta_X) \circ \sigma_{\mathsf{P}(X)} &= \mu^{\mathsf{P}}_X \circ \mathsf{PP}(\xi_X) \circ \mathsf{P}(\sigma_X) \circ \sigma_{\mathsf{P}(X)} \\ &= \mathsf{P}(\xi_X) \circ \mu^{\mathsf{P}}_{\mathsf{T}(X)} \circ \mathsf{P}(\sigma_X) \circ \sigma_{\mathsf{P}(X)} \\ &= \mathsf{P}(\xi_X) \circ \sigma_X \circ \mathsf{T}(\mu^{\mathsf{P}}_X) = \zeta_X \circ \mathsf{T}(\mu^{\mathsf{P}}_X).\end{aligned}$$

Now we observe:

$$\mu^{\mathsf{P}}_X \circ \mathsf{P}(\zeta_X) = \mu^{\mathsf{P}}_X \circ \mathsf{PP}(\xi_X) \circ \mathsf{P}(\sigma_X) = \mathsf{P}(\xi_X) \circ \mu^{\mathsf{P}}_{\mathsf{T}(X)} \circ \mathsf{P}(\sigma_X) = \vartheta.$$

Hence we conclude from Proposition 1.2.15 that $(\mathsf{P}(X), \vartheta)$ is a **PT**-algebra.

As an additional remark we would like to point out that the structure map ϑ occurs as the second column of the diagram appearing in the proof of Proposition 1.2.15.

Exercises

1.2.1. If $\mathsf{F} \dashv \mathsf{G}$ is an adjoint pair of functors, then show that the multiplication μ of the associated monad has the form $\mu = 1_{\mathsf{G}} \star \varepsilon \star 1_{\mathsf{F}}$ where ε is the counit of $\mathsf{F} \dashv \mathsf{G}$.

1.2.2. Let P and T be endofunctors of a category C. Further, let $\eta^{\mathsf{P}} \colon \mathrm{id}_{\mathsf{C}} \to \mathsf{P}$ and $\eta^{\mathsf{T}} \colon \mathrm{id}_{\mathsf{C}} \to \mathsf{T}$ be natural transformations. Show that:

(a) $\big((\eta^{\mathsf{P}} \star 1_{\mathsf{T}}) \circ \eta^{\mathsf{T}}\big) \star 1_{\mathsf{P}} \star \eta^{\mathsf{T}} = \eta^{\mathsf{P}} * \eta^{\mathsf{T}} \star 1_{\mathsf{P}} \star \eta^{\mathsf{T}} = \big(\eta^{\mathsf{P}} \star \eta^{\mathsf{T}} \star 1_{\mathsf{PT}}\big) \circ \big(1_{\mathsf{P}} \star \eta^{\mathsf{T}}\big)$.
(b) $\eta^{\mathsf{P}} \star 1_{\mathsf{T}} \star \big((1_{\mathsf{P}} \star \eta^{\mathsf{T}}) \circ \eta^{\mathsf{P}}\big) = \eta^{\mathsf{P}} \star 1_{\mathsf{T}} \star \eta^{\mathsf{P}} \star \eta^{\mathsf{T}} = \big(1_{\mathsf{PT}} \star \eta^{\mathsf{P}} \star \eta^{\mathsf{T}}\big) \circ \big(\eta^{\mathsf{P}} \star 1_{\mathsf{T}}\big)$.

1.2.3. Let $\mathbf{P} = (\mathsf{P}, \eta, \mu)$ be a monad on a category C. Show that the natural transformation $\sigma \colon \mathsf{PP} \to \mathsf{PP}$ with $\sigma_X = \mathsf{P}(\eta_X) \circ \mu_X$ is a distributive law of the functor P over the monad $\mathbf{P}$ (cf. Definition 1.2.4).

1.3 The Category of Preordered Sets

A *preorder* on a set X is a binary relation $\leq$ on X satisfying the following well-known axioms

– if $x \in X$, then $x \leq x$, (Reflexivity)
– if $x \leq y$ and $y \leq z$, then $x \leq z$. (Transitivity)

A *preordered set* is a pair $(X, \leq)$ where X is a set and $\leq$ is a preorder on X. Two elements x and y of a preordered set are called *equivalent* if $x \leq y$ *and* $y \leq x$.

If $(X, \leq)$ is a preordered set, then the dual preorder $\leq^{op}$ on X is determined by

$$x \leq^{op} y \iff y \leq x.$$

Obviously, $(X, \leq^{op})$ is again a preordered set. In this context the *principle of duality* can be described as follows:

If P is a property (notion) in $(X, \leq)$, then the *dual* property (notion) P^{op} in $(X, \leq)$ is the property (notion) P phrased in $(X, \leq^{op})$.

Definition 1.3.1. Let $(X, \leq)$ and $(Y, \leq)$ be preordered sets. A map $X \xrightarrow{f} Y$ is *isotone*, if $x_1 \leq x_2$ implies $f(x_1) \leq f(x_2)$ for all $x_1, x_2 \in X$. A map $X \xrightarrow{f} Y$ is *antitone* if f is isotone from $(X, \leq)$ to $(Y, \leq^{op})$. A pair $X \overset{f}{\underset{g}{\rightrightarrows}} Y$ of isotone maps f and g is called *naturally equivalent* if $f(x) \leq g(x)$ and $g(x) \leq f(x)$ for all $x \in X$.

If the context is clear, we may drop the notation for the preorder of a preordered set and simply write X. In the case of dual preorders this approach leads to the rule to write X^{op} instead of $(X, \leq^{op})$.

Obviously preordered sets and isotone maps form a category denoted by `Preord`. Since preorders on X are preserved under arbitrary intersections of binary relations on X, it follows immediately that `Preord` is a topological category over `Set` (cf. [2]). Hence `Preord` is a complete and cocomplete category (cf. [2, 21.16 Theorem]). Moreover, the hom-sets of `Preord` can always be enriched by a preorder — i.e. if $X \overset{f}{\underset{g}{\rightrightarrows}} Y$ are isotone maps, then $f \leq g \iff f(x) \leq g(x)$ for all $x \in X$. Therefore it is not difficult to prove that `Preord` is cartesian closed. Hence the categorical product in `Preord` induces the structure of a symmetric and monoidal closed category on `Preord`. In particular, magmas of `Preord` are preordered groupoids — these are triples $(X, \leq, m_X)$ such that $(X, \leq)$ is a preordered set and the binary operation m_X is isotone in each variable separately. Since `Preord` is cocomplete and cartesian closed, it follows immediately from Corollary 1.1.5 that for every preordered set there exists a free preordered groupoid. In detail, this results means that for every preordered set $(X, \leq)$ there exists a preordered groupoid $(X^\sharp, \leq^\sharp, m_{X^\sharp})$ and an isotone map $X \xrightarrow{\eta_X} X^\sharp$ such that for any further preordered groupoid $(Y, \leq, \cdot)$ and isotone map $X \xrightarrow{f} Y$ there exists a unique isotone groupoid homomorphism $(X^\sharp, \leq^\sharp, m_{X^\sharp}) \xrightarrow{f^\sharp} (Y, \leq, \cdot)$ making the following diagram commutative:

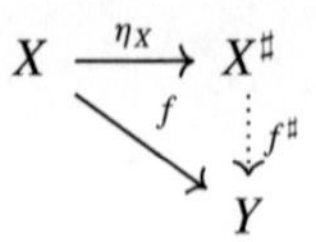

The corresponding monad on `Preord` is called the *term monad* $\mathbf{T}$ on `Preord` (of a single binary operator symbol). In particular, the Eilenberg–Moore category $\texttt{Preord}^{\mathbf{T}}$ is isomorphic to the category of preordered groupoids.

The preorder on hom-sets of `Preord` also expresses a second feature of `Preord` — namely the concept of adjoint pairs of isotone maps. In order to fix some notation we recall here its definition in terms of `Preord`.

A pair $X \overset{f}{\underset{g}{\rightleftarrows}} Y$ of isotone maps is said to be *adjoint* (i.e. forms a *covariant Galois connection*) if (f, g) satisfies the following condition

$$1_X \leq g \circ f \quad \text{and} \quad f \circ g \leq 1_Y. \tag{AD}$$

In this context we use the notation $f \dashv g$ and call f a *left adjoint* map of g and g a *right adjoint* map of f.

Obviously a pair $X \overset{f}{\underset{g}{\rightleftarrows}} Y$ of isotone maps is adjoint if and only if the relation

$$f(x) \leq y \iff x \leq g(y)$$

holds for all $x \in X$ and $y \in Y$. It is easily seen that left adjointness and right adjointness are dual notions, because $Y \xrightarrow{g} X$ is right adjoint to $X \xrightarrow{f} Y$ if and only if $Y^{op} \xrightarrow{g} X^{op}$ is left adjoint to $X^{op} \xrightarrow{f} Y^{op}$. Moreover, adjoint pairs of isotone maps are uniquely determined by each other up to a natural equivalence — this means

- If $f \dashv g$ and $f \dashv g'$, then g and g' are naturally equivalent.
- If $f \dashv g$ and $f' \dashv g$, then f and f' are naturally equivalent.

Hence this observation motivates the requirement that equivalent elements are equal and leads to the following terminology. A preorder is *antisymmetric* if $x \leq y$ and $y \leq x$ imply $x = y$. An antisymmetric preorder is also called a *partial order*, and a preordered set with a partial order is said to be a *partially ordered set* (*poset* for short).

As we will see later, large parts of the theory of complete preordered sets are independent from the antisymmetry axiom. But for those readers who prefer to work with partial orders we insert the following result.

Corollary 1.3.2. *Let* $X \underset{g}{\overset{f}{\rightleftarrows}} Y$ *be a pair of isotone maps. If either* $f \dashv g$ *or* $g \dashv f$ *holds, then the following properties hold*:

(i) *If* Y *is a poset, then* $f = f \circ g \circ f$.
(ii) *If* X *is a poset, then* $g = g \circ f \circ g$.

The aim of the following considerations is to show how completeness of preordered sets is based on adjointness. We begin with some terminology. A subset U of a preordered set X is said to be *downclosed* if whenever $y \in X$ has the property that there exists an $x \in U$ with $y \leq x$, then y is an element of U — i.e.

$$U = \{y \in X \mid \exists x \in U : \ y \leq x\}.$$

The empty set is a special example of a downclosed subset. The set of all downclosed subsets of X is denoted by $\mathsf{Dwn}(X)$ and is provided with the set-inclusion. Then $(\mathsf{Dwn}(X), \subseteq)$ is a preordered set, $\subseteq$ is a partial order on $\mathsf{Dwn}(X)$, and the map $X \xrightarrow{\eta_X} \mathsf{Dwn}(X)$ defined by $\eta_X(x) = \downarrow x = \{y \in X \mid y \leq x\}$ for $x \in X$ is isotone. Finally, the *downclosed hull* of an arbitrary subset A of X is given as follows:

$$\downarrow A = \{z \in X \mid \exists a \in A : \ z \leq a\} = \bigcup_{a \in A} \eta_X(a).$$

Lemma 1.3.3. *Let* X *be a preordered set and* 1_X *be the identity map of* X. *An isotone map* $\mathsf{Dwn}(X) \xrightarrow{\xi} X$ *is left adjoint to* η_X *(i.e.* $\xi \dashv \eta_X$*) if and only if* 1_X *and* $\xi \circ \eta_X$ *are naturally equivalent.*

Proof. It is sufficient to prove $1_{\mathsf{Dwn}(X)} \subseteq \eta_X \circ \xi \iff 1_X \leq \xi \circ \eta_X$. The necessity of $1_X \leq \xi \circ \eta_X$ for $1_{\mathsf{Dwn}(X)} \subseteq \eta_X \circ \xi$ follows from

$$\eta_X(x) \subseteq \eta_X\big(\xi(\eta_X(x))\big) \iff x \le \xi(\eta_X(x)), \qquad x \in X.$$

On the other hand, let us assume that $1_X \le \xi \circ \eta_X$ holds. If A is a downclosed subset of X, then $A = \bigcup_{a\in A} \eta_X(a) \subseteq \bigcup_{a\in A} \eta_X(\xi(\eta_X(a))) \subseteq \eta_X\big(\xi(A)\big)$ follows from the isotonicity of ξ. Hence $1_X \le \xi \circ \eta_X$ is also sufficient for $1_{\mathsf{Dwn}(X)} \subseteq \eta_X \circ \xi$. □

In order to give a further characterization under which condition η_X has a left adjoint map we need some more terminology. Let $(X, \le)$ be a preordered set and A be a subset of X. An element $x \in X$ is called an *upper bound* of A if $a \le x$ for all $a \in A$ — i.e. $A \subseteq \eta_X(x)$. An upper bound x of A is called a *join* of A if $x \le z$ for all upper bounds of z of A. If a subset A has joins, then the joins of A are uniquely determined by A up to equivalence — i.e. all joins of A are equivalent elements.

Theorem 1.3.4. *Let $(X, \le)$ be a preordered set. Then $X \xrightarrow{\eta_X} \mathsf{Dwn}(X)$ has a left adjoint map if and only if every subset of X has a join.*

Proof. Let $\mathsf{Dwn}(X) \xrightarrow{\xi} X$ be a left adjoint map of η_X. If A is a subset of X and $\downarrow A$ its downclosed hull, then $\xi(\downarrow A)$ is a join of A. On the other hand, let $\mathbb{J}(A)$ be the set of all joins of a subset A. If every subset of X has a join, then $\mathbb{J}(A)$ is nonempty for all $A \in \mathsf{Dwn}(X)$. Using the axiom of choice we can select a choice function ξ of $\{\mathbb{J}(A) \mid A \in \mathsf{Dwn}(X)\}$ — i.e. $\xi(A) \in \mathbb{J}(A)$ for each $A \in \mathsf{Dwn}(X)$. If $A \subseteq B$, then $\xi(A)$ is a join of A and $\xi(B)$ is an upper bound of A. Hence $\xi(A) \le \xi(B)$ follows — i.e. ξ is isotone. By definition x is equivalent to $\xi(\eta_X(x))$ for each $x \in X$. Hence Lemma 1.3.3 implies that ξ is left adjoint to η_X. □

Motivated by the previous theorem we introduce the following terminology. A preordered set X is *join-complete* if every subset of X has a join. A join of the empty subset is called a *universal lower bound* of X, while a join of X is called a *universal upper bound* of X. Based on this terminology we can reformulate Theorem 1.3.4 as follows. A preordered set X is join-complete if and only if the isotone map $X \xrightarrow{\eta_X} \mathsf{Dwn}(X)$ has a left adjoint map. Since left adjoint maps of a given isotone map are naturally equivalent, in the case of join-complete preordered sets X a left adjoint map of η_X will always be denoted by $\sup_X$ and will also be called the *formation of arbitrary joins* in X.

The dual concept of downclosed subsets are *upclosed* subsets and the dual notions of upper bound and join are *lower bound* and *meet*. For later applications we briefly recall some notation. The partially ordered set of all upclosed subsets of a preordered set X is denoted by $\mathsf{Up}(X)$ and is provided with the containment relation $\subseteq^{op}$. The isotone map $X \xrightarrow{\eta_X^*} \mathsf{Up}(X)$ is determined by $\eta_X^*(x) = \uparrow x = \{z \in X \mid x \le z\}$ for each $x \in X$. Then a preordered set X is *meet-complete* (i.e. all subsets have meets) if and only if η_X^* has a right adjoint map $\inf_X$ which is also called the *formation of arbitrary meets* in X.

The next proposition reveals that join-completeness and meet-completeness are equivalent concepts.

Proposition 1.3.5. *A preordered set X is join-complete if and only X is meet-complete.*

Proof. Let us assume that X is join-complete. We fix a subset A of X and consider the set B of all lower bounds of A. Let y be a join of B. Since every element of A is an upper bound of B, y is a lower bound of A. In particular, $b \leq y$ for all $b \in B$. Hence y is a meet of A. The converse direction follows by the principle of duality. □

Motivated by the previous proposition in general we will speak of *complete* preordered sets where we mean either join-completeness or meet-completeness. An antisymmetric, complete preordered set is also called a *complete lattice*. In this context $\sup_X$ is denoted by $\bigvee$ and $\inf_X$ by $\bigwedge$.

In the following considerations we are interested in the extension of isotone maps to join-preserving maps, which will lead to a further monad on `Preord`. First we make a simple observation.

Lemma 1.3.6. *Let X and Y be complete preordered sets, and $X \xrightarrow{f} Y$ be an isotone map. Then for every $A \in \mathsf{Dwn}(X)$ the following relation holds:*

$$\sup_Y(\downarrow f(A)) \leq f(\sup_X(A)).$$

Proof. Let A be a downclosed subset of X. Since $A \subseteq \eta_X(\sup_X(A))$, we infer from the isotonicity of f that $\downarrow f(A) \subseteq \eta_Y(f(\sup_X(A)))$ holds. Hence

$$\sup_Y(\downarrow f(A)) \leq \sup_Y(\eta_Y(f(\sup_X(A)))) \leq f(\sup_X(A)). \qquad \square$$

Motivated by the previous lemma we introduce the following terminology. Let X and Y be complete preordered sets. An isotone map $X \xrightarrow{f} Y$ is *join-preserving* if for every downclosed subset A of X the following inequality holds:

$$f(\sup_X(A)) \leq \sup_Y(\downarrow f(A)). \tag{1.23}$$

A characterization of join-preserving maps is given in the next theorem.

Theorem 1.3.7. *Let X and Y be complete preordered sets. An isotone $X \xrightarrow{f} Y$ is join-preserving if and only if f has a right adjoint map.*

Proof. Since X is complete, we introduce an isotone map $Y \xrightarrow{g} X$ by

$$g(y) = \sup_X\big(f^{-1}(\eta_Y(y))\big) = \sup_X\{x \in X \mid f(x) \leq y\}, \qquad y \in Y. \tag{1.24}$$

Since $f(\eta_X(x)) \subseteq \eta_Y(f(x))$, the relation

$$x \leq \sup_X(\eta_X(x)) \leq \sup_X\big(f^{-1}(\eta_Y(f(x)))\big) = g(f(x))$$

follows for all $x \in X$. If f is join-preserving, we conclude from (1.23) that

$$f(g(y)) = f\big(\sup_X(f^{-1}(\eta_Y(y)))\big) \le \sup_Y\big(\downarrow f\big(f^{-1}(\eta_Y(y))\big)\big) \le \sup_Y(\eta_Y(y)) \le y$$

holds for all $y \in Y$. Hence g is a right adjoint map to f.
On the other hand, let A be a downclosed subset of X. If g is a right adjoint map of f, then we infer from the relation

$$f(a) \le \sup_Y(\downarrow f(A)), \qquad a \in A$$

that $g\big(\sup_Y(\downarrow f(A))\big)$ is an upper bound of A. Hence $\sup_X(A) \le g\big(\sup_Y(\downarrow f(A))\big)$ and a fortiori $f\big(\sup_X(A)\big) \le \sup_Y(\downarrow f(A))\big)$ follows — i.e. f is join-preserving. □

Corollary 1.3.8. *Let X and Y be complete preordered sets. An isotone map $X \xrightarrow{f} Y$ is meet-preserving if and only if f has a left adjoint map.*

Proof. The assertion follows from Theorem 1.3.7 by the principle of duality. □

Since arbitrary unions of downclosed subsets of a preordered set X are again downclosed, the partially ordered set $(\mathsf{Dwn}(X), \subseteq)$ is obviously a complete lattice. If Y is now an arbitrary complete preordered set, we are interested in the extension of isotone maps $X \xrightarrow{h} Y$ to join-preserving maps from $\mathsf{Dwn}(X)$ to Y. As a preparation we first prove an interesting property.

Lemma 1.3.9. *Let X and Y be preordered sets and $X \xrightarrow{h} Y$ be an isotone map. Further, let $\mathsf{Dwn}(X) \underset{g}{\overset{f}{\rightleftarrows}} Y$ be a pair of adjoint isotone maps such that f is left adjoint to g. Then h and $f \circ \eta_X$ are naturally equivalent if and only if g has the form $g(y) = \{x \in X \mid h(x) \le y\}$ for each $y \in Y$.*

Proof. First we observe that h and $f \circ \eta_X$ are natural equivalent if and only if the equivalence $h(x) \le y \iff (f \circ \eta_X)(x) \le y$ holds for all $x \in X$ and for all $y \in Y$. Hence the assertion follows from

$$(f \circ \eta_X)(x) \le y \iff \eta_X(x) \subseteq g(y) \iff x \in g(y)$$

where $x \in X$ and $y \in Y$. □

Theorem 1.3.10. *Let X be a preordered set and Y be a complete preordered set. For every isotone map $X \xrightarrow{h} Y$ there exists a join-preserving map $\mathsf{Dwn}(X) \xrightarrow{h^\sharp} Y$ such that h and $h^\sharp \circ \eta_X$ are naturally equivalent. Moreover, $h^\sharp$ is unique up to a natural equivalence.*

Proof. The uniqueness of $h^\sharp$ up to a natural equivalence follows immediately from Theorem 1.3.7 and Lemma 1.3.9. On the other hand, since the map $Y \xrightarrow{g} \mathsf{Dwn}(X)$ defined by $g(y) = \{x \in X \mid h(x) \le y\}$ is meet-preserving, Corollary 1.3.8 implies that g has a left adjoint map $\mathsf{Dwn}(X) \xrightarrow{h^\sharp} Y$. Finally, by Lemma 1.3.9 the maps $h^\sharp \circ \eta_X$ and h are naturally equivalent. □

If in the previous theorem we now assume that the preorder on Y is antisymmetric — this means that Y is a complete lattice, then Theorem 1.3.10 can be reformulated as follows.

Corollary 1.3.11. *Let X be a preordered set and Y be a complete lattice. Then for every isotone map $X \xrightarrow{h} Y$ there exists a unique join-preserving map $\mathsf{Dwn}(X) \xrightarrow{h^\sharp} Y$ such that the following diagram is commutative:*

$$\begin{array}{ccc} X & \xrightarrow{\eta_X} & \mathsf{Dwn}(X) \\ & \searrow^{h} & \vdots\, h^\sharp \\ & & Y \end{array} \qquad (1.25)$$

Obviously $h^\sharp$ is determined by $h^\sharp(A) = \bigvee h(A)$ for each $A \in \mathsf{Dwn}(X)$.

1.3.1 The Monad of Downclosed Sets

Corollary 1.3.11 permits an equivalent categorical formulation. For this purpose we introduce the category `Sup` of complete lattices and join-preserving maps. On this background Corollary 1.3.11 means that the forgetful functor $\mathsf{U}\colon \texttt{Sup} \to \texttt{Preord}$ has a left adjoint functor $\mathsf{F}\colon \texttt{Preord} \to \texttt{Sup}$. In particular, F is determined by:

$$\mathsf{F}(X) = \mathsf{Dwn}(X), \quad X \xrightarrow{h} Y, \quad \mathsf{Dwn}(X) \xrightarrow{\mathsf{F}(h)} \mathsf{Dwn}(Y), \quad \mathsf{F}(h) = (\eta_Y \circ h)^\sharp.$$

The monad on `Preord` corresponding to $\mathsf{F} \dashv \mathsf{U}$ (cf. Sect. 1.2) has the form $\mathbf{Dwn} = (\mathsf{Dwn}, \eta, \mu)$, where

- The endofunctor Dwn assigns $\mathsf{Dwn}(X)$ to a preordered set X and acts on isotone maps $X \xrightarrow{h} Y$ as follows:

$$\mathsf{Dwn}(h)(A) = (\eta_Y \circ h)^\sharp(A) = \bigcup(\eta_Y \circ h)(A) = \bigcup_{a \in A} \downarrow h(a) = \downarrow h(A),\ A \in \mathsf{Dwn}(X).$$

- The unit has the form $\eta_X(x) = \downarrow x$, and the X-component of the multiplication μ coincides with the arbitrary union of downclosed sets — i.e. $\mu_X(\mathbb{A}) = \bigcup \mathbb{A}$ for all $\mathbb{A} \in \mathsf{Dwn}(\mathsf{Dwn}(X))$.

The monad **Dwn** is called the *monad of downclosed sets* on `Preord`. Obviously, the endofunctor Dwn of **Dwn** preserves the order on hom-sets of `Preord`. Hence Dwn has the interesting property that Dwn preserves adjoint pairs of isotone maps.

If $((X, \leq), \xi)$ is now a **Dwn**-algebra, then the unit axiom implies (cf. Lemma 1.3.3 and Theorem 1.3.4) that $(X, \leq)$ is complete and $\leq$ is antisymmetric — i.e. the underlying preordered set of a **Dwn**-algebra is always a complete lattice, and the structure map coincides with the formation of arbitrary joins. On the other hand, if X is a complete lattice, then $\mathsf{Dwn}(X) \xrightarrow{\sup_X} X$ is left adjoint to $X \xrightarrow{\eta_X} \mathsf{Dwn}(X)$ (cf. Lemma 1.3.3) and consequently join-preserving (cf. Theorem 1.3.7). Hence the associativity axiom of $(X, \sup_X)$ follows — i.e. every complete lattice is a **Dwn**-algebra. Moreover **Dwn**-homomorphisms and join-preserving maps between complete lattices are the same things. Hence the Eilenberg–Moore category of **Dwn** coincides with `Sup`.

1.3.2 The Composite Monad DwnT

In this subsection we show that the composition of the term monad **T** on `Preord` (of a single binary operator symbol) with the monad of downclosed sets exists in the sense of some distributive law $\sigma : \mathsf{TDwn} \to \mathsf{DwnT}$. First we fix a preordered groupoid $(X, \leq, \cdot)$ and define an isotone binary operation $\boxdot$ on $\mathsf{Dwn}(X)$ as follows:

$$A \boxdot B = \{z \in X \mid \exists a \in A,\ \exists b \in B : z \leq a \cdot b\}, \qquad A, B \in \mathsf{Dwn}(X). \quad (1.26)$$

Then $\boxdot$ is called the *Minkowski multiplication* on $\mathsf{Dwn}(X)$ induced by the multiplication of X. In particular, $(\mathsf{Dwn}(X), \subseteq, \boxdot)$ is again a partially order groupoid. If h is an isotone groupoid homomorphism, then it is easily seen that $\mathsf{Dwn}(h)$ is again an isotone groupoid homomorphism with respect to the respective Minkowski multiplications. Hence we conclude from Theorem 1.2.7 that there exists a distributive law

$$\sigma : \mathsf{TDwn} \to \mathsf{DwnT}$$

of the endofunctor Dwn of `Preord` over the term monad **T** (of a single binary operator symbol) on `Preord`. In particular, if X is a preordered set, then the construction of the X-component $\mathsf{TDwn}(X) \xrightarrow{\sigma_X} \mathsf{DwnT}(X)$ of the "swapper map" can be reproduced as follows.

Let $\mathsf{T}(X)$ be the free preordered groupoid generated by X with the corresponding "embedding" $X \xrightarrow{\eta_X^{\mathsf{T}}} \mathsf{T}(X)$, and let $\mathsf{TDwn}(X)$ be the free preordered groupoid generated by $\mathsf{Dwn}(X)$. Further, let $\mathsf{Dwn}(\mathsf{T}(X))$ be the partially order groupoid provided with the Minkowski multiplication induced by the multiplication of $\mathsf{T}(X)$. Then $\mathsf{TDwn}(X) \xrightarrow{\sigma_X} \mathsf{Dwn}(\mathsf{T}(X)) = \mathsf{DwnT}(X)$ is given by the unique extension of $\mathsf{Dwn}(X) \xrightarrow{\mathsf{Dwn}(\eta_X^{\mathsf{T}})} \mathsf{Dwn}(\mathsf{T}(X))$ to an isotone groupoid homomorphism.

Moreover, if $(X, \leq, \cdot)$ is a preordered groupoid, then it is easily seen that the following relations hold:

$$\eta_X(x \cdot z) = \downarrow(x \cdot z) = (\downarrow x) \boxdot (\downarrow z) = \eta_X(x) \boxdot \eta_X(z),$$
$$\mu_X(\mathbb{A}) \boxdot \mu_X(\mathbb{B}) = \{z \in X \mid \exists\, A \in \mathbb{A},\ \exists B \in \mathbb{B},\ \exists a \in A,\ \exists b \in B : z \leq a \cdot b\}$$
$$= \bigcup_{A \in \mathbb{A},\, B \in \mathbb{B}} A \boxdot B = \mu_X(\mathbb{A} \boxdot \mathbb{B}).$$

Hence we conclude from Corollary 1.2.9 and Corollary 1.2.11 that we can apply Beck's Theorem (cf. Theorem 1.2.12) to our setting. So, the composition of the term monad on `Preord` (of a single binary operator symbol) with the monad of down-closed sets exists. As we have seen, here the Minkowski multiplication plays a rôle of fundamental importance.

Exercises

1.3.1. A *closure operator* on a preordered set $(X, \leq)$ is an isotone map $X \xrightarrow{c} X$ satisfying the properties $x \leq c(x)$ and $c(c(x)) \leq c(x)$ for all $x \in X$.

(a) Let $X \underset{g}{\overset{f}{\rightleftarrows}} Y$ be a pair of adjoint maps such that $f \dashv g$ holds. Show that $c = g \circ f$ is a closure operator on X.
(b) Let c be a closure operator on $(X, \leq)$. On $c(X)$ we consider the preorder $\leq$ inherited from $(X, \leq)$. Now, let $c(X) \xrightarrow{\iota} X$ be the inclusion map and $X \xrightarrow{q} c(X)$ be the isotone map determined by c as map onto its range — i.e. $q(x) = c(x)$ for all $x \in X$. Show that $q \dashv \iota$ holds.

Comment. If we identify preordered sets with thin and small categories, then closure operators and monads are equivalent concepts.

1.3.2. Let X and Y be posets and $X \underset{g}{\overset{f}{\rightleftarrows}} Y$ be a pair of isotone maps such that $f \dashv g$ or $g \dashv f$ holds.

(a) Show that the following assertions are valid (cf. Corollary 1.3.2):
  (i) $f \circ g = 1_Y \iff f$ is surjective $\iff g$ is injective.
  (ii) $g \circ f = 1_X \iff f$ is injective $\iff g$ is surjective.
  (iii) f is bijective $\iff$ ($f \dashv g$ *and* $g \dashv f$) $\iff g$ is bijective.
(b) Show that f is an order isomorphism if and only if f is bijective.

Warning. In the general case a bijective isotone map may not be an order isomorphism.

1.3.3. (Cf. [75]) Let $\mathsf{Up}(X)$ be the complete lattice of all upclosed subsets of a preordered set X. Since the partial order on $\mathsf{Up}(X)$ is given by the containment relation $\subseteq^{op}$, meets in $\mathsf{Up}(X)$ are given by unions of upclosed subsets. The *upclosed* *hull* of a subset A of X has the form $\uparrow A = \{x \in X \mid \exists a \in A : a \leq x\}$. Further, we recall that $\eta^*_X(x) = \uparrow x$ for all $x \in X$. In this context the upclosed set $\uparrow x$ is also called the *upsegment* of x. After having fixed this notation we complete the object function $X \longmapsto \mathsf{Up}(X)$ to an endofunctor Up of `Preord` as follows:

$$X \xrightarrow{f} Y, \quad \mathsf{Up}(X) \xrightarrow{\mathsf{Up}(f)} \mathsf{Up}(Y), \quad \mathsf{Up}(f)(A) = \uparrow f(A), \quad A \in \mathsf{Up}(X).$$

Show that the following assertions are valid:

(a) The triple $\mathbf{Up} = (\mathsf{Up}, \eta^*, \mu^*)$ is a monad on `Preord` where the X-component $\mathsf{Up}(\mathsf{Up}(X)) \xrightarrow{\mu^*_X} \mathsf{Up}(X)$ of the multiplication μ^* coincides with the formation of arbitrary unions of upclosed subsets.
(b) The Eilenberg–Moore category of $\mathbf{Up}$ coincides with the category `Inf` of complete lattices and meet-preserving maps.
(c) There exists a distributive law $\sigma : \mathsf{Up}\mathsf{Dwn} \to \mathsf{Dwn}\mathsf{Up}$ of the endofunctor Dwn over the monad $\mathbf{Up}$. In particular, for some preordered set X compute explicitly the X-component σ_X of σ.
(Hint: Use the fact that Dwn preserves adjoint pairs of isotone maps and apply Theorem 1.2.7).
(d) The composite monad **DwnUp** exists.
(Hint: Corollaries 1.2.9 and 1.2.11 and Beck's Theorem.)
(e) The Eilenberg–Moore category of **DwnUp** is isomorphic to the category `CD` of completely distributive complete lattices with join- and meet-preserving maps.
(Hint: Theorem 1.2.16.)

Notes

The history of preorders is not easily traceable, but preorders were already occurring in the literature as far back as 1937 under the name "quasi-orderings" (cf. [13]). Since then, preordered sets have played a remarkable rôle in various areas of mathematics — especially in category theory (see e.g. [34, 36, 73]). How adjoint pairs of maps, respectively pairs of adjoint functors, have emerged is a complex topic, and we will not attempt to give a precise historical account here (cf. [36, 73]). We only mention two important developments in the theory

- In 1944 O. Ore was the first to give a comprehensive study of adjoint pairs of antitone maps (i.e. contravariant Galois connections) (cf. [87]).
- In 1958 D.M. Kan introduced the important concept of adjoint functors (cf. [62]).

If preordered sets are viewed as 2-enriched categories (cf. [63]), then the internal characterization of complete preordered sets by the requirement that the covariant (resp. contravariant) 2-enriched Yoneda embedding (cf. [15, 16]) has a left (resp. right) adjoint map goes back to C.J. Mikkelsen 1976 (cf. [78], where an internal characterization of complete ordered object is given). The extension of isotone maps to join-preserving (resp. meet-preserving) maps is a standard construction in order theory which is evidently independent from the antisymmetry axiom.

The investigation of the interrelationship between adjoint situations and monads goes back to P. Huber 1961, S. Eilenberg, J.C. Moore and H. Kleisli 1965 (cf. [26, 57, 66]). Huber showed that any adjoint situation gives rise to a standard construction, which is a comonad in contemporary language, while Eilenberg and Moore, and Kleisli give two different constructions showing that every monad is induced by an adjoint pair of functors. In this book we have only focused on the Eilenberg–Moore category and not on the Kleisli category.

The composition of monads in the sense of distributive laws (cf. [10]) depends essentially on the star composition of natural transformations — a construction which seems to go back to C. Ehresmann 1960 (cf. [25]), even though in 1958 various pieces of this construction had already appeared in the appendix of [38] under the title *Cinq règles de calcul fonctoriel*. For this reason the star composition is sometimes also called the *Godement product* (cf. [15, p. 13]). A graphical explanation of the Interchange Law (cf. [47, 73]) is given in Appendix A.3, where a graphical representation of a mixed application of the (vertical) composition and star composition is also developed. In this context, the equation in Lemma 1.2.13 and its application in the proof of Beck's Theorem (cf. Theorem 1.2.12), including the related graphical representation, seem to appear explicitly for the first time in [27].

The introduction of monoidal categories goes back to J. Bénabou and S. Mac Lane 1963 (cf. [12, 72]). This concept forms a perfect algebraic framework for the definition of magmas, semigroups and monoids. The existence of free magmas in the context of monoidal biclosed categories has recently been established by P. Eklund, U. Höhle and J. Kortelainen (cf. [28]). This proof is based on the free algebra algorithm formulated by J. Adámek 1974 (cf. [1, 3]).

Chapter 2
Fundamentals of Quantales

The aim of this chapter is to give a survey of the most important properties of the theory of (pre)quantales from the perspective of the category `Sup` of complete lattices and join-preserving maps. We therefore begin with a detailed account of the categorical properties of `Sup`.

As a first important observation we point out that the category of prequantales is the Eilenberg–Moore category of the composition of the term monad of a single binary operator symbol on `Preord` with the monad of downclosed sets. This result also explains the rôle of the Minkowski multiplication in the theory of (pre)quantales. As a corollary of this fact we obtain that the category of prequantales is algebraic.

The second important observation is the fact that the Eilenberg–Moore category of the composition of the monad of involutive lattices on `Sup` with the term monad of a single binary operator symbol on `Sup` coincides with the category of involutive prequantales.

Quantales are associative prequantales. In this context the concept of prime elements of quantales opens the door for the topological representation of quantales. Left-sided idempotent quantales and balanced, bisymmetric quantales play a dominant rôle in these constructions. In both cases the non-commutativity forces the absence of a unit. Moreover, the non-commutativity requires many-valued topologies (cf. [49]) for the topological representation. In particular, semi-unital, bisymmetric and spatial quantales are represented by six-valued topological spaces where the quantale of six elements coincides with quantization of the two-chain — i.e. the tensor product of the left-sided, non-commutative, idempotent three-chain with the right-sided, non-commutative, idempotent three-chain.

As an application of these constructions we treat the problem of defining the spectrum of a C^*-algebra and constructing its topological representation. This approach leads to the concept of the non-commutative Gelfand topology, which coincides with the usual Gelfand topology in the commutative case.

Finally, we try to give a coherent account of the law of double negation in the framework of quantales. Hence we describe the fundamental properties of Frobenius

P. Eklund et al., *Semigroups in Complete Lattices*, Developments in Mathematics 54, https://doi.org/10.1007/978-3-319-78948-4_2

quantales, including a class of examples which arise from completely distributive lattices. We also study complete MV-algebras as divisible and commutative Frobenius quantales. Without mentioning the concept of semi-simplicity we pay special attention to the MacNeille completion of MV-algebras. As a special result we prove that every infinite, simple and complete MV-algebra is isomorphic to the real unit interval provided with the **Łukasiewicz** arithmetic conjunction.

2.1 The Category Sup

In Sect. 1.3.1 we have already encountered the category Sup of complete lattices and join-preserving maps. In this section we will give a detailed account of its categorical properties and related topics.

First of all, since Sup is the Eilenberg–Moore category of the monad of down-closed sets, the completeness of Sup is inherited by the completeness of Preord. With regard to cocompleteness, Theorem 1.3.7 and Corollary 1.3.8 suggest that on Sup there exists a self-duality expressed by the following contravariant endofunctor $\mathsf{S}\colon \mathtt{Sup} \to \mathtt{Sup}$:

$$\mathsf{S}(X) = X^{op} = (X, \leq^{op}), \quad X \xrightarrow{f} Y, \quad \mathsf{S}(f) = f^{\vdash}, \quad Y^{op} \xrightarrow{f^{\vdash}} X^{op}$$

where $f^{\vdash}$ is the right adjoint map of f and has the following explicit form (see (1.24)):

$$f^{\vdash}(y) = \bigvee\{x \in X \mid f(x) \leq y\}, \qquad y \in Y.$$

Hence the cocompleteness of Sup follows from the completeness of Sup and the existence of the self-duality S. Obviously Sup has a zero object given by the lattice consisting of a single element. For the convenience of the reader we briefly describe the construction of coproducts and coequalizers in Sup.

Let $\mathbb{F} = \{X_i \mid i \in I\}$ be a family of complete lattices and $\prod_{i\in I} X_i$ be the product of $\mathbb{F}$ (in Sup). Since the projections $\prod_{i\in I} X_i \xrightarrow{\pi_j} X_j$ $(j \in I)$ are meet-preserving, the coproduct of $\mathbb{F}$ is also given by $\prod_{i\in I} X_i$, but the corresponding coproduct injections $X_j \xrightarrow{q_j} \prod_{i\in I} X_i$ are determined by:

$$q_j(x_j) = (z_i)_{i\in I}, \quad z_i = \begin{cases} x_j, & i = j, \\ \bot, & i \neq j, \end{cases} \qquad x_j \in X_j$$

where $\bot$ denotes the universal lower bound in X_i.

If $R \overset{f}{\underset{g}{\rightrightarrows}} X$ is a pair of arrows in Sup, then we can explicitly construct the coequalizer of f and g as follows. We put $Z = \{x \in X \mid f^{\vdash}(x) = g^{\vdash}(x)\}$ and define $X \xrightarrow{\pi} Z$ by $\pi(x) = \bigwedge\{z \in Z \mid x \leq z\}$ for all $x \in X$. Obviously $\pi \circ f = \pi \circ g$ holds. In order to verify the universal property, we consider a further arrow $X \xrightarrow{h} U$ with $h \circ f = h \circ g$ and define $Z \xrightarrow{h^*} U$ by $h^*(z) = \bigwedge\{u \in U \mid z \leq h^{\vdash}(u)\}$ for all $z \in Z$. Since

$$h(x) \leq u \iff x \leq h^{\vdash}(u) \iff \pi(x) \leq h^{\vdash}(u) \iff h^*(\pi(x)) \leq u,$$

the relation $h(x) = (h^* \circ \pi)(x)$ holds for all $x \in X$. The uniqueness of h^* follows from the surjectivity of π.

In the next step we investigate the rôle of epimorphisms in Sup. We begin with a simple, but interesting property.

Lemma 2.1.1. *Let $X \xrightarrow{f} Y$ be an epimorphism in* Sup*. Then f, viewed as a map, is surjective.*

Proof. Let $C_2 = \{0, 1\}$ be the 2-chain with the usual order. Then every element $y \in Y$ can be identified with a join-preserving map $Y \xrightarrow{\lambda_y} C_2$ defined by:

$$\lambda_y(z) = \begin{cases} 1, & z \not\leq y, \\ 0, & z \leq y, \end{cases} \qquad z \in Y.$$

Since $f(x) \leq y$ if and only if $f(x) \leq f(f^{\vdash}(y))$ we obtain the following relation for all $y \in Y$:

$$\lambda_y \circ f = \lambda_{f(f^{\vdash}(y))} \circ f.$$

Since f is an epimorphism, the relation $y = f(f^{\vdash}(y))$ follows for all $y \in Y$. Hence f is surjective. □

Theorem 2.1.2. *Let $X \xrightarrow{h} Y$ be an epimorphism in* Sup *and $R \underset{q}{\overset{p}{\rightrightarrows}} X$ be the kernel pair of h. Then the diagram*

$$\begin{array}{ccc} R & \xrightarrow{q} & X \\ {\scriptstyle p}\downarrow & & \downarrow{\scriptstyle h} \\ X & \xrightarrow{h} & Y \end{array} \tag{2.1}$$

is a pullback and a pushout square — i.e. a pulation square.

Proof. By definition, the diagram (2.1) is a pullback square. In order to show that (2.1) is also a pushout square we proceed as follows.
Since pullback squares in Sup are computed at the level of Preord, we conclude from $h \circ h^{\vdash} \circ h = h$ (cf. Corollary 1.3.2) that there exist isotone maps $X \underset{\zeta}{\overset{\xi}{\rightrightarrows}} R$ such that the relations $p \circ \xi = h^{\vdash} \circ h = q \circ \zeta$ and $q \circ \xi = 1_X = p \circ \zeta$ hold in Preord.
Now we consider a further complete lattice Z and a pair $X \underset{l}{\overset{k}{\rightrightarrows}} Z$ of join-preserving maps with $k \circ p = l \circ q$. By Corollary 1.3.2 again, we have the relation

$$k \circ h^{\vdash} = k \circ h^{\vdash} \circ h \circ h^{\vdash} = k \circ p \circ \xi \circ h^{\vdash} = l \circ q \circ \xi \circ h^{\vdash} = l \circ h^{\vdash}$$

and we can introduce an isotone map $Y \xrightarrow{\pi} Z$ given by $\pi = k \circ h^{\vdash} = l \circ h^{\vdash}$. Now we observe:

$$k = k \circ p \circ \zeta = l \circ q \circ \zeta = l \circ h^{\vdash} \circ h = \pi \circ h = k \circ h^{\vdash} \circ h = k \circ p \circ \xi = l \circ q \circ \xi = l.$$

Hence we have verified $k = \pi \circ h = l$. Since h is surjective (cf. Lemma 2.1.1), π is uniquely determined by the previous relation. Further, we show that $h \circ l^{\vdash}$ is the right adjoint map of π. Referring again to the surjectivity of h we obtain:

$$h \circ l^{\vdash} \circ \pi = h \circ l^{\vdash} \circ l \circ h^{\vdash} \geq h \circ h^{\vdash} = 1_Y.$$

On the other hand we observe that

$$\pi \circ h \circ l^{\vdash} = k \circ h^{\vdash} \circ h \circ l^{\vdash} = k \circ p \circ \xi \circ l^{\vdash} = l \circ q \circ \xi \circ l^{\vdash} = l \circ l^{\vdash} \leq 1_Z.$$

Hence π is join-preserving (Theorem 1.3.7), and so the universal property of a pushout square is established. □

By the previous theorem every epimorphism in Sup is regular and can be understood as a surjective join-preserving map. If we apply the self-duality of Sup to Theorem 2.1.2, then we conclude from Theorem 2.1.2 that the amalgamation property holds in Sup — this means that for any monomorphism $Y \xrightarrow{m} X$ in Sup the pushout square

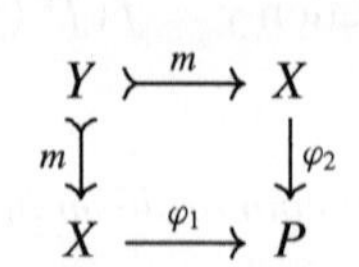

is also a pullback square.

2.1.1 Closure Operators and Quotient Objects in Sup

Let X be a complete lattice and $\mathsf{CL}(X)$ be the set of all closure operators on X (cf. Exercise 1.3.1). On $\mathsf{CL}(X)$ we introduce a partial order as follows

$$c_1 \leq c_2 \iff c_1(x) \leq c_2(x) \text{ for all } x \in X.$$

Then $(\mathsf{CL}(X), \leq)$ is again a complete lattice, and meets in $\mathsf{CL}(X)$ are computed pointwisely.

Let a closure operator c on X be given. Then on the range $c(X)$ of c we consider always the partial order inherited from X. Thus $c(X)$ is again a complete lattice and the join of a subset A of $c(X)$ is the c-closure of the join of A in X — i.e. $\bigvee A = c(\bigvee A)$. Moreover, we conclude from Exercise 1.3.1 (b) and Theorem 1.3.7 that the map c viewed as map onto its range — i.e. $X \xrightarrow{\pi} c(X)$ with $\pi(x) = c(x)$ for all $x \in X$ — is join-preserving. Hence $(c(X), \pi)$ can be understood as a quotient of X in Sup.

Further, every join-preserving map $X \xrightarrow{h} Y$ induces a closure operator c_h on X by $c_h = h^{\vdash} \circ h$. In this context c_h is also called the *closure operator associated with* h. We show that Sup has the (epi,mono)-factorization property (cf. [47]).

As a first step we prove a factorization lemma.

Lemma 2.1.3. *Let $X \xrightarrow{\pi} Z$ be an epimorphism in* Sup *and $X \xrightarrow{h} Y$ be an arbitrary join-preserving map with the associated closure operators c_π and c_h on X. Then h factors through π — i.e. the commutativity of the following diagram holds:*

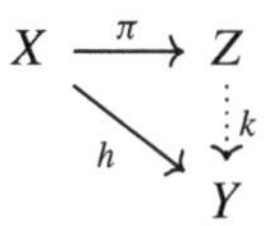

if and only if the relation $c_\pi \leq c_h$ is valid.

Proof. Since adjoint situations compose, the necessity of $c_\pi \leq c_h$ follows from $h = k \circ \pi$ and the relation $\pi^{\vdash} \circ \pi \leq \pi^{\vdash} \circ k^{\vdash} \circ k \circ \pi = h^{\vdash} \circ h$. On the other hand, let us assume $c_\pi \leq c_h$ and observe that $c_\pi \leq c_h$ is equivalent to $c_h = c_h \circ c_\pi$. Now we define an isotone map $Z \xrightarrow{k} Y$ by $k = h \circ \pi^{\vdash}$ and show that $\pi \circ h^{\vdash}$ is the right adjoint map of k. Since π is surjective (cf. Lemma 2.1.1), the relation $1_Z \leq (\pi \circ h^{\vdash}) \circ k$ is obvious. Further, we obtain (cf. Corollary 1.3.2):

$$k \circ (\pi \circ h^{\vdash}) = h \circ c_h \circ c_\pi \circ h^{\vdash} = h \circ c_h \circ h^{\vdash} = h \circ h^{\vdash} \leq 1_Y.$$

Hence k is join-preserving and the following relation holds:

$$k \circ \pi = h \circ c_h \circ c_\pi = h \circ c_h = h.$$

So, h factors through π. □

It follows immediately from Lemma 2.1.3 and the previous constructions that quotient objects of a complete lattice X in the sense of Sup can be identified with closure operators on X and vice versa. Hence the class of quotient objects of a complete lattice X is a partially ordered set and order isomorphic to the complete lattice $(\mathsf{CL}(X), \leq)$.

Finally, every join-preserving map $X \xrightarrow{h} Y$ has a decomposition

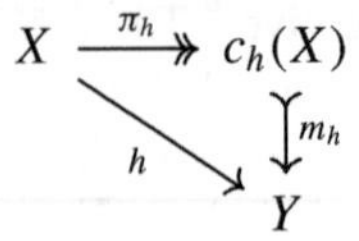

into an epimorphism π_h determined by c_h and followed by a monomorphism m_h defined on $c_h(X)$ by $m_h(x) = h(x)$ for all $x \in c_h(X)$. Obviously, this decomposition is unique up to isomorphism.

We finish this subsection with an example of a quotient construction given by the MacNeille completion.

Example 2.1.4. (a) Let X be a preordered set, $\mathsf{Dwn}(X)$ be the complete lattice of all downclosed subsets of X and $\mathsf{Up}(X)$ be the complete lattice of all upclosed subsets of X. We introduce an adjoint pair $\mathsf{Dwn}(X) \underset{\mathsf{L}}{\overset{\mathsf{U}}{\rightleftarrows}} \mathsf{Up}(X)$ of isotone maps by

$$\mathsf{U}(A) = \{\text{all upper bounds of } A\}$$

and

$$\mathsf{L}(B) = \{\text{all lower bounds of } B\},$$

where $A \in \mathsf{Dwn}(X)$ and $B \in \mathsf{Up}(X)$. Obviously, U is left adjoint to L. Hence $\mathsf{C} = \mathsf{L} \circ \mathsf{U}$ is a closure operator on $\mathsf{Dwn}(X)$ and the quotient $\mathsf{Dwn}(X) \xrightarrow{\pi_{\mathsf{C}}} \widehat{X}$ of $\mathsf{Dwn}(X)$ w.r.t. C is called the *MacNeille completion* of X. Since $\downarrow x = \eta_X(x)$ is C-closed (i.e. $\mathsf{C}(\downarrow x) = \downarrow x$) for all $x \in X$, the MacNeille completion $\widehat{X}$ satisfies the following well-known property:

$$\bigcap_{b \in \mathsf{U}(A)} \eta_X(b) = \mathsf{C}(A) = \bigvee_{x \in A} \eta_X(a), \qquad A \in \mathsf{Dwn}(X). \tag{2.2}$$

We show now in what sense the MacNeille completion is uniquely determined up to order isomorphism. For this purpose we choose a further complete lattice Y and an isotone map $X \xrightarrow{j} Y$ such that the following properties hold for all $x_1, x_2 \in X$ and $y \in Y$:

- $j(x_1) \le j(x_2) \implies x_1 \le x_2$,
- $\bigwedge\{j(x) \mid x \in X,\ y \le j(x)\} = y = \bigvee\{j(x) \mid x \in X,\ j(x) \le y\}$.

In a first step we consider the extension of the isotone map j to a join-preserving map $\mathsf{Dwn}(X) \xrightarrow{j^\sharp} Y$ (cf. Corollary 1.3.11). Obviously $j^\sharp$ is surjective, and the closure operator $c_{j^\sharp}$ associated with $j^\sharp$ has the following form:

$$c_{j^\sharp}(A) = \{x \in X \mid j(x) \le \bigvee j(A)\}, \qquad A \in \mathsf{Dwn}(X).$$

By the previous properties the relations

$$\{x \in X \mid j^\sharp(A) \le j(x)\} = \mathsf{U}(A) \quad \text{and} \quad \mathsf{L}(\mathsf{U}(A)) = c_{j^\sharp}(A), \qquad A \in \mathsf{Dwn}(X)$$

hold. Hence the closure operators C and $c_{j^\sharp}$ coincide — this means that $\widehat{X}$ and Y are order isomorphic.

(b) If Y is a further preordered set and $X \underset{g}{\overset{f}{\rightleftarrows}} Y$ is a pair of adjoint isotone maps with $f \dashv g$, then f has a unique extension to a join-preserving map $\widehat{X} \xrightarrow{\widehat{f}} \widehat{Y}$ such that the following diagram is commutative:

$$\begin{array}{ccc} X & \xrightarrow{f} & Y \\ {\scriptstyle \eta_X}\downarrow & & \downarrow{\scriptstyle \eta_Y} \\ \widehat{X} & \xrightarrow{\widehat{f}} & \widehat{Y} \end{array}$$

The uniqueness of $\widehat{f}$ follows immediately from (2.2). In order to verify the existence of $\widehat{f}$ we proceed as follows. Since the functor Dwn preserves the order on the respective hom-sets of $\texttt{Preord}$, $\mathsf{Dwn}(f)$ is left adjoint to $\mathsf{Dwn}(g)$. Further, we consider the join-preserving map $\mathsf{Dwn}(X) \xrightarrow{h} \widehat{Y}$ determined by the diagram:

$$\begin{array}{ccc} \mathsf{Dwn}(X) & \xrightarrow{\mathsf{Dwn}(f)} & \mathsf{Dwn}(Y) \\ & \searrow^{h} & \downarrow{\scriptstyle \pi_{\mathsf{C}}} \\ & & \widehat{Y} \end{array}$$

Since $1_X \le g \circ f$, we obtain the following relation for every $A \in \mathsf{Dwn}(X)$:

$$A \subseteq \downarrow g(f(A)) \subseteq \downarrow g\big(\mathsf{L}(\mathsf{U}(f(A)))\big) = \mathsf{Dwn}(g)\big(\pi_{\mathsf{C}}(\mathsf{Dwn}(f)(A))\big) = c_h(A).$$

Hence we conclude from Lemma 2.1.3 that h factors through the MacNeille completion of X — i.e. the commutativity of the following diagram:

$$\begin{array}{ccc} \mathsf{Dwn}(X) & \xrightarrow{\mathsf{Dwn}(f)} & \mathsf{Dwn}(Y) \\ {\scriptstyle \pi_{\mathsf{C}}}\downarrow & \searrow^{h} & \downarrow{\scriptstyle \pi_{\mathsf{C}}} \\ \widehat{X} & \xrightarrow{\widehat{f}} & \widehat{Y} \end{array}$$

Evidently, $\widehat{f}\big(\eta_X(x)\big) = \pi_{\mathsf{C}}\big(\mathsf{Dwn}(f)(\eta_X(x))\big) = \eta_Y(f(x))$ holds for all $x \in X$.

2.1.2 *The Tensor Product in* **Sup**

From the algebraic point of view one of the most important properties of $\texttt{Sup}$ is the existence of a tensor product turning $\texttt{Sup}$ into a symmetric monoidal closed category.

First we begin with some terminology and an important example related to contravariant Galois connections. Let X, Y and Z be complete lattices. A map $X \times Y \xrightarrow{b} Z$ is called a *bimorphism* in Sup (cf. [6]) if b is join-preserving in each variable separately — i.e. for all $y \in Y$ and $x \in X$ the maps $X \xrightarrow{b(\cdot,y)} Z$ and $Y \xrightarrow{b(x,\cdot)} Z$ are join-preserving. In the next example we construct an important bimorphism which already appeared in [106].

Example 2.1.5. Let X and Y be arbitrary complete lattices. We start by recalling that a pair (f, f^+) of maps $X \xrightarrow{f} Y$ and $Y \xrightarrow{f^+} X$ is a *contravariant Galois connection* between X and Y if for all $x \in X$ and $y \in Y$ the equivalence

$$y \leq f(x) \iff x \leq f^+(y)$$

holds (cf. [14, 35]) — i.e. $X \underset{f^+}{\overset{f}{\rightleftarrows}} Y^{op}$ is an adjoint pair of isotone maps such that $f \dashv f^+$ holds. Since X and Y are complete lattices, it follows from Theorem 1.3.7 that an antitone map $X \xrightarrow{f} Y$ is the left part of a contravariant Galois connection if and only if f is join-reversing. — i.e. $f(\bigvee A) = \bigwedge f(A)$ for all $A \subseteq X$. In this case the right part f^+ is given by (cf. (1.24))

$$f^+(y) = \bigvee\{x \in X \mid y \leq f(x)\}, \qquad y \in Y. \tag{2.3}$$

After this preparation we introduce the set $\mathscr{G}(X, Y)$ of all join-reversing maps $X \xrightarrow{f} Y$ provided with the following partial order:

$$f_1 \leq f_2 \iff f_1(x) \leq f_2(x) \quad \text{for all } x \in X. \tag{2.4}$$

Obviously, $(\mathscr{G}(X, Y), \leq)$ is a complete lattice in which meets (but in general not joins) are computed pointwisely. In particular, the universal upper bound $\underline{\top}$ and the universal lower bound $\underline{\perp}$ in $\mathscr{G}(X, Y)$ have the following form:

$$\underline{\top}(x) = \top \quad \text{and} \quad \underline{\perp}(x) = \begin{cases} \top, & x = \perp, \\ \perp, & x \neq \perp, \end{cases} \qquad x \in X.$$

In this context (2.3) determines an order isomorphism $\mathscr{G}(X, Y) \xrightarrow{(\cdot)^+} \mathscr{G}(Y, X)$. Now we introduce a map $X \times Y \xrightarrow{\beta} \mathscr{G}(X, Y)$ as follows:

$$[\beta(x, y)](x') = \begin{cases} \top, & \text{if } x' = \perp, \\ y, & \text{if } \perp \neq x' \leq x, \\ \perp, & \text{if } x' \nleq x, \end{cases} \qquad x, x' \in X, \ y \in Y. \tag{2.5}$$

Obviously, $\beta(x, y) \in \mathcal{G}(X, Y)$, and the relation $\beta(x, \bot) = \beta(\bot, y) = \underline{\bot}$ holds for all $x \in X$ and $y \in Y$. We show that β is a bimorphism. If we fix the left argument of β, then $\beta(x, \cdot)$ is evidently join-preserving. Hence we fix the right argument $y \in Y$ and observe $\beta(\bot, y) = \underline{\bot}$. Therefore it is sufficient to consider a nonempty subset $A \subseteq X$. If $f \in \mathcal{G}(X, Y)$ and $x \in X$, then

$$\beta(x, y) \le f \iff y \le f(x). \tag{2.6}$$

Hence an element $f \in \mathcal{G}(X, Y)$ is an upper bound of $\{\beta(x, y) \mid x \in A\}$ if and only if $y \le \bigwedge_{x \in A} f(x) = f(\bigvee A)$, and so $\beta\big(\bigvee A, y\big) \le f$ follows — this means that $\bigvee_{x \in A} \beta(x, y) = \beta\big(\bigvee A, y\big)$ holds.

Moreover, β satisfies the following important properties:

$$\beta(\top, \top) = \underline{\top} \tag{2.7}$$

$$\bigwedge_{i \in I} \beta(x_i, y_i) = \beta\big(\bigwedge_{i \in I} x_i, \bigwedge_{i \in I} y_i\big) \tag{2.8}$$

$$f = \bigvee_{x \in X} \beta(x, f(x)) \quad \text{for any } f \in \mathcal{G}(X, Y). \tag{2.9}$$

The properties (2.7) and (2.8) mean that β is *meet-preserving*, while the property (2.9) expresses the fact that $\{\beta(x, y) \mid x \in X, y \in Y\}$ forms a *join-basis* of $\mathcal{G}(X, Y)$.

Remark 2.1.6. Let X, Y and Z be complete lattices. If $X \times Y \xrightarrow{b} Z$ is a bimorphism, then for all $z \in Z$ the maps $X \xrightarrow{b_z} Y$ and $Y \xrightarrow{b_z^+} X$ given by

$$b_z(x) = \bigvee\{y \in Y \mid b(x, y) \le z\} \quad \text{and} \quad b_z^+(y) = \bigvee\{x \in X \mid b(x, y) \le z\} \tag{2.10}$$

satisfy the chain of equivalences

$$y \le b_z(x) \iff b(x, y) \le z \iff x \le b_z^+(y).$$

Hence (b_z, b_z^+) is a contravariant Galois connection between X and Y. In particular, $b_z \in \mathcal{G}(X, Y)$ and $b_z^+ \in \mathcal{G}(Y, X)$.

Now we give a definition which will play a fundamental rôle in this book.

Definition 2.1.7. Let X and Y be complete lattices. A pair (t, T) is called a *tensor product* of X and Y if T is a complete lattice and $X \times Y \xrightarrow{t} T$ is a bimorphism such that the following universal property holds:

For every bimorphism $X \times Y \xrightarrow{b} Z$ there exists a unique join-preserving map $T \xrightarrow{h_b} Z$ making the following diagram commutative:

$$\begin{array}{ccc} X \times Y & \xrightarrow{t} & T \\ & \searrow^{b} & \vdots\, h_b \\ & & Z \end{array} \tag{2.11}$$

In this context t is also called the *universal bimorphism*.

It follows immediately from the universal property in Definition 2.1.7 that a tensor product of two complete lattices is always unique up to isomorphism in the sense of Sup. Thus we speak of *the* tensor product. In what follows we will show that the tensor product always exists.

The next theorem is a special case of a general construction in [6] (cf. Corollary of Proposition 5 in [6]) where we maintain the notation from Example 2.1.5 and Remark 2.1.6

Theorem 2.1.8. *If X and Y are complete lattices, then $(\beta, \mathcal{G}(X, Y))$ is the tensor product of X and Y in* Sup.

Proof. Let Z be a complete lattice and $X \times Y \xrightarrow{b} Z$ be a bimorphism. We have to show that there exists a unique join-preserving map $\mathcal{G}(X, Y) \xrightarrow{h_b} Z$ making the diagram (2.11) commutative. The uniqueness of h_b follows immediately from (2.9), the commutativity of (2.11) and the fact that h_b is join-preserving. In order to verify the existence of h_b we define a map $\mathcal{G}(X, Y) \xrightarrow{h_b} Z$ as follows:

$$h_b(f) = \bigvee_{i \in I} b(x_i, y_i), \tag{2.12}$$

where $\bigvee_{i \in I} \beta(x_i, y_i) = f$ is a representation of $f \in \mathcal{G}(X, Y)$ as a join of some elements of the join-basis $\{\beta(x, y) \mid x \in X, y \in Y\}$. If we can show that h_b is well defined by (2.12) — this means that the value $h_b(f)$ is independent from the chosen representation — then it is easily seen that h_b is join-preserving and makes the diagram (2.11) commutative. Hence it is sufficient to show that h_b is well defined. For this purpose we choose a further representation $\bigvee_{j \in J} \beta(x_j, y_j) = f$ of f. Because of the law of antisymmetry it is sufficient to establish the following relation

$$\bigvee_{j \in J} b(x_j, y_j) \le \bigvee_{i \in I} b(x_i, y_i). \tag{2.13}$$

First we put $z = \bigvee_{i \in I} b(x_i, y_i)$. Since b is a bimorphism, we define an element $b_z \in \mathcal{G}(X, Y)$ by (cf. Remark 2.1.6):

$$b_z(x) = \bigvee \{y \in Y \mid b(x, y) \le z\}, \qquad x \in X.$$

Since $y_i \le b_z(x_i)$, it follows from the definition of β (cf. Example 2.1.5) that the relation $\beta(x_i, y_i) \le b_z$ holds for all $i \in I$ — i.e. $f = \bigvee_{i \in I} \beta(x_i, y_i) \le b_z$. Now we make use of the relation

$$\bigvee_{j \in J} \beta(x_j, y_j) = f \le b_z$$

and obtain $y_j \le b_z(x_j)$ for all $j \in J$. Then $b(x_j, y_j) \le z$ follows from Remark 2.1.6 for all $j \in J$. Hence (2.13) is verified. □

Addition and Comment. If we use the representation of $f \in \mathcal{G}(X, Y)$ given in (2.9), then the unique extension of a bimorphism $X \times Y \xrightarrow{b} Z$ to a join-preserving map $\mathcal{G}(X, Y) \xrightarrow{h_b} Z$ (cf. Theorem 2.1.8) can be expressed, among other ways, as follows:

$$h_b(f) = \bigvee_{x \in X} b(x, f(x)), \qquad f \in \mathcal{G}(X, Y). \tag{2.14}$$

This formulation will play a special rôle in the proofs of Lemmas 2.1.13 and 2.1.14.

Motivated by the previous theorem we introduce the following:

Notation. The tensor product of two complete lattices X and Y is denoted by $X \otimes Y$. By abuse of notation the universal bimorphism $X \times Y \xrightarrow{\beta} X \otimes Y$ is also denoted by $\otimes$. Instead of $\beta(x, y)$ we write $x \otimes y$. In this context every element of $f \in X \otimes Y$ is called a *tensor*, and tensors of the special type $x \otimes y$ are called *elementary tensors*.

Before we proceed we briefly sketch some simple properties of the tensor calculus. Obviously, $x \otimes \bot = \bot \otimes y$ and $\top \otimes \top$ are the bottom and the top of $X \otimes Y$, respectively (cf. Example 2.1.5). If $f \in X \otimes Y$, then for all $x \in X$ and $y \in Y$

$$x \otimes y \leq f \iff y \leq f(x). \tag{2.15}$$

In particular, if $x \neq \bot$ and $y \neq \bot$, then for all $x' \in X$ and $y' \in Y$

$$x \otimes y \leq x' \otimes y' \iff x \leq x' \text{ and } y \leq y' \tag{2.16}$$

holds. Moreover, it follows immediately from (2.7) – (2.9) that the universal bimorphism $\otimes$ is meet-preserving and every tensor is a join of an appropriate family of elementary tensors — e.g.

$$f = \bigvee_{x \in X} x \otimes f(x), \qquad f \in X \otimes Y = \mathcal{G}(X, Y). \tag{2.17}$$

Further, the maps $X \xrightarrow{j_X} X \otimes Y$ and $Y \xrightarrow{j_Y} X \otimes Y$ determined by

$$j_X(x) = x \otimes \top, \quad j_Y(y) = \top \otimes y, \qquad x \in X,\ y \in Y, \tag{2.18}$$

are meet-preserving and join-preserving embeddings provided X and Y have at least two different elements.

We continue with two examples which will play a special rôle in a later context.

Example 2.1.9. Let X be an arbitrary set and $\mathcal{P}(X)$ be the power set of X with the usual partial order. Then $\mathcal{P}(X)$ is a complete lattice. Further, let M be an arbitrary complete lattice. Then the set M^X of all maps $X \xrightarrow{f} M$ is again a complete lattice w.r.t. the partial order:

$$f \le g \iff f(x) \le g(x) \quad \text{for all } x \in X.$$

We will refer to M^X as the *M-valued power set of X*.

If we now restrict a join-reversing map $\mathscr{P}(X) \xrightarrow{F} M$ to the atoms of $\mathscr{P}(X)$, then we obtain an order isomorphism $\mathscr{P}(X) \otimes M \xrightarrow{\Phi} M^X$. In particular, Φ and Φ^{-1} are given by:

$$\big(\Phi(F)\big)(x) = F(\{x\}) \quad \text{and} \quad \big(\Phi^{-1}(f)\big)(A) = \bigwedge_{x \in A} f(x),$$

where $F \in \mathscr{P}(X) \otimes M$, $f \in M^X$, $x \in X$ and $A \in \mathscr{P}(X)$. Since the tensor product is unique up to isomorphism, M^X is also the tensor product of the power set $\mathscr{P}(X)$ with M. In this context it is interesting to see what the corresponding universal bimorphism looks like. Referring to the definition of β in Example 2.1.5 it is easily seen that the following relation holds for all $A \in \mathscr{P}(X)$ and all $m \in M$:

$$\big(m \cdot 1_A\big)(x) := \begin{Bmatrix} m, & x \in A \\ \perp, & x \notin A \end{Bmatrix} = \Phi\big(\beta(A, m)\big)(x), \qquad x \in X. \tag{2.19}$$

We emphasize that the dot in $m \cdot 1_A$ does not stand for any binary operation (it is simply part of the symbol). Hence, in this context, the universal bimorphism has the form $(A, m) \longmapsto m \cdot 1_A$. It is interesting to note that maps of the type $m \cdot 1_A$ play a significant rôle in the study of many-valued structures, where their meaning remains obscure. However, from the perspective of the category Sup these maps are simply *elementary tensors* of the *tensor product* of the power set with some complete lattice M.

The next example is related to probability theory and nonnegative probability distribution functions.

Example 2.1.10. Let $[0, +\infty]^{op}$ be the nonnegative extended real line provided with the dual order, and let $[0, 1]$ be the real unit interval. Then the tensor product $[0, +\infty]^{op} \otimes [0, 1]$ is the complete lattice of all meet-preserving maps

$$[0, +\infty] \xrightarrow{G} [0, 1]$$

— these are all nonnegative, isotone, right-continuous probability distribution functions. Obviously, for $x \in [0, +\infty]$ and $\alpha \in [0, 1]$ the elementary tensor $x \otimes \alpha$ has the following form (cf. Example 2.1.5):

$$(x \otimes \alpha)(r) = \begin{cases} 1, & r = +\infty, \\ \alpha, & x \le r \ne +\infty, \\ 0, & 0 \le r < x, \end{cases} \qquad r \in [0 + \infty].$$

If at every point of discontinuity we now move the right-limit point to the left-limit point, then this procedure leads to an order isomorphism

$$[0, +\infty]^{op} \otimes [0, 1] \xrightarrow{\Psi} \Delta^+$$

where Δ^+ is the complete lattice of all nonnegative, isotone, left-continuous probability distribution functions — these are all join-preserving maps $[0, +\infty] \xrightarrow{F} [0, 1]$.
Explicitly Ψ is given by:

$$\big(\Psi(G)\big)(r) = \sup\{G(x) \mid x < r\}, \qquad r \in [0, +\infty],\ G \in [0, +\infty]^{op} \otimes [0, 1]. \tag{2.20}$$

Hence Δ^+ is also the tensor product of $[0, +\infty]^{op}$ with $[0, 1]$ in Sup. Since

$$\big(\alpha \cdot H_x\big)(r) = \big(\Psi(x \otimes \alpha)\big)(r) = \begin{cases} 0, & r \leq x, \\ \alpha, & x < r, \end{cases} \qquad r \in [0, +\infty],$$

the corresponding universal bimorphism has the form $(x, \alpha) \longmapsto \alpha \cdot H_x$, where H_x denotes the left-continuous unit step function at x. From a probabilistic point of view, the *elementary tensor* $\alpha \cdot H_x$ describes a *random variable* taking the value x with probability α and the value $+\infty$ with probability $1 - \alpha$. In this sense the tensor product Δ^+ plays an important rôle in the theory of probabilistic metric spaces (cf. [104]). In Examples 2.3.5 and 2.3.36 we will describe certain monoidal structures on Δ^+.

The next example can be regarded as a refinement of Example 2.1.9.

Example 2.1.11. We maintain the notation from Example 2.1.9 and consider a topological space (X, τ) — this means that X is a set and τ is a subset of the power set $\mathscr{P}(X)$ which is closed under finite intersections and arbitrary unions. Hence τ is a complete lattice with respect to the partial order inherited from $\mathscr{P}(X)$. In particular, the embedding $\tau \overset{\iota_\tau}{\hookrightarrow} \mathscr{P}(X)$ is join-preserving.

Let x be an element of X. An *open neighborhood* of x is a set $U \in \tau$ such that $x \in U$. The set of all open neighborhoods of x is denoted by $\mathscr{N}_x$.

Further, let M be an arbitrary complete lattice. A map $X \xrightarrow{f} M$ is called *lower semicontinuous* if the following condition holds for all $x \in X$ (cf. [54]):

$$f(x) = \bigvee_{U \in \mathscr{N}_x} \big(\bigwedge_{y \in U} f(y)\big).$$

It is not difficult to verify that a map $X \xrightarrow{f} M$ is lower semicontinuous if and only if there exists an index set I and subsets of the form $\{U_i \mid i \in I\} \subseteq \tau$ and $\{m_i \mid i \in I\} \subseteq M$ such that the relation

$$f = \bigvee_{i \in I} m_i \cdot 1_{U_i}$$

holds.

Now we consider the join-preserving embedding $\tau \overset{\iota_\tau}{\hookrightarrow} \mathscr{P}(X)$ and the bimorphism $\tau \times M \xrightarrow{b} \mathscr{P}(X) \otimes M$ determined by $b(U, m) = U \otimes m$ for all $U \in \tau$ and $m \in M$. Then we conclude from the universal property of the tensor product that there exists a unique join-preserving map from $\tau \otimes M$ to $\mathscr{P}(X) \otimes M$, denoted by $\iota_\tau \otimes 1_M$, such that $b(U, m) = (\iota_\tau \otimes 1_M)(U \otimes m)$ holds for all $U \in \tau$ and $m \in M$.[1]

Further, we compose $\iota_\tau \otimes 1_M$ with the order isomorphism $\mathscr{P}(X) \otimes M \xrightarrow{\Phi} M^X$ given in Example 2.1.9 and denote it by Φ_τ — i.e. $\Phi_\tau = \Phi \circ (\iota_\tau \otimes 1_M)$. Then it is easily seen that Φ_τ acts on elementary tensors $U \otimes m \in \tau \otimes M$ as follows:

$$\Phi_\tau(U \otimes m) = m \cdot 1_U.$$

Consequently, if $F \in \tau \otimes M$, then by (2.17), $\Phi_\tau(F)$ is explicitly given by:

$$\Phi_\tau(F) = \bigvee_{U \in \tau} F(U) \cdot 1_U$$

and the range of Φ_τ coincides with the set $\mathrm{LSC}(X, M)$ of all lower semicontinuous maps, which is a complete sublattice of M^X.

Since Φ_τ is join-preserving, Φ_τ has a right adjoint map $\mathrm{LSC}(X, M) \xrightarrow{\Phi_\tau^\vdash} \tau \otimes M$. Because of the following chain of equivalences

$$\Phi_\tau(F) \le f \iff (\forall U \in \tau,\ F(U) \cdot 1_U \le f) \iff (\forall U \in \tau,\ F(U) \le \bigwedge_{x \in U} f(x)),$$

the right adjoint map of Φ_τ can be expressed by the formula:

$$\big(\Phi_\tau^\vdash(f)\big)(U) = \bigwedge_{x \in U} f(x), \qquad f \in \mathrm{LSC}(X, M),\ U \in \tau.$$

Obviously, the relation $\Phi_\tau \circ \Phi_\tau^\vdash = 1_{\mathrm{LSC}(X,M)}$ follows from the surjectivity of Φ_τ (cf. Exercise 1.3.2 (a)). Hence Φ_τ is injective (i.e. bijective in this context) if and only if $\Phi_\tau^\vdash$ is left adjoint to Φ_τ (see again Exercise 1.3.2 (a)). This observation motivates the question: under which condition does the relation $\Phi_\tau^\vdash \dashv \Phi_\tau$ hold? We show that the continuity of the complete lattice M (cf. [36, 37]) is a sufficient condition.

ASSERTION. If M is a continuous lattice, then $\Phi_\tau^\vdash \dashv \Phi_\tau$ holds.

In fact, let us choose $F \in \tau \otimes M$ and $f \in \mathrm{LSC}(X, M)$. Since Φ_τ is surjective, the implication $\Phi_\tau^\vdash(f) \le F \implies f \le \Phi_\tau(F)$ is evident. On the other hand, in order to verify the converse implication, we assume $f \le \Phi_\tau(F)$ and so

[1] For the general definition of the tensor product of two join-preserving maps, see p. 59.

$$\big(\Phi_\tau^{\vdash}(f)\big)(U) = \bigwedge_{x\in U} f(x) \leq \bigwedge_{x\in U} \Phi_\tau(F)(x) = \bigwedge_{x\in U}\big(\bigvee_{V\in\mathcal{N}_x} F(V)\big)$$

for each $U \in \tau$. Since $\{F(V) \mid V \in \mathcal{N}_x\}$ is directed for each $x \in U$, we conclude from the continuity of M (cf. [36, 37]) that the following relation holds for each $U \in \tau$:

$$\big(\Phi_\tau^{\vdash}(f)\big)(U) \leq \bigvee_{\xi\in\prod_{x\in U}\mathcal{N}_x} \big(\bigwedge_{x\in U} F(\xi(x))\big) = \bigvee_{\xi\in\prod_{x\in U}\mathcal{N}_x} F\big(\bigcup_{x\in U}\xi(x)\big) \leq F(U),$$

where we have used the property $U \subseteq \bigcup_{x\in U}\xi(x)$.

To sum up we have established the following important

FACT I. *If M is continuous, then the complete lattice of all lower semicontinuous M-valued maps is the tensor product of τ with M — i.e.* $\mathrm{LSC}(X, M) \cong \tau \otimes M$.

For more details on the tensor product $\tau \otimes M$, we refer to Sect. 3.2 in [43].

Now we return to our general train of thought and construct a *monoidal* structure on Sup which transforms Sup into a monoidal closed category. We begin with the definition of a bifunctor Sup × Sup → Sup, which we also denote by ⊗. The action of ⊗ on

– an object (X, Y) is given by $X \otimes Y$ (the complete lattice of all join-reversing maps $X \xrightarrow{f} Y$);
– a morphism $(X_1, Y_1) \xrightarrow{(h,k)} (X_2, Y_2)$ is given by the *tensor product* of h and k, which is the unique join-preserving map $h \otimes k$ making the following diagram commutative:

$$\begin{array}{ccc} X_1 \times Y_1 & \xrightarrow{\otimes} & X_1 \otimes Y_1 \\ {\scriptstyle h\times k}\downarrow & & \vdots{\scriptstyle h\otimes k} \\ X_2 \times Y_2 & \xrightarrow{\otimes} & X_2 \otimes Y_2 \end{array}$$

Before we show the associativity of the bifunctor ⊗, we prove that for each complete lattice X the endofunctor $X \otimes \_$ has a right adjoint functor. Therefore for each pair (X, Y) of complete lattices we first introduce the *complete lattice* $[X, Y]$ *of all join-preserving maps* $X \xrightarrow{f} Y$ ordered pointwisely — i.e.

$$f_1 \leq f_2 \iff f_1(x) \leq f_2(x) \quad \text{for all } x \in X.$$

On this basis we define an endofunctor $\hom_X$ of Sup as follows:

$$\hom_X(Y) = [X, Y], \quad Y_1 \xrightarrow{h} Y_2, \quad [X, Y_1] \xrightarrow{\hom_X(h)} [X, Y_2], \quad \hom_X(h)(f) = h \circ f.$$

In this context it is interesting to note that $[X, Y^{op}]^{op}$ coincides with the tensor product $X \otimes Y$.

Theorem 2.1.12. *For every complete lattice X the functor* $\hom_X$ *is right adjoint to* $X \otimes \_$.

Proof. Let X be a fixed complete lattice. Then for every further complete lattice Y there exists a join-preserving map $Y \xrightarrow{\eta_Y} [X, X \otimes Y]$ defined by $\eta_Y(y)(x) = x \otimes y$ for all $x \in X$ and $y \in Y$.

Evidently $\eta = (\eta_Y)_Y$ is a natural transformation from $\mathrm{id}_{\mathtt{Sup}}$ to $\hom_X \circ (X \otimes \_)$. We show that η is the unit of the adjoint situation $(X \otimes \_) \dashv \hom_X$ — i.e. for every complete lattice Z and every join-preserving map $Y \xrightarrow{g} [X, Z]$ there exists a unique join-preserving map $X \otimes Y \xrightarrow{\ulcorner g \urcorner} Z$ making the following diagram commutative:

$$\begin{array}{ccc} Y & \xrightarrow{\eta_Y} & [X, X \otimes Y] \\ {\scriptstyle g}\downarrow & \swarrow_{\hom_X(\ulcorner g \urcorner)} & \\ [X, Z] & & \end{array}$$

First we notice that $Y \xrightarrow{g} [X, Z]$ can be identified with a bimorphism $X \times Y \xrightarrow{b_g} Z$ according to the formula

$$b_g(x, y) = g(y)(x), \qquad x \in X,\ y \in Y.$$

Then the previous diagram is equivalent to the following relation:

$$\hom_X(\ulcorner g \urcorner)\big(\eta_Y(y)\big)(x) = \ulcorner g \urcorner(x \otimes y) = g(y)(x) = b_g(x, y). \tag{2.21}$$

Hence $\ulcorner g \urcorner$ coincides with the unique extension of b_g to a join-preserving map from $X \otimes Y$ to Z, whose existence is ensured by the universal property of the tensor product in Sup. □

Comment. Let $\varepsilon = (\varepsilon_Y)_Y$ be the counit of the adjoint situation $X \otimes \_ \dashv \hom_X$. Then $X \otimes [X, Y] \xrightarrow{\varepsilon_Y} Y$ attains the following form at elementary tensors:

$$\varepsilon_Y(x \otimes f) = f(x), \qquad x \in X,\ f \in [X, Y].$$

Therefore ε_Y is also called the evaluation arrow of $[X, Y]$.

The aim of the following considerations is to show that the bifunctor $\otimes$ induces the structure of a symmetric, monoidal category on Sup.

Lemma 2.1.13. (Associativity) *Let X, Y and Z be complete lattices. There exists a unique order isomorphism $(X \otimes Y) \otimes Z \xrightarrow{a_{XYZ}} X \otimes (Y \otimes Z)$ satisfying the following condition*:

$$a_{XYZ}((x \otimes y) \otimes z) = x \otimes (y \otimes z), \qquad x \in X,\ y \in Y,\ z \in Z. \tag{2.22}$$

Proof. For every $z \in Z$ we define a bimorphism $X \times Y \xrightarrow{b_z} X \otimes (Y \otimes Z)$ by

$$b_z(x, y) = x \otimes (y \otimes z), \qquad (x, y) \in X \times Y.$$

By the universal property of the tensor product there exists a unique join-preserving map $X \otimes Y \xrightarrow{h_z} X \otimes (Y \otimes Z)$ making the following diagram commutative:

$$\begin{array}{ccc} X \times Y & \xrightarrow{\ \otimes\ } & X \otimes Y \\ {\scriptstyle b_z}\downarrow & \swarrow {\scriptstyle h_z} & \\ X \otimes (Y \otimes Z) & & \end{array}$$

Now we define a map $(X \otimes Y) \times Z \xrightarrow{\xi} X \otimes (Y \otimes Z)$ by $\xi(f, z) = h_z(f)$ and show that ξ is again a bimorphism. Since h_z is join-preserving, it is sufficient to prove that ξ is join-preserving in its second variable. For this purpose we choose a subset C of Z, put $z_0 = \bigvee C$ and derive the following relation from (2.14):

$$\begin{aligned} \xi(f, z_0) = h_{z_0}(f) &= \bigvee_{x \in X} b_{z_0}(x, f(x)) = \bigvee_{x \in X} x \otimes (f(x) \otimes z_0) = \bigvee_{x \in X,\, z \in C} x \otimes (f(x) \otimes z) \\ &= \bigvee_{z \in C,\, x \in X} b_z(x, f(x)) = \bigvee_{z \in C} h_z(f) = \bigvee_{z \in C} \xi(f, z). \end{aligned}$$

Hence ξ is a bimorphism. Referring again to the universal property of the tensor product, we obtain a unique join-preserving map $(X \otimes Y) \otimes Z \xrightarrow{a_{XYZ}} X \otimes (Y \otimes Z)$ with $a_{XYZ}(f \otimes z) = \xi(f, z) = h_z(f)$ for all $f \in X \otimes Y$ and all $z \in Z$. In particular, $a_{XYZ}((x \otimes y) \otimes z) = h_z(x \otimes y) = x \otimes (y \otimes z)$ for all $x \in X$, $y \in Y$ and $z \in Z$, and so a_{XYZ} satisfies (2.22).

By analogy with the previous argumentation there exists a unique join-preserving map

$$X \otimes (Y \otimes Z) \xrightarrow{b_{XYZ}} (X \otimes Y) \otimes Z$$

such that $b_{XYZ}(x \otimes (y \otimes z)) = (x \otimes y) \otimes z$ for all $x \in X$, $y \in Y$ and $z \in Z$. Since every tensor is the join of an appropriate family of elementary tensors, we conclude from

$$b_{XYZ}\big(a_{XYZ}((x \otimes y) \otimes z)\big) = (x \otimes y) \otimes z \quad \text{and} \quad a_{XYZ}\big(b_{XYZ}(x \otimes (y \otimes z))\big) = x \otimes (y \otimes z)$$

that $1_{(X \otimes Y) \otimes Z} = b_{XYZ} \circ a_{XYZ}$ and $1_{X \otimes (Y \otimes Z)} = a_{XYZ} \circ b_{XYZ}$ holds. Hence a_{XYZ} is an order isomorphism. □

It is easily seen that $a = (a_{XYZ})_{XYZ}$ is a natural transformation

$$a \colon \otimes \circ (\otimes \times \mathrm{id}_{\mathsf{Sup}}) \to \otimes \circ (\mathrm{id}_{\mathsf{Sup}} \times \otimes).$$

Since every tensor is a join of elementary tensors, it follows immediately from Lemma 2.1.13 that the natural isomorphism a satisfies the pentagonal diagram (cf. (1) in Sect. 1.1). Hence the bifunctor $\otimes$ is associative.

Lemma 2.1.14. (Unit object) *Let X and Y be complete lattices and $\mathbb{1} = \{0, 1\}$ be the chain with two elements. Then there exist order isomorphisms $\mathbb{1} \otimes Y \xrightarrow{\ell_Y} Y$ and $X \otimes \mathbb{1} \xrightarrow{r_X} X$ such that the diagram*

$$\begin{array}{ccc} (X \otimes \mathbb{1}) \otimes Y & \xrightarrow{\quad a_{X\mathbb{1}Y} \quad} & X \otimes (\mathbb{1} \otimes Y) \\ & {\scriptstyle r_X \otimes 1_Y} \searrow \quad \swarrow {\scriptstyle 1_X \otimes \ell_Y} & \\ & X \otimes Y & \end{array} \tag{2.23}$$

is commutative. If $X = Y = \mathbb{1}$, then $r_{\mathbb{1}}$ and $\ell_{\mathbb{1}}$ coincide — i.e. $r_{\mathbb{1}} = \ell_{\mathbb{1}}$.

Proof. By (2.17) the tensors $f \in X \otimes \mathbb{1}$ and $g \in \mathbb{1} \otimes Y$ have the form:

$$f = x_0 \otimes 1 \qquad \text{where } x_0 = \bigvee\{x \in X \mid f(x) = 1\},$$
$$g = 1 \otimes y_0 \qquad \text{where } y_0 = g(1).$$

Hence the order isomorphisms r_X and ℓ_Y are defined by $r_X(x \otimes 1) = x$ and $\ell_Y(1 \otimes y) = y$. It is easily seen that the diagram (2.23) is commutative and the relation $r_{\mathbb{1}} = \ell_{\mathbb{1}}$ holds. □

Obviously, the maps ℓ_X and r_X are components of two natural isomorphisms $\ell \colon \mathbb{1} \otimes \_ \to \mathrm{id}_{\mathsf{Sup}}$ and $r \colon \_ \otimes \mathbb{1} \to \mathrm{id}_{\mathsf{Sup}}$. In this context Lemma 2.1.14 assures the validity of coherence axiom (2) in Sect. 1.1.

The commutativity of the tensor product follows immediately from its universal property by interchanging the variables.

Lemma 2.1.15. (Commutativity) *Let X and Y be complete lattices. Then there exists a unique order isomorphism $X \otimes Y \xrightarrow{c_{XY}} Y \otimes X$ satisfying the condition:*

$$c_{XY}(x \otimes y) = y \otimes x \tag{2.24}$$

for all $x \in X$ and $y \in Y$. Moreover, the following diagram is commutative:

$$\begin{array}{ccc} X \otimes \mathbb{1} & \xrightarrow{\quad c_{X\mathbb{1}} \quad} & \mathbb{1} \otimes X \\ & {\scriptstyle r_X} \searrow \quad \swarrow {\scriptstyle \ell_X} & \\ & X & \end{array} \tag{2.25}$$

Proof. Let π_X and π_Y be the respective projections of the product $X \times Y$ in Sup. By the universal property of the tensor product there exists a unique join-preserving map $X \otimes Y \xrightarrow{c_{XY}} Y \otimes X$ making the following diagram commutative:

$$\begin{array}{ccc} X \times Y & \xrightarrow{\ \otimes\ } & X \otimes Y \\ {\scriptstyle \langle \pi_Y, \pi_X \rangle} \downarrow & & \vdots\, {\scriptstyle c_{XY}} \\ Y \times X & \xrightarrow{\ \otimes\ } & Y \otimes X \end{array}$$

Obviously c_{XY} satisfies the desired properties. □

Addition. In order to have a concrete perception of c_{XY} we point out that c_{XY} coincides with the order isomorphism

$$X \otimes Y = \mathscr{G}(X, Y) \xrightarrow{(\cdot)^+} \mathscr{G}(Y, X) = Y \otimes X$$

mentioned in Example 2.1.5 (cf. (2.3)).

Let $\mathsf{F}\colon \mathtt{Sup} \times \mathtt{Sup} \to \mathtt{Sup} \times \mathtt{Sup}$ be the endofunctor of $\mathtt{Sup} \times \mathtt{Sup}$ interchanging the "components" of $\mathtt{Sup} \times \mathtt{Sup}$. Then $c = (c_{XY})_{XY}$ is a natural transformation $c\colon \otimes \to \otimes \circ \mathsf{F}$. It follows immediately from Lemma 2.1.15 that c satisfies the coherence axioms (3) and (4) in Sect. 1.1.

Moreover, since c is a natural isomorphism, the functors $\_ \otimes X$ and $\hom_X$ also form an adjoint situation $\_ \otimes X \dashv \hom_X$ for each complete lattice X. If $\varepsilon = (\varepsilon_Y)_Y$ is the counit of $(X \otimes \_) \dashv \hom_X$, then the counit $(\mathrm{ev}_Y)_Y$ of $\_ \otimes X \dashv \hom_X$ obviously has the form $\mathrm{ev}_Y = \varepsilon_Y \circ c_{[X,Y]X}$ — i.e. $\mathrm{ev}_Y(f \otimes x) = f(x)$. In this context we introduce the following terminology.

If $Z \otimes X \xrightarrow{\xi} Y$ is a join-preserving map, then the join-preserving map $Z \xrightarrow{\ulcorner \xi \urcorner} [X, Y]$ satisfying the property $\xi = \mathrm{ev}_Y \circ (\ulcorner \xi \urcorner \otimes 1_X)$ is called the *monoidal adjoint map* of ξ.

We summarize the previous results as follows.

Fact II. *The septuple* $(\mathtt{Sup}, \otimes, \mathbb{1}, a, c, \ell, r)$ *is a symmetric and monoidal closed category. If the self-duality of* $\mathtt{Sup}$ *determined by the construction of right adjoint maps is added, then the octuple* $(\mathtt{Sup}, \otimes, \mathbb{1}, a, c, \ell, r, {}^{op})$ *is even a star-autonomous category* (for a full discussion of star-autonomous categories, we refer to [8, 9]).

As an immediate corollary of Theorem 1.1.4 and the previous Fact II we obtain the following result.

Corollary 2.1.16. *The tensor product in* $\mathtt{Sup}$ *preserves direct limits.*

We finish this subsection with a discussion of some important lattice-theoretical properties of the tensor product. In particular, we show that the tensor product preserves continuity, complete distributivity, compactness of continuous lattices and the property of being algebraic. The fact that the tensor product also preserves the structure of frames will be postponed to Sect. 2.4, where we will study idempotent quantales in more detail.

Recall that, given a complete lattice X and $x_1, x_2 \in X$, we say that x_1 is *way below* x_2, and we write $x_1 \ll x_2$, if for any directed set $D \subseteq X$ with $x_2 \leq \bigvee D$ there is a

$d \in D$ such that $x_1 \leq d$. An element $x \in X$ is said to be *compact* if $x \ll x$. Hence a complete lattice X is continuous (cf. [36, 37]) if and only if $\ll$ is approximating, i.e. $x = \bigvee\{x' \in X \mid x' \ll x\}$ for all $x \in X$. A complete lattice is *algebraic* if every element of X is a join of a family of compact elements of X. Every algebraic lattice is continuous. A complete lattice is *compact* if its universal upper bound is compact.

Following Raney (see [96, Definition 3 and Theorem 1]), we say that $x_1 \in X$ is *totally below* $x_2 \in X$, and we write $x_1 \lhd x_2$, if for any subset $A \subseteq X$ with $x_2 \leq \bigvee A$ there is an $a \in A$ such that $x_1 \leq a$. Hence a complete lattice X is completely distributive if and only if $\lhd$ is approximating, i.e. $x = \bigvee\{x' \in X \mid x' \lhd x\}$ for all $x \in X$.

Theorem 2.1.17. *Let X and Y be complete lattices with at least two different elements and $X \otimes Y$ be its tensor product. Then the following equivalences hold*:

(i) *$X \otimes Y$ is continuous if and only if X and Y are continuous.*
(ii) *$X \otimes Y$ is completely distributive if and only if X and Y are completely distributive.*
(iii) *$X \otimes Y$ is continuous and compact if and only if X and Y are continuous and compact.*

The proof of Theorem 2.1.17 is based on a sequence of lemmas. Since, in general, joins in the tensor product are not computed pointwisely, we first give an explicit description of directed joins. For this purpose we recall that a complete lattice X is *meet-continuous* if for every $x \in X$ and for every directed subset D of X the relation

$$x \wedge (\bigvee D) = \bigvee_{d \in D} (x \wedge d)$$

holds. Obviously, continuous lattices are meet-continuous.

Lemma 2.1.18. *Let X be a continuous lattice and Y be a meet-continuous lattice. If $\{g_i \mid i \in I\}$ is a directed subset of $X \otimes Y$, then the join of $\{g_i \mid i \in I\}$ has the following form*:

$$\Big(\bigvee_{i \in I} g_i\Big)(x) = \bigwedge_{x' \ll x} \Big(\bigvee_{i \in I} g_i(x')\Big), \qquad x \in X.$$

Proof. For every directed subset $\{g_i \mid i \in I\}$ of $X \otimes Y$ we define an antitone map $X \xrightarrow{g_0} Y$ by

$$g_0(x) = \bigvee_{i \in I} g_i(x), \qquad x \in X.$$

Since Y is meet-continuous, g_0 is *finite* join-reversing. In fact, the following holds:

$$g_0(x_1) \wedge g_0(x_2) = \bigvee_{i,j \in I} g_i(x_1) \wedge g_j(x_2) = \bigvee_{i \in I} g_i(x_1) \wedge g_i(x_2) = g_0(x_1 \vee x_2).$$

Now we smooth g_0 and define a map $X \xrightarrow{\widehat{g_0}} Y$ by $\widehat{g_0}(x) = \bigwedge_{x' \ll x} g_0(x')$ for each $x \in X$. We show that $\widehat{g_0}$ is join-reversing. Since $\{x' \in X \mid x' \ll \bot\} = \{\bot\}$, the value

$\widehat{g_0}(\perp)$ coincides with $\top$. Hence it is sufficient to consider a nonempty subset A of X. Since X is continuous and therefore the way below relation is approximating, for every $x \ll \bigvee A$ there exist finitely many

$$x_k \in \bigcup_{a\in A}\{x \in X \mid x \ll a\}, \qquad k = 1, \dots, n$$

such that the relation $x \le \bigvee_{k=1}^{n} x_k$ holds. Now we apply the property that g_0 is finite join-reversing and obtain:

$$\bigwedge_{a\in A} \widehat{g_0}(a) \le \bigwedge_{k=1}^{n} g_0(x_k) = g_0\Big(\bigvee_{k=1}^{n} x_k\Big) \le g_0(x).$$

Hence the relation

$$\bigwedge_{a\in A} \widehat{g_0}(a) \le \bigwedge_{x\ll\bigvee A} g_0(x) = \widehat{g_0}\big(\bigvee A\big)$$

follows — i.e. $\widehat{g_0}$ is join-reversing. Referring again to the continuity of X it follows immediately from the construction that $\widehat{g_0}$ is the join of $\{g_i \mid i \in I\}$. □

Lemma 2.1.19. *Let X and Y be complete lattices, and let X be completely distributive. If $\{g_i \mid i \in I\}$ is a subset of $X \otimes Y$, then the join of $\{g_i \mid i \in I\}$ has the following form*:

$$\big(\bigvee_{i\in I} g_i\big)(x) = \bigwedge_{x'\lhd x} \big(\bigvee_{i\in I} g_i(x')\big), \qquad x \in X.$$

Proof. For every subset $\{g_i \mid i \in I\}$ of $X \otimes Y$ we define a map $X \xrightarrow{f} Y$ by

$$f(x) = \bigwedge_{x'\lhd x} \big(\bigvee_{i\in I} g_i(x')\big), \qquad x \in X.$$

Since $\{x' \in X \mid x' \lhd \perp\} = \varnothing$, the relation $f(\perp) = \top$ is obvious. Further, since the totally below relation $\lhd$ is approximating, for every nonempty subset A of X the property

$$\bigcup_{a\in A}\{x' \in X \mid x' \lhd a\} = \{x' \in X \mid x' \lhd \bigvee A\}$$

holds. Hence

$$f(\bigvee A) = \bigwedge_{a\in A} \bigwedge_{x'\lhd a} \big(\bigvee_{i\in I} g_i(x')\big) = \bigwedge_{a\in A} f(a)$$

follows — i.e. f is join-reversing. Finally, if we apply again the property that $\lhd$ is approximating in X, then we conclude immediately from the construction of f that f is the join of $\{g_i \mid i \in I\}$. Hence the assertion is verified. □

Lemma 2.1.20. *Let X be a continuous lattice with $x, a \in X$ and Y be a meet-continuous lattice with $y, b \in Y$.*

(a) *If $a \neq \perp$ and $b \neq \perp$, then $a \ll x$ and $b \ll y$ if and only if $a \otimes b \ll x \otimes y$.*

(b) *If $x \neq \bot$ and $y \neq \bot$, then x and y are compact if and only if $x \otimes y$ is compact.*

Proof. Since (b) is a special case of (a), we only prove (a). We therefore assume $\bot \neq a \ll x$ and $\bot \neq b \ll y$ and consider a directed subset $\{g_i \mid i \in I\}$ of $X \otimes Y$ with $x \otimes y \leq \bigvee_{i \in I} g_i$. Since $y \leq \big(\bigvee_{i \in I} g_i\big)(x)$, the relation $y \leq \bigvee_{i \in I} g_i(a)$ follows from Lemma 2.1.18. Since $\{g_i(a) \mid i \in I\}$ is a directed subset of Y and b is way below y, there exists an $i \in I$ with $b \leq g_i(a)$ — i.e. $a \otimes b \leq g_i$. Hence $a \otimes b \ll x \otimes y$. On the other hand, we assume $a \neq \bot, b \neq \bot$ and $a \otimes b \ll x \otimes y$. Then we consider a directed subset U of X with $x \leq \bigvee U$. Hence $\{u \otimes \top \mid u \in U\}$ is a directed subset of $X \otimes Y$ and the relation $x \otimes y \leq \bigvee\{u \otimes \top \mid u \in U\}$ holds. Since $a \otimes b$ is way below $x \otimes y$, there exists a $u \in U$ with $a \otimes b \leq u \otimes \top$. In particular, $a \leq u$ follows from (2.16). Hence a is way below x. Analogously we show $b \ll y$. □

Lemma 2.1.21. *Let X be a completely distributive lattice and Y be an arbitrary complete lattice. Further, let $x, a \in X$ and $y, b \in Y$. If $a \neq \bot$ and $b \neq \bot$, then $a \lhd x$ and $b \lhd y$ if and only if $a \otimes b \lhd x \otimes y$.*

Proof. If we replace Lemma 2.1.18 by Lemma 2.1.19 and $\ll$ by $\lhd$, then the proof of Lemma 2.1.20 (a) can be repeated verbatim. □

Proof of Theorem 2.1.17. In order to verify (i) we proceed as follows. Since the embeddings $X \xrightarrow{j_X} X \otimes Y$ and $Y \xrightarrow{j_Y} X \otimes Y$ preserve arbitrary meets and joins (cf. (2.18)), we infer from the definition of continuity given supra that the continuity of X and the continuity of Y are necessary for the continuity of $X \otimes Y$. On the other hand, if X and Y are continuous, then the continuity of $X \otimes Y$ follows from Lemma 2.1.20 (a) and the fact that every tensor is the join of an appropriate family of elementary tensors.

If in the previous argumentation we replace continuity by complete distributivity and Lemma 2.1.20 (a) by Lemma 2.1.21, then we obtain a proof for (ii).

Finally, the equivalence in (iii) follows from (i) and Lemma 2.1.20 (b). □

Proposition 2.1.22. *If X and Y are algebraic lattices, then the tensor product $X \otimes Y$ is again algebraic.*

Proof. The assertion follows from Lemma 2.1.20 (b) and the fact that every tensor is a join of an appropriate family of elementary tensors. □

As an immediate corollary of Theorem 2.1.17 and Example 2.1.11 we obtain the following result.

Corollary 2.1.23. *Let (X, τ) be a topological space and M be a continuous lattice. Then the following equivalences hold:*

(i) $\mathrm{LSC}(X, M)$ *is continuous if and only if τ is continuous.*
(ii) $\mathrm{LSC}(X, M)$ *is completely distributive if and only if τ and M are completely distributive.*
(iii) $\mathrm{LSC}(X, M)$ *is continuous and compact if and only if τ and M are continuous and compact.*

2.1.3 The Tensor Product on Completely Distributive Lattices

Let CD be the category of completely distributive lattices with join- and meet-preserving maps (see also Exercise 1.3.3 (e)). Hence CD is a subcategory of Sup. Motivated by Theorem 2.1.17 (ii) we investigate the restriction of the tensor product of Sup to CD.

By (2.7) and (2.8) it follows immediately from Theorem 2.1.8 that the universal bimorphism $(x, y) \longmapsto x \otimes y$ is meet-preserving. Hence every join- and meet-preserving map $X \otimes Y \xrightarrow{h} Z$ induces a meet-preserving bimorphism $X \times Y \xrightarrow{b} Z$ by

$$b(x, y) = h(x \otimes y), \qquad x \in X,\ y \in Y.$$

This observation motivates the following:

Question 2.1.24. Does every meet-preserving bimorphism $X \times Y \xrightarrow{b} Z$ have a unique extension to a meet- and join-preserving map $X \otimes Y \xrightarrow{h} Z$ such that the diagram

$$\begin{array}{ccc} X \times Y & \xrightarrow{\ \otimes\ } & X \otimes Y \\ & \searrow^{b} & \vdots\, h \\ & & Z \end{array}$$

is commutative?

The next example shows that the previous question does not always have a positive answer.

Counterexample Let $\mathbb{B}$ be a complete and atomless Boolean algebra. Then the binary meet operation $\mathbb{B} \times \mathbb{B} \xrightarrow{\wedge} \mathbb{B}$ is a meet-preserving bimorphism. We show that the unique join-preserving map $\mathbb{B} \otimes \mathbb{B} \xrightarrow{w} \mathbb{B}$ making the diagram

$$\begin{array}{ccc} \mathbb{B} \times \mathbb{B} & \xrightarrow{\ \otimes\ } & \mathbb{B} \otimes \mathbb{B} \\ & \searrow^{\wedge} & \vdots\, w \\ & & \mathbb{B} \end{array}$$

commutative is *not* meet-preserving.
In the following we denote the complement of $x \in \mathbb{B}$ by x' and introduce a subset I of $\mathbb{B}$ as follows:

$$I = \{x \in \mathbb{B} \mid x \neq \bot \text{ and } x' \neq \bot\}.$$

For every $x \in I$ we define a join-reversing map $\mathbb{B} \xrightarrow{g_x} \mathbb{B}$ by

$$g_x(z) = \begin{cases} \bot, & z \not\leq x \text{ and } z \not\leq x', \\ x, & z \leq x \text{ and } z \neq \bot, \\ x', & z \leq x' \text{ and } z \neq \bot, \\ \top, & z = \bot. \end{cases} \qquad z \in \mathbb{B}$$

It is not difficult to confirm the relation $g_x = (x \otimes x) \vee (x' \otimes x')$ (cf. Exercise 2.1.6). Now we refer to the definition of w and observe:

$$w(g_x) = w(x \otimes x) \vee w(x' \otimes x') = (x \wedge x) \vee (x' \wedge x') = \top, \qquad x \in I.$$

Hence $\bigwedge_{x \in I} w(g_x) = \top$ follows. On the other hand, the meet of $\{g_x \mid x \in I\}$ (which is computed pointwisely) coincides with the universal lower bound in $\mathbb{B} \otimes \mathbb{B}$. In fact, if z is an element of $\mathbb{B}$ with $z \neq \bot$, then we apply the property that $\mathbb{B}$ is atomless and choose elements $u, v \in \mathbb{B}$ satisfying the following conditions:

$$u \neq \bot, \quad v \neq \bot, \quad u \wedge v = \bot, \quad u \vee v = z.$$

Now we define $x_0 = u \vee z'$ and observe that $x_0' = v$. Since $z \not\leq x_0$ and $z \not\leq x_0'$, the relation $g_{x_0}(z) = \bot$ follows.
Since w is join-preserving, we have $w\big(\bigwedge_{x \in I} g_x\big) = \bot$. Thus w is not meet-preserving.

The next theorem shows that under the complete distributivity of all three complete lattices X, Y and Z the previous question has a positive answer. For this purpose we recall that complete distributivity is preserved under the tensor product in $\mathtt{Sup}$ (cf. Theorem 2.1.17 (ii) in Sect. 2.1.2).

Theorem 2.1.25. *Let X, Y and Z be three completely distributive lattices, and let $X \times Y \xrightarrow{b} Z$ be a bimorphism in* Sup *which also preserves arbitrary meets. Then there exists a unique join- and meet-preserving map $X \otimes Y \xrightarrow{h_b} Z$ making the following diagram commutative*:

$$\begin{array}{ccc} X \times Y & \xrightarrow{\ \otimes\ } & X \otimes Y \\ & \underset{b}{\searrow} & \Big\downarrow{\scriptstyle h_b} \\ & & Z \end{array} \qquad \text{(D)}$$

Proof. By Theorem 2.1.8 we know that there exists a unique join-preserving map $X \otimes Y \xrightarrow{h_b} Z$ making the diagram (D) commutative. In particular, h_b is determined by (cf. (2.12))

$$h_b(f) = \bigvee_{i \in I} b(x_i, y_i), \qquad f = \bigvee_{i \in I} x_i \otimes y_i \in X \otimes Y.$$

We have to show that h_b preserves meets. Since b is meet-preserving, the relation $b(\top, \top) = \top$ holds. Hence h_b preserves the respective universal upper bounds —

i.e. $h_b(\top \otimes \top) = \top$. Now we choose a nonempty subset $\{f_i \mid i \in I\}$ of $X \otimes Y$ and use the fact that every tensor f_i is a join of an appropriate family of elementary tensors — i.e. $f_i = \bigvee_{j \in J_i}(x_{(i,j)} \otimes y_{(i,j)})$. Since b is meet-preserving, we derive the following relation from the complete distributivity of $X \otimes Y$ and Z:

$$\begin{aligned}
\bigwedge_{i \in I} h_b(f_i) &= \bigwedge_{i \in I}\Big(\bigvee_{j \in J_i} h_b(x_{(i,j)} \otimes y_{(i,j)})\Big) \\
&= \bigvee_{(j_i)_{i \in I} \in \prod_{i \in I} J_i} \Big(\bigwedge_{i \in I} b(x_{(i,j_i)}, y_{(i,j_i)})\Big) \\
&= \bigvee_{(j_i)_{i \in I} \in \prod_{i \in I} J_i} b\Big(\big(\bigwedge_{i \in I} x_{(i,j_i)}\big), \big(\bigwedge_{i \in I} y_{(i,j_i)}\big)\Big) \\
&= h_b\Big(\bigvee_{(j_i)_{i \in I} \in \prod_{i \in I} J_i} \big(\bigwedge_{i \in I} x_{(i,j_i)}\big) \otimes \big(\bigwedge_{i \in I} y_{(i,j_i)}\big)\Big) \\
&= h_b\Big(\bigvee_{(j_i)_{i \in I} \in \prod_{i \in I} J_i} \big(\bigwedge_{i \in I}(x_{(i,j_i)} \otimes y_{(i,j_i)})\big)\Big) \\
&= h_b\Big(\bigwedge_{i \in I}\big(\bigvee_{j \in J_i}(x_{(i,j)} \otimes y_{(i,j)})\big)\Big) = h_b\big(\bigwedge_{i \in I} f_i\big).
\end{aligned}$$

Hence the assertion is verified. □

As an immediate corollary of Theorems 2.1.17 (ii) and 2.1.25 we obtain that the tensor product $\otimes$ in Sup induces a *symmetric monoidal structure* on CD. In fact, since every pair of morphisms $X_1 \xrightarrow{h_1} Y_1$ and $X_2 \xrightarrow{h_2} Y_2$ in CD induces a meet-preserving bimorphism $X_1 \times X_2 \xrightarrow{b} Y_1 \otimes Y_2$ by:

$$b(x_1, x_2) = h_1(x_1) \otimes h_2(x_2), \qquad x_1 \in X_1,\ x_2 \in X_2$$

(where we have used the fact that the universal bimorphism $(x, y) \longmapsto x \otimes y$ is meet-preserving), there exists a bifunctor $\mathrm{CD} \times \mathrm{CD} \to \mathrm{CD}$ which assigns a meet and join-preserving map $X_1 \otimes X_2 \xrightarrow{h_1 \otimes h_2} Y_1 \otimes Y_2$ to each pair of morphisms $X_1 \xrightarrow{h_1} Y_1$ and $X_2 \xrightarrow{h_2} Y_2$ such that the following diagram is commutative:

$$\begin{array}{ccc}
X_1 \times X_2 & \xrightarrow{\ \otimes\ } & X_1 \otimes X_2 \\
{\scriptstyle h_1 \times h_2}\downarrow & & \vdots\, {\scriptstyle h_1 \otimes h_2} \\
Y_1 \times Y_2 & \xrightarrow{\ \otimes\ } & Y_1 \otimes Y_2
\end{array}$$

Obviously, the validity of the pentagonal diagram, the unit object $\mathbb{1} = \{0, 1\}$ and the coherence axioms of a symmetry c is inherited by Sup. Hence $(\mathrm{CD}, \otimes, \mathbb{1}, \ell, r, c)$ is a symmetric and monoidal category. In contrast to Sup we conjecture that $(\mathrm{CD}, \otimes, \mathbb{1}, \ell, r, c)$ is not a (monoidal) closed category.

Exercises

2.1.1. Let X be a complete lattice. Show that every idempotent and join-preserving map $X \xrightarrow{e} X$ splits — i.e. there exists a complete lattice Y and join-preserving maps $X \xrightarrow{r} Y$ and $Y \xrightarrow{s} X$ such that $r \circ s = 1_Y$ and $e = s \circ r$.
(Hint: Consider the quotient w.r.t. the closure operator associated with e.)

2.1.2. Let $X \xrightarrow{f} Z$ be a join-preserving map. Show that the pullback of an epimorphism along f is again an epimorphism.

2.1.3. Let X, Y and Z be complete lattices. For $x \in X$, $y \in Y$ and $z \in Z$ let us view $(x \otimes y) \otimes z$ as a join-preserving map from $X \otimes Y$ to Z^{op}. Show that the join-preserving map $x \otimes (y \otimes z)$ from X to $[Y, Z^{op}]$ is the monoidal adjoint map of $(x \otimes y) \otimes z$.

2.1.4. Let X and Y be complete lattices. Further, for each $x, y \in X$ let us consider a tensor f_{xy} of $X \otimes Y$ having the form $f_{xy} = (x \otimes \top) \vee (\top \otimes y)$. Show:

(a)

$$f_{xy}(x') = \begin{cases} \top, & \text{if } x' \le x, \\ y, & \text{if } x' \not\le x, \end{cases} \qquad x' \in X.$$

(b) $u \otimes v \le f_{xy}$ if and only if $u \le x$ or $v \le y$.
(c) $f_{x_1 y_1} \wedge f_{x_2 y_2} = f_{(x_1 \wedge x_2)(y_1 \wedge y_2)} \vee (x_1 \otimes y_2) \vee (x_2 \otimes y_1)$.
(d) $\bigwedge_{i \in I} f_{x_i y_i} = \bigvee_{A \subseteq I} \big((\bigwedge_{i \in A} x_i) \otimes (\bigwedge_{i \notin A} y_i) \big)$.
(e) $(\bigwedge_{i \in I} f_{x_i y_i}) \wedge (u \otimes v) = \bigvee_{A \subseteq I} \big((u \wedge (\bigwedge_{i \in A} x_i)) \otimes (v \wedge (\bigwedge_{i \notin A} y_i)) \big)$.
(Hint to (c) – (d): Use (b) and the fact that every tensor is a join of an appropriate family of elementary tensors. Further, apply the meet preservation of the universal bimorphism.)

2.1.5. Let X be a complete lattice and $C_3 = \{\bot, a, \top\}$ be the three-chain. Show that the tensor product $X \otimes C_3$ is order isomorphic to the complete lattice

$$Z = \{(x, y) \in X \times X \mid x \le y\}$$

provided with the partial order $(x_1, y_1) \le (x_2, y_2) \iff x_1 \le x_2$ and $y_1 \le y_2$.
(Hint: Consider the following bimorphism $X \times C_3 \xrightarrow{b} Z$ determined by:

$$b(x, \top) = (x, x), \quad b(x, a) = (\bot, x), \quad b(x, \bot) = (\bot, \bot), \quad x \in X,$$

and use Exercise 2.1.6.)

2.1.6. Let X and Y be complete lattices. If $x_1, x_2 \in X$ and $y_1, y_2 \in Y$, then show that the tensor $f = (x_1 \otimes y_1) \vee (x_2 \otimes y_2)$ has the following explicit form:

$$f(z) = \begin{cases} y_1 \vee y_2, & z \le x_1 \wedge x_2, \\ y_1, & z \le x_1,\ z \not\le x_2, \\ y_2, & z \not\le x_1,\ z \le x_2, \\ y_1 \wedge y_2, & z \not\le x_1,\ z \not\le x_2,\ z \le x_1 \vee x_2, \\ \bot, & z \not\le x_1 \vee x_2, \end{cases} \qquad z \in X \setminus \{\bot\}.$$

2.1.7. Let (X, τ) be a topological space, M be a complete lattice, and $\mathrm{LSC}(X, M)$ be the complete lattice of all M-valued lower semicontinuous maps from X to M. If τ and M are algebraic, show that $\mathrm{LSC}(X, M)$ is also algebraic.

2.1.8. Let X be a complete lattice and $f \in X \otimes X$. Show that the pair (f, f) is a contravariant Galois connection if and only if the relation $x \le f(f(x))$ holds for all $x \in X$.

2.1.9. Let $X_1 \xrightarrow{h} X_2$ and $Y_1 \xrightarrow{k} Y_2$ be join-preserving maps. Show that the right adjoint map $(h \otimes k)^{\vdash}$ of $X_1 \otimes Y_1 \xrightarrow{h \otimes k} X_2 \otimes Y_2$ has the following form:

$$(h \otimes k)^{\vdash}(f) = k^{\vdash} \circ f \circ h, \qquad f \in X_2 \otimes Y_2$$

where $k^{\vdash}$ is the right adjoint map of k.

2.1.10. Let X, Y and Z be complete lattices, and let $X \times Y \xrightarrow{b} Z$ be a bimorphism. With every $z \in Z$ we associate a tensor $b_z \in X \otimes Y$ defined by (cf. Remark 2.1.6):

$$b_z(x) = \bigvee \{y \in Y \mid b(x, y) \le z\}.$$

If $X \otimes Y \xrightarrow{h} Z$ is the unique join-preserving map extending b (i.e. $b(x, y) = h(x \otimes y)$, $x \in X$, $y \in Y$), then show that the following relation holds:

$$h(f) = \bigwedge \{z \in Z \mid f \le b_z\}, \quad f \in X \otimes Y.$$

2.1.11. Let X, Y and Z be complete lattices. Show that the following chain of isomorphisms hold:

$$[X \otimes Y, Z] \cong [Z^{op}, [X, Y^{op}]] \cong [X, [Z^{op}, Y^{op}]] \cong [X, [Y, Z]] \cong \mathrm{Bimorph}(X \times Y, Z).$$

Comment. The previous exercise was inspired by a personal communication to the authors by M. Barr.

2.2 Prequantales: The Basis of Quantales

A pair $(X, *)$ is a *prequantale* if X is a complete lattice and $X \times X \xrightarrow{*} X$ is a bimorphism of Sup — i.e. $X \times X \xrightarrow{*} X$ is a map in Set and is join-preserving in each variable separately. Instead of $*(x, y)$ we also write $x * y$ for $x, y \in X$. If $(X, *)$ is a prequantale, then $*$ is called the *multiplication* of X.

By the universal property of the tensor product in Sup the multiplication $*$ of a prequantale X can be identified with a binary operation in the sense of Sup — i.e. a join-preserving map $X \otimes X \xrightarrow{\circledast} X$. In this context, the commutative diagram (2.11) attains the following form

$$\begin{array}{ccc} X \times X & \xrightarrow{\otimes} & X \otimes X \\ & {\scriptstyle *}\searrow & \downarrow{\scriptstyle \circledast} \\ & & X \end{array}$$

where $\otimes$ denotes the universal bimorphism. Hence *prequantales* and *magmas* in Sup are equivalent concepts, and we will not distinguish between them.

Since in the case of $x \neq \perp$ and $y \neq \perp$ every pair (x, y) can be identified with its corresponding elementary tensor $x \otimes y$ (cf. (2.16)), the previous diagram suggests the following simplification of notation.

Notation. Let $(X, *)$ be a prequantale with its corresponding binary operation $\circledast$. Then we will also write $x \circledast y$ for $\circledast(x \otimes y)$ where $x, y \in X$. Since now $x * y$ and $x \circledast y$ coincide, it will depend on the context which kind of notation we will prefer.

Since the category Sup shares various properties with the category of abelian groups — e.g. zero object, biproducts, (epi-mono)-factorization property and a tensor product producing a symmetric and monoidal closed structure (see Sect. 2.1), a prequantale can therefore be viewed as a "non-associative ring-like object" — the formation of joins plays the rôle of addition, while $*$ is the ring multiplication.

Before we proceed, we first present an interesting class of examples of prequantales induced by non-associative algebras.

Example 2.2.1. Let $\mathfrak{g} = (V, [\_, \_])$ be a finite-dimensional real or complex *Lie algebra* (cf. [45, Definition 2.36 on p. 53]). Then V is a finite-dimensional real or complex vector space, and $V \times V \xrightarrow{[\_,\_]} V$ is a bilinear map which is *alternating* (i.e. $[a, a] = 0$ for each $a \in V$) and satisfies the *Jacobi identity*:

$$[a, [b, c]] + [b, [c, a]] + [c, [a, b]] = 0, \quad a, b, c \in V.$$

The map $[\_, \_]$ is also called *Lie bracket*.

Now let $\mathbb{X}(V)$ be the lattice of all linear subspaces U of V ordered by set-inclusion. Obviously, meets in $\mathbb{X}(V)$ are intersections of subspaces, and joins in $\mathbb{X}(V)$ are given by the linear hull of unions of subspaces.

The Lie bracket induces a binary operation on $\mathbb{X}(V)$ as follows:

$$U * W = \text{ linear hull of } \{[a,b] \mid a \in U,\, b \in W\}, \qquad U, W \in \mathbb{X}(V).$$

Since the Lie bracket is bilinear, the operation $*$ is join-preserving in each variable separately. Hence $(\mathbb{X}(V), *)$ is a prequantale.

As a first simple property we mention the unitalization of prequantales. In this context we recall that a prequantale $(X, *)$ is unital if $(X, *)$ has a unit e — i.e. $x * e = x = e * x$ for all $x \in X$. Referring to Sect. 1.1 the unitalization of prequantales always exists (for details, see Exercise 2.2.2).

A *homomorphism* between prequantales[2] $(X, *)$ and $(Y, *)$ is a homomorphism between magmas in Sup; this means that $X \xrightarrow{h} Y$ is join-preserving and also preserves the respective multiplications — i.e. $h(x_1 * x_2) = h(x_1) * h(x_2)$ for all $x_1, x_2 \in X$. A homomorphism h is *unital* if h preserves the respective units e_X and e_Y — i.e. $h(e_X) = e_Y$.

In order to give a characterization of homomorphisms we need some more terminology. Let $(X, *)$ and $(Y, *)$ be prequantales. An isotone map $(X, \leq) \xrightarrow{h} (Y, \leq)$ is said to be *closed* if the relation $h(x_1) * h(x_2) \leq h(x_1 * x_2)$ holds for all $x_1, x_2 \in X$.

Proposition 2.2.2. *Let $(X, *)$ and $(Y, *)$ be prequantales. Further, let $X \xrightarrow{h} Y$ be a join-preserving map and $Y \xrightarrow{h^{\vdash}} X$ its right adjoint map. Then the following assertions are equivalent*:

(i) *h and $h^{\vdash}$ are closed maps.*
(ii) *h is a homomorphism.*

Proof. If $h^{\vdash}$ is closed, then $x_1 * x_2 \leq h^{\vdash}(h(x_1)) * h^{\vdash}(h(x_2)) \leq h^{\vdash}(h(x_1) * h(x_2))$ for all $x_1, x_2 \in X$. Hence $h(x_1 * x_2) \leq h(x_1) * h(x_2)$ follows. If h is also closed, then h is a homomorphism. On the other hand, if h is a homomorphism, then we obtain $h\big(h^{\vdash}(y_1) * h^{\vdash}(y_2)\big) = h\big(h^{\vdash}(y_1)\big) * h\big(h^{\vdash}(y_2)\big) \leq y_1 * y_2$ for all $y_1, y_2 \in Y$. Hence the closedness of $h^{\vdash}$ follows. □

Prequantales and homomorphisms form a category denoted by Pq. In the following considerations we will focus on the categorical properties of prequantales.

As a first important property we show that Pq is isomorphic to the Eilenberg–Moore category of the composite monad **DwnT**, where **Dwn** is the monad of down-closed sets and **T** is the term monad of a single binary operator symbol on Preord (cf. Sect. 1.3.2). In this sense prequantales have deep roots in the category of preordered sets.

Since a prequantale $(X, *)$ has two structures — on one hand a preordered groupoid $(X, \leq, *)$ and on the other hand a complete lattice $(X, \leq)$, we identify

[2] In order to avoid a cumbersome notation the multiplication of prequantales is denoted by $*$ on both sides.

$(X, \leq, *)$ with a structure map $\mathsf{T}(X, \leq) \xrightarrow{\xi_1} (X, \leq)$ in the sense of $\mathbf{T}$ and the formation sup_X of arbitrary joins with a structure map $(\mathsf{Dwn}(X), \subseteq) \xrightarrow{\xi_2} (X, \leq)$ in the sense of **Dwn**. If we now consider the Minkowski multiplication $\boxdot$ on $\mathsf{Dwn}(X)$ induced by $*$, which is intrinsically related to the distributive law of Dwn over $\mathbf{T}$ (see Sect. 1.3.2), then it is easily seen that the distributive law between the multiplication and the formation of arbitrary joins in a prequantale means the commutativity of the diagram

$$\begin{array}{ccc} \mathsf{Dwn}(X) \times \mathsf{Dwn}(X) & \xrightarrow{\mathrm{sup}_X \times \mathrm{sup}_X} & X \times X \\ \boxdot \big\downarrow & & \big\downarrow * \\ \mathsf{Dwn}(X) & \xrightarrow{\mathrm{sup}_X} & X \end{array} \tag{2.26}$$

— i.e. ξ_2 is a $\mathbf{T}$-homomorphism. Hence we conclude from Proposition 1.2.15 and the subsequent comment that $\xi_2 \circ \mathsf{Dwn}(\xi_1)$ is a structure map of a **DwnT**-algebra on $(X, \leq)$.

On the other hand, given a **DwnT**-algebra $((X, \leq), \xi)$, ξ can be decomposed into a $\mathbf{T}$-algebra $\big((X, \leq), \xi_1\big)$ and a **Dwn**-algebra $\big((X, \leq), \xi_2\big)$ such that the following relations hold (cf. Theorem 1.2.14):

$$\xi = \xi_2 \circ \mathsf{Dwn}(\xi_1) \quad \text{and} \quad \xi \circ \sigma_X = \xi_1 \circ \mathsf{T}(\xi_2). \tag{2.27}$$

Hence $(X, \leq)$ is a complete lattice, and $\big((X, \leq), \xi_1\big)$ can be identified with a preordered groupoid $(X, \leq, *)$. Moreover, the second property in (2.27) means that sup_X is a groupoid homomorphism (cf. (2.26)). Hence $*$ satisfies the required distributive law — this means that $*$ is a bimorphism in `Sup`.

Since **DwnT**-homomorphisms are join-preserving maps which are simultaneously groupoid homomorphisms (cf. Theorem 1.2.16), we can summarize the previous results as follows.

FACT I. *The category* `Pq` *of prequantales is isomorphic to the Eilenberg–Moore category of the composite monad* **DwnT** *on* `Preord`.

Because of the previous fact, completeness of `Pq` is inherited from `Preord`. In particular, limits in `Pq` are computed at the level of `Preord`.

Since it appears that `Pq` does not have a self-duality, the investigation of colimits is more complicated than what we have encountered in `Sup`. For example, we leave the characterization of epimorphisms in `Pq` as an open problem to the reader. Therefore we turn directly to regular epimorphisms. For this purpose we need an adaptation of closure operators to the situation given by prequantales.

Definition 2.2.3. Let $(X, *)$ be a prequantale. A closure operator c on X is called a *nucleus* on $(X, *)$ if c is compatible with the multiplication $*$ in the following sense:

$$c(x_1) * c(x_2) \leq c(x_1 * x_2), \qquad x_1, x_2 \in X.$$

Example 2.2.4. Let $X \xrightarrow{h} Y$ be a homomorphism between prequantales $(X, *)$ and $(Y, *)$. If $c_h = h^{\vdash} \circ h$ is the closure operator on X associated with h, then we infer from Proposition 2.2.2 that c_h is always a nucleus on $(X, *)$.

By analogy with closure operators, the set $\mathsf{N}(X, *)$ of all nuclei on $(X, *)$ is a complete lattice with respect to the partial order defined by

$$c_1 \le c_2 \iff c_1(x) \le c_2(x) \text{ for all } x \in X.$$

In particular, meets in $\mathsf{N}(X, *)$ are again computed pointwisely.

The first important observation is that the range of every nucleus c on a prequantale $(X, *)$ gives rise to a prequantale $(c(X), \star)$. In fact, the multiplication $\star$ is given by

$$c(x_1) \star c(x_2) = c(c(x_1) * c(x_2)) = c(x_1 * x_2), \qquad x_1, x_2 \in X. \tag{2.28}$$

Obviously, $(c(X), \star)$ is a prequantale and the quotient map $X \xrightarrow{\pi_c} c(X)$ is a surjective homomorphism in $\mathtt{Pq}$. We call $(c(X), \star)$ the *regular quotient* of $(X, *)$ w.r.t. c and will justify this terminology in the following. We begin with the construction of coequalizers in $\mathtt{Pq}$. As a first step we present an analogue of the factorization Lemma 2.1.3.

Lemma 2.2.5. *Let $X \xrightarrow{\pi} Z$ be a surjective homomorphism and $X \xrightarrow{h} Y$ be an arbitrary homomorphism between prequantales with the associated nuclei c_π and c_h on X. Then h factors through π — i.e. the commutativity of the following diagram holds*:

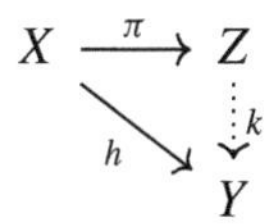

if and only if the relation $c_\pi \le c_h$ is valid.

Proof. The necessity of $c_\pi \le c_h$ is evident. On the other hand, if $c_\pi \le c_h$, then we conclude from the proof of Lemma 2.1.3 that $h \circ \pi^{\vdash}$ is left adjoint to $\pi \circ h^{\vdash}$. By Proposition 2.2.2 both isotone maps $h \circ \pi^{\vdash}$ and $\pi \circ h^{\vdash}$ are closed. Hence by Theorem 1.3.7 and again by Proposition 2.2.2 the map $k = h \circ \pi^{\vdash}$ is a homomorphism between prequantales. Obviously $h = k \circ \pi$ holds — i.e. h factors through π. □

Theorem 2.2.6. *The coequalizer of every pair of parallel homomorphism in $\mathtt{Pq}$ exists.*

Proof. Let $R \underset{h_2}{\overset{h_1}{\rightrightarrows}} X$ be a pair of parallel homomorphisms between prequantales and $\mathbb{F}$ be the set of all nuclei c on $(X, *)$ satisfying the following condition:

$$c(h_1(r)) = c(h_2(r)), \qquad r \in R.$$

Since the meet of nuclei is computed pointwisely, the meet c_0 of $\mathbb{F}$ belongs to $\mathbb{F}$. Further, let $(c_0(X), \star)$ be the regular quotient of $(X, *)$ w.r.t. c_0 and $X \xrightarrow{\pi_{c_0}} c_0(X)$ be the corresponding quotient map. Then $\pi_{c_0} \circ h_1 = \pi_{c_0} \circ h_2$ follows. If $X \xrightarrow{k} Z$ is a further homomorphism with $k \circ h_1 = k \circ h_2$, then the associated nucleus c_k is an element of $\mathbb{F}$. Then we infer from Lemma 2.2.5 that k factors through π_{c_0}. Since π_{c_0} is surjective, the factorizing homomorphism is unique. Hence $\big(\pi_{c_0}, (c_0(X), \star)\big)$ is the coequalizer of $R \overset{h_1}{\underset{h_2}{\rightrightarrows}} X$. □

The proof of the previous theorem implies that every coequalizer in Pq can be viewed as a surjective homomorphism. The next theorem reveals that every surjective homomorphism in Pq is the coequalizer of its kernel pair.

Theorem 2.2.7. *Let $X \xrightarrow{h} Y$ be a surjective homomorphism between prequantales and $R \overset{p}{\underset{q}{\rightrightarrows}} X$ be the kernel pair of h. Then the diagram*

$$\begin{array}{ccc} R & \xrightarrow{q} & X \\ {\scriptstyle p}\downarrow & & \downarrow{\scriptstyle h} \\ X & \xrightarrow{h} & Y \end{array} \tag{2.29}$$

is a pullback and a pushout square — i.e. a pulation square in Pq.

Proof. By definition the diagram (2.29) is a pullback square. Further, we choose homomorphisms $X \overset{k}{\underset{l}{\rightrightarrows}} Z$ with $k \circ p = l \circ q$. Then we conclude from the proof of Theorem 2.1.2 that the following properties hold:

$$k = l \quad \text{and} \quad k \circ h^{\vdash} \dashv h \circ k^{\vdash}.$$

Since $k \circ h^{\vdash}$ and $h \circ k^{\vdash}$ are closed maps, $Y \xrightarrow{k \circ h^{\vdash}} Z$ is a homomorphism from $(Y, *)$ to $(Z, *)$. In particular, $(k \circ h^{\vdash}) \circ h = k$ holds (cf. Proof of Theorem 2.1.2). Hence the universal property of pushouts in Pq follows. □

We can summarize the previous results as follows.

FACT II. *Regular epimorphisms of* Pq *and surjective homomorphisms between prequantales are equivalent concepts.*

Since the forgetful functor from `Preord` to the category `Set` (of sets and maps) has a left adjoint functor, we derive the following important result from the previous Facts I, II and Theorem 2.2.6.

Corollary 2.2.8. *The category* Pq *of prequantales is algebraic.*

It follows from the general theory of algebraic categories that Pq has the (regular epi, mono)-factorization property and coproducts (cf. [47, Corollary 32.13 and Theorem 32.14]). For the convenience of the reader we will give here direct proofs of these properties.

Since the associated closure operator of every homomorphism between prequantales is a nucleus (cf. Example 2.2.4), it follows immediately from Lemma 2.2.5 that every homomorphism $(X, *) \xrightarrow{h} (Y, *)$ has a unique decomposition (up to isomorphism)

$$\begin{array}{ccc} X & \xrightarrow{\pi_{c_h}} & c_h(X) \\ & \searrow^{h} & \downarrow{\scriptstyle m_h} \\ & & Y \end{array}$$

into an regular epimorphism π_{c_h} determined by the nucleus c_h and followed by a monomorphism m_h defined by $m_h(x) = h(x)$ for all $x \in c_h(X)$. In particular, the partially ordered set of all regular quotient objects of a given prequantale $(X, *)$ is order isomorphic to the complete lattice $\mathsf{N}(X, *)$ of all nuclei on $(X, *)$.

As a preparation of the construction of coproducts in $\mathtt{Pq}$, we conclude from Fact II in Sect. 2.1 and Corollary 1.1.5 that for every complete lattice X the free magma $(X^\sharp, m_{X^\sharp})$ exists. We record this result as follows.

Theorem 2.2.9. (Free prequantales) *Let X be a complete lattice. Then there exists a prequantale $(X^\sharp, *_\sharp)$ and a join-preserving map $X \xrightarrow{\eta_X} X^\sharp$ such that for every further prequantale $(Y, *)$ and for every further join-preserving map $X \xrightarrow{h} Y$ there exists a unique homomorphism $(X^\sharp, *_\sharp) \xrightarrow{h^\sharp} (Y, *)$ such that the following diagram is commutative:*

$$\begin{array}{ccc} X & \xrightarrow{\eta_X} & X^\sharp \\ & \searrow^{h} & \vdots{\scriptstyle h^\sharp} \\ & & Y \end{array}$$

Let $\{(X_i, *_i) \mid i \in I\}$ be a family of prequantales. As a first step we form the coproduct $C = \sqcup_{i \in I} X_i$ of the family $\{X_i \mid i \in I\}$ in $\mathtt{Sup}$ with the corresponding coproduct injections $X_i \xrightarrow{q_i} C$. Subsequently we consider the free prequantale $(C^\sharp, *_\sharp)$ generated by the complete lattice C. Now we introduce the set $\mathbb{F}$ of all nuclei c on $(C^\sharp, *_\sharp)$ satisfying the following condition:

$$c\big((\eta_C \circ q_i)(x_i) *_\sharp (\eta_C \circ q_i)(y_i)\big) = c\big((\eta_C \circ q_i)(x_i *_i y_i)\big)$$

for all $x_i, y_i \in X_i$ and for all $i \in I$. Finally, let c_0 be the meet of $\mathbb{F}$ in $\mathsf{N}(C^\sharp, *_\sharp)$. Then c_0 is again an element of $\mathbb{F}$.

Now we construct the regular quotient $(X_0, *)$ of $(C^\sharp, *_\sharp)$ and the corresponding join-preserving quotient map $C^\sharp \xrightarrow{\pi_{c_0}} X_0$. Finally, for every $i \in I$ we introduce a join-preserving map $X_i \xrightarrow{\rho_i} X_0$ by $\rho_i = \pi_{c_0} \circ \eta_C \circ q_i$.

It follows from the construction of $X_0 = c_0(C^\sharp)$ that $X_i \xrightarrow{\rho_i} X_0$ is a homomorphism for all $i \in I$. We show that $(X_0, *)$ is the coproduct of $\mathbb{F}$ with the coproduct injections ρ_i where i is varying in I. For this purpose we consider a further prequantale $(Y, *)$ and homomorphisms $X_i \xrightarrow{h_i} Y$ for all $i \in I$. We have to show that there exists a unique homomorphism $X_0 \xrightarrow{h_0} Y$ such that $h_0 \circ \rho_i = h_i$ holds for all $i \in I$.

(Uniqueness) Since $h_0 \circ \pi_{c_0} \circ \eta_C \circ q_i = h_i = (\sqcup_{i\in I} h_i) \circ q_i$, we conclude from the universal property of coproducts in Sup that the relation $h_0 \circ \pi_{c_0} \circ \eta_C = \sqcup_{i\in I} h_i$ holds. Hence $h_0 \circ \pi_{c_0} = (\sqcup_{i\in I} h_i)^\sharp$ follows from Theorem 2.2.9. Since π_{c_0} is surjective, the uniqueness of h_0 follows.

(Existence) With regard to the existence of h_0, the previous argumentation suggests to verify the factorization of $(\sqcup_{i\in I} h_i)^\sharp$ through π_{c_0}. If $\widetilde{c}$ is the nucleus associated with $(\sqcup_{i\in I} h_i)^\sharp$, then we infer from Lemma 2.2.5 that it is sufficient to show $c_0 \le \widetilde{c}$. For this purpose we fix $j \in I$ and choose $x_j, y_j \in X_j$. Now we observe:

$$\begin{aligned}
&(\sqcup_{i\in I} h_i)^\sharp(\eta_C(q_j(x_j)) *_\sharp \eta_C(q_j(y_j)))\\
&\quad = ((\sqcup_{i\in I} h_i)^\sharp(\eta_C(q_j(x_j)))) * ((\sqcup_{i\in I} h_i)^\sharp(\eta_C(q_j(y_j))))\\
&\quad = h_j(x_j) * h_j(y_j)\\
&\quad = h_j(x_j *_j y_j)\\
&\quad = (\sqcup_{i\in I} h_i)^\sharp(\eta_C(q_j(x_j *_j y_j))).
\end{aligned}$$

Hence $\widetilde{c}$ is an element of $\mathbb{F}$, and $c_0 \le \widetilde{c}$ follows.

As the reader might have noticed, the construction of coproducts of prequantales is not simple, but rather complicated in Pq. Therefore it would be most desirable to have a method at our disposal which simplifies the situation under certain restrictions. In Sect. 2.5 we will see that the tensor product will play such a rôle (cf. Theorem 2.5.9).

2.2.1 *Left- and Right-Implication*

In this subsection we prefer to understand prequantales as magmas $(X, \circledast)$ in Sup and make repeated use of the self-duality in Sup. First we are interested in the right adjoint map of binary operations in Sup.

Theorem 2.2.10. *Let $(X, \circledast)$ be a magma in* Sup. *Then the right adjoint map $X^{op} \xrightarrow{\Gamma_\circledast} (X \otimes X)^{op} = [X, X^{op}]$ of the binary operation $\circledast$ has the following form:*

$$[\Gamma_\circledast(z)](x) = x \searrow z := \bigvee\{y \in X \mid x * y \le z\}, \quad z, x \in X. \tag{R}$$

Proof. Let us recall that the tensor product $X \otimes X$ in Sup is the complete lattice of all join-reversing maps $X \xrightarrow{g} X$. Since every tensor is an appropriate join of elementary tensors, for every $z \in X$ we have to compute the join of the following subset of $X \otimes X$:

$$A_z = \{x \otimes y \in X \otimes X \mid x \circledast y \le z\}.$$

Further, we recall $x \circledast y = x * y$ and the equivalence $x \otimes y \le g \iff y \le g(x)$ (cf. (2.15)). Thus we have to determine the meet of the following subset B_z of $X \otimes X$:

$$B_z = \{g \in X \otimes X \mid x * y \le z \implies y \le g(x)\}.$$

Obviously, every $g \in B_z$ satisfies the property $\bigvee\{y \in X \mid x * y \le z\} \le g(x)$ for all $x \in X$. Hence the correspondence $x \longmapsto \bigvee\{y \in X \mid x * y \le z\}$ is the smallest join-reversing map contained in B_z. Thus (R) is verified. □

Since Sup is a symmetric and monoidal closed category, every binary operation $X \otimes X \xrightarrow{\circledast} X$ induces a join-preserving map $X \otimes X^{op} \xrightarrow{\ominus} X^{op}$ by

$$\ominus = \varepsilon_{X^{op}} \circ (1_X \otimes \Gamma_{\circledast}), \tag{2.30}$$

where $X \otimes [X, X^{op}] \xrightarrow{\varepsilon_{X^{op}}} X^{op}$ is the evaluation arrow of $X \otimes \_ \dashv \hom_X$. If we evaluate $\ominus$ *at elementary tensors* of $X \otimes X^{op}$, then we conclude from (2.30) and Theorem 2.2.10 that the relation

$$\ominus(x \otimes z) = x \searrow z, \qquad x, z \in X \tag{2.31}$$

holds. Therefore $\ominus$ is also called the *right-implication* of $\circledast$ (respectively of $*$).

In order to define the left-implication of $\circledast$ we proceed as follows. Since Sup has a symmetry c (cf. Lemma 2.1.15), we first recall that the *transposed multiplication* of $\circledast$ is determined by:

$$\circledast^{\tau} = \circledast \circ c_{XX} \quad \text{— i.e. } x_1 *^{\tau} x_2 = x_2 * x_1, \qquad x_1, x_2 \in X.$$

Then the *left-implication* $X^{op} \otimes X \xrightarrow{\ominus} X^{op}$ of $\circledast$ is defined as the composition of $c_{X^{op}X}$ with the right-implication of the *transposed multiplication* $\circledast^{\tau}$. Hence $\ominus$ is explicitly given by

$$\ominus = \mathrm{ev}_{X^{op}} \circ (\Gamma_{\circledast^{\tau}} \otimes 1_X). \tag{2.32}$$

In this context, the evaluation of $\ominus$ at elementary tensors now attains the following form:

$$\ominus(z \otimes y) = z \swarrow y := \bigvee\{x \in X \mid x * y \le z\}, \qquad z, y \in X. \tag{2.33}$$

As we will see later, the right- and left-implication will play a basic rôle in the theory of modules in Sup (cf. Sect. 3.1). In particular, the right- (left-)implication coincides with the formation of right- (left-)residuals (cf. [14, p. 325]).

A *left unit* of a prequantale $(X, *)$ (resp. magma $(X, \circledast)$ in Sup) is a join-preserving map $\mathbb{1} \xrightarrow{e} X$ such that the diagram

$$\begin{array}{ccc} \mathbb{1} \otimes X & \xrightarrow{e \otimes 1_X} & X \otimes X \\ & \searrow^{\ell_X} & \downarrow \circledast \\ & & X \end{array}$$

is commutative. If we identify the join-preserving map e with its value at 1, then this means $e * x = x$ for all $x \in X$. The concept of a *right unit* is defined analogously, via the commutativity of the diagram:

$$\begin{array}{ccc} X \otimes \mathbb{1} & \xrightarrow{1_X \otimes e} & X \otimes X \\ & \searrow^{r_X} & \downarrow \circledast \\ & & X \end{array}$$

In fact, e is the unit of a magma $(X, \circledast)$ in Sup (cf. Sect. 1.1) if and only if e is a left unit and a right unit.

After these preliminaries we will show that a left (resp. right) unit can be characterized by the right-implication (resp. left-implication) of the corresponding multiplication.

Theorem 2.2.11. *Let $(X, \circledast)$ be a magma in Sup and $\ominus$ be the right-implication of $\circledast$. A join-preserving map $\mathbb{1} \xrightarrow{e} X$ is a left unit of $(X, \circledast)$ if and only if the following diagram is commutative*:

$$\begin{array}{ccc} \mathbb{1} \otimes X^{op} & \xrightarrow{e \otimes 1_{X^{op}}} & X \otimes X^{op} \\ & \searrow^{\ell_{X^{op}}} & \downarrow \ominus \\ & & X^{op} \end{array}$$

Proof. By the definition of the natural isomorphism ℓ (see the proof of Lemma 2.1.14) the diagram

$$\begin{array}{ccc} \mathbb{1} \otimes X^{op} & & \\ {\scriptstyle 1_{\mathbb{1}} \otimes \ell_X^{\vdash}} \downarrow & \searrow^{\ell_{X^{op}}} & \\ \mathbb{1} \otimes [\mathbb{1}, X^{op}] & \xrightarrow{\varepsilon_{X^{op}}} & X^{op} \end{array}$$

is commutative. Further, the right adjoint map of $\mathbb{1} \otimes X \xrightarrow{e \otimes 1_X} X \otimes X$ is the evaluation at e — i.e. $m \longmapsto m(e)$ for all $m \in [X, X^{op}]$ (see Exercise 2.1.9). Hence the diagram

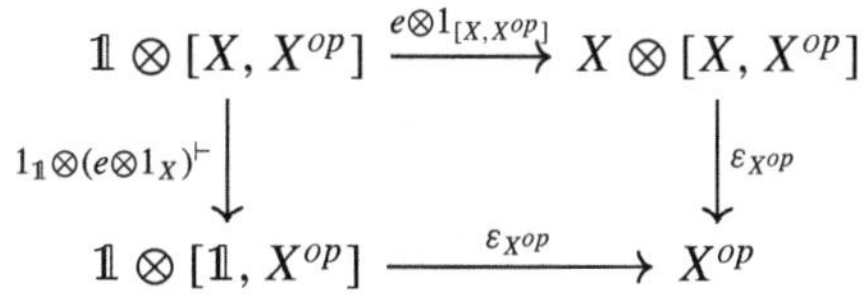

is also commutative. Since $\circledast^{\vdash}$ coincides with $\Gamma_{\circledast}$, we now put the previous diagrams together and observe

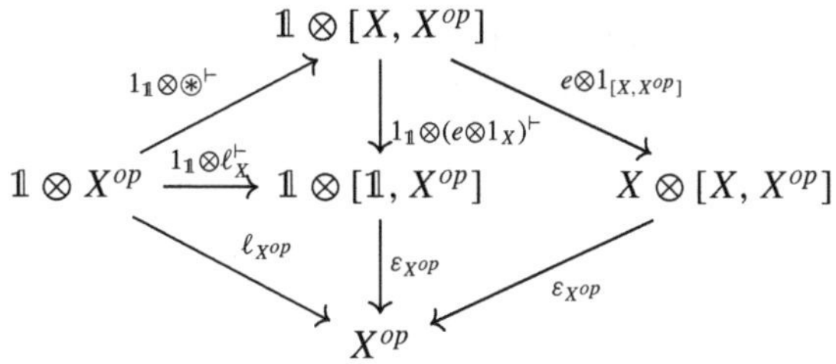

Hence the assertion follows. □

Corollary 2.2.12. *Let* $(X, \circledast)$ *be a magma in* `Sup` *and* $\Leftarrow$ *be the left-implication of* $\circledast$. *A join-preserving map* $\mathbb{1} \xrightarrow{e} X$ *is a right unit of* $(X, \circledast)$ *if and only if the following diagram is commutative*:

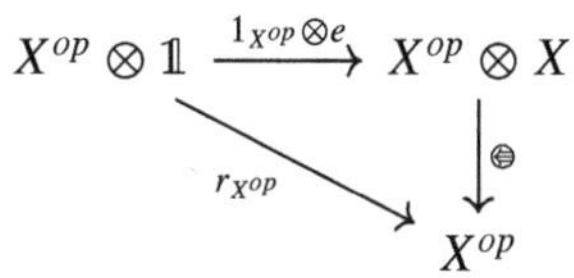

Proof. Since e is a right unit of $(X, \circledast)$ if and only if e is a left unit of the transposed magma $(X, \circledast^{\tau})$, the assertion follows immediately from the construction of $\Leftarrow$ and Theorem 2.2.11. □

We finish this subsection with an example describing the MacNeille completion of residuated preordered groupoids.

Example 2.2.13. A preordered groupoid $(X, \cdot, \leq)$ is called *residuated* if there exists two further binary operations $\searrow$ and $\swarrow$ on X (in the sense of `Set`) satisfying the following condition for all $x, y, z \in X$:

$$y \leq x \searrow z \iff x \cdot y \leq z \iff x \leq z \swarrow y. \tag{RR}$$

By (R), (2.31) and (2.33), the underlying groupoid of any prequantale is residuated. On the other hand, we will show that the MacNeille completion of any residuated preordered groupoid is a prequantale such that the corresponding "embedding" preserves the respective multiplications and formations of residuals. For this purpose we maintain the notation from Example 2.1.4 (a).

Let $(X, \leq, *, \searrow, \swarrow)$ be a residuated groupoid. In order to prove that the MacNeille completion $\widehat{X}$ of the underlying preordered set $(X, \leq)$ is a prequantale it is sufficient to show that the closure operator $\mathsf{C} = \mathsf{L} \circ \mathsf{U}$ is a nucleus with respect to the Minkowski multiplication $\boxdot$ on $\mathsf{Dwn}(X)$ induced by $*$. If A and B are down-closed subsets of X, then we choose $u \in \mathsf{C}(A)$, $v \in \mathsf{C}(B)$ and $z \in \mathsf{U}(A \boxdot B)$. Since $a * b \leq z$ holds for all $a \in A$ and $b \in B$, we infer from (RR) that $a \searrow z$ is an upper bound of B for all $a \in A$. Hence $v \leq a \searrow z$ follows for all $a \in A$. Applying (RR) again we obtain that $z \swarrow v$ is an upper bound of A. In particular, we have $u \leq z \swarrow v$. So, $u * v \leq z$ follows — i.e. $\mathsf{C}(A) \boxdot \mathsf{C}(B) \subseteq \mathsf{C}(A \boxdot B)$.

Obviously, η_X preserves the respective multiplications. Further, we conclude from (RR) that the following equivalences hold:

$$\downarrow x \boxdot A \subseteq \downarrow y \iff A \subseteq \downarrow(x \searrow y),$$
$$A \boxdot \downarrow x \subseteq \downarrow y \iff A \subseteq \downarrow(y \swarrow x).$$

Hence η_X also preserves the respective formation of residuals.

As a trivial corollary of the previous example we obtain that every prequantale $(X, *)$ is a regular quotient of the complete lattice $\mathsf{Dwn}(X)$ of all downclosed subsets of X provided with the Minkowski multiplication induced by $*$.

2.2.2 *Involutive Prequantales*

In this subsection we enrich prequantales with an involution which is also an anti-homomorphism. For this purpose we first briefly review the category of involutive lattices.

Let X be a complete lattice. An isotone map $X \xrightarrow{\iota_X} X$ is called an involution on X if $\iota_X \circ \iota_X = 1_X$. Since involutions are always automorphisms, every involution on a complete lattice is join- and meet-preserving. A pair (X, ι_X) is called an *involutive* lattice if X is a complete lattice and ι_X is an involution on X. Let (X, ι_X) and (Y, ι_Y) be involutive lattices. A join-preserving map $X \xrightarrow{h} Y$ is said to be *involutive* if the relation $h \circ \iota_X = \iota_Y \circ h$ holds. Obviously, involutive lattices and involutive join-preserving maps form a category, which we denote by $\mathbb{I}\mathtt{Sup}$.

Lemma 2.2.14. *Let (X, ι_X) and (Y, ι_X) be involutive lattices and $X \xrightarrow{h} Y$ be a join-preserving map. Then $(X, \iota_X) \xrightarrow{h} (Y, \iota_Y)$ is a morphism in $\mathbb{I}\mathtt{Sup}$ if and only if its right adjoint $(Y^{op}, \iota_Y) \xrightarrow{h^{\vdash}} (X^{op}, \iota_X)$ is a morphism in $\mathbb{I}\mathtt{Sup}$.*

Proof. Let $X \xrightarrow{h} Y$ be an involutive and join-preserving map. Since the following chain of equivalences holds for all $x \in X$ and $y \in Y$:

$$x \leq \iota_X\big(h^{\vdash}(\iota_Y(y))\big) \iff h(x) = \iota_Y\big(h(\iota_X(x))\big) \leq y \iff x \leq h^{\vdash}(y),$$

the right adjoint map $h^{\vdash}$ of h is involutive. The converse implication follows by an application of the principle of duality. □

Obviously, the self-duality of $\mathbb{I}\mathtt{Sup}$ follows immediately from Lemma 2.2.14. A further property is the simple fact that every complete lattice gives rise to an involutive lattice $X \times X$ provided with the involution $\iota_{X\times X}$ defined by

$$\iota_{X\times X}(x_1, x_2) = (x_2, x_1), \qquad (x_1, x_2) \in X \times X.$$

We will show that $(X \times X, \iota_{X\times X})$ is the *free involutive lattice generated by* X where the corresponding embedding $X \xrightarrow{\eta_X} X \times X$ is given as follows:

$$\eta_X(x) = (x, \bot), \qquad x \in X.$$

Proposition 2.2.15. *Let* X *be a complete lattice,* (Y, ι_Y) *be an involutive lattice and* $X \xrightarrow{h} Y$ *be a join-preserving map. Then there exists a unique involutive and join-preserving map* $X \times X \xrightarrow{\ulcorner h \urcorner} Y$ *making the following diagram commutative*:

$$\begin{array}{ccc} X & \xrightarrow{\eta_X} & X \times X \\ & {\scriptstyle h}\searrow & \vdots\, {\scriptstyle \ulcorner h \urcorner} \\ & & Y \end{array} \qquad \text{(I)}$$

Proof. (a) (Uniqueness). Since $\ulcorner h \urcorner$ is involutive, the commutativity of diagram (I) means:

$$\iota_Y(h(x)) = (\iota_Y \circ \ulcorner h \urcorner)(x, \bot) = \ulcorner h \urcorner(\bot, x), \qquad x \in X.$$

Subsequently, we use the property that $\ulcorner h \urcorner$ is join-preserving. Hence we obtain:

$$h(x_1) \vee \iota_Y(h(x_2)) = \ulcorner h \urcorner(x_1, x_2), \qquad (x_1, x_2) \in X \times X. \tag{2.34}$$

(b) (Existence) We define $\ulcorner h \urcorner$ by (2.34). Then $\ulcorner h \urcorner$ is join-preserving and the diagram (I) is commutative. Further, the following relation holds for all $(x_1, x_2) \in X \times X$:

$$\iota_Y(\ulcorner h \urcorner(x_1, x_2)) = \iota_Y(h(x_1)) \vee h(x_2) = \ulcorner h \urcorner(x_2, x_1) = \ulcorner h \urcorner(\iota_{X\times X}(x_1, x_2)).$$

□

As an immediate corollary of Proposition 2.2.15 we obtain that the forgetful functor $\mathsf{U}\colon \mathbb{I}\mathtt{Sup} \to \mathtt{Sup}$ has a left adjoint functor $\mathsf{F}\colon \mathtt{Sup} \to \mathbb{I}\mathtt{Sup}$. The monad $\mathbf{I}$ induced by this adjoint situation $\mathsf{F} \dashv \mathsf{U}$ is called the *monad of involutive lattices* (for details, see Exercise 2.2.5).

Definition 2.2.16. (a) Let $(X, *)$ and $(Y, *)$ be prequantales. A join-preserving map $X \xrightarrow{h} Y$ is called an *anti-homomorphism* if $h(x_1 * x_2) = h(x_2) * h(x_1)$ holds for all

$x_1, x_2 \in X$. Hence an anti-homomorphism means a homomorphism from X to Y where Y is provided with the *transposed* multiplication of Y.
(b) A triple $(X, *, \iota_X)$ is an *involutive prequantale* if (X, ι_X) is an involutive lattice, $(X, *)$ is a prequantale and the involution ι_X is an anti-homomorphism.
(c) Let $(X, *, \iota_X)$ and $(Y, *, \iota_Y)$ be involutive prequantales. A join-preserving map $X \xrightarrow{h} Y$ is an *involutive homomorphism* if h is a homomorphism between prequantales and an involutive, join-preserving map between involutive lattices.

Since prequantales are magmas in Sup, involutive prequantales are involutive magmas in Sup. Further, free magmas in Sup generated by complete lattices X are always denoted by $(X^\sharp, \circledast_\sharp)$. In this context we also refer to the $\mathsf{F}_\otimes$-term monad $\mathbf{T} = (\mathsf{T}, \eta, \mu)$ (of a single binary operator symbol) on Sup and recall the construction of the $\mathsf{F}_\otimes$-term functor T (see Sect. 1.1.1 and Appendix A.1 in the special case $\mathsf{C}_0 = \mathsf{Sup}$):

$$\mathsf{T}(X) = X^\sharp, \quad X \xrightarrow{h} Y, \quad \mathsf{T}(X) \xrightarrow{\mathsf{T}(h)} \mathsf{T}(Y), \quad \mathsf{T}(h) = (\eta_Y \circ h)^\sharp.$$

The next theorem and the subsequent corollaries can be understood as a confirmation that there exists a distributive law of the endofunctor T over the monad $\mathbf{I}$ of involutive lattices.

Theorem 2.2.17. *Let (X, ι_X) be an involutive lattice and $(X^\sharp, \circledast_\sharp)$ be the free magma in Pq generated by the complete lattice X with the corresponding join-preserving embedding $X \xrightarrow{\eta_X} X^\sharp$. On $X^\sharp$ there exists a unique isotone involution $X^\sharp \xrightarrow{\iota_{X^\sharp}} X^\sharp$ satisfying the following conditions*:

(A1) $\iota_{X^\sharp} \circ \eta_X = \eta_X \circ \iota_X$.
(A2) $(X^\sharp, \circledast_\sharp, \iota_{X^\sharp})$ *is an involutive magma.*

Proof. Let (X, ι_X) be an involutive lattice, and $(\circledast_\sharp)^\tau$ be the transposed multiplication of $X^\sharp$. Then we conclude from Theorem 2.2.9 that there exists a unique homomorphism

$$(X^\sharp, \circledast_\sharp) \xrightarrow{\iota_{X^\sharp}} (X^\sharp, (\circledast_\sharp)^\tau)$$

satisfying (A1). Obviously $\iota_{X^\sharp}$ is an anti-homomorphism from $(X^\sharp, \circledast_\sharp)$ to $(X^\sharp, \circledast_\sharp)$. Since $\iota_{X^\sharp}$ is also a homomorphism from $\big(X^\sharp, (\circledast_\sharp)^\tau\big)$ to $(X^\sharp, \circledast_\sharp)$, it follows from the uniqueness of the extension of join-preserving maps to homomorphisms on free prequantales (see again Theorem 2.2.9) that $\iota_{X^\sharp} \circ \iota_{X^\sharp} = 1_{X^\sharp}$ holds — i.e. $\iota_{X^\sharp}$ is an involution on $X^\sharp$. Hence condition (A2) is verified. □

The next corollary reveals that the involutive magma $(X^\sharp, \circledast_\sharp, \iota_{X^\sharp})$ constructed in Theorem 2.2.17 (i.e. $\iota_{X^\sharp}$ satisfies (A1)) is the free involutive magma generated by the involutive lattice (X, ι_X).

Corollary 2.2.18. *Let (X, ι_X) be an involutive lattice, $(X^\sharp, \circledast_\sharp)$ be the free magma generated by X, and let $\iota_{X^\sharp}$ be the involution on $(X^\sharp, \circledast_\sharp)$ satisfying the conditions*

(A1) *and* (A2) *in Theorem 2.2.17. Further, let* $(Y, *, \iota_Y)$ *be an involutive prequantale and* $X \xrightarrow{h} Y$ *be an involutive and join-preserving map. Then there exists a unique involutive homomorphism* $X^\sharp \xrightarrow{h^\sharp} Y$ *making the following diagram commutative*

$$\begin{array}{ccc} X & \xrightarrow{\eta_X} & X^\sharp \\ & {\scriptstyle h}\searrow & \vdots\, {\scriptstyle h^\sharp} \\ & & Y \end{array}$$

Proof. Let $h^\sharp$ be the unique extension of h to a homomorphism from $(X^\sharp, \circledast_\sharp)$ to $(Y, *)$. We have to show that $h^\sharp$ is involutive. Obviously, $h^\sharp$ is also a homomorphism from $\big(X^\sharp, (\circledast_\sharp)^\tau\big)$ to $(Y, *^\tau)$. Hence $(X^\sharp, \circledast_\sharp) \underset{\iota_Y \circ h^\sharp}{\overset{h^\sharp \circ \iota_{X^\sharp}}{\rightrightarrows}} (Y, *^\tau)$ is a pair of homomorphisms. Now we invoke (A1) and obtain:

$$h^\sharp \circ \iota_{X^\sharp} \circ \eta_X = h^\sharp \circ \eta_X \circ \iota_X = h \circ \iota_X = \iota_Y \circ h = \iota_Y \circ h^\sharp \circ \eta_X .$$

Since the extension to homomorphisms is unique, $h^\sharp \circ \iota_{X^\sharp}$ and $\iota_Y \circ h^\sharp$ coincide — i.e. $h^\sharp$ is involutive. □

Corollary 2.2.19. *Let* (X, ι_X) *and* (Y, ι_Y) *be involutive lattices, and let*

$$X^\sharp = \mathsf{T}(X) \xrightarrow{\iota_{X^\sharp}} \mathsf{T}(X) = X^\sharp \quad \text{and} \quad Y^\sharp = \mathsf{T}(Y) \xrightarrow{\iota_{Y^\sharp}} \mathsf{T}(Y) = Y^\sharp$$

be the involutions satisfying conditions (A1) *and* (A2) *in Theorem 2.2.17. If* $X \xrightarrow{h} Y$ *is a morphism in* $\mathbb{I}\mathtt{Sup}$, *then* $\mathsf{T}(X) \xrightarrow{\mathsf{T}(h)} \mathsf{T}(Y)$ *is not only a homomorphism between prequantales, but also a morphism of* $\mathbb{I}\mathtt{Sup}$.

Proof. By (A1) the map η_Y is involutive and a fortiori join-preserving. Hence in the case of $\eta_Y \circ h$ the hypothesis of Corollary 2.2.18 is satisfied. Since $\mathsf{T}(h)$ coincides with $(\eta_Y \circ h)^\sharp$, the assertion follows from Corollary 2.2.18. □

If we now combine Corollary 2.2.19 with Theorem 1.2.7, then it is evident that there exists a distributive law $\sigma : \mathsf{IT} \to \mathsf{TI}$ of the $\mathsf{F}_\otimes$-term functor T over the monad $\mathbf{I}$ of involutive lattices.

In the next step we verify that the composite monad **TI** exists in the sense of σ. For this purpose we have to confirm the assumptions of Beck's Theorem (cf. Theorem 1.2.12). Obviously, the assertion (i) of Lemma 1.2.8 follows immediately from condition (A1) (cf. Theorem 2.2.17). In fact, if (X, ι_X) is an involutive lattice, then (A1) means that the X-component of the unit of the $\mathsf{F}_\otimes$-term monad **T** is involutive. The assertion (i) of Lemma 1.2.10 in the setting of the monads **I** and **T** is the subject of the next proposition.

Proposition 2.2.20. *Let* (X, ι_X) *be an involutive lattice, and let* $(X^\sharp, \circledast_\sharp, \iota_{X^\sharp})$ *and* $\big((X^\sharp)^\sharp, (\circledast_\sharp)_\sharp, \iota_{(X^\sharp)^\sharp}\big)$ *be the involutive magmas constructed in Theorem 2.2.17.*

Then the X-component $(X^\sharp)^\sharp = \mathsf{TT}(X) \xrightarrow{\mu_X} \mathsf{T}(X) = X^\sharp$ of the multiplication of the $\mathsf{F}_\otimes$-term monad $\mathbf{T}$ is involutive.

Proof. First we recall that the X-component $\mathsf{T}(\mathsf{T}(X)) \xrightarrow{\mu_X} \mathsf{T}(X)$ of the multiplication of $\mathbf{T}$ is also a homomorphism w.r.t. the transposed multiplications $\big((\circledast_\sharp)_\sharp\big)^\tau$ and $(\circledast_\sharp)^\tau$. Hence

$$\big((X^\sharp)^\sharp, (\circledast_\sharp)_\sharp\big) \underset{\iota_{X^\sharp} \circ \mu_X}{\overset{\mu_X \circ \iota_{(X^\sharp)^\sharp}}{\rightrightarrows}} \big(X^\sharp, (\circledast_\sharp)^\tau\big)$$

are homomorphisms. Since $\mu_X \circ \eta_{\mathsf{T}(X)} = 1_{X^\sharp}$, we infer from (A1) that the following relation holds:

$$\iota_{X^\sharp} \circ \mu_X \circ \eta_{\mathsf{T}(X)} = \mu_X \circ \eta_{\mathsf{T}(X)} \circ \iota_{X^\sharp} = \mu_X \circ \iota_{(X^\sharp)^\sharp} \circ \eta_{\mathsf{T}(X)}.$$

Since the extension of a join-preserving map to a homomorphism is unique, the homomorphisms $\iota^\sharp_{X^\sharp} \circ \mu_X$ and $\mu_X \circ \iota_{(X^\sharp)^\sharp}$ necessarily coincide. Thus μ_X is involutive. □

The previous results show that we have verified the assumptions of Beck's Theorem. Thus the composite monad **TI** of the monad of involutive lattices on $\mathtt{Sup}$ with the $\mathsf{F}_\otimes$-term monad (of a single binary operator symbol) on $\mathtt{Sup}$ exists.

Moreover, it is not difficult to verify that **TI**-algebras (X, ξ) and involutive prequantales can be identified with each other (cf. Exercise 2.2.6). Hence the Eilenberg–Moore category $\mathtt{Sup}^{\mathbf{TI}}$ is isomorphic to the category $\mathbb{I}\mathtt{Pq}$ of involutive prequantales with involutive homomorphisms. Since $\mathtt{Sup}$ is complete, $\mathbb{I}\mathtt{Pq}$ is also complete. In order to verify the existence of colimits in $\mathbb{I}\mathtt{Pq}$ we need some more terminology.

Definition 2.2.21. Let $(X, *, \iota_X)$ be an involutive prequantale. A closure operator c on X is called an *involutive nucleus* if c is a nucleus on $(X, *)$ and satisfies the additional condition:

$$\iota_X(c(x)) \le c(\iota_X(x)), \qquad x \in X. \tag{IN}$$

It is easily seen that every involutive nucleus c on an involutive prequantale $(X, *, \iota_X)$ satisfies the property $\iota_X(c(x)) = c(\iota_X(x))$ for all $x \in X$. Hence the regular quotient of an involutive prequantale with respect to an involutive nucleus is again an involutive prequantale. By Lemma 2.2.14 the closure operator associated with an involutive homomorphism is always an involutive nucleus. Since right adjoint maps of involutive homomorphisms are obviously involutive, the results from Lemma 2.2.5, Theorems 2.2.6 and 2.2.7 can be transferred to $\mathbb{I}\mathtt{Pq}$ — this means that

- the category $\mathbb{I}\mathtt{Pq}$ has coequalizers,
- the regular epimorphisms in $\mathbb{I}\mathtt{Pq}$ are precisely the surjective and involutive homomorphisms,
- the category $\mathbb{I}\mathtt{Pq}$ has the (regular epi, mono)-factorization property.

Morover, if we combine Proposition 2.2.15 with Corollary 2.2.18, then we obtain that every complete lattice X generates a *free involutive prequantale*

$$\big((X \times X)^\sharp, \circledast_\sharp, \iota_{(X\times X)^\sharp}\big).$$

In fact, for every involutive prequantale $(Y, *, \iota_Y)$ and for every join-preserving map $X \xrightarrow{h} Y$ there exists a unique involutive homomorphism $(X \times X)^\sharp \xrightarrow{h^\sharp} Y$ such that $h^\sharp\big(\eta_{X\times X}(x, \bot)\big) = h(x)$ holds for all $x \in X$. If we put

$$\ulcorner h \urcorner(x_1, x_2) = h(x_1) \vee \iota_Y\big(h(x_2)\big), \qquad (x_1, x_2) \in X \times X,$$

then $h^\sharp$ coincides with the extension $(\ulcorner h \urcorner)^\sharp$ of $\ulcorner h \urcorner$ to a homomorphism on the free prequantale $((X \times X)^\sharp, \circledast_\sharp)$ generated by $X \times X$ (cf. Corollary 2.2.18).

Thus the construction of coproducts also transfers from $\mathtt{Pq}$ to $\mathbb{I}\mathtt{Pq}$. So, $\mathbb{I}\mathtt{Pq}$ is a cocomplete category.

Since the forgetful functor from $\mathtt{Sup}$ to $\mathtt{Set}$ has a left adjoint functor, we derive the following important property from the previous results.

FACT III. *The category of involutive prequantales is algebraic.*

We finish this subsection with the observation that in involutive prequantales $(X, *, \iota_X)$ the right-implication and left-implication are mutually determined by each other. Referring to (R), (2.31) and (2.33), the relations

$$\iota_X(x_1 \searrow x_2) = \iota_X(x_2) \swarrow \iota_X(x_1) \quad \text{and} \quad \iota_X(x_2 \swarrow x_1) = \iota_X(x_1) \searrow \iota_X(x_2) \qquad (2.35)$$

hold for all $x_1, x_2 \in X$. Hence these properties can be expressed by the following commutative diagrams:

$$\begin{array}{ccccc} X^{op} \otimes X & & \xrightarrow{\quad c_{X^{op}X} \quad} & & X \otimes X^{op} \\ {\scriptstyle \iota_X \otimes \iota_X}\downarrow & & & & \downarrow{\scriptstyle \ominus} \\ X^{op} \otimes X & \xrightarrow{\ominus} & X^{op} & \xleftarrow{\iota_X} & X^{op} \end{array} \qquad \begin{array}{ccccc} X^{op} \otimes X & & \xrightarrow{\quad c_{X^{op}X} \quad} & & X \otimes X^{op} \\ {\scriptstyle \ominus}\downarrow & & & & \downarrow{\scriptstyle \iota_X \otimes \iota_X} \\ X^{op} & \xrightarrow{\iota_X} & X^{op} & \xleftarrow{\ominus} & X \otimes X^{op} \end{array}$$

Exercises

2.2.1. Confirm that there exist precisely 20 prequantale structures on the three-chain $C_3 = \{\bot, a, \top\}$ given by the following multiplication tables:

(1)

$*$	$\bot$	a	$\top$
$\bot$	$\bot$	$\bot$	$\bot$
a	$\bot$	$\bot$	$\bot$
$\top$	$\bot$	$\bot$	$\bot$

(2)

$*$	$\bot$	a	$\top$
$\bot$	$\bot$	$\bot$	$\bot$
a	$\bot$	$\bot$	$\bot$
$\top$	$\bot$	$\bot$	a

(3)

$*$	$\bot$	a	$\top$
$\bot$	$\bot$	$\bot$	$\bot$
a	$\bot$	$\bot$	a
$\top$	$\bot$	$\bot$	a

(4)

$\ast$	$\bot$	a	$\top$
$\bot$	$\bot$	$\bot$	$\bot$
a	$\bot$	$\bot$	$\bot$
$\top$	$\bot$	a	a

(5)

$\ast$	$\bot$	a	$\top$
$\bot$	$\bot$	$\bot$	$\bot$
a	$\bot$	$\bot$	a
$\top$	$\bot$	a	a

(6)

$\ast$	$\bot$	a	$\top$
$\bot$	$\bot$	$\bot$	$\bot$
a	$\bot$	$\bot$	$\bot$
$\top$	$\bot$	$\bot$	$\top$

(7)

$\ast$	$\bot$	a	$\top$
$\bot$	$\bot$	$\bot$	$\bot$
a	$\bot$	$\bot$	a
$\top$	$\bot$	$\bot$	$\top$

(8)

$\ast$	$\bot$	a	$\top$
$\bot$	$\bot$	$\bot$	$\bot$
a	$\bot$	$\bot$	$\bot$
$\top$	$\bot$	a	$\top$

(9)

$\ast$	$\bot$	a	$\top$
$\bot$	$\bot$	$\bot$	$\bot$
a	$\bot$	$\bot$	a
$\top$	$\bot$	a	$\top$

(10)

$\ast$	$\bot$	a	$\top$
$\bot$	$\bot$	$\bot$	$\bot$
a	$\bot$	$\bot$	$\top$
$\top$	$\bot$	$\bot$	$\top$

(11)

$\ast$	$\bot$	a	$\top$
$\bot$	$\bot$	$\bot$	$\bot$
a	$\bot$	$\bot$	$\bot$
$\top$	$\bot$	$\top$	$\top$

(12)

$\ast$	$\bot$	a	$\top$
$\bot$	$\bot$	$\bot$	$\bot$
a	$\bot$	$\bot$	a
$\top$	$\bot$	$\top$	$\top$

(13)

$\ast$	$\bot$	a	$\top$
$\bot$	$\bot$	$\bot$	$\bot$
a	$\bot$	$\bot$	$\top$
$\top$	$\bot$	a	$\top$

(14)

$\ast$	$\bot$	a	$\top$
$\bot$	$\bot$	$\bot$	$\bot$
a	$\bot$	$\bot$	$\top$
$\top$	$\bot$	$\top$	$\top$

(15)

$\ast$	$\bot$	a	$\top$
$\bot$	$\bot$	$\bot$	$\bot$
a	$\bot$	a	a
$\top$	$\bot$	a	a

(16)

$\ast$	$\bot$	a	$\top$
$\bot$	$\bot$	$\bot$	$\bot$
a	$\bot$	a	a
$\top$	$\bot$	a	$\top$

(17)

$\ast$	$\bot$	a	$\top$
$\bot$	$\bot$	$\bot$	$\bot$
a	$\bot$	a	$\top$
$\top$	$\bot$	a	$\top$

(18)

$\ast$	$\bot$	a	$\top$
$\bot$	$\bot$	$\bot$	$\bot$
a	$\bot$	a	a
$\top$	$\bot$	$\top$	$\top$

(19)

$\ast$	$\bot$	a	$\top$
$\bot$	$\bot$	$\bot$	$\bot$
a	$\bot$	a	$\top$
$\top$	$\bot$	$\top$	$\top$

(20)

$\ast$	$\bot$	a	$\top$
$\bot$	$\bot$	$\bot$	$\bot$
a	$\bot$	$\top$	$\top$
$\top$	$\bot$	$\top$	$\top$

2.2.2. If $(X, \ast)$ is a prequantale and $\mathbb{1}$ is the two-chain $\{0, 1\}$, then define a binary operation $\widehat{\ast}$ on $\widehat{X} = X \times \mathbb{1}$ as follows:

$$(x,0)\,\widehat{\ast}\,(y,0) = (x \ast y, 0), \qquad (x,1)\,\widehat{\ast}\,(y,1) = ((x \ast y) \vee x \vee y, 1),$$
$$(x,1)\,\widehat{\ast}\,(y,0) = ((x \ast y) \vee y, 0), \qquad (y,0)\,\widehat{\ast}\,(x,1) = ((y \ast x) \vee y, 0).$$

(a) Show that $(\widehat{X}, \widehat{\ast})$ is a unital prequantale with unit $(\bot, 1)$.
(b) Show that $(\widehat{X}, \widehat{\ast})$ is the unitalization of $(X, \ast)$.
(c) Let $\ast^\tau$ be the transposed multiplication of $\ast$. Show that $\widehat{\ast}^\tau = \widehat{\ast^\tau}$ holds.

2.2.3. Let $(X, *)$ be a prequantale and c be a closure operator on X. Show that c is a nucleus if and only if c satisfies the following additional properties for all $x, y \in X$:

$$x * c(y) \leq c(x * y) \quad \text{and} \quad c(x) * y \leq c(x * y).$$

2.2.4. Let $(X, *)$ be a prequantale. An isotone map $X \xrightarrow{p} X$ is called a *prenucleus* if p satisfies the following conditions for all $x, y \in X$ (cf. [94]):

$$x \leq p(x), \quad x * p(y) \leq p(x * y) \quad \text{and} \quad x * p(y) \leq p(x * y).$$

If now p is a prenucleus on X, then show that the map $X \xrightarrow{p_0} X$ defined by

$$p_0(x) = \bigwedge\{z \in X \mid x \leq z,\ p(z) \leq z\}, \qquad x \in X$$

is a nucleus on $(X, *)$.
(Hint: For $x, u \in X$ with $p(u) \leq u$ first verify the relations $p(x \searrow u) \leq x \searrow u$ and $p(u \swarrow x) \leq u \swarrow x$, and then use Exercise 2.2.3.)

2.2.5. Let $\mathbf{I} = (\mathsf{I}, \eta^{\mathsf{I}}, \mu^{\mathsf{I}})$ be the monad of involutive lattices on Sup.

(a) Show that the data I, η^{I} and μ^{I} of $\mathbf{I}$ are explicitly given as follows:

$$\mathsf{I}(X) = X \times X, \quad \eta^{\mathsf{I}}_X(x) = (x, \perp), \quad \mu^{\mathsf{I}}_X\big((x_1, x_2), (x_3, x_4)\big) = (x_1 \vee x_4, x_2 \vee x_3)$$

for all $x, x_1, x_2, x_3, x_4 \in X$.

(b) Let $X \xrightarrow{\iota} X$ and $\mathsf{I}(X) = X \times X \xrightarrow{\xi} X$ be join-preserving maps such that $\xi(x, \perp) = x$ and $\xi(\perp, x) = \iota(x)$ for all $x \in X$. Show that (X, ξ) is an $\mathbf{I}$-algebra if and only if ι is involutive.
(Hint: Prove that $\mathsf{II}(X) \xrightarrow{\mathsf{I}(\xi)} \mathsf{I}(X)$ is defined for all $((x_1, x_2), (x_3, x_4)) \in \mathsf{II}(X)$ by

$$\mathsf{I}(\xi)((x_1, x_2), (x_3, x_4)) = (\xi(x_1, x_2), \xi(x_3, x_4))$$

2.2.6. Let $\mathbf{I}$ be the monad of involutive lattices, $\mathbf{T} = (\mathsf{T}, \eta, \mu)$ be the monad of a single binary operator symbol on Sup, and let X be a complete lattice provided with two algebra structures: (X, ξ_1) is an $\mathbf{I}$-algebra and (X, ξ_2) is a $\mathbf{T}$-algebra such that ξ_2 is an involutive homomorphism — i.e. $\big(\mathsf{TI}(X), \xi_2 \circ \mathsf{T}(\xi_1)\big)$ is a $\mathbf{TI}$-algebra and the relation $\xi_2 \circ \mathsf{T}(\xi_1) \circ \sigma_X = \xi_1 \circ \mathsf{I}(\xi_2)$ holds (cf. Theorem 1.2.14 and Proposition 1.2.15). Further, let $X \xrightarrow{\iota} X$ and $X \otimes X \xrightarrow{\circledast} X$ be determined by (cf. Exercise 2.2.5 (b)):

$$\xi_2 \circ \circledast_\sharp \circ (\eta_X \otimes \eta_X) = \circledast, \quad \text{and} \quad \xi_1(\perp, x) = \iota(x), \qquad x \in X.$$

Show that $(X, \circledast, \iota)$ is an involutive magma in Sup — i.e. an involutive prequantale. On the other hand, show that every involutive prequantale $(X, *, \iota)$ gives rise to a $\mathbf{TI}$-algebra.

(Hint: Conclude from Corollary 2.2.18 that the extension of 1_X to a homomorphism $\mathsf{T}(X) \xrightarrow{\xi_2} X$ is involutive.)

2.3 Quantales: The Axiom of Associativity

A prequantale $(X, *)$ is a *quantale* if the multiplication $*$ is associative — i.e.

$$(x * y) * z = x * (y * z), \quad x, y, z \in X. \qquad \text{(Associativity)}$$

With regard to Lemma 2.1.13 the associativity axiom of quantales means that quantales are *semigroups in* Sup. Hence, because of similarities between Sup and the category of abelian groups, quantales can be understood as ring-like objects (see also the Introduction in [59]).

Quantales obviously form a full subcategory of the category of prequantales which we will denote by Quant. We show now that Quant is even a reflective subcategory of Pq.

Theorem 2.3.1. *Let $(X, *)$ be a prequantale. Then there exists a quantale $(X^{\flat}, *^{\flat})$ and a homomorphism $X \xrightarrow{\eta_X^{\flat}} X^{\flat}$ such that for every quantale $(Y, *)$ and for every homomorphism $X \xrightarrow{h} Y$ there exists a unique homomorphism $X^{\flat} \xrightarrow{h^{\flat}} Y$ making the following diagram commutative*:

$$\begin{array}{ccc} X & \xrightarrow{\eta_X^{\flat}} & X^{\flat} \\ & \underset{h}{\searrow} & \downarrow{\scriptstyle h^{\flat}} \\ & & Y \end{array}$$

Proof. Let $\mathbb{F}$ be the set of all nuclei c on a given prequantale $(X, *)$ satisfying the following condition:

$$c\big((x * y) * z\big) = c\big(x * (y * z)\big), \qquad x, y, z \in X.$$

Now we compute the meet c_0 of $\mathbb{F}$ in the sense of the complete lattice $\mathsf{N}(X, *)$ of all nuclei on $(X, *)$. Then c_0 stays inside $\mathbb{F}$ — i.e. $c_0 \in \mathbb{F}$ — and the regular quotient of $(X, *)$ w.r.t. c_0 is a quantale, which we denote by $(X^{\flat}, *^{\flat})$. In this context $X \xrightarrow{\eta_X^{\flat}} X^{\flat}$ is the corresponding quotient map — i.e. $\eta_X^{\flat}(x) = c_0(x)$, $x \in X$. If $(Y, *)$ is a further quantale and $X \xrightarrow{h} Y$ is a homomorphism, then the associativity of the multiplication in Y implies that the nucleus c_h associated with h is an element of $\mathbb{F}$. Hence $c_0 \leq c_h$ follows — i.e. h factors through $\eta^{\flat}$ (cf. Lemma 2.2.5). □

The quantale $(X^{\flat}, *^{\flat})$ constructed in Theorem 2.3.1 is called the *associative reflection* of the prequantale $(X, *)$. Obviously, the associative reflection of the free prequantale generated by a complete lattice X (cf. Theorem 2.2.9) is the *free quantale* generated by X.

It follows immediately from Theorem 2.3.1 that the cocompleteness of Pq is transmitted to Quant. The completeness of Quant is evident. In particular, limits in Quant are computed at the level of Sup. Finally, regular epimorphisms in Quant are surjective homomorphisms and coincide with the coequalizer of their kernel pair. Since the forgetful functor from Sup to Set has a right adjoint functor, the category of quantales is obviously algebraic.

A quantale is *unital* if it has a unit. Hence unital quantales are *monoids in* Sup. A typical example of a unital quantale can be given as follows.

Example 2.3.2. Let X be a complete lattice and $[X, X]$ be the set of all join-preserving self-maps $X \xrightarrow{f} X$ provided with the partial order defined by

$$f_1 \le f_2 \iff f_1(x) \le f_2(x) \quad \text{for all } x \in X.$$

Then $([X, X], \le)$ is a complete lattice. On $[X, X]$ we consider the composition of maps as multiplication — i.e. $f_1 * f_2 = f_1 \circ f_2$. Then $([X, X], \circ)$ is a unital quantale with the unit given by the identity 1_X of X.

As a further important class of unital quantales we point out (cf. Sect. 1.1.2) that for every complete lattice X the free unital quantale generated by X exists.

A subset of a quantale $(X, *)$ is a *subquantale* of $(X, *)$ if the inclusion map is a homomorphism. A unital quantale is called *integral* if the unit coincides with the universal upper bound $\top$. If e is the unit of a unital quantale $(X, *)$, then the subquantale $\{x \in X \mid x \le e\}$ is always an integral quantale.

Since in the case of symmetric and monoidal closed categories with underlying cocomplete category the unitalization of semigroups always exists (cf. Sect. 1.1), we mainly focus here on non-unital quantales.

Let X be a complete lattice with the trivial multiplication $*$ — i.e. $x * y = \bot$ for all $x, y \in X$. Then the pair $(X, *)$ is a quantale — the so-called *trivial* quantale. Hence a quantale $(X, *)$ is said to be *non-trivial* if $\top * \top \neq \bot$ holds. Every non-trivial quantale contains at least two different elements. A trivial quantale $(X, *)$ is non-unital provided that X has at least two different elements. In order to understand the rôle of the universal upper bound, especially in non-unital quantales, we introduce the following terminology.

An element x of a quantale $(X, *)$ is *left-sided* (resp. *right-sided*) if $\top * x \le x$ (resp. $x * \top \le x$). An element x of a quantale $(X, *)$ is *two-sided* if it is left-sided and right-sided. A quantale $(X, *)$ is *left-sided* (resp. *right-sided*) if every element of X is left-sided (resp. right-sided). A quantale $(X, *)$ is *two-sided* if every element of X is two-sided.

Obviously, left-sidedness and right-sidedness are transposed notions. In fact, a quantale $(X, *)$ is right-sided if and only if the *transposed* quantale $(X, *^{\tau})$ (i.e. X provided with the transposed multiplication) is left-sided. Hence at various places it is sufficient to restrict our interest, for example, to left-sided quantales (cf. Sect. 2.4).

Every integral quantale is two-sided, but not vice versa. If a complete lattice X has at least two different elements, then a counterexample is given by the trivial quantale $(X, *)$, which is two-sided, but not unital as we have seen above.

As a first important result we note that the *tensor product* of quantales $(X, *)$ and $(Y, *)$ is again a quantale (cf. Sect. 1.1.3). Referring to Lemmas 2.1.13, 2.1.15 and Theorem 1.1.9, it is easily seen that the multiplication $\star$ of the tensor product $X \otimes Y$ is determined on elementary tensors as follows:

$$(x_1 \otimes y_1) \star (x_2 \otimes y_2) = (x_1 * x_2) \otimes (y_1 * y_2), \qquad x_1, x_2 \in X,\ y_1, y_2 \in Y. \quad (2.36)$$

If e_X and e_Y are the respective units of $(X, *)$ and $(Y, *)$, then the elementary tensor $e_X \otimes e_Y$ is obviously the unit of $(X \otimes Y, \star)$. The tensor product of quantales is non-trivial if and only if its factors are non-trivial. Since $\top \otimes \top$ is the universal upper bound in $X \otimes Y$, left-sidedness (resp. right-sidedness) of tensor products can be characterized as follows.

Proposition 2.3.3. *Let $(X, *)$ and $(Y, *)$ be non-trivial quantales. Then the tensor product $(X \otimes Y, \star)$ is left-sided (right-sided) if and only if $(X, *)$ and $(Y, *)$ are left-sided (right-sided).*

Proof. If $(X \otimes Y, \star)$ is left-sided, then $\top * x \le x$ follows from $\top * \top \ne \bot$ and the relation

$$(\top * x) \otimes (\top * \top) = (\top \otimes \top) \star (x \otimes \top) \le x \otimes \top.$$

Hence $(X, *)$ is left-sided. Analogously we treat the quantale $(Y, *)$. On the other hand, if $(X, *)$ and $(Y, *)$ are left-sided, then we observe

$$(\top \otimes \top) \star (x \otimes y) = (\top * x) \otimes (\top * y) \le x \otimes y.$$

Since every tensor is a join of an appropriate family of elementary tensors, $(X \otimes Y, \star)$ is left-sided. □

We continue with some interesting examples of quantales occurring in system theory and probability theory (cf. the probabilistic metric spaces in [104]).

Example 2.3.4. Let $(X, \cdot)$ be a semigroup (in Set) and $\mathscr{P}(X)$ be the power set of X. On $\mathscr{P}(X)$ we consider the set-inclusion as the partial order and the Minkowski multiplication

$$A \boxdot B = \{a \cdot b \mid a \in A,\ b \in B\}, \qquad A, B \in \mathscr{P}(X).$$

Then $(\mathscr{P}(X), \boxdot)$ is a quantale. Let $(\mathfrak{Q}, *)$ be a further arbitrary quantale. By (2.36) the multiplication $\star$ of the tensor product $\mathscr{P}(X) \otimes \mathfrak{Q}$ is determined on elementary tensors by

$$(A \otimes \alpha) \star (B \otimes \beta) = (A \boxdot B) \otimes (\alpha * \beta), \qquad A, B \in \mathscr{P}(X),\ \alpha, \beta \in \mathfrak{Q}.$$

Referring to Example 2.1.9 we know that $\mathscr{P}(X) \otimes \mathfrak{Q}$ is order isomorphic to $\mathfrak{Q}^X$, and the respective order isomorphism $\mathscr{P}(X) \otimes \mathfrak{Q} \xrightarrow{\Phi} \mathfrak{Q}^X$ has the following form:

$$\big(\Phi(F)\big)(x) = F(\{x\}), \qquad x \in X, \ F \in \mathscr{P}(X) \otimes \mathfrak{Q}.$$

Obviously, for every $x \in X$ and $\alpha \in \mathfrak{Q}$ the relation $\Phi(\{x\} \otimes \alpha) = \alpha \cdot 1_{\{x\}}$ holds, where the map $\alpha \cdot 1_{\{x\}}$ is given by (Example 2.1.9):

$$\big(\alpha \cdot 1_{\{x\}}\big)(z) = \begin{cases} \alpha, & z = x, \\ \bot, & z \neq x. \end{cases}$$

Since $\otimes$ is a bimorphism, it follows from (2.17) that $\{\{x\} \otimes \alpha \mid x \in X, \ \alpha \in \mathfrak{Q}\}$ is also a join-basis of $\mathscr{P}(X) \otimes \mathfrak{Q}$. Hence for every tensor $F \in \mathscr{P}(X) \otimes \mathfrak{Q}$ there exists a unique map $X \xrightarrow{f} \mathfrak{Q}$ satisfying the condition

$$F = \bigvee_{x \in X} \{x\} \otimes f(x). \tag{2.37}$$

Now we apply Φ and infer from (2.37) that the relation

$$\Phi(F) = \bigvee_{x \in X} f(x) \cdot 1_{\{x\}} = f \tag{2.38}$$

holds. Hence the unique map f in (2.38) coincides with the value of the order isomorphism Φ at the tensor F — i.e. $\Phi(F) = f$. In order to fix terminology we call the map f the *reduced representation* of the tensor $F \in \mathscr{P}(X) \otimes \mathfrak{Q}$.

Now we multiply two tensors F and G of $\mathscr{P}(X) \otimes \mathfrak{Q}$ in the sense of the tensor product of $(\mathscr{P}(X), \boxdot)$ with $(\mathfrak{Q}, *)$ and obtain the following situation for their respective reduced representations $f, g \in \mathfrak{Q}^X$:

$$\begin{aligned}
\big(\bigvee_{x \in X} \{x\} \otimes f(x)\big) \star \big(\bigvee_{y \in X} \{y\} \otimes g(y)\big) &= \bigvee_{x, y \in X} (\{x\} \otimes f(x)) \star (\{y\} \otimes g(y)) \\
&= \bigvee_{x, y \in X} \{x \cdot y\} \otimes (f(x) * g(y)) \\
&= \bigvee_{z \in X} \{z\} \otimes \big(\bigvee_{x \cdot y = z} f(x) * g(y)\big).
\end{aligned}$$

Further, let $\odot$ be the binary operation on $\mathfrak{Q}^X$ determined by the semigroup operation $\cdot$ according to Zadeh's extension principle[3] — i.e.

$$(f \odot g)(z) = \bigvee \{f(x) * g(y) \mid x \cdot y = z\}, \qquad z \in X, \ f, g \in \mathfrak{Q}^X.$$

[3] In his original papers [113] L.A. Zadeh only used a frame as an underlying quantale, which means that $*$ coincides with the binary meet operation $\wedge$.

Then we conclude from (2.37) and (2.38) that the reduced representation of the tensor product $F \star G$ coincides with $f \odot g$. Hence the tensor product of the quantale $(\mathscr{P}(X), \boxdot)$ with the quantale $(\mathfrak{Q}, *)$ is isomorphic to the quantale given by the $\mathfrak{Q}$-valued power set $\mathfrak{Q}^X$ provided with the multiplication $\odot$ defined according to Zadeh's extension principle.

Example 2.3.5. On the complete lattice $[0, +\infty]^{op}$ we consider the usual addition of the nonnegative extended real line. Then $([0, +\infty]^{op}, +)$ is a unital quantale. Further, let $*$ be a binary operation on the real unit interval $[0, 1]$ such that $([0, 1], *)$ is also a quantale. Now we consider the tensor product of $[0, +\infty]^{op}$ with $[0, 1]$ given by the complete lattice Δ^+ of all nonnegative left-continuous probability distribution functions. Since $(x, \alpha) \longmapsto \alpha \cdot H_x$ is the corresponding universal bimorphism (cf. Example 2.1.10), the product $f \star g$ of $f, g \in \Delta^+$ according to the multiplication of the tensor product of $([0, +\infty]^{op}, +)$ with $([0, 1], *)$ has the following form:

$$\begin{aligned} f \star g &= \Big(\bigvee_{x \in [0,+\infty]} f(x) \cdot H_x \Big) \star \Big(\bigvee_{y \in [0,+\infty]} g(y) \cdot H_y \Big) \\ &= \bigvee_{x,y \in [0,+\infty]} (f(x) \cdot H_x) \star (g(y) \cdot H_y) \\ &= \bigvee_{x,y \in [0,+\infty]} (f(x) * g(y)) \cdot H_{x+y}. \end{aligned}$$

Thus $f \star g$ is explicitly given by:

$$(f \star g)(z) = \bigvee_{x+y=z} f(x) * g(y), \qquad z \in [0, +\infty]. \tag{2.39}$$

We will return to this situation in Sect. 2.3.3 when we consider the special case in which $*$ is given by a left-continuous t-norm (cf. Example 2.3.36).

In the next theorem we give a characterization of the law of associativity.

Theorem 2.3.6. *Let $(X, *)$ be a prequantale viewed as magma $(X, \circledast)$ in* Sup. *Further, let $\overset{\leftarrow}{\ominus}$ and $\overset{\rightarrow}{\ominus}$ be the corresponding left- and right-implication. Then the following assertions are equivalent:*

(i) *$(X, *)$ is a quantale — i.e. $(X, \circledast)$ is a semigroup in* Sup.
(ii) *The following diagram is commutative:*

$$\begin{array}{ccccc} X \otimes (X \otimes X^{op}) & & \xleftarrow{\quad a_{XXX^{op}} \quad} & & (X \otimes X) \otimes X^{op} \\ & & & & \big\downarrow {\scriptstyle c_{XX} \otimes 1_{X^{op}}} \\ {\scriptstyle 1_X \otimes \overset{\leftarrow}{\ominus}} \big\downarrow & & & & (X \otimes X) \otimes X^{op} \\ & & & & \big\downarrow {\scriptstyle \circledast \otimes 1_{X^{op}}} \\ X \otimes X^{op} & \xrightarrow{\ \overset{\leftarrow}{\ominus}\ } & X^{op} & \xleftarrow{\ \overset{\leftarrow}{\ominus}\ } & X \otimes X^{op} \end{array}$$

where c is the symmetry and a is the natural isomorphism expressing the associativity of the tensor product in Sup.

(iii) *The following diagram is commutative*:

$$
\begin{array}{ccccc}
(X^{op} \otimes X) \otimes X & & \xleftarrow{\quad a_{X^{op}XX} \quad} & & X^{op} \otimes (X \otimes X) \\
\Big\downarrow & & & & \Big\downarrow {\scriptstyle 1_{X^{op}} \otimes c_{XX}} \\
{\scriptstyle \ominus \otimes 1_X} & & & & X^{op} \otimes (X \otimes X) \\
\Big\downarrow & & & & \Big\downarrow {\scriptstyle 1_{X^{op}} \otimes \circledast} \\
X^{op} \otimes X & \xrightarrow{\ \ominus\ } & X^{op} & \xleftarrow{\ \ominus\ } & X^{op} \otimes X
\end{array}
$$

(iv) *The following diagram is commutative*:

$$
\begin{array}{ccccc}
(X \otimes X^{op}) \otimes X & & \xrightarrow{\quad a_{XX^{op}X} \quad} & & X \otimes (X^{op} \otimes X) \\
{\scriptstyle \ominus \otimes 1_X} \Big\downarrow & & & & \Big\downarrow {\scriptstyle 1_X \otimes \ominus} \\
X^{op} \otimes X & \xrightarrow{\ \ominus\ } & X^{op} & \xleftarrow{\ \ominus\ } & X \otimes X^{op}
\end{array}
$$

Proof. Let $\searrow$ and $\swarrow$ be the restrictions of the right- and left-implication to elementary tensors. Then for all $x, y, z \in X$ the following chain of equivalences holds (cf. (R), (2.31) and (2.33)):

$$y \le x \searrow z \iff x * y \le z \iff x \le z \swarrow y.$$

In order to verify the statement of the theorem it is sufficient to show that the associativity of the multiplication $*$ is equivalent to each of the following properties for all $x, y, z \in X$:

$$
\begin{cases}
y \searrow (x \searrow z) = (x * y) \searrow z, \\
(z \swarrow x) \swarrow y = z \swarrow (y * x), \\
(x \searrow y) \swarrow z = x \searrow (y \swarrow z).
\end{cases}
\tag{E}
$$

In a first step we verify the equivalence of the following assertions:

(a) $(x * y) * z \le x * (y * z)$ for all $x, y, z \in X$.
(b) $y \searrow (x \searrow z) \le (x * y) \searrow z$ for all $x, y, z \in X$.
(c) $z \swarrow (y * x) \le (z \swarrow x) \swarrow y$ for all $x, y, z \in X$.
(d) $(x \searrow y) \swarrow z \le x \searrow (y \swarrow z)$ for all $x, y, z \in X$.

The implication (a) $\Longrightarrow$ (b) follows from the relation:

$$(x * y) * \big(y \searrow (x \searrow z)\big) \le x * \big(y * \big(y \searrow (x \searrow z)\big)\big) \le x * (x \searrow z) \le z.$$

The verification of (b) $\Longrightarrow$ (c) is based on the observation:

$$x \leq y \searrow (y * x) \leq y \searrow \big((z \swarrow (y * x)) \searrow z\big) \leq \big((z \swarrow (y * x)) * y\big) \searrow z.$$

Further, the implication (c) $\Longrightarrow$ (d) follows from the relation

$$x \leq y \swarrow (x \searrow y) \leq y \swarrow \big(((x \searrow y) \swarrow z) * z\big) \leq (y \swarrow z) \swarrow \big((x \searrow y) \swarrow z\big).$$

Finally, the verification of (d) $\Longrightarrow$ (a) makes use of the observation:

$$y \leq (y * z) \swarrow z \leq \big(x \searrow (x * (y * z))\big) \swarrow z \leq x \searrow \big((x * (y * z)) \swarrow z\big).$$

If in (a) – (d) we replace the multiplication $*$ of X by its transposed multiplication $*^{\tau}$, then in a second step we notice that the equivalences between (a) – (d) remain valid with respect to the dual order $\leq^{op}$ of $\leq$. Hence the equivalence of the associativity of $*$ to each of the properties in (E) follows from the antisymmetry axiom. $\square$

Since the Minkowski multiplication induced by an associative binary operation is again associative, we have the following associative version of Example 2.2.13.

Example 2.3.7. Let $(X, \leq, *, \searrow, \swarrow)$ be a residuated preordered semigroup — i.e. a residuated preordered groupoid with an associative multiplication $*$. Then the MacNeille completion of $(X, \leq, *, \searrow, \swarrow)$ is a quantale.

Finally, we restrict our interest to completely distributive lattices. In this situation quantales whose multiplication is in addition meet-preserving can be understood as semigroups in CD (cf. Theorem 2.1.25).

The next theorem shows that semigroups in CD have an intrinsic "diagonal map". For this purpose we add the following terminology.

A pair (X, n) is a *cosemigroup in* Sup if X is a complete lattice and $X \xrightarrow{n} X \otimes X$ is a join-preserving map such that the following diagram is commutative:

$$\begin{array}{ccccc} (X \otimes X) \otimes X & & \xrightarrow{\quad a_{XXX} \quad} & & X \otimes (X \otimes X) \\ {\scriptstyle n \otimes 1_X}\big\uparrow & & & & \big\uparrow{\scriptstyle 1_X \otimes n} \\ X \otimes X & \xleftarrow{\quad n \quad} & X & \xrightarrow{\quad n \quad} & X \otimes X \end{array} \qquad \text{(CS)}$$

Theorem 2.3.8. *Let (X, m) be a semigroup in CD and $X \xrightarrow{m^{\dashv}} X \otimes X$ be the left adjoint map of m. Then $(X, m^{\dashv})$ is a cosemigroup in Sup.*

Proof. Let $m^{\dashv}$ be the left adjoint map of m. Since $m \otimes 1_X$ and $1_X \otimes m$ are morphisms in CD, their respective left adjoint maps $(m \otimes 1_X)^{\dashv}$ and $(1_X \otimes m)^{\dashv}$ exist. Now we first observe: $(m \otimes 1_X)^{\dashv} = m^{\dashv} \otimes 1_X$ and $(1_X \otimes m)^{\dashv} = 1_X \otimes m^{\dashv}$. In fact, since $1_X \leq m \circ m^{\dashv}$ and $m^{\dashv} \circ m \leq 1_{X \otimes X}$, the following relations hold:

$$\begin{aligned} 1_{X \otimes X} &\leq (m \otimes 1_X) \circ (m^{\dashv} \otimes 1_X) \quad \text{and} \quad (m^{\dashv} \otimes 1_X) \circ (m \otimes 1_X) \leq 1_{(X \otimes X) \otimes X}, \\ 1_{X \otimes X} &\leq (1_X \otimes m) \circ (1_X \otimes m^{\dashv}) \quad \text{and} \quad (1_X \otimes m^{\dashv}) \circ (1_X \otimes m) \leq 1_{X \otimes (X \otimes X)}. \end{aligned}$$

Now we derive the following relation from the associativity of m:

$$\begin{aligned} a_{XXX}^{\dashv} \circ (1_X \otimes m^{\dashv}) \circ m^{\dashv} &= (m \circ (1_X \otimes m) \circ a_{XXX})^{\dashv} \\ &= (m \circ (m \otimes 1_X))^{\dashv} = (m^{\dashv} \otimes 1_X) \circ m^{\dashv}. \end{aligned}$$

Since $a_{XXX}^{\dashv}$ coincides with a_{XXX}^{-1}, the diagram (CS) is commutative. □

Motivated by the previous theorem we introduce the following terminology. If (X, m) is a semigroup in CD, then the left adjoint $m^{\dashv}$ of m is also called the *comultiplication of* (X, m).

Example 2.3.9. Let X be a completely distributive lattice and $\wedge$ be the binary meet operation on X. Then $\wedge$ is a meet-preserving bimorphism which determines a meet- and join-preserving map $X \otimes X \xrightarrow{\circledcirc_{\wedge}} X$ given explicitly by (cf. (2.17) and Theorem 2.1.25):

$$\circledcirc_{\wedge}(f) = \bigvee_{u \in X} u \wedge f(u), \qquad f \in X \otimes X.$$

Hence $(X, \circledcirc_{\wedge})$ is a monoid in CD. The corresponding comultiplication $\circledcirc_{\wedge}^{\dashv}$ has the following form:

$$\circledcirc_{\wedge}^{\dashv}(x) = \bigvee_{z \lhd x} z \otimes z, \qquad x \in X, \tag{2.40}$$

where $\lhd$ is the totally below relation in X.
In fact, if $f \in X \otimes X$ and $x \in X$, then the relation $x \leq \circledcirc_{\wedge}(f)$ holds if and only if for every element $z \in X$ with $z \lhd x$ there exists a $u \in X$ with the property $z \leq u \wedge f(u)$. Since f is antitone, this implies that the relation $z \leq f(z)$ holds for $z \lhd x$ — i.e. $\bigvee_{z \lhd x} z \otimes z \leq f$. Since $\lhd$ is approximating and $\circledcirc_{\wedge}$ is join-preserving, we obtain $x = \circledcirc_{\wedge}\big(\bigvee_{z \lhd x} z \otimes z\big)$. Hence formula (2.40) follows.

2.3.1 Balanced Quantales

In the following we give some further properties of non-unital quantales. First we need a bit more terminology.

Definition 2.3.10. Let $(X, *)$ be a quantale and $\top$ be the universal upper bound in X. Then $(X, *)$ is called

- *balanced* if $\top$ of X is idempotent — i.e. $\top * \top = \top$.
- *semi-unital* if the relations $x \leq x * \top$ and $x \leq \top * x$ hold for all $x \in X$.
- *strictly left-sided* if $\top$ is a left unit — i.e. $\top * x = x$ for all $x \in X$.
- *strictly right-sided* if $\top$ is a right unit — i.e. $x * \top = x$ for all $x \in X$.
- *idempotent* if every element x of X is *idempotent* — i.e. $x * x = x$.

Every semi-unital quantale is balanced, and every strictly left-sided (right-sided) quantale is left-sided (right-sided). Further, every semi-unital and left-sided (right-sided) quantale is strictly left-sided (right-sided), but not vice versa (cf. Exercise 2.3.1 (l)). A quantale is integral if and only if it is strictly left-sided and strictly right-sided. Finally, every unital or idempotent quantale is semi-unital, and every idempotent and two-sided quantale is integral.

If a quantale $(X, *)$ is not semi-unital, then it is not difficult to see that $(X, *)$ can be extended to a semi-unital quantale $(\overline{X}, \overline{*})$ as follows:

- The complete lattice $\overline{X}$ arises from X by adding an improper element ω — i.e. $\overline{X} = X \cup \{\omega\}$ with $\omega \notin X$.
- ω acts as a universal upper bound in $\overline{X}$.
- The semigroup operation $\overline{*}$ is defined by:

$$\omega \,\overline{*}\, \omega = \omega, \quad x \,\overline{*}\, y = x * y, \quad \omega \,\overline{*}\, x = (\top * x) \vee x, \quad x \,\overline{*}\, \omega = (x * \top) \vee x, \quad x, y \in X,$$

where $\top$ is the universal upper bound in X. The quantale $(\overline{X}, \overline{*})$ is semi-unital and is called the *semi-unitalization* of $(X, *)$. Obviously, $(X, *)$ is a subquantale of its semi-unitalization. Moreover, $(\overline{X}, \overline{*})$ is strictly left-sided (resp. right-sided) if and only if $(X, *)$ is left-sided (resp. right-sided). Consequently, $(\overline{X}, \overline{*})$ is an integral quantale if and only if $(X, *)$ is two-sided.

Since every quantale on the chain C_2 with two elements is two-sided, it might be meaningful to include the following example.

Example 2.3.11. On the three-chain C_3 there exist precisely two structures of an integral quantale which appear as the semi-unitalization of quantales on C_2. In fact, the semi-unitalization of the trivial quantale on C_2 is (9) in Exercise 2.2.1 and coincides therefore with the three-valued MV-algebra (cf. Sect. 2.7), while the second quantale structure on C_2 is the binary meet on C_2 and its semi-unitalization coincides with the binary meet on C_3 (see (16) in Exercise 2.2.1).

Referring to Example 2.3.26 infra it is hopeless to expect that the tensor product preserves the structure of idempotent quantales. However, the tensor product does preserve the remaining properties formulated in Definition 2.3.10. Since every tensor is a join of elementary tensors and the relations

$$(\top * x) \otimes (\top * y) = (\top \otimes \top) \star (x \otimes y) \quad \text{and} \quad (x * \top) \otimes (y * \top) = (x \otimes y) \star (\top \otimes \top)$$

hold in any tensor product of quantales $(X, *)$ and $(Y, *)$, the following lemma can easily be verified.

Lemma 2.3.12. *The following assertions are valid*:

(i) *The tensor product of balanced quantales is balanced.*
(ii) *The tensor product of semi-unital quantales is semi-unital.*
(iii) *The tensor product of strictly left-sided* (*right-sided*) *quantales is strictly left-sided* (*right-sided*).

If the underlying lattices of the respective quantales contain at least two different elements, then the converse of the assertions in Lemma 2.3.12 is true.

Definition 2.3.13. Let $(X, *)$ and $(Y, *)$ be quantales. A homomorphism $X \xrightarrow{h} Y$ is *strong* if h preserves the respective universal upper bounds — i.e. $h(\top) = \top$.

Every surjective homomorphism is strong. In the context of C^*-algebras we will see that irreducible representations and strong homomorphisms are closely related to each other (cf. Exercise 3.2.8 (a)). Moreover, we have the following result.

Proposition 2.3.14. *Let $(X \otimes Y, \star)$ be the tensor product of the quantales $(X, *)$ and $(Y, *)$. The following assertions hold:*

(i) *If $(Y, *)$ is balanced, then the embedding $X \xrightarrow{j_X} X \otimes Y$* (cf. (2.18)) *is a strong homomorphism.*
(ii) *If $(X, *)$ is balanced, then the embedding $Y \xrightarrow{j_Y} X \otimes Y$* (cf. (2.18)) *is a strong homomorphism.*
(iii) *If $(X, *)$ is non-trivial and $X \xrightarrow{j_X} X \otimes Y$ is a strong homomorphism, then $(Y, *)$ is balanced.*
(iv) *If $(Y, *)$ is non-trivial and $Y \xrightarrow{j_Y} X \otimes Y$ is a strong homomorphism, then $(X, *)$ is balanced.*

Proof. By definition j_X preserves the universal upper bounds. Since $(Y, *)$ is balanced, the homomorphism property of j_X follows from

$$j_X(x_1 * x_2) = (x_1 * x_2) \otimes \top = (x_1 * x_2) \otimes (\top * \top) = j_X(x_1) \star j_X(x_2).$$

Analogously we verify the assertion (ii). With regard to (iii) the homomorphism property of j_X implies

$$(\top * \top) \otimes \top = j_X(\top * \top) = j_X(\top) \star j_X(\top) = (\top * \top) \otimes (\top * \top).$$

Since $(X, *)$ is non-trivial, $(Y, *)$ is balanced. Analogously we verify (iv). □

The previous proposition motivates the introduction of the category `BQuant` of balanced quantales and strong homomorphisms. Obviously `BQuant` is a subcategory of `Quant`. By Lemma 2.3.12 (i) the tensor product of `Sup` induces a monoidal structure on `BQuant` — i.e. `BQuant` is a monoidal category. It is not difficult to verify that `BQuant` is a complete category.

In order to prove the cocompleteness of `BQuant` we proceed as follows.

Theorem 2.3.15. *The category* `BQuant` *has set-indexed coproducts.*

Proof. Obviously, the quantale $\mathbb{1} = \{0, 1\}$ with $1 * 1 = 1$ (i.e. the unit object of the tensor product in `Sup` (cf. Lemma 2.1.14)) is the initial object of `BQuant`. On the other hand, given a nonempty system of balanced quantales $\{(X_i, *) \mid i \in I\}$, we first consider the coproduct $(X_0, *)$ (in the sense of `Quant`) of $\{(X_i, *) \mid i \in I\}$ with the coproduct injections $X_i \xrightarrow{q_i} X_0$. Now we construct the following regular quotient of $(X_0, *)$. For this purpose let $\mathbb{F}$ be the set of all nuclei c on $(X_0, *)$ satisfying the conditions

$$c(\top * \top) = \top \quad \text{and} \quad c(q_i(\top)) = \top, \qquad i \in I.$$

Then the meet c_0 of $\mathbb{F}$ is again an element of $\mathbb{F}$. Let $(Z, \star)$ be the regular quotient of $(X_0, *)$ w.r.t. c_0 and $X_0 \xrightarrow{\pi_{c_0}} Z$ be the quotient homomorphism. Obviously, $(Z, \star)$ is a balanced quantale and $r_i = \pi_{c_0} \circ q_i$ is a strong homomorphism for all $i \in I$. We show that $\big((Z, \star), (r_i)_{i\in I}\big)$ is the coproduct of $\{(X_i, *) \mid i \in I\}$ in the sense of `BQuant`.

Let $(Y, *)$ be a further balanced quantale and $(h_i)_{i\in I}$ be a system of strong homomorphisms $X_i \xrightarrow{h_i} Y$. Then there exists a unique homomorphism $X_0 \xrightarrow{h_0} Y$ such that $h_0 \circ q_i = h_i$ holds for all $i \in I$. Since

$$h_0(q_i(\top)) = h_i(\top) = \top, \qquad i \in I, \tag{2.41}$$

h_0 is a strong homomorphism. It is sufficient to show that h_0 factors through π_{c_0}. Since $(Y, *)$ is a balanced quantale, we conclude from (2.41) that the associated nucleus c_{h_0} of h_0 is an element of $\mathbb{F}$. Hence $c_0 \le c_{h_0}$ follows, and we infer from Lemma 2.2.5 that h_0 factors through π_{c_0}. □

Proposition 2.3.16. *The category* `BQuant` *has set-indexed coequalizers.*

Proof. Since quotients of balanced quantales are again balanced, coequalizers in `BQuant` are computed at the level of `Quant`, where we recall that `Quant` is cocomplete. □

We summarize the previous results as follows.

FACT I. *The category* `BQuant` *is a complete and cocomplete, monoidal category.*

2.3.2 Prime Elements of Quantales

Let $(X, *)$ be a quantale and $\top$ be the universal upper bound of X. An element $p \in X$ is called *prime* in $(X, *)$ if $p \neq \top$ and the following implication holds for all $x_1, x_2 \in X$:

$$x_1 * x_2 \le p \quad \Longrightarrow \quad x_1 * \top \le p \ \text{ or } \ \top * x_2 \le p. \qquad \text{(Prime Property)}$$

In the case of integral quantales the previous definition coincides with the terminology introduced by G. Birkhoff in [14].

Since the implication $(x_1 * \top \le p \ \text{or} \ \top * x_2 \le p \Longrightarrow x_1 * x_2 \le p)$ always holds, the primality can also be expressed by the following equivalence:

$$x_1 * x_2 \leq p \iff x_1 * \top \leq p \text{ or } \top * x_2 \leq p. \qquad \text{(Prime Property}')$$

Lemma 2.3.17. *Let p be a prime element in a quantale $(X, *)$. If $\top * \top \not\leq p$, then $\top \searrow p$ and $p \swarrow \top$ are again prime elements in $(X, *)$.*

Proof. Let p be a prime element in $(X, *)$ with the property $\top * \top \not\leq p$. Obviously, the relations $\top \searrow p \neq \top$ and $p \swarrow \top \neq \top$ hold. In order to verify the primality of $\top \searrow p$ we choose $x_1, x_2 \in X$ with $x_1 * x_2 \leq \top \searrow p$. Since $*$ is associative, we obtain:

$$\top * (x_1 * x_2) = (\top * x_1) * x_2 \leq p.$$

Hence the primality of p and again the associativity of $*$ imply:

$$\top * (x_1 * \top) = (\top * x_1) * \top \leq p \quad \text{or} \quad \top * (\top * x_2) = (\top * \top) * x_2 \leq \top * x_2 \leq p.$$

Thus $x_1 * \top \leq \top \searrow p$ or $\top * x_2 \leq \top \searrow p$ — this means that $\top \searrow p$ is prime in $(X, *)$. Analogously we show that $p \swarrow \top$ is prime in $(X, *)$. □

In order to understand the rôle of prime elements we need some additional terminology. A quantale $(X, *)$ is *semi-integral* if the relation $x_1 * \top * x_2 \leq x_1 * x_2$ holds for all $x_1, x_2 \in X$. Every integral or left-sided (right-sided) quantale is semi-integral.

Lemma 2.3.18. *Let $(X, *)$ be a quantale and let $\odot$ be the multiplication defined by:*

$$x_1 \odot x_2 = x_1 * \top * x_2, \qquad x_1, x_2 \in X. \tag{2.42}$$

Then $(X, \odot)$ is a semi-integral quantale.

Proof. The assertion follows immediately from the associativity of $*$ and the following relation

$$x_1 \odot \top \odot x_2 = x_1 * \top * \top * \top * x_2 \leq x_1 * \top * x_2 = x_1 \odot x_2, \qquad x_1, x_2 \in X.$$

□

The quantale $(X, \odot)$ constructed in Lemma 2.3.18 is called the *semi-integral regularization* of $(X, *)$. It is not difficult to see that the semi-integral regularization gives rise to an endofunctor of the category `BQuant` (cf. Sect. 2.3.1).

The next proposition characterizes the semi-integrality of the tensor product.

Proposition 2.3.19. *Let $(X, *)$ and $(Y, *)$ be quantales. Then the following assertions are equivalent:*

(i) *The tensor product $(X \otimes Y, \star)$ is semi-integral.*
(ii) *One of the following properties occurs:*

  (1) *$(X, *)$ or $(Y, *)$ satisfies the condition $\top * \top * \top = \bot$.*

(2) $(X, *)$ *and* $(Y, *)$ *are semi-integral.*

Proof. Since (1) is equivalent to $(\top \otimes \top) \star (\top \otimes \top) \star (\top \otimes \top) = \bot$, the implication (ii) $\Longrightarrow$ (i) is evident. On the other hand, let us assume that $(X \otimes Y, \star)$ is semi-integral and $\top * \top * \top \neq \bot$ holds in both quantales $(X, *)$ and $(Y, *)$. Then we conclude from

$$\begin{aligned}(\top * \top * \top) \otimes (y_1 * \top * y_2) &= (\top \otimes y_1) \star (\top \otimes \top) \star (\top \otimes y_2)\\ &\leq (\top \otimes y_1) \star (\top \otimes y_2)\\ &= (\top * \top) \otimes (y_1 * y_2)\end{aligned}$$

that $y_1 * \top * y_2 \leq y_1 * y_2$ holds for all $y_1, y_2 \in Y$ — i.e. $(Y, *)$ is semi-integral. Analogously we verify the semi-integrality of $(X, *)$. □

Definition 2.3.20. A left-sided (resp. right-sided) element x of a quantale $(X, *)$ is called *maximal* if $x \neq \top$ and for every further left-sided (resp. right-sided) element $z \in X$ with the properties $z \neq \top$ and $x \leq z$ the relation $x = z$ follows.

Theorem 2.3.21. *In every semi-integral quantale* $(X, *)$ *the following properties hold*:

(i) *Every maximal left-sided element is prime in* $(X, *)$.
(ii) *Every maximal right-sided element is prime in* $(X, *)$.

Proof. We verify property (i). For this purpose we fix a maximal left-sided element p of $(X, *)$ and choose $x, y \in X$ with $x * y \leq p$. If $\top * y \not\leq p$, then the maximality of p implies $(\top * y) \vee p = \top$. Now we use the semi-integrality of $(X, *)$ and the left-sidedness of p and obtain:

$$x * \top = (x * \top * y) \vee (x * p) \leq (x * y) \vee (\top * p) \leq p.$$

Hence p is prime in $(X, *)$. Property (ii) can be verified analogously. □

Comment. Prime elements of left-sided (resp. right-sided) quantales are not necessarily maximal (cf. Exercise 2.3.10 (b)).

The prime elements of the tensor product of quantales can be classified in terms of the prime elements of its respective factors, provided that the left factor is strictly left-sided and the right factor is strictly right-sided.

As a first step we show how prime elements of the respective factors induce prime elements of the tensor product.

Lemma 2.3.22. *Let* $(X, *)$ *and* $(Y, *)$ *be quantales. If* x *is a prime element in* $(X, *)$ *and* y *is a prime element in* $(Y, *)$, *then* $(x \otimes \top) \vee (\top \otimes y)$ *is a prime element in the tensor product* $(X \otimes Y, \star)$.

Proof. Let $x \in X$ and $y \in Y$ be prime elements in $(X, *)$ and $(Y, *)$, respectively. Further, we put $p = (x \otimes \top) \vee (\top \otimes y)$ and observe that the join-reversing map p evaluated at $z \in X$ attains the following value (cf. Exercise 2.1.6):

$$p(z) = \begin{cases} \top, & \text{if } z \leq x, \\ y, & \text{if } z \not\leq x. \end{cases}$$

Now let us assume that $p = (x \otimes \top) \vee (\top \otimes y)$ is not a prime element in $(X \otimes Y, \star)$. Then there exist tensors $f, g \in X \otimes Y$ satisfying the conditions

$$f \star g \leq p, \quad f \star (\top \otimes \top) \not\leq p, \quad (\top \otimes \top) \star g \not\leq p$$

which can be rewritten as follows (cf. (2.36)):

$$\begin{aligned} f \star g &= \bigvee_{z_1, z_2 \in X} (z_1 * z_2) \otimes (f(z_1) * g(z_2)) \leq p, \\ f \star (\top \otimes \top) &= \bigvee_{z \in X} (z * \top) \otimes (f(z) * \top) \not\leq p, \\ (\top \otimes \top) \star g &= \bigvee_{z \in X} (\top * z) \otimes (\top * g(z)) \not\leq p. \end{aligned}$$

Hence there exist elements $z_1, z_2 \in X$ with the properties:

$$z_1 * \top \not\leq x, \quad f(z_1) * \top \not\leq y \qquad \text{and} \qquad \top * z_2 \not\leq x, \quad \top * g(z_2) \not\leq y.$$

Since both x and y are prime, we obtain $z_1 * z_2 \not\leq x$ and $f(z_1) * g(z_2) \not\leq y$. Referring again to the special form of p we conclude from $(z_1 * z_2) \otimes (f(z_1) * g(z_2)) \leq p$ that $p(z_1 * z_2) = \top$ holds. Thus $z_1 * z_2 \leq x$ follows, which is a contradiction to $z_1 * z_2 \not\leq x$. So, the assumption is false and $p = (x \otimes \top) \vee (\top \otimes y)$ is prime in $(X \otimes Y, \star)$. □

In a second step we show that the right adjoint map of a strong homomorphism preserves prime elements. We begin with a preliminary property.

Lemma 2.3.23. *Let $(X, *)$ and $(Y, *)$ be quantales. Further, let $X \xrightarrow{h} Y$ be a homomorphism and $Y \xrightarrow{h^{\vdash}} X$ be its right adjoint map. If q is a prime element in $(Y, *)$ with the property $h^{\vdash}(q) \neq \top$, then $h^{\vdash}(q)$ is a prime element in $(X, *)$.*

Proof. Let q be a prime element in $(Y, *)$ with $h^{\vdash}(q) \neq \top$. We choose $x_1, x_2 \in X$ with $x_1 * x_2 \leq h^{\vdash}(q)$. Then $h(x_1) * h(x_2) \leq q$ follows, and the primality of q implies $h(x_1) * \top \leq q$ or $\top * h(x_2) \leq q$. Since $h(\top) \leq \top$, we obtain $h(x_1 * \top) \leq q$ or $h(\top * x_2) \leq q$ — i.e. $x_1 * \top \leq h^{\vdash}(q)$ or $\top * x_2 \leq h^{\vdash}(q)$. Hence $h^{\vdash}(q)$ is prime in $(X, *)$. □

Corollary 2.3.24. *The right adjoint map of a strong homomorphism preserves prime elements.*

Proof. Let $X \xrightarrow{h} Y$ be a strong homomorphism and $Y \xrightarrow{h^{\vdash}} X$ be its right adjoint map. If q is prime in $(Y, *)$, then it follows from Lemma 2.3.23 that it is sufficient to prove $h^{\vdash}(q) \neq \top$. In fact, since h is strong, $h(h^{\vdash}(q)) \leq q \neq \top$ implies $h^{\vdash}(q) \neq \top$. □

Having made these preparations, we record the following important fact.

Theorem 2.3.25. *Let $(X, *)$ be a strictly left-sided quantale and $(Y, *)$ be a strictly right-sided quantale. If p is a prime element of the tensor product $(X \otimes Y, \star)$, then there exist prime elements x in $(X, *)$ and y in $(Y, *)$ such that p has the following form*

$$p = (x \otimes \top) \vee (\top \otimes y).$$

Proof. Let $X \xrightarrow{j_X} X \otimes Y$ and $Y \xrightarrow{j_Y} X \otimes Y$ be the embeddings defined in (2.18). Since $(X, *)$ and $(Y, *)$ are balanced, j_X and j_Y are strong homomorphisms. Now we fix a prime element $p \in X \otimes Y$. By Corollary 2.3.24 the elements $x = j_X^{\vdash}(p)$ and $y = j_Y^{\vdash}(p)$ are prime elements in $(X, *)$ and $(Y, *)$, respectively. In particular, $j_X(x) \vee j_Y(y) \leq p$ follows. On the other hand, let us choose an elementary tensor $u \otimes v \leq p$. Since $(X, *)$ is strictly left-sided and $(Y, *)$ is strictly right-sided, $u \otimes v$ has the form $u \otimes v = (\top \otimes v) \star (u \otimes \top)$. Hence the primality of p implies:

$$(\top \otimes v) \star (\top \otimes \top) \leq p \quad \text{or} \quad (\top \otimes \top) \star (u \otimes \top) \leq p.$$

Since v is strictly right-sided and u is strictly left-sided, the previous relation is equivalent to the statement:

$$\top \otimes (v * \top) = \top \otimes v \leq p \quad \text{or} \quad (\top * u) \otimes \top = u \otimes \top \leq p.$$

Thus $v \leq y$ or $u \leq x$, and consequently $u \otimes v \leq (x \otimes \top) \vee (\top \otimes y)$ holds. Since every tensor is a join of elementary tensors, the relation $p = (x \otimes \top) \vee (\top \otimes y)$ follows. Hence the assertion is verified. □

As we have seen, Lemma 2.3.22 and Theorem 2.3.25 give a complete description of prime elements of the tensor product of quantales provided its left factor is strictly left-sided and its right factor is strictly right-sided. We illustrate this result with the following example, which will also play a crucial rôle in some later contexts — e.g. in the topological representation of semi-unital quantales (cf. Sect. 2.5.2).

Example 2.3.26. Let $C_3 = \{\bot, a, \top\}$ be the three-chain. On C_3 we consider two non-commutative, idempotent and associative multiplications given by the multiplication tables (see (17) and (18) in Exercise 2.2.1):

$*_\ell$	$\bot$	a	$\top$
$\bot$	$\bot$	$\bot$	$\bot$
a	$\bot$	a	$\top$
$\top$	$\bot$	a	$\top$

$*_r$	$\bot$	a	$\top$
$\bot$	$\bot$	$\bot$	$\bot$
a	$\bot$	a	a
$\top$	$\bot$	$\top$	$\top$

Then $C_3^\ell = (C_3, *_\ell)$ is a left-sided, non-commutative, idempotent quantale, while $C_3^r = (C_3, *_r)$ is a right-sided, non-commutative, idempotent quantale. As a next step we compute the tensor product $C_3^\ell \otimes C_3^r = (C_3 \otimes C_3, \star)$ and observe that $C_3 \otimes C_3$ consists of six elements

$$\top = \top \otimes \top, \quad \alpha = (a \otimes \top) \vee (\top \otimes a), \quad \lambda = a \otimes \top, \quad \rho = \top \otimes a, \quad \beta = a \otimes a,$$
$$\bot = \bot \otimes \bot = \bot \otimes a = a \otimes \bot.$$

The order-theoretic structure of $C_3 \otimes C_3$ can be visualized by the following Hasse diagram:

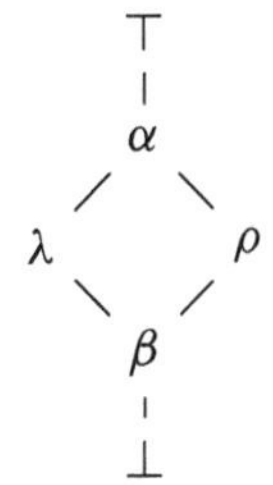

and the multiplication table has the form (see also Example 8.1 in [32])

$\star$	$\bot$	β	λ	ρ	α	$\top$
$\bot$	$\bot$	$\bot$	$\bot$	$\bot$	$\bot$	$\bot$
β	$\bot$	β	β	ρ	ρ	ρ
λ	$\bot$	λ	λ	$\top$	$\top$	$\top$
ρ	$\bot$	β	β	ρ	ρ	ρ
α	$\bot$	λ	λ	$\top$	$\top$	$\top$
$\top$	$\bot$	λ	λ	$\top$	$\top$	$\top$

The unique non-elementary tensor α is the unique non-idempotent element in $(C_3 \otimes C_3, \star)$. Further, the elementary tensor β is neither left-sided nor right-sided, while λ is left-sided, but not right-sided and ρ is right-sided, but not left-sided. Moreover, the tensor product $(C_3 \otimes C_3, \star)$ is semi-integral (cf. Proposition 2.3.19).

Since $\bot$ and a are the prime elements of C_3^ℓ, respectively C_3^r, we conclude from Lemma 2.3.22 and Theorem 2.3.25 that $(C_3 \otimes C_3, \star)$ has four prime elements, namely, α, λ, ρ and $\bot$.

In the following we construct the "spatial reflection" of quantales. We begin with a simple property.

Lemma 2.3.27. *Let $(X, *)$ and $(Y, *)$ be quantales. Further, let $X \xrightarrow{h} Y$ be a surjective homomorphism and $Y \xrightarrow{h^\vdash} X$ be its right adjoint map. If p is a prime element in $(X, *)$ with the property $h^\vdash(h(p)) = p$, then $h(p)$ is a prime element in $(Y, *)$.*

Proof. Let p be a prime element in $(X, *)$ satisfying the condition $h^\vdash(h(p)) = p$. Since $h^\vdash(\top) = \top$, we first notice that $h(p) \neq \top$. Now we choose $u, v \in Y$ with $u * v \leq h(p)$. Since $h^\vdash$ is a closed map (cf. Proposition 2.2.2), the relation

$$h^{\vdash}(u) * h^{\vdash}(v) \leq h^{\vdash}(h(p)) = p$$

follows. We apply the primality of p and obtain $h^{\vdash}(u) * \top \leq p$ or $\top * h^{\vdash}(v) \leq p$. Finally, $u * \top \leq h(p)$ or $\top * v \leq h(p)$ follows form the surjectivity of h. Hence $h(p)$ is prime in $(Y, *)$. □

Lemma 2.3.28. *Let $(X, *)$ be a quantale. Then the map $X \xrightarrow{c_s} X$ defined by*

$$c_s(x) = \bigwedge\{p \in X \mid p \text{ prime in } (X, *), x \leq p\}, \quad x \in X \tag{S}$$

is a nucleus on $(X, *)$.

Proof. The map c_s determined by (S) is obviously a closure operator on $(X, \leq)$. In order to prove that c_s is a nucleus we fix elements $x, y \in X$ and choose an arbitrary prime element p in $(X, *)$ with $x * y \leq p$. If $\top * \top \leq p$, then $c_s(x) * c_s(y) \leq p$ is obvious. Hence we assume $\top * \top \not\leq p$. Because of the primality of p we obtain $x * \top \leq p$ or $\top * y \leq p$. Without loss of generality we can assume $x * \top \leq p$, which means $x \leq p \swarrow \top$. By Lemma 2.3.17 the element $p \swarrow \top$ is prime. Hence the relation

$$c_s(x) * c_s(y) \leq (p \swarrow \top) * \top \leq p$$

follows — this means that c_s is a nucleus. □

Corollary 2.3.29. *Let $(X, *)$ be a quantale and c_s be the nucleus on $(X, *)$ determined by* (S)*. Further, let $(c_s(X), \star)$ be the regular quotient of $(X, *)$ w.r.t. c_s. Then every element of the quantale $(c_s(X), \star)$ is a meet of prime elements of $(c_s(X), \star)$.*

Proof. Let $X \xrightarrow{\pi_{c_s}} c_s(X)$ be the corresponding quotient homomorphism. Since all prime elements p of $(X, *)$ are fixpoints of c_s (i.e. $p = \pi_{c_s}^{\vdash}(\pi_{c_s}(p)))$ and π_{c_s} is surjective, we conclude from Lemma 2.3.27 that every prime element in $(X, *)$ is also a prime element in $(c_s(X), \star)$. Hence the construction of c_s ensures that every element of $c_s(X)$ is a meet of prime elements in $(c_s(X), \star)$. □

Motivated by Corollary 2.3.29 we introduce the following terminology. A quantale $(X, *)$ is called *spatial* if the set of prime elements is order generating — i.e. every element $x \in X \setminus \{\top\}$ is a meet of prime elements of $(X, *)$. A simple example of a spatial, non-idempotent, but semi-integral quantale is given by the tensor product $C_3^{\ell} \otimes C_3^{r}$ (cf. Example 2.3.26).

Moreover, the regular quotient of $(X, *)$ constructed in Lemma 2.3.28 and Corollary 2.3.29 is called the *spatial reflection of* $(X, *)$.

The next theorem justifies the chosen terminology.

Theorem 2.3.30. *Let $(X, *)$ be a quantale and $(Y, *)$ be a spatial quantale. Further, let $(X_s, \star)$ be the spatial reflection of $(X, *)$ with the quotient homomorphism η_X. For every homomorphism $X \xrightarrow{h} Y$ there exists a unique homomorphism $X_s \xrightarrow{\ulcorner h \urcorner} Y$ making the following diagram commutative:*

$$\begin{array}{ccc} X & \xrightarrow{\;h\;} & Y \\ {\scriptstyle \eta_X}\big\downarrow & \nearrow_{\ulcorner h \urcorner} & \\ X_s & & \end{array}$$

Proof. Let c_s be the nucleus determined by (S). In order to show that h factors (uniquely) through η_X it is sufficient to verify that the nucleus $c_h = h^{\vdash} \circ h$ associated with h satisfies the condition $c_s \le c_h$ (cf. Lemma 2.2.5). Without loss of generality, we can choose $x \in X$ with $c_h(x) \neq \top$. Since $(Y, *)$ is spatial and the right adjoint map $h^{\vdash}$ of h is meet-preserving, we derive the following relation from Lemma 2.3.23:

$$\begin{aligned} c_h(x) = h^{\vdash}(h(x)) &= \bigwedge\{h^{\vdash}(q) \mid h(x) \le q,\ q \text{ prime in } (Y, *)\} \\ &= \bigwedge\{p \mid c_h(x) \le p,\ p \text{ prime in } (X, *)\}. \end{aligned}$$

Hence $c_h(x) = c_s(c_h(x))$ follows for all $x \in X$ — i.e. $c_s \le c_h$. □

Let $\mathtt{SpQ}$ be the category of spatial quantales and homomorphisms. Then Theorem 2.3.30 means that $\mathtt{SpQ}$ is a reflective subcategory of the category of quantales.

2.3.3 Involutive Quantales

A triple $(X, *, \iota_X)$ is an *involutive quantale* if $(X, *)$ is a quantale and $(X, *, \iota_X)$ is an involutive prequantale. In order to simplify the notation we sometimes also denote the involution ι_X of an involutive quantale by $'$. An element x of X is *hermitian* if x is a fixed point of the involution — i.e. $x' = x$. Obviously, the universal bounds of X are always hermitian.

The next example shows that every complete lattice with an order-reversing involution gives rise to an involutive and unital quantale (cf. Example (5) in [93]).

Example 2.3.31. Let X be a complete lattice and $([X, X], \circ, 1_X)$ be the (non-commutative) unital quantale constructed in Example 2.3.2. Further, we assume that there exists an *order-reversing involution* $x \longmapsto x^{\perp}$ on X — this is an antitone self-map of X with the property $(x^{\perp})^{\perp} = x$ for all $x \in X$. Since X is complete, for every order-reversing involution on X the relation

$$\bigvee A = ((\bigvee A)^{\perp})^{\perp} \ge \big(\bigwedge_{x \in A} x^{\perp}\big)^{\perp} \ge \bigvee A, \qquad A \subseteq X$$

holds. Hence, for all subsets A of X, an order-reversing involution always satisfies the following properties:

$$\big(\bigvee A\big)^{\perp} = \bigwedge_{x \in A} x^{\perp} \quad \text{and} \quad \big(\bigwedge A\big)^{\perp} = \bigvee_{x \in A} x^{\perp}. \qquad \text{(De Morgan's Laws)}$$

On this basis we now construct an order-preserving involution $'$ on $[X, X]$ as follows. Let $X \xrightarrow{f} X$ be a join-preserving map. We define $X \xrightarrow{f'} X$ by $f'(x) = (f^{\vdash}(x^{\perp}))^{\perp}$ for all $x \in X$, where $f^{\vdash}$ is the right adjoint map of f. Obviously f' is join-preserving. Since the chain of equivalences

$$f'(x) \leq y \iff y^{\perp} \leq f^{\vdash}(x^{\perp}) \iff f(y^{\perp}) \leq x^{\perp} \iff x \leq (f(y^{\perp}))^{\perp}$$

holds for each $x, y \in X$, we obtain $(f')^{\vdash}(y) = (f(y^{\perp}))^{\perp}$ for each $y \in X$. Hence $f'' = f$ follows. In fact, the correspondence $f \longmapsto f'$ is an order-preserving involution on $[X, X]$. Since adjoint situations compose, for all $f, g \in [X, X]$ the relation $(f \circ g)' = g' \circ f'$ is valid—i.e. $f \longmapsto f'$ is an anti-homomorphism on $([X, X], \circ)$. To sum up, $([X, X], \circ, 1_X, ')$ is an involutive and unital quantale.

Comment. An interesting example of a complete lattice with an order-reversing involution is the complete lattice of all closed subspaces of a Hilbert space provided with the formation of orthogonal complements.

An element x of an involutive and unital quantale $(X, *, e, ')$ is called *unitary* if $x * x' = x' * x = e$. Obviously unitary elements form a group. Hence there exists a functor from the category of involutive and unital quantales to the category of groups which has a left adjoint functor determined by the power set functor. In this sense we obtain further examples of involutive and unital quantales (see also Exercise 2.3.12).

Before we continue, we return to Example 2.3.31 and give a characterization of unitary join-preserving maps.

Example 2.3.32. Let X be a complete lattice equipped with an order-reversing involution $x \longmapsto x^{\perp}$. Further, let $([X, X], \circ, 1_X, ')$ be the involutive and unital quantale constructed in Example 2.3.31. A join-preserving map $X \xrightarrow{f} X$ is unitary if and only if f is bijective and preserves the given order-reversing involution — i.e. $f(x^{\perp}) = f(x)^{\perp}$ for all $x \in X$.

In fact, if $f \circ f' = 1_X$ and $f' \circ f = 1_X$ hold, then f and its right adjoint map $f^{\vdash}$ are obviously surjective. Hence we infer from Exercise 1.3.2 (a) that f is bijective, and $f^{\vdash}$ coincides with f^{-1}. Since $f' \circ f = 1_X$, we now obtain

$$f(x)^{\perp} = f\big(f^{\vdash}\big(f(x)^{\perp}\big)\big) = f\big(\big(f'(f(x))\big)^{\perp}\big) = f(x^{\perp})$$

for all $x \in X$ — i.e. f preserves the order-reversing involution.

On the other hand, if a join-preserving map $X \xrightarrow{f} X$ is bijective, then again $f^{\vdash} = f^{-1}$ follows. If we now assume that f also preserves the given order-reversing involution $x \longmapsto x^{\perp}$, then f is obviously unitary.

Since the universal upper bound of an involutive quantale is always hermitian, the semi-integral regularization of involutive quantales (cf. Lemma 2.3.18) is again an involutive quantale.

The next theorem explains in what sense the tensor product generates involutive quantales. For this purpose we first refer to the embeddings $X \xrightarrow{j_X} X \otimes Y$ and $Y \xrightarrow{j_Y} X \otimes Y$ defined in (2.18).

Theorem 2.3.33. *Let $(X, *)$ be a strictly left-sided quantale and $(Y, *)$ be a strictly right-sided quantale. Further, let $X \underset{\vartheta_Y}{\overset{\vartheta_X}{\rightleftarrows}} Y$ be a pair of join-preserving anti-homomorphisms with the property $\vartheta_Y \circ \vartheta_X = 1_X$ and $\vartheta_X \circ \vartheta_Y = 1_Y$. Then there exists a unique isotone involution on $(X \otimes Y, \star)$ such that $(X \otimes Y, \star, \iota)$ is an involutive quantale and the following diagram is commutative:*

$$\begin{array}{ccccc} X \otimes Y & \xrightarrow{\iota} & X \otimes Y & \xrightarrow{\iota} & X \otimes Y \\ {\scriptstyle j_X}\uparrow & & {\scriptstyle j_Y}\uparrow & & \uparrow{\scriptstyle j_X} \\ X & \xrightarrow{\vartheta_X} & Y & \xrightarrow{\vartheta_Y} & X \end{array} \tag{I}$$

Proof. (Uniqueness) Let ι be an involution on the tensor product $X \otimes Y$ such that $(X \otimes Y, \star, \iota)$ is an involutive quantale and the diagram (I) is commutative. Since $(X, *)$ is strictly left-sided and $(Y, *)$ is strictly right-sided, the relation

$$\begin{aligned} \iota(x \otimes y) = \iota\big((\top \otimes y) \star (x \otimes \top)\big) &= \iota(x \otimes \top) \star \iota(\top \otimes y) \\ &= j_Y(\vartheta_X(x)) \star j_X(\vartheta_Y(y)) \\ &= (\top \otimes \vartheta_X(x)) \star (\vartheta_Y(y) \otimes \top) \\ &= \vartheta_Y(y) \otimes \vartheta_X(x) \end{aligned}$$

follows for all $x \in X$ and $y \in Y$. Since ι is join-preserving and every tensor is a join of elementary tensors, ι is uniquely determined by (I).
(Existence). Since ϑ_X and ϑ_Y are join-preserving, we conclude from the universal property of the tensor product that there exists a unique join-preserving map $X \otimes Y \xrightarrow{\iota} X \otimes Y$ satisfying the following condition:

$$\iota(x \otimes y) = \vartheta_Y(y) \otimes \vartheta_X(x), \qquad x \in X,\ y \in Y. \tag{2.43}$$

We show that ι is an anti-homomorphism and an involution. For this purpose it is sufficient to consider elementary tensors only. Therefore we choose $x, u \in X$ and $y, v \in Y$ and observe:

$$\begin{aligned} \iota\big((x \otimes y) \star (u \otimes v)\big) &= \iota\big((x * u) \otimes (y * v)\big) \\ &= \vartheta_Y(y * v) \otimes \vartheta_X(x * u) \\ &= \big(\vartheta_Y(v) * \vartheta_Y(y)\big) \otimes \big(\vartheta_X(u) * \vartheta_X(x)\big) \\ &= \big(\vartheta_Y(v) \otimes \vartheta_X(u)\big) \star \big(\vartheta_Y(y) \otimes \vartheta_X(x)\big) \\ &= \iota(u \otimes v) \star \iota(x \otimes y). \end{aligned}$$

Hence ι is an anti-homomorphism. Since $\vartheta_X \circ \vartheta_Y = 1_Y$ and $\vartheta_Y \circ \vartheta_X = 1_X$, the property $\iota \circ \iota = 1_X$ is obvious. Thus $(X \otimes Y, \star, \iota)$ is an involutive quantale.
Since both anti-homomorphisms ϑ_X and ϑ_Y preserve the universal upper bounds, we obtain for all $x \in X$:

$$\iota(j_X(x)) = \iota(x \otimes \top) = \vartheta_Y(\top) \otimes \vartheta_X(x) = j_Y(\vartheta_X(x)).$$

Analogously, we verify $\iota \circ j_Y = j_X \circ \vartheta_Y$. Hence the commutativity of diagram (I) follows. □

Example 2.3.34. Let C_3^ℓ be the left-sided, non-commutative and idempotent chain of three elements and C_3^r be the right-sided, non-commutative and idempotent chain of three elements (cf. Example 2.3.26). Then we observe that the identity map of C_3 gives rise to a pair of anti-homomorphisms between C_3^ℓ and C_3^r. Hence we conclude from Theorem 2.3.33 that there exists a unique involution $'$ on $C_3 \otimes C_3$ satisfying the following conditions:

(i) $(C_3 \otimes C_3, \star, ')$ is an involutive quantale.
(ii) $(a \otimes \top)' = \top \otimes a$ and $(\top \otimes a)' = a \otimes \top$.

In particular, the tensors $(a \otimes \top) \vee (\top \otimes a)$ and $a \otimes a$ are hermitian. Moreover, the involutive quantale $(C_3 \otimes C_3, \star, ')$ is isomorphic to the semi-integral regularization of $([C_3, C_3], \circ, 1_{C_3}, ')$ (cf. Example 2.3.31).

We finish this subsection with a brief discussion related to the commutativity of quantales. A quantale $(X, *)$ is *commutative* if $(X, *, 1_X)$ is an involutive quantale — this means $x_1 * x_2 = x_2 * x_1$ for all $x_1, x_2 \in X$. Hence a quantale is commutative if and only if its multiplication coincides with its transposed multiplication.

Theorem 2.3.35. *The coproduct of two commutative and unital quantales coincides with their tensor product.*

Proof. Let $(X, *, e_X)$ and $(Y, *, e_Y)$ be unital and commutative quantales, and let $(X \otimes Y, \star)$ be their tensor product. We recall that the unit e and the multiplication $\star$ of the tensor product have the following form (cf. (2.36):

$$e = e_X \otimes e_Y \quad \text{and} \quad (x \otimes y) \star (u \otimes v) = (x * u) \otimes (y * v), \qquad x, u \in X, \quad y, v \in Y.$$

Then $(X \otimes Y, \star, e)$ is a commutative and unital quantale, and the join-preserving maps $X \xrightarrow{q_X} X \otimes Y$ and $Y \xrightarrow{q_Y} X \otimes Y$ defined by

$$q_X(x) = x \otimes e_Y, \quad q_Y(y) = e_X \otimes y, \qquad x \in X, \ y \in Y$$

are obviously unital homomorphisms. We show that $((X \otimes Y, \star), (q_X, q_Y))$ is the coproduct of $(X, *)$ and $(Y, *)$ in the category of commutative and unital quantales. For this purpose let $(Z, *, e_Z)$ be a further commutative and unital quantale with unital homomorphisms $X \xrightarrow{h_X} Z$ and $Y \xrightarrow{h_Y} Z$. We have to show that there exists unique unital homomorphism $X \otimes Y \xrightarrow{k} Z$ with $h_X = k \circ q_X$ and $h_Y = k \circ q_Y$. Obviously, every elementary tensor $x \otimes y$ of $X \otimes Y$ can be written as:

$$x \otimes y = (x \otimes e_Y) \star (e_X \otimes y) = (e_X \otimes y) \star (x \otimes e_Y).$$

If we now assume the existence of a unital homomorphism $X \otimes Y \xrightarrow{k} Z$ with the properties $h_X = k \circ q_X$ and $h_Y = k \circ q_Y$, then we obtain

$$k(x \otimes y) = k\big((x \otimes e_Y) \star (e_X \otimes y)\big) = k(x \otimes e_Y) * k(e_X \otimes y) = h_X(x) * h_Y(y).$$

Since k is join-preserving, k is uniquely determined by h_X and h_Y.
In order to verify the existence of k we apply the universal property of the tensor product in Sup and introduce k by:

$$k(x \otimes y) = h_X(x) * h_Y(y), \qquad x \in X,\ y \in Y.$$

Since the multiplication of Z is associative and commutative, we obtain:

$$\begin{aligned} k\big((x \otimes y) \star (u \otimes v)\big) &= k\big((x * u) \otimes (y * v)\big) \\ &= h_X(x * u) * h_Y(y * v) \\ &= h_X(x) * h_X(u) * h_Y(y) * h_Y(v) \\ &= h_X(x) * h_Y(y) * h_X(u) * h_X(v) \\ &= k(x \otimes y) * k(u \otimes v). \end{aligned}$$

Hence k is a homomorphism. Finally, it is easily seen that k is unital and satisfies the relations $h_X = k \circ q_X$ and $h_Y = k \circ q_Y$. □

Example 2.3.36. A binary operation $*$ (in the sense of Set) on the real unit interval $[0, 1]$ is called a *left-continuous t-norm* if $([0, 1], *)$ is a commutative and integral quantale (cf. [67]). Motivated by the need for a triangle inequality for probabilistic metric spaces B. Schweizer and A. Sklar formulated a binary operation τ_* on the complete lattice Δ^+ of all left-continuous probability distribution function as follows (cf. [104]):

$$\big(\tau_*(f, g)\big)(r) = \sup\{f(x) * g(y) \mid x + y = r\}, \qquad r \in [0, +\infty].$$

It follows immediately from (2.39) that τ_* coincides with $\star$. Hence we conclude from Theorem 2.3.35 that (Δ^+, τ_*) is the coproduct of the commutative and unital quantales $([0, +\infty]^{op}, +)$ and $([0, 1], *)$.

2.3.4 Topological Semigroups as Quantales on [0, 1]

In this subsection we prove the Mostert–Shields theorem (cf. [79, 89]) in the case of [0, 1], which can be stated as follows:

> If a topological semigroup on [0, 1] is an integral quantale, then the semigroup multiplication is commutative. More specifically, every non-idempotent topological semigroup on [0, 1] with the boundary points 0 and 1 as zero and unit element is an ordinal sum whose summands are isomorphic either to the usual *product* or to *Łukasiewicz arithmetic conjunction* given by:
> $$x *_{Ł} y = \max(x + y - 1, 0), \qquad x, y \in [0, 1]$$
> or to a restriction of the binary *minimum*.

Throughout this subsection $([0, 1], *)$ will always be an integral quantale. First we note that $*$ is continuous (i.e. $([0, 1], *)$ is a topological semigroup) if and only if $*$ is nonempty meet-preserving in each variable separately. If $*$ is continuous, then $([0, 1], *)$ is called a *continuous quantale*. In this sense, every continuous quantale is integral.

Lemma 2.3.37. *In every continuous quantale* $([0, 1], *)$ *the following properties hold*:

(i) *For* $n \in \mathbb{N}$ *and* $x \in [0, 1]$ *there exists a* $z \in [0, 1]$ *such that* x *coincides with the* n^{th} *power of* z *w.r.t.* $*$ *— i.e.* $x = z^n := z * \overset{n}{\cdots} * z$.
(ii) *If* $x \leq y$, *then there exist* $u, v \in [0, 1]$ *with* $x = u * y = y * v$.

Proof. Since 1 is the unit and [0, 1] is a connected topological space, the properties (i) and (ii) follow from the continuity of $*$ and the intermediate mean value theorem. □

As an immediate corollary of Lemma 2.3.37 (i) we obtain that every continuous quantale has roots — i.e. for every $n \in \mathbb{N}$ there exists a unique isotone map $x \longmapsto x^{1/n}$ satisfying the following conditions:

(R1) $z^n \leq x \iff z \leq x^{1/n}$.
(R2) $x = \left(x^{1/n}\right)^n$ for all $x \in [0, 1]$.

Moreover, the assertion (ii) in Lemma 2.3.37 implies:

$$x * \left(x \searrow (x \wedge y)\right) = x \wedge y = \left((x \wedge y) \swarrow x\right) * x, \qquad x, y \in [0, 1]. \tag{2.44}$$

With regard to cancellation properties we have the following situation.

Corollary 2.3.38. *Let* $([0, 1], *)$ *be a continuous quantale. If* 1 *and* 0 *are the only idempotent elements of* [0, 1], *then for all* $x, y, z \in [0, 1]$ *the following implications hold*:

$$0 \neq x * y = x * z \implies y = z \quad \textit{and} \quad 0 \neq y * x = z * x \implies y = z.$$

Proof. If $0 \neq x * y = x * z$, then we infer from (2.44) that:

$$0 \neq x * y * \big(y \searrow (y \wedge z)\big) = x * (y \wedge z) = x * y = x * z.$$

Since $*$ is nonempty meet-preserving in each variable separately, the element

$$e = \bigwedge_{n \in \mathbb{N}} \big(y \searrow (y \wedge z)\big)^n$$

is idempotent and satisfies the property $0 \neq x * y * e = x * y$. Hence $y \searrow (y \wedge z)$ necessarily coincides with 1 — i.e. $y \leq z$. If we interchange the rôle of y and z, then we also obtain $z \leq y$ — i.e. $y = z$. Analogously we verify the second implication. □

Corollary 2.3.39. *Let* $([0, 1], *)$ *be a continuous quantale. If* 0 *and* 1 *are the only idempotent elements of* $[0, 1]$*, then for all* $x, y \in (0, 1)$ *and for all* $n, m \in \mathbb{N}$ *the following properties hold*:

(i) $0 \neq x^m = x^n \implies m = n$.
(ii) $0 \neq x^n = y^n \implies x = y$.
(iii) $\big(x^{1/n}\big)^{1/m} = x^{1/(n \cdot m)}$.

Proof. Without loss of generality, we can assume $n \leq m$ and consider the case $0 \neq x^n = x^m$ where $x \neq 1$. Then $x^{m-n} = 1$ follows from Corollary 2.3.38. Hence we obtain $n = m$. Now we choose $x, y \in (0, 1]$ with $0 \neq x^n = y^n$ and assume, without loss of generality, $x \leq y$. Then $0 \neq y^n * (y \searrow x) = y^{n-1} * x \geq x^n$ follows from Lemma 2.3.37 (ii). Since $x^n = y^n$, the previous relation means $0 \neq y^n * (y \searrow x) = y^n$. Now we introduce an idempotent element e of $[0, 1]$ by $e = \bigwedge_{m \in \mathbb{N}} (y \searrow x)^m$ and observe that $0 \neq y^n * e = y^n$. Hence e necessarily coincides with 1 — i.e. $x = y$. Finally, referring to (R2), it is easily seen that the relation

$$\big((x^{1/n})^{1/m}\big)^{n \cdot m} = x = \big(x^{1/(n \cdot m)}\big)^{n \cdot m}$$

holds. Hence (iii) follows immediately from (ii). □

Theorem 2.3.40. *Let* $([0, 1], *)$ *be a continuous quantale such that the relation* $0 < x * x < x$ *holds for all* $x \in (0, 1)$*. Then* $([0, 1], *)$ *is isomorphic to the quantale* $([0, +\infty]^{op}, +)$.

Proof. We fix $x \in (0, 1)$ and choose natural numbers $i, j, m, n \in \mathbb{N}$ with the property $i \cdot n = m \cdot j$. Then we obtain from Corollary 2.3.39 (iii):

$$\big(x^{1/n}\big)^m = \big(\big(x^{1/n}\big)^{1/j}\big)^{j \cdot m} = \big(x^{1/(n \cdot j)}\big)^{m \cdot j} = \big(x^{1/(n \cdot j)}\big)^{i \cdot n} = \big(x^{1/j}\big)^i.$$

Thus we can assign a unique element $x^r \in [0, 1]$ to each rational number r in $[0, +\infty]$ according to the following rule:

$$x^r = \big(x^{1/n}\big)^m \quad \text{where} \quad r = \tfrac{m}{n}. \tag{2.45}$$

Obviously $x^r * x^s = x^{r+s}$ holds for all rational numbers $r, s \in (0, +\infty)$. Now we extend the antitone correspondence $r \longmapsto x^r$ to a join-preserving map $[0, +\infty]^{op} \xrightarrow{\varphi} [0, 1]$ by:

$$\varphi(0) = 1, \quad \varphi(+\infty) = 0, \quad \varphi(\alpha) = \sup\{x^r \mid \alpha < r\}, \qquad \alpha \in (0, +\infty).$$

It is easily seen that φ also preserves the multiplications $+$ and $*$, respectively. Hence φ is a unital homomorphism.

We show that $[0, +\infty]^{op} \xrightarrow{\varphi} [0, 1]$ is also meet-preserving. Since 0 and 1 are the only idempotent elements in $[0, 1]$, we first notice that $\sup_{n \in \mathbb{N}} x^{1/n} = 1$. Now we conclude from

$$x^r * x^{1/n} = x^{r+\frac{1}{n}} \leq x^{r'} \leq x^r, \qquad r < \alpha < r' < r + \tfrac{1}{n}$$

that for all $n \in \mathbb{N}$ the relation

$$(\inf\{x^r \mid r < \alpha\}) * x^{1/n} \leq \varphi(\alpha) \leq \inf\{x^r \mid r < \alpha\}$$

holds. Hence

$$\varphi(\alpha) = \inf_{r < \alpha} x^r$$

follows for all $\alpha \in (0, +\infty)$. To sum up, φ is continuous. Then, by the intermediate value theorem, φ is surjective.

In order to verify the injectivity of φ we proceed as follows. Since rational numbers of $[0, +\infty]$ are dense in $[0, +\infty]$, it is sufficient to verify the implication:

$$\left(x^{1/j}\right)^i = \left(x^{1/n}\right)^m \implies \tfrac{i}{j} = \tfrac{m}{n}. \tag{2.46}$$

If $\left(x^{1/j}\right)^i = \left(x^{1/n}\right)^m$ holds, then we derive the following relation from (R2) and Corollary 2.3.39 (iii):

$$\left(x^{1/(j \cdot n)}\right)^{i \cdot n} = \left(x^{1/j}\right)^i = \left(x^{1/n}\right)^m = \left(x^{1/(n \cdot j)}\right)^{m \cdot j}.$$

Since all powers x^k of x are different from 0, we infer from Corollary 2.3.39 (i) that $i \cdot n = m \cdot j$ holds. Hence (2.46) is verified.

To sum up, we have shown that φ is an isomorphism of unital quantales. In particular, $*$ is commutative. □

In the next step of the classification we need quantales with nilpotent elements. Here we consider the interval $[0, 2]$ provided with the dual order and the multiplication $\oplus$ defined by:

$$x \oplus y = \min(x + y, 2), \qquad x, y \in [0, 2].$$

Then $([0,2]^{op}, \oplus)$ is a unital quantale with unit 0 and zero element 2. Since $1 \oplus 1 = 2$, the element 1 is nilpotent.

Theorem 2.3.41. *Let* $([0,1], *)$ *be a continuous quantale satisfying the condition* $x * x < x$ *for all* $x \in (0,1)$. *If there exists an element* $x_0 \in (0,1)$ *with* $x_0 * x_0 = 0$, *then* $([0,1], *)$ *is isomorphic to* $([0,2]^{op}, \oplus)$.

Proof. Let x_0 be an element in $(0,1)$ with $x_0 * x_0 = 0$. Then we put

$$z = \sup\{x \in [0,1] \mid x * x = 0\}$$

and observe that $z * z = 0$ and $z^r \neq 0$ holds for all $r \in (0,2)$ (cf. (2.45)). Hence the essential arguments from the proof of Theorem 2.3.40 carry over, and there exists an order isomorphism $[0,2]^{op} \xrightarrow{\psi} [0,1]$ defined by:

$$\psi(0) = 1, \quad \psi(2) = 0, \quad \psi(\alpha) = \sup\{z^r \mid \alpha < r\}, \qquad \alpha \in (0,2).$$

We show that ψ is a homomorphism. For this purpose we distinguish the following cases. If $\alpha + \beta < 2$, then $\psi(\alpha \oplus \beta) = \psi(\alpha) * \psi(\beta)$ follows from the definition of z^r with $0 < r < 2$ (cf. (2.45)). If $2 \leq \alpha + \beta$, then we obtain:

$$\psi(\alpha) * \psi(\beta) \leq \sup\{z^r \mid 2 < r\} \leq z^2 = 0.$$

Hence $0 = \psi(2) = \psi(\alpha \oplus \beta) = \psi(\alpha) * \psi(\beta)$ follows. To sum up we have verified that ψ is an isomorphism between unital quantales. □

Since the map $\alpha \longmapsto \ln(-\alpha)$ is a unital homomorphism from $([0,+\infty]^{op}, +)$ to $[0,1]$ provided with the usual *product* and the map $\alpha \longmapsto 1 - \frac{\alpha}{2}$ is a unital homomorphism from $([0,2]^{op}, \oplus)$ to $[0,1]$ provided with Łukasiewicz arithmetic conjunction, we can summarize the results of Theorems 2.3.40 and 2.3.41 as follows.

FACT II. *Let* $([0,1], *)$ *be a continuous quantale with the property* $x * x < x$ *for all* $x \in (0,1)$. *Then* $([0,1], *)$ *is isomorphic to* $([0,1], \star)$, *where* $\star$ *is either the usual product or the Łukasiewicz arithmetic conjunction.*

Now we complete the classification when we consider an arbitrary continuous quantale $([0,1], *)$. Since idempotent elements play a special rôle, we distinguish the following cases:
(1) If every element of $[0,1]$ is idempotent, then $*$ coincides with the binary minimum.
(2) If there exists at least one non-idempotent element of $[0,1]$, then we define $I = \{x \in [0,1] \mid x * x = x\}$ and observe that I is a compact subset of $[0,1]$. Hence its complement, $\{x \in [0,1] \mid x * x < x\}$, is an open, nonempty subset and can be read as a countable union of open intervals (a_n, b_n). Then the restriction of $*$ to $[a_n, b_n]$ is either isomorphic to the usual product or to Łukasiewicz arithmetic conjunction. If the pair (x, y) is not contained in some rectangle $[a_n, b_n] \times [a_n, b_n]$, then $*$ operates on (x, y) as the minimum — i.e. $x * y = \min(x, y)$. Hence we can summarize

the situation as follows. A continuous quantale having at least one non-idempotent element is the ordinal sum of those summands which are isomorphic either to the usual product, or to Łukasiewicz arithmetic conjunction or to a restriction of the binary minimum. In this sense we have a complete classification of all continuous quantales. In particular, every continuous quantale is commutative, and consequently the multiplication is a continuous t-norm (cf. [67]).

Exercises

2.3.1. (See Exercise 2.2.1) Let C_3 be chain with three elements. Prove:

(a) There exist precisely 12 quantale structures on the three-chain C_3, namely (1)–(2), (6)–(9) and (15)–(20).
(b) 6 of them are semi-unital, namely (9) and (16)–(20).
(c) 3 of them are unital, namely (9), (16) and (19).
(d) 8 of them are commutative, namely (1)–(2), (6), (9) and (15)–(16) and (19)–(20).
(e) 4 of them are idempotent, namely (16)–(19).
(f) 9 of them are left-sided, namely (1)–(2), (6)–(9) and (15)–(17).
(g) 9 of them are right-sided, namely (1)–(2), (6)–(9), (15)–(16) and (18).
(h) 8 of them are two-sided, namely (1)–(2), (6)–(9) and (15)–(16).
(i) 9 of them are balanced, namely (6)–(9) and (16)–(20).
(k) 2 of them are integral (cf. Example 2.3.11), namely (9) and (16).
(l) 2 of them are non-commutative and two-sided, namely (7) and (8). More precisely, (8) is strictly left-sided and not semi-unital, while (7) is strictly right-sided and not semi-unital.
(m) One of them is non-commutative, left-sided and not right-sided (17), and another one is non-commutative, right-sided and not left-sided (18).

Comment. As a corollary of the previous observations we record the following statements:

(i) Every unital quantale on the three-chain C_3 is commutative.
(ii) There exists a unique left-sided and not right-sided quantale on C_3 which is necessarily idempotent.
(iii) There exists a unique right-sided and not left-sided quantale on C_3 which is necessarily idempotent.
(iv) There exist two non-commutative and two-sided quantales on C_3.

2.3.2. Show that the associative reflection of the eight prequantales (3)–(5) and (10)–(14) introduced in Exercise 2.2.1 have the following form:

(i) The associative reflection of (3)–(5) are isomorphic to the trivial quantale on the two-chain C_2.
(ii) The associative reflection of (10)–(14) coincide with the quantale consisting of a single element.

2.3.3. Determine all integral quantales on the chain with four elements.
(Hint: Use Exercise 2.3.1 (h) and apply the semi-unitalization.)

2.3.4. Let C_2 be the chain with two elements. Show that on a Boolean algebra with four elements there exist three structures of a unital quantale up to isomorphism. More precisely, there exists a unique structure of an integral quantale and two structures of a non-integral and unital quantale occuring as the unitalization of quantales on C_2 (cf. Exercise 2.2.2).

2.3.5. Let $[0, 1]$ be the real unit interval and $*$ be the binary geometric mean — i.e. $x * y = \sqrt{x \cdot y}$. Show that $([0, 1], *)$ is a prequantale and compute its associative reflection.
(Hint: If c is a nucleus on $([0, 1], *)$ such that $c\big((x_1 * x_2) * x_3\big) = c\big(x_1 * (x_2 * x_3)\big)$ holds for all $x_1, x_2, x_3 \in [0, 1]$, then c also satisfies the property $x^{2/3} \le c(x)$ for all $x \in (0, 1]$. For this purpose fix $x \in (0, 1]$ and consider the case $x_1 = 1$ and $x_2 = x_3 = x^{4/3}$.)

2.3.6. Let $\mathbb{N}$ be the set of natural numbers and $\mathbb{1}$ be the unit object of `Sup` (cf. Lemma 2.1.14). Show that the free quantale generated by $\mathbb{1}$ in the sense of `Sup` is isomorphic to the power set $\mathscr{P}(\mathbb{N})$ of $\mathbb{N}$ provided with the Minkowski addition.

2.3.7. Let $(X, *, e, ')$ be an involutive and unital quantale, and let us recall that the universal bounds are always hermitian. Show that the unit e of X is also hermitian.

2.3.8. Show that the category of commutative quantales is a reflective subcategory of the category `Quant` of quantales.

2.3.9. Construct a continuous quantale $([0, 1], *)$ which satisfies the following properties:

(a) $\frac{1}{2}$ is the unique idempotent element $x \in (0, 1)$.
(b) $\frac{1}{2} < x * x$ for all $x \in (\frac{1}{2}, 1)$.
(c) There exists an element $x \in (0, 1)$ with $x * x = 0$.

2.3.10. Let X be a complete lattice with at least two different elements. Further, let $*_\ell$ and $*_r$ be binary operations on X (in the sense of `Set`) given by

$$x *_\ell y = \begin{cases} \bot, & x = \bot, \\ y, & x \neq \bot, \end{cases} \quad \text{and} \quad x *_r y = \begin{cases} \bot, & y = \bot, \\ x, & y \neq \bot, \end{cases} \qquad x, y \in X.$$

Show

(a) $(X, *_\ell)$ is a left-sided and idempotent quantale, while $(X, *_r)$ is a right-sided and idempotent quantale.
(b) The set of prime elements in both quantales $X^\ell = (X, *_\ell)$ and $X^r = (X, *_r)$ coincides with $X \setminus \{\top\}$. What are the prime elements of $X^\ell \otimes X^r$?

(c) All the elementary tensors in the tensor product $X^\ell \otimes X^r$ are idempotent elements. $X^\ell \otimes X^r$ is idempotent if and only if X is isomorphic to the unit object of the tensor product.
(d) A tensor $f \in X^\ell \otimes X^r$ is left-sided if and only if there exists a $u \in X$ such that $f = u \otimes \top$.
(Hint: $u = \bigvee\{x \in X \mid f(x) \neq \bot\}$.)
(e) A tensor $g \in X^\ell \otimes X^r$ is right-sided if and only if there exists a $v \in X$ such that $g = \top \otimes v$.
(Hint: $v = \bigvee\{g(x) \mid x \neq \bot\}$.)
(f) Conclude from (d) and (e) that the subquantale $\mathbb{L}(X^\ell \otimes X^r)$ of all left-sided tensors is isomorphic to X^ℓ and the subquantale $\mathbb{R}(X^\ell \otimes X^r)$ of all right-sided tensors is isomorphic to X^r.
(g) The subquantale $\mathbb{I}(X^\ell \otimes X^r)$ of all two-sided tensors coincides with the quantale consisting only of the universal bounds of $X^\ell \otimes X^r$ — i.e. $\{\bot, \top\}$.
(h) Conclude from (d), (e) and (g) that $X^\ell \otimes X^r$ is left-sided (resp. right-sided) if and only if $X = \{\bot, \top\}$.

2.3.11. Let $[0, 1]$ be the real unit interval provided with the order-reversing involution determined by $x^\perp = 1 - x$ for all $x \in [0, 1]$. Further, let $([[0, 1], [0, 1]], \circ, 1_{[0,1]}, ')$ be the involutive and unital quantale of all join-preserving maps $[0, 1] \xrightarrow{f} [0, 1]$ (cf. Example 2.3.31). Show that the group of all unitary join-preserving maps in $[[0, 1], [0, 1]]$ is isomorphic to the group of all join-preserving and bijective maps $[0, \frac{1}{2}] \xrightarrow{g} [0, \frac{1}{2}]$.
(Hint: Use Example 2.3.32 and for every $[0, \frac{1}{2}] \xrightarrow{g} [0, \frac{1}{2}]$ consider the map f determined by: $f(x) = g(x)$ for all $x \in [0, \frac{1}{2}]$ and $f(x) = 1 - g(1 - x)$ for all $x \in (\frac{1}{2}, 1]$.)

2.3.12. For every topological group $(X, \cdot, \tau)$ with the multiplication $\cdot$ show that the complete lattice τ of open subsets is an involutive quantale $(\tau, *, ')$ with respect to the following operations:

$$U * V = \{x \cdot y \mid x \in U, y \in V\} \quad \text{and} \quad U' = \{x^{-1} \mid x \in U\}, \qquad U, V \in \tau.$$

2.3.13. Show that the quantales (7) and (8) in Exercise 2.2.1 are semigroups in CD — i.e. the corresponding binary operation in Sup is meet-preserving. Moreover, compute the comultiplication of (7) and (8).

2.4 Idempotent Quantales and Frames

As a first general property we show that the idempotency of the tensor product entails the idempotency of its factors.

Lemma 2.4.1. *Let $(X, *)$ and $(Y, *)$ be quantales such that X and Y contain at least two different elements. If $(X \otimes Y, \star)$ is idempotent, then both factors $(X, *)$ and $(Y, *)$ are idempotent.*

Proof. If $(X \otimes Y, \star)$ is idempotent, then $(X \otimes Y, \star)$ is balanced. Since X and Y contain at least two different elements, the quantales $(X, *)$ and $(Y, *)$ are also balanced. Hence the embeddings j_X and j_Y are homomorphisms (cf. Proposition 2.3.14 (i) and (ii)), which implies that $(X, *)$ and $(Y, *)$ are idempotent. □

Comment. The converse of the previous lemma does not hold (see Example 2.3.26 for a counterexample). A complete characterization of the idempotency of the tensor product will be given in Theorem 2.4.11 infra.

In the following considerations we investigate the impact of left- and right-sidedness on the theory of idempotent quantales. Obviously, left-sided (resp. right-sided) and idempotent quantales are always strictly left-sided (resp. right-sided). Under the hypothesis of left-sidedness (resp. right-sidedness) we show now that the tensor product preserves the structure of idempotent quantales.

Corollary 2.4.2. *If $(X, *)$ and $(Y, *)$ are left-sided (resp. right-sided) and idempotent quantales, then their tensor product $(X \otimes Y, \star)$ is also left-sided (resp. right-sided) and idempotent.*

Proof. The left-sidedness of $(X \otimes Y, \star)$ is evident (cf. Lemma 2.3.12 (iii)). Further, for every tensor $f \in X \otimes Y$ the following relation holds:

$$f = \bigvee_{x \in X} (x * x) \otimes (f(x) * f(x)) = \bigvee_{x \in X} (x \otimes f(x)) \star (x \otimes f(x)) \leq f \star f.$$

Since $f \star f \leq f$ follows from the left-sideness of $(X \otimes Y, \star)$, we conclude that the tensor product $(X \otimes Y, \star)$ is idempotent. □

If X and Y contain at least two different elements, then we conclude from Proposition 2.3.3 and Lemma 2.4.1 that the converse statement of Corollary 2.4.2 is also true.

Lemma 2.4.3. *Let $(X, *)$ be an idempotent quantale. Then the following properties hold:*

(i) *If $(X, *)$ is left-sided, then $(X, *)$ is left-symmetric — i.e.*

$$x * y * z = y * x * z, \qquad x, y, z \in X.$$

(ii) *If $(X, *)$ is two-sided, then $(X, *)$ is an integral and commutative quantale.*

Proof. Let $(X, *)$ be a left-sided and idempotent quantale. Then we obtain:

$$x * y * z = (x * y * z) * (x * y * z) \leq \top * y * \top * x * \top * z \leq y * x * z, \qquad x, y, z \in X.$$

Interchanging the rôle of x and y, the left-symmetry of $(X, *)$ follows.
If $(X, *)$ is two-sided, then the universal upper bound $\top$ is the unit of $(X, *)$ — i.e. $(X, *)$ is integral. The commutativity of $(X, *)$ follows from (i). □

Since in any integral, commutative and idempotent quantale the multiplication coincides with the binary meet operation, two-sided and idempotent quantales receive a special name and are called *frames* (cf. [58]). Typical examples of frames are given by lattices of open subsets of topological spaces.

Further, it is important to note that left-sided, idempotent and unital quantales are always frames. Hence the non-commutativity of a left-sided and idempotent quantale is characterized by the absence of a unit. This fundamental observation leads to the motivation to study non-unital quantales in more detail.

Before we proceed, we first present a non-trivial class of left-sided, idempotent and non-commutative quantales arising from non-commutative unital C^*-algebras. For the convenience of the reader we begin with the axioms of a C^*algebra.

A Banach algebra $\mathfrak{A} = (A, +, \cdot)$ with unit e (cf. [60, 77]) is a *unital C^*-algebras* if A is provided with a conjugate-linear map $a \longmapsto a^*$ of A into itself satisfying the following conditions:

(C1) $(a^*)^* = a$ for all $a \in A$.
(C2) $(a \cdot b)^* = b^* \cdot a^*$ for all $a, b \in A$.
(C3) $\|a^* \cdot a\| = \|a\|^2$ for all $a \in A$.

Sometimes the conjugate-linear map $a \longmapsto a^*$ is called the *involution* of $\mathfrak{A}$. In this context condition (C3) is also known as the *C^*-property*.

Because of (C1), (C3) and the submultiplicativity of the norm the involution $a \longmapsto a^*$ is always an isometry.

A *closed left ideal* of a C^*-algebra $\mathfrak{A} = (A, \cdot, +)$ is a closed linear subspace I of A satisfying the additional property:

$$a \in A, \quad x \in I \quad \Longrightarrow \quad a \cdot x \in I.$$

Morphisms between C^*-algebras are *$*$-homomorphisms* — these are algebra homomorphisms $A \xrightarrow{\pi} B$ satisfying the property $\pi(a^*) = \pi(a)^*$ for all $a \in A$. Consequently $*$-homomorphisms are algebra homomorphisms preserving the corresponding involutions. It follows from the spectral theory of self-adjoint elements of A and the C^*-property that a $*$-homomorphism π always satisfies the condition (cf. [5, p. 12])

$$\|\pi(a)\| \le \|a\|, \qquad a \in A.$$

Hence $*$-homomorphisms are continuous.

Further, we recall that the multiplication of closed left ideals of C^*-algebras is idempotent. Since these constructions make essential use of spectral theory in Banach algebras, we first summarize the fundamental properties of the continuous function calculus. For details the reader is referred to standard references in Functional Analysis — e.g. [77]. Here we only recall the spectrum of self-adjoint elements of unital C^*-algebras A. If $x \in A$ is *self-adjoint* (i.e. $x^* = x$), then in this case the *spectrum* $\sigma(x)$ of x is always a compact subset of the real line and is given by

$$\sigma(x) = \{\lambda \in \mathbb{R} \mid \lambda e - x \text{ is not invertible}\},$$

where e denotes the unit of A.

Theorem 2.4.4. *Let A be a unital C^*-algebra, x be a self-adjoint element of A and let $\sigma(x)$ be the spectrum of x. Further let $\mathscr{C}(\sigma(x))$ be the commutative C^*-algebra of all continuous functions $\sigma(x) \xrightarrow{f} \mathbb{C}$ and $\sigma(x) \overset{\iota}{\hookrightarrow} \mathbb{C}$ be the inclusion map. Then there exists a unique $*$-homomorphism $\mathscr{C}(\sigma(x)) \xrightarrow{\Phi} A$ satisfying the condition $\Phi(\iota) = x$. Moreover, Φ satisfies the following properties:*

(i) *Φ is an isometry.*
(ii) *For every $f \in \mathscr{C}(\sigma(x))$ the spectrum of $\Phi(f)$ coincides with the image of $\sigma(x)$ under f — i.e. $\sigma(\Phi(f)) = f(\sigma(x))$.*

Comment. The first part of Theorem 2.4.4, including property (i), is the *(continuous) function calculus* of self-adjoint elements (cf. [60, p. 239 and p. 271]), while property (ii) is also called the *spectral mapping theorem*.

Corollary 2.4.5. *Let $(A, \cdot, \|\ \|)$ be a unital C^*-algebra with unit e, and let I be a closed left ideal of A. Then $I \subseteq I * I$ holds where*

$$I * I = \text{top.closure}(\text{lin.hull}\,\{a \cdot b \mid a, b \in I\}).$$

Proof. We distinguish the following cases.
(a) Let x be a self-adjoint element of I and $\sigma(x)$ be the spectrum of x. Further, let $\mathscr{C}(\sigma(x)) \xrightarrow{\Phi} A$ be the unique $*$-homomorphism satisfying the conditions of Theorem 2.4.4. Then we consider a sequence $(f_n)_{n\in\mathbb{N}}$ of continuous functions $\sigma(x) \xrightarrow{f_n} \mathbb{C}$ defined by

$$f_n(t) = \tfrac{nt^2}{1+nt^2}, \qquad t \in \sigma(x),\ n \in \mathbb{N}.$$

Since $\Phi(1) = e$ and $\iota^2(t) = t^2$ for all $t \in \sigma(x)$, we obtain that there exists a sequence $(e_n)_{n\in\mathbb{N}}$ of elements $e_n \in A$ such that

$$e_n = \Phi(f_n) = nx^2(e + nx^2)^{-1},$$

where e is the unit of A. We show that e_n belongs to I. Since f_n is bounded ($\sigma(x)$ is compact) and $f_n(0) = 0$ holds, f_n can uniformly be approximated by polynomials p having the following form $p(t) = \sum_{k=1}^{m} a_k t^k$. Since x is self-adjoint and Φ is an isometry, e_n is the limit of self-adjoint elements of the form

$$\Phi(p) = \sum_{k=1}^{m} a_k x^k.$$

Hence e_n is self-adjoint, and $\Phi(p)$ is an element of I where we have used the ideal property of I. Now the closedness of I ensures that the self-adjoint element e_n is also an element of I. Finally, we invoke the spectral mapping theorem and obtain:

$$\sigma(e - e_n) = (1 - f_n)(\sigma(x)) \subseteq [0, 1).$$

Since the norm of $e - e_n$ coincides with its spectral radius, we have $\|e - e_n\| \leq 1$. Now we use the C^*-property of C^*-algebras and again the property that Φ is an isometry. Then the relation

$$\begin{aligned}\|x \cdot e_n - x\|^2 &= \|(e_n - e) \cdot x \cdot x \cdot (e_n - e)\| \\ &\leq \|x^2 \cdot (e_n - e)\| = \|x^2 \cdot (e - e_n)\| \\ &= \sup_{t \in \sigma(x)} t^2\big(1 - \tfrac{nt^2}{1+nt^2}\big) \leq \tfrac{1}{n}\end{aligned}$$

follows — this means $\lim_{n\to\infty} x \cdot e_n = x$. Thus $x \in I * I$ holds by definition of $*$.

(b) Now we choose an arbitrary element x of I. Since I is a left ideal, $x^* \cdot x$ is contained in I. In particular, $x^* \cdot x$ is self-adjoint. Hence $\lim_{n\to\infty} x^* \cdot x \cdot e_n = x^* \cdot x$ follows from (a). Moreover, we observe:

$$\|x \cdot e_n - x\|^2 = \|(e_n - e) \cdot x^* \cdot x \cdot (e_n - e)\| \leq \|x^* \cdot x \cdot (e_n - e)\|.$$

The previous relation implies $\lim_{n\to\infty} x \cdot e_n = x$. Thus we have verified $x \in I * I$. □

Corollary 2.4.6. *Let $(A, \cdot, \|\ \|)$ be a unital C^*-algebra. Further, let I be a closed left ideal of A and $I' = \{x^* \mid x \in I\}$ be its adjoint closed right ideal. Then the relation $I \subseteq I * I'$ holds, where $I * I' =$ top.closure (lin.hull $\{a \cdot b \mid a \in I,\ b \in I'\}$).*

Proof. Since the sequence $(e_n)_{n\in\mathbb{N}}$ constructed in the proof of Corollary 2.4.5 consists of self-adjoint elements of I, the assertion follows from the proof of Corollary 2.4.5. □

As an immediate corollary of Corollary 2.4.6 we obtain the following well-known result (cf. [5]).

Corollary 2.4.7. *Every closed two-sided ideal I of a unital C^*-algebra is self-adjoint — i.e. $I = I'$.*

Example 2.4.8. Let $(A, \cdot, \|\ \|)$ be a unital C^*-algebra and $\mathbb{L}(A)$ be the set of all closed left ideals I of A ordered by set-inclusion. Then $\mathbb{L}(A)$ is a complete lattice. In particular, meets in $\mathbb{L}(A)$ are computed by intersections, while joins in $\mathbb{L}(A)$ are formed by the closure of the linear hull of unions. Further, we consider the ideal multiplication $*$ on $\mathbb{L}(A)$ — i.e.

$$I * J = \text{top.closure (lin. hull} \{a \cdot b \mid a \in I,\ b \in J\}, \qquad I, J \in \mathbb{L}(A).$$

Then it is not difficult to verify that $(\mathbb{L}(A), *)$ is a quantale. Further, $(\mathbb{L}(A), *)$ is a left-sided quantale and also an idempotent quantale by Corollary 2.4.5. Since $(\mathbb{L}(A), *)$ is semi-integral, maximal left ideals are prime (cf. Theorem 2.3.21). Hence we conclude from [61, Theorem 10.2.10(iii) on p. 733] that every closed left ideal is a meet of prime elements. Thus $(\mathbb{L}(A), *)$ is a spatial quantale.
In this context it is well-known that a unital C^*-algebra is commutative if and only if each closed left ideal is a two-sided ideal (cf. [60, Exercise 4.6.29 on p. 292]). Hence $(\mathbb{L}(A), *)$ is a non-commutative, idempotent and left-sided quantale if and only if the underlying C^*-algebra $(A, \cdot, \| \ \|)$ is non-commutative — e.g. the C^*-algebra of all bounded, linear operators on a Hilbert space.
If the underlying unital C^*-algebra $(A, \cdot, \| \ \|)$ is commutative (i.e. every closed left ideal is two-sided), then we conclude from Lemma 2.4.3 (ii) that $(\mathbb{L}(A), *)$ is a frame. In this case $\mathbb{L}(A)$ is isomorphic to the *Gelfand topology* on the spectrum of A (cf. [77]).

We continue our train of thought with a categorical framework of left-sided and idempotent quantales. First we note that a strong homomorphism between frames is a frame homomorphism — i.e. an arbitrary join and finite meet-preserving map. Now we introduce the following terminology and notation.

Let `LQuant` be the category of left-sided and idempotent quantales and strong homomorphisms. Obviously, `LQuant` is a full subcategory of `BQuant` (cf. Sect. 2.3.1). Further, let `Frm` be the category of frames and frame homomorphisms. Then `Frm` is evidently a full subcategory of `LQuant`.

It is easily seen that `LQuant` is complete. Moreover, `LQuant` satisfies the following important property.

Theorem 2.4.9. *The category* `LQuant` *is a reflective subcategory of* `BQuant`.

Proof. Let $(X, *)$ be a balanced quantale. We have to prove that there exists a left-sided and idempotent quantale $(X^\flat, *^\flat)$ and a strong homomorphism $X \xrightarrow{\eta_X^\flat} X^\flat$ such that for every further left-sided and idempotent quantale $(Y, *)$ and every further strong homomorphism $X \xrightarrow{h} Y$ there exists a unique strong homomorphism $X^\flat \xrightarrow{h^\flat} Y$ making the following diagram commutative:

$$\begin{array}{ccc} X & \xrightarrow{\eta_X^\flat} & X^\flat \\ {\scriptstyle h}\downarrow & \swarrow {\scriptstyle h^\flat} & \\ Y & & \end{array}$$

Let $\mathbb{F}$ be the set of all nuclei c on $(X, *)$ satisfying the following condition:

$$c(\top * x) = c(x) = c(x * x), \qquad x \in X.$$

Further, let c_0 be the meet of $\mathbb{F}$. Since the meet of nuclei is computed pointwisely, c_0 is contained in $\mathbb{F}$. Now let $(X^\flat, *^\flat) = (c_0(X), \star)$ be the regular quotient of $(X, *)$

w.r.t. c_0. Then $(X^\flat, *^\flat)$ is a left-sided and idempotent quantale, and the quotient map $X \xrightarrow{\eta_X^\flat} X^\flat$ is a strong homomorphism. If $(Y, *)$ is a further left-sided and idempotent quantale and $X \xrightarrow{h} Y$ is a further strong homomorphism, then the nucleus $c_h = h^\vdash \circ h$ associated with h is contained in $\mathbb{F}$. Hence $c_0 \le c_h$ follows, and h factors uniquely through $\eta_X^\flat$. □

As an immediate corollary of Theorem 2.3.15, Proposition 2.3.16 and Theorem 2.4.9 we obtain that `LQuant` is cocomplete. We can summarize the previous results as follows.

FACT I. *The category* `LQuant` *of left-sided and idempotent quantales is a complete, cocomplete and monoidal category.*

The next proposition is a first step on the way to giving a complete characterization of idempotent tensor products of quantales.

Proposition 2.4.10. *Let $(X, *)$ and $(Y, *)$ be idempotent quantales. If $(X, *)$ or $(Y, *)$ is a frame, then their tensor product $(X \otimes Y, \star)$ is idempotent.*

Proof. Let $(X, *)$ and $(Y, *)$ be idempotent quantales. We fix a tensor $f \in X \otimes Y$. Then the relation $f \le f \star f$ is obvious (cf. the proof of Corollary 2.4.2). In order to verify $f \star f \le f$ we distinguish the following cases.
Case 1. Let $(X, *)$ be two-sided — i.e. a frame. Since f is antitone and $(Y, *)$ is idempotent, we obtain for all $x, z \in X$:

$$(x \otimes f(x)) \star (z \otimes f(z)) = (x \wedge z) \otimes (f(x) * f(z)) \le (x \wedge z) \otimes f(x \wedge z) \le f.$$

Case 2. Let $(Y, *)$ be two-sided — i.e. a frame. Since f is join-reversing and $(X, *)$ is idempotent, we obtain for all $x, z \in X$:

$$(x \otimes f(x)) \star (z \otimes f(z)) = (x * z) \otimes (f(x) \wedge f(z)) \le (x \vee z) \otimes f(x \vee z) \le f.$$

Hence $f \star f \le f$ follows. □

Warning. The tensor product of a frame with an arbitrary idempotent quantale is in general neither left-sided nor right-sided.

Theorem 2.4.11. *Let $(X, *)$ and $(Y, *)$ be quantales such that X and Y contain at least two different elements. Then the following assertions are equivalent:*

(i) *The tensor product $(X \otimes Y, \star)$ is idempotent.*
(ii) *$(X, *)$ and $(Y, *)$ are idempotent and one of the following cases occurs:*

  (1) *$(X, *)$ or $(Y, *)$ is a frame.*
  (2) *$(X, *)$ and $(Y, *)$ are left-sided.*
  (3) *$(X, *)$ and $(Y, *)$ are right-sided.*

Proof. The implication (ii) $\Longrightarrow$ (i) follows immediately from Corollary 2.4.2 and Proposition 2.4.10. In order to verify (i) $\Longrightarrow$ (ii) the idempotency of $(X, *)$ and $(Y, *)$ has already been established in Lemma 2.4.1. Without loss of generality, we can assume that $(X, *)$ is not two-sided and show now that one of the cases (1)–(3) occurs. For this purpose we choose $x \in X$ and $y \in Y$ and consider the tensor $f_{xy} = (x \otimes \top) \vee (\top \otimes y)$. Obviously, f_{xy} has the explicit form (see Exercise 2.1.4 (a)):

$$f_{xy}(z) = \begin{cases} \top, & z \le x, \\ y, & z \not\le x, \end{cases} \qquad z \in X.$$

Further, we observe:

$$\begin{aligned} f_{xy} \star f_{xy} &\ge \big((x \otimes \top) \star (\top \otimes y)\big) \vee \big((\top \otimes y) \star (x \otimes \top)\big) \\ &= \big((x * \top) \otimes (\top * y)\big) \vee \big((\top * x) \otimes (y * \top)\big). \end{aligned}$$

Since f_{xy} is idempotent, the relation

$$\big((x * \top) \otimes (\top * y)\big) \vee \big((\top * x) \otimes (y * \top)\big) \le f_{xy} \tag{2.47}$$

follows for all $x \in X$ and for all $y \in Y$. Now we distinguish the following cases:
Case 1. Let $(X, *)$ be neither left-sided nor right-sided. Then there exist elements $x_1, x_2 \in X$ such that $\top * x_1 \not\le x_1$ and $x_2 * \top \not\le x_2$. Now the relations

$$y * \top \le y = f_{x_1 y}(\top * x_1) \quad \text{and} \quad \top * y \le y = f_{x_2 y}(x_2 * \top), \quad y \in Y$$

follow from (2.47). Hence $(Y, *)$ is two-sided (i.e. a frame), and (1) occurs.
Case 2. Let $(X, *)$ be left-sided and not right-sided. Then there exists an $x \in X$ such that $x * \top \not\le x$. Now we infer from (2.47) that

$$\top * y \le y = f_{xy}(x * \top)$$

holds for all $y \in Y$. Hence $(Y, *)$ is also left-sided, and (2) occurs.
If $(X, *)$ is right-sided and not left-sided, then by analogy with *Case 2* the quantale $(Y, *)$ is right-sided, and (3) occurs. □

The next theorem shows that in special cases the coproduct of left-sided and idempotent quantales can be expressed by their tensor product. Since idempotent quantales are balanced, we know already that the embeddings $X \xrightarrow{j_X} X \otimes Y$ and $Y \xrightarrow{j_Y} X \otimes Y$ are strong homomorphisms (cf. Proposition 2.3.14 (i) and (ii)).

Theorem 2.4.12. *Let $(X, *)$ be a left-sided, idempotent quantale, and let $(Y, *)$ be a two-sided, idempotent quantale (i.e. a frame). Then $(X \otimes Y, \star)$ with the embeddings j_X and j_Y as coproduct injections is the coproduct of $(X, *)$ and $(Y, *)$ in* `LQuant`.

Proof. Let $(X, *)$ and $(Y, *)$ be idempotent quantales such that $(X, *)$ is left-sided and $(Y, *)$ is two-sided. By Corollary 2.4.2 we already know that their tensor product $(X \otimes Y, \star)$ is an object of LQuant. In order to confirm that the universal property of coproducts holds for $(X \otimes Y, \star)$ with the coproduct injections j_X and j_Y we choose a further left-sided and idempotent quantale $(Z, *)$, and further strong homomorphisms $X \xrightarrow{h_X} Z$ and $Y \xrightarrow{h_Y} Z$. We show that there exists a unique strong homomorphism $X \otimes Y \xrightarrow{k} Z$ with $h_X = k \circ j_X$ and $h_Y = k \circ j_Y$.

(a) (Uniqueness) Since $(X, *)$ is strictly left-sided and $(Y, *)$ is strictly right-sided, the universal upper bound is a left-unit in $(X, *)$ and a right-unit in $(Y, *)$. Hence the relation

$$(\top \otimes y) \star (x \otimes \top) = (\top * x) \otimes (y * \top) = x \otimes y$$

holds for all $x \in X$ and $y \in Y$. If $X \otimes Y \xrightarrow{k} Z$ is a strong homomorphism satisfying the conditions $h_X = k \circ j_X$ and $h_Y = k \circ j_Y$, then we obtain:

$$k(x \otimes y) = k(\top \otimes y) * k(x \otimes \top) = h_Y(y) * h_X(x), \qquad x \in X,\ y \in Y. \tag{2.48}$$

Since every tensor is a join of an appropriate family of elementary tensors, k is uniquely determined.

(b) (Existence) Let $Z \otimes Z \xrightarrow{\circledast} Z$ be the binary operation corresponding to the multiplication $*$ in Z. Motivated by (2.48) we now define a join-preserving map $X \otimes Y \xrightarrow{k} Z$ by $k = \circledast \circ c_{ZZ} \circ (h_X \otimes h_Y)$, where c_{ZZ} is the Z-component of the symmetry in Sup (cf. Lemma 2.1.15). We show that k is a strong homomorphism. Since $k(\top \otimes \top) = h_Y(\top) * h_X(\top) = \top * \top = \top$, the map k preserves the respective universal upper bounds. Further, we use the left-symmetry of $(Z, *)$ (cf. Lemma 2.4.3 (i)) and observe:

$$\begin{aligned} k\big((x_1 \otimes y_1) \star (x_2 \otimes y_2)\big) &= k\big((x_1 * x_2) \otimes (y_1 * y_2)\big) \\ &= h_Y(y_1) * h_Y(y_2) * h_X(x_1) * h_X(x_2) \\ &= \big(h_Y(y_1) * h_X(x_1)\big) * \big(h_Y(y_2) * h_X(x_2)\big) \\ &= k(x_1 \otimes y_1) * k(x_2 \otimes y_2). \end{aligned}$$

Since every tensor is a join of elementary tensors, the map k is a strong homomorphism. Finally, for $x \in X$ and $y \in Y$ we observe:

$$\begin{aligned} &k(j_X(x)) = k(x \otimes \top) = \top * h_X(x) = h_X(x) \\ &k(j_Y(y)) = k(\top \otimes y) = h_Y(y) * \top = h_Y(y) * h_Y(\top) = h_Y(y * \top) = h_Y(y). \end{aligned}$$

Hence k satisfies the desired properties. □

Corollary 2.4.13. ([110]) *If X and Y are frames, then their tensor product $X \otimes Y$ is the coproduct of X and Y in* Frm.

The previous corollary implies that the coproduct of a family of frames in the sense of `Frm` is the direct limit of their finite tensor products (cf. [59]).

In the next subsection we describe the relationship between quantales and frames.

2.4.1 Localic Quotients

A nucleus c on a prequantale $(X, *)$ is called *localic* if c satisfies the following additional property:

$$c(x_1) * c(x_2) \leq c(x_1) \wedge c(x_2) \leq c(x_1 * x_2), \qquad x_1, x_2 \in X. \tag{L}$$

Since $c(c(x_1) \wedge c(x_2)) = c(x_1) \wedge c(x_2)$, condition (L) is equivalent to

$$c(x_1) \wedge c(x_2) = c(x_1 * x_2), \qquad x_1, x_2 \in X. \tag{L$'$}$$

The following properties are now obvious.

Lemma 2.4.14. *Let $(X, *)$ be a prequantale and $(Y, \wedge)$ be a frame. If $X \xrightarrow{h} Y$ is a homomorphism, then the associated nucleus c_h (i.e. $c_h = h^{\vdash} \circ h$) is localic.*

Corollary 2.4.15. *Let $(X, *)$ be a prequantale and c be a nucleus on $(X, *)$. Then the following assertions are equivalent:*

(i) *The regular quotient $(c(X), \star)$ of $(X, *)$ w.r.t. c is localic — i.e. $\star$ coincides with the binary meet operation. In particular, $c(X)$ is a frame.*
(ii) *The nucleus c is localic.*

Example 2.4.16. Let R be a commutative ring with unit and $(\mathbb{I}(R), *)$ be the unital and commutative quantale of all ideals of R provided with the ideal multiplication $*$. In particular, the unit coincides with R, which is the universal upper bound of $\mathbb{I}(R)$. Further, $r(I)$ denotes the radical of each ideal I of R — i.e.

$$r(I) = \{x \in R \mid \exists n \in \mathbb{N} : x^n \in I\}.$$

It is well-known that the correspondence $I \longmapsto r(I)$ is a closure operator r on $\mathbb{I}(R)$ satisfying the additional condition:

$$r(I * J) = r(I) \cap r(J), \qquad I, J \in \mathbb{I}(R).$$

Hence r is a localic nucleus, and the localic quotient of $\mathbb{I}(R)$ w.r.t. r coincides with the frame of all open subsets of the Zariski spectrum of R.

Theorem 2.4.17. *The spatial reflection of any integral quantale is a frame.*

Proof. Let c_s be the nucleus on $(X, *)$ defined by (S) in Lemma 2.3.28. Referring to Corollary 2.3.29 and Theorem 2.3.30 it is sufficient to prove that c_s is localic.
Since $(X, *)$ is integral, the relation $c_s(x_1 * x_2) \leq c_s(x_1) \wedge c_s(x_2)$ holds for all $x_1, x_2 \in X$. On the other hand, let us choose an arbitrary prime element p of $(X, *)$ with $x_1 * x_2 \leq p$. Using again the integrality of $(X, *)$ we obtain $x_1 \leq p$ or $x_2 \leq p$. Hence $c_s(x_1) \wedge c_s(x_2) \leq p$ follows. Since p is an arbitrary prime element with $x_1 * x_2 \leq p$, the relation $c_s(x_1) \wedge c_s(x_2) \leq c_s(x_1 * x_2)$ holds. □

As an immediate corollary of Theorem 2.4.17 we obtain the following result due to J. Rosický (cf. [102, Corollary 1 on p. 392]).

Corollary 2.4.18. *Every spatial and integral quantale is a frame.*

Theorem 2.4.19. *The category* `Frm` *is a reflective subcategory of* `LQuant`.

Proof. Let $(X, *)$ be a left-sided and idempotent quantale. We have to prove that there exists a frame $(X^\flat, \wedge)$ and a strong homomorphism $X \xrightarrow{\eta_X^\flat} X^\flat$ such that for every further frame $(Y, \wedge)$ and strong homomorphism $X \xrightarrow{h} Y$ there exists a unique frame homomorphism $X^\flat \xrightarrow{h^\flat} Y$ making the following diagram commutative:

$$\begin{array}{ccc} X & \xrightarrow{\eta_X^\flat} & X^\flat \\ {\scriptstyle h}\downarrow & \swarrow{\scriptstyle h^\flat} & \\ Y & & \end{array} \tag{2.49}$$

In comparison to the proof of Theorem 2.4.9 we will here apply a similar proof strategy. Therefore let $\mathbb{F}$ be the set of all nuclei c on $(X, *)$ satisfying the condition:

$$c(x * \top) = c(x), \qquad x \in X.$$

Further, let c_0 be the meet of $\mathbb{F}$. Since the meet of nuclei is computed pointwisely, c_0 is contained in $\mathbb{F}$. Then the regular quotient of $(X, *)$ w.r.t. c_0 is an idempotent and two-sided quantale $(X^\flat, \wedge) = (c_0(X), \star)$ — i.e. a frame (cf. Lemma 2.4.3 (ii)). Obviously, the quotient map $X \xrightarrow{\eta_X^\flat} X^\flat$ is a strong homomorphism. If $(Y, \wedge)$ is a further frame and $X \xrightarrow{h} Y$ is a further strong homomorphism, then the nucleus $c_h = h^\vdash \circ h$ associated with h is contained in $\mathbb{F}$. Hence $c_0 \leq c_h$ follows, and h factors uniquely through $\eta_X^\flat$. □

Since adjoint situations compose, we conclude from Theorems 2.4.9 and 2.4.19 that `Frm` is a reflective subcategory of `BQuant`, and the action of the corresponding reflector at a balanced quantale $(X, *)$ is called the *localic reflection* of $(X, *)$. Hence `Frm` is a cocomplete category. The completeness of `Frm` is evident.

2.4.2 Topological Representation of Left-Sided and Idempotent Quantales

In the case of left-sided quantales $(X, *)$ we begin with the observation that an element $p \in X \setminus \{\top\}$ is *prime* if and only if the following implication holds for all $x_1, x_2 \in X$:

$$x_1 * x_2 \leq p \quad \Longrightarrow \quad x_1 * \top \leq p \text{ or } x_2 \leq p. \qquad \text{(Prime Property}'')$$

Since the implication $(x_1 * \top \leq p \text{ or } x_2 \leq p \Longrightarrow x_1 * x_2 \leq p)$ holds in any left-sided quantale, the previous prime property can also expressed by the following equivalence

$$x_1 * x_2 \leq p \quad \Longleftrightarrow \quad x_1 * \top \leq p \text{ or } x_2 \leq p.$$

Further, let $C_3^\ell = (C_3, *_\ell)$ be the idempotent, left-sided and non-commutative chain with three elements (cf. Example 2.3.26) — i.e. $C_3 = \{\bot, a, \top\}$ and the multiplication table of $*_\ell$ is given by

$*_\ell$	$\bot$	a	$\top$
$\bot$	$\bot$	$\bot$	$\bot$
a	$\bot$	a	$\top$
$\top$	$\bot$	a	$\top$

First we show that prime elements of left-sided and idempotent quantales and strong homomorphisms with codomain C_3^ℓ are equivalent concepts (cf. [51]).

We begin with the observation that in an arbitrary quantale strong homomorphisms with codomain C_3^ℓ always induce prime elements.

Lemma 2.4.20. *Let $(X, *)$ be a quantale and $X \xrightarrow{h} C_3^\ell$ be a strong homomorphism. Then the element $p \in X$ defined by $p = \bigvee\{x \in X \mid h(x) \leq a\}$ is a prime element in $(X, *)$.*

Proof. Since h is join-preserving and strong, the element p is different from the universal upper bound of X. In order to show that p is prime we choose $x_1, x_2 \in X$ with $x_1 * x_2 \leq p$. If $x_2 \not\leq p$, then $h(x_2) = \top$, and we observe:

$$h(x_1 * \top) = h(x_1) *_\ell \top = h(x_1) *_\ell h(x_2) = h(x_1 * x_2) \leq h(p) \leq a.$$

Hence $x_1 * \top \leq p$. □

Theorem 2.4.21. *Let $(X, *)$ be a left-sided and idempotent quantale. Then for every prime element p in $(X, *)$ there exists a unique strong homomorphism $X \xrightarrow{h} C_3^\ell$ satisfying the condition*

$$p = \bigvee\{x \in X \mid h(x) \leq a\}. \tag{2.50}$$

Proof. (a) (Uniqueness) Let $X \xrightarrow{h} C_3^\ell$ be an arbitrary strong homomorphism satisfying (2.50). Since h is join-preserving, the equivalence $x \le p \iff h(x) \le a$ holds for all $x \in X$. Now we distinguish the following cases:

- If $x * \top \le p$, then $h(x) *_\ell \top = h(x * \top) \le h(p) \le a$. Hence $h(x) = \bot$.
- If $x * \top \not\le p$ and $x \le p$, then $h(x) *_\ell \top = h(x * \top) = \top$ and $h(x) \le a$. Hence $h(x) = a$.
- If $x \not\le p$, then $h(x) = \top$.

Since $(X, *)$ is semi-unital, it follows from the previous cases that h is uniquely determined by (2.50).
(b) (Existence) Let p be a prime element. Motivated by the uniqueness proof we define a map $X \xrightarrow{h_p} C_3^\ell$ by:

$$h_p(x) = \begin{cases} \bot, & x * \top \le p, \\ a, & x * \top \not\le p \text{ and } x \le p, \\ \top, & x \not\le p, \end{cases} \qquad x \in X. \tag{2.51}$$

Obviously, h_p is join-preserving. Since $(X, *)$ is semi-unital, h_p satisfies property (2.50) — i.e. $p = \bigvee\{x \in X \mid h_p(x) \le a\}$. Therefore we have only to show that h_p is a strong homomorphism. Since $p \ne \top$, the relation $h_p(\top) = \top$ holds. In order to verify $h_p(x_1 * x_2) = h_p(x_1) *_\ell h_p(x_2)$ we distinguish the following cases:

- If $h_p(x_1 * x_2) = \bot$, then $x_1 * x_2 * \top \le p$. Since p is prime, we have either $x_1 * \top \le p$ or $x_2 * \top \le p$ — i.e. either $h_p(x_1) = \bot$ or $h_p(x_2) = \bot$. Hence $h_p(x_1) *_\ell h_p(x_2) = \bot$.
- If $h_p(x_1 * x_2) = a$, then $x_1 * x_2 * \top \not\le p$ and $x_1 * x_2 \le p$. Because of the prime property we obtain $x_2 \le p$, $x_1 * \top \not\le p$ and $x_2 * \top \not\le p$. Hence the relations $a \le h_p(x_1)$, $h_p(x_2) = a$ and $h_p(x_1) *_\ell h_p(x_2) = a$ follow.
- If $h_p(x_1 * x_2) = \top$, then $x_1 * x_2 \not\le p$. Hence $x_1 * \top \not\le p$ and $x_2 \not\le p$ hold, because of the prime property. This means $a \le h_p(x_1)$ and $h_p(x_2) = \top$, and so $h_p(x_1) *_\ell h_p(x_2) = \top$. □

Since it is possible to identify prime elements of left-sided and idempotent quantales with strong three-valued homomorphisms (cf. Theorem 2.4.21), this observation motivates the introduction of the basic properties of three-valued topological spaces.

Let again $C_3^\ell = (C_3, *_\ell)$ be the left-sided, idempotent and non-commutative chain with three elements and X be an arbitrary set. Then we extend pointwisely the partial order and the multiplication in C_3^ℓ to the set $(C_3^\ell)^X$ of all maps $X \xrightarrow{f} C_3^\ell$ — i.e.

$$f \le g \iff f(x) \le g(x) \quad \text{for all } x \in X,$$
$$(f * g)(x) = f(x) *_\ell g(x), \quad x \in X.$$

Obviously, $((C_3^\ell)^X, *)$ is again a left-sided and idempotent quantale.

Definition 2.4.22. Let X be a set. A subset τ of $(C_3^\ell)^X$ is called a *three-valued topology* on X if X satisfies the following properties (see also [49, p. 150]):

(O1) τ is closed under arbitrary joins in the sense of $(C_3^\ell)^X$ — i.e. the inclusion map $\tau \overset{\iota}{\hookrightarrow} (C_3^\ell)^X$ is join-preserving.
(O2) The universal upper bound $\underline{\top}$ of $(C_3^\ell)^X$ belongs to τ.
(O3) If $f, g \in \tau$, then $f * g \in \tau$.

If τ is a three-valued topology on X, then the pair (X, τ) is called a *three-valued topological space*.
Let (X, τ) and (Y, σ) be three-valued topological spaces. A map $X \xrightarrow{\varphi} Y$ is called *continuous* if for all $g \in \sigma$ the composition $g \circ \varphi$ is an element of τ.

Three-valued topological spaces and continuous maps form a category denoted by $\mathtt{Top}(3_\ell)$. Since every three-valued topology is a left-sided and idempotent quantale, there exists an object function

$$|\mathtt{Top}(3_\ell)| \to |\mathtt{LQuant}|$$

sending each three-valued topological space to its underlying three-valued topology. This object function can be completed to a functor $\mathsf{F}\colon \mathtt{Top}(3_\ell) \to \mathtt{LQuant}^{op}$ as follows:

$$(X, \tau) \xrightarrow{\varphi} (Y, \sigma), \quad \sigma \xrightarrow{\mathsf{F}(\varphi)} \tau, \quad \mathsf{F}(\varphi) = h_\varphi \quad \text{where} \quad h_\varphi(g) = g \circ \varphi, \quad g \in \sigma.$$

In fact, since φ is continuous, it is easily seen that h_φ is a strong homomorphism.

The aim of the following considerations is to show that $\mathsf{F}\colon \mathtt{Top}(3_\ell) \to \mathtt{LQuant}^{op}$ has a right adjoint functor $\mathtt{Pt}_{3_\ell}\colon \mathtt{LQuant}^{op} \to \mathtt{Top}(3_\ell)$. For this purpose we change our notation and denote now a left-sided and idempotent quantale by $(\mathscr{Q}, *)$ and by $\mathsf{pt}_{3_\ell}(\mathscr{Q})$ the set of all strong homomorphisms $\mathscr{Q} \xrightarrow{h} C_3^\ell$. On $\mathsf{pt}_{3_\ell}(\mathscr{Q})$ we introduce a three-valued topology $\tau_\mathscr{Q}$ as follows. First, for each $q \in \mathscr{Q}$, we define a map $\mathsf{pt}_{3_\ell}(\mathscr{Q}) \xrightarrow{\mathbb{A}_q} C_3^\ell$ by

$$\mathbb{A}_q(h) = h(q), \qquad h \in \mathsf{pt}_{3_\ell}(\mathscr{Q}).$$

Since all $h \in \mathsf{pt}_{3_\ell}(\mathscr{Q})$ are strong homomorphisms, it is easily seen that

$$\tau_\mathscr{Q} = \{\mathbb{A}_q \mid q \in \mathscr{Q}\}$$

is a three-valued topology on $\mathsf{pt}_{3_\ell}(\mathscr{Q})$. Further, let k be a morphism in $\mathtt{LQuant}$ — i.e. a strong homomorphism $\mathscr{P} \xrightarrow{k} \mathscr{Q}$. Then k induces a map $\mathsf{pt}_{3_\ell}(\mathscr{Q}) \xrightarrow{\varphi_k} \mathsf{pt}_{3_\ell}(\mathscr{P})$ by

$$\varphi_k(h) = h \circ k, \qquad h \in \mathsf{pt}_{3_\ell}(\mathscr{Q}). \tag{2.52}$$

Since

$$\mathbb{A}_p \circ \varphi_k = \mathbb{A}_{k(p)}, \qquad p \in \mathscr{P}, \tag{2.53}$$

φ_k is continuous. Hence we can introduce a functor $\mathsf{Pt}3_\ell : \mathtt{LQuant}^{op} \to \mathtt{Top}(3_\ell)$ as follows:

$$\mathsf{Pt}3_\ell(\mathcal{Q}, *) = (\mathsf{pt}3_\ell(\mathcal{Q}), \tau_{\mathcal{Q}}),$$

$$\mathcal{P} \xrightarrow{k} \mathcal{Q}, \quad \mathsf{Pt}3_\ell(\mathcal{Q}, *) \xrightarrow{\mathsf{Pt}3_\ell(k)} \mathsf{Pt}3_\ell(\mathcal{P}, *), \quad \mathsf{Pt}3_\ell(k) = \varphi_k.$$

Theorem 2.4.23. *The functor* $\mathsf{Pt}3_\ell$ *is right adjoint to* $\mathsf{F}\colon \mathtt{Top}(3_\ell) \to \mathtt{LQuant}^{op}$.

Proof. Let (X, τ) be a three-valued topological space. Since for every $x \in X$ the evaluation at x induces a strong homomorphism $\tau \xrightarrow{\delta_x} C_3^\ell$ by

$$\delta_x(f) = f(x), \qquad f \in \tau,$$

there exists a natural transformation $\eta\colon \mathrm{id}_{\mathtt{Top}(3_\ell)} \to \mathsf{Pt}3_\ell\,\mathsf{F}$ defined as follows:

$$\eta_{(X,\tau)}(x) = \delta_x, \qquad x \in X. \tag{2.54}$$

In fact, $\eta_{(X,\tau)}$ is continuous, because $\mathbb{A}_f(\delta_x) = f(x)$ for all $x \in X$ and all $f \in \tau$. Since $\mathtt{LQuant}^{op}$ is the dual category of $\mathtt{LQuant}$, we have to show that the following property is valid:

> For every left-sided and idempotent quantale $(\mathcal{Q}, *)$ and for every continuous map $(X, \tau) \xrightarrow{\varphi} (\mathsf{pt}3_\ell(\mathcal{Q}), \tau_{\mathcal{Q}})$ there exists a unique strong homomorphism $\mathcal{Q} \xrightarrow{\ulcorner\varphi\urcorner} \tau$ such that the following diagram is commutative:
>
> $$\begin{array}{ccc} X & \xrightarrow{\eta_{(X,\tau)}} & \mathsf{pt}3_\ell(\mathsf{F}(X,\tau)) \\ & {\scriptstyle\varphi}\searrow & \Big\downarrow {\scriptstyle \mathsf{Pt}3_\ell(\ulcorner\varphi\urcorner)} \\ & & \mathsf{pt}3_\ell(\mathcal{Q}) \end{array} \tag{2.55}$$

(Uniqueness) Let $\mathcal{Q} \xrightarrow{\ulcorner\varphi\urcorner} \tau$ be an arbitrary strong homomorphism making the diagram (2.55) commutative. Then for all $x \in X$ and $q \in \mathcal{Q}$ the following holds:

$$\begin{aligned} \big(\varphi(x)\big)(q) &= \Big(\mathsf{Pt}3_\ell(\ulcorner\varphi\urcorner)\big(\eta_{(X,\tau)}(x)\big)\Big)(q) \\ &= \big(\mathsf{Pt}3_\ell(\ulcorner\varphi\urcorner)(\delta_x)\big)(q) \\ &= \big(\delta_x \circ \ulcorner\varphi\urcorner\big)(q) \\ &= \big(\ulcorner\varphi\urcorner(q)\big)(x). \end{aligned}$$

Hence $\mathcal{Q} \xrightarrow{\ulcorner\varphi\urcorner} \tau$ is uniquely determined by φ and the commutativity of (2.55).
(Existence) Motivated by the uniqueness proof we define a map $\mathcal{Q} \xrightarrow{k} (C_3^\ell)^X$ by

$$\big(k(q)\big)(x) = \big(\varphi(x)\big)(q), \qquad q \in \mathcal{Q},\ x \in X$$

and observe that $k(q) = \mathbb{A}_q \circ \varphi$ holds for all $q \in \mathcal{Q}$. Since φ is continuous, k is a strong homomorphism from $\mathcal{Q}$ to τ. Referring again to the definition of k we obtain:

$$\big(\mathsf{Pt}_{\ell_3}(k)\big)(\delta_x)(q) = \delta_x \circ k(q) = \big(k(q)\big)(x) = \varphi(x)(q), \qquad x \in X,\ q \in \mathcal{Q}.$$

Hence k makes the diagram (2.55) commutative. □

Comment. Since the value of a strong homomorphism $\mathcal{Q} \xrightarrow{h} C_3^\ell$ at a two-sided element q of $\mathcal{Q}$ is always two-valued (i.e. $h(q) \in \{\bot, \top\}$), the restriction of the adjunction of Theorem 2.4.23 to the category $\mathtt{Frm}^{op} = \mathtt{Loc}$ of locales and to the category $\mathtt{Top}(2) \cong \mathtt{Top}$ of 2-valued topological spaces coincides with the well-known adjunction between locales and topological spaces (cf. [58, p. 42]). In this sense, left-sided and idempotent quantales can be viewed as *non-commutative frames* and the adjunction in Theorem 2.4.23 can be considered as a *non-commutative extension* of the adjunction between $\mathtt{Frm}^{op}$ and $\mathtt{Top}$.

Let $\mathcal{Q}$ be a left-sided and idempotent quantale. Since the $\mathcal{Q}$-component of the counit ε of the adjoint situation $\mathsf{F} \dashv \mathsf{Pt}_{3_\ell}$ is the unique arrow $\mathsf{FPt}_{3_\ell}(\mathcal{Q}) \to \mathcal{Q}$ in the sense of $\mathtt{LQuant}^{op}$ — i.e. the unique strong homomorphism $\mathcal{Q} \xrightarrow{\varepsilon_{\mathcal{Q}}} \tau_{\mathcal{Q}}$ such that the diagram

$$\begin{array}{ccc}
\mathsf{Pt}_{3_\ell}(\mathcal{Q}) & \xrightarrow{\eta_{\mathsf{Pt}_{3_\ell}(\mathcal{Q})}} & \mathsf{Pt}_{3_\ell}\mathsf{FPt}_{3_\ell}(\mathcal{Q}) \\
{\scriptstyle 1_{\mathsf{Pt}_{3_\ell}(\mathcal{Q})}}\big\downarrow & \swarrow {\scriptstyle \mathsf{Pt}_{3_\ell}(\varepsilon_{\mathcal{Q}})} & \\
\mathsf{Pt}_{3_\ell}(\mathcal{Q}) & &
\end{array}$$

is commutative, we conclude from the proof of Theorem 2.4.23 that $\varepsilon_{\mathcal{Q}}$ has the form $\big(\varepsilon_{\mathcal{Q}}(q)\big)(h) = h(q)$ where $q \in \mathcal{Q}$ and $h \in \mathsf{Pt}_{3_\ell}(\mathcal{Q})$ — i.e.

$$\varepsilon_{\mathcal{Q}}(q) = \mathbb{A}_q, \qquad q \in \mathcal{Q}.$$

We are now interested in those left-sided and idempotent quantales $\mathcal{Q}$ for which the counit at $\mathcal{Q}$ (i.e. $\varepsilon_{\mathcal{Q}}$) is an isomorphism.

Theorem 2.4.24. *A left-sided and idempotent quantale $(\mathcal{Q}, *)$ is spatial if and only if strong homomorphisms $\mathcal{Q} \xrightarrow{h} C_3^\ell$ separate elements of $\mathcal{Q}$ — i.e. for $q_1, q_2 \in \mathcal{Q}$ with $q_1 \neq q_2$ there exists a strong homomorphism $\mathcal{Q} \xrightarrow{h} C_3^\ell$ with $h(q_1) \neq h(q_2)$.*

Proof. (a) Let $(\mathcal{Q}, *)$ be a spatial, left-sided and idempotent quantale. Then for $q_1, q_2 \in \mathcal{Q}$ with $q_1 \not\leq q_2$ there exists a prime element $p \in \mathcal{Q}$ with $q_2 \leq p$ but $q_1 \not\leq p$. Now let $\mathcal{Q} \xrightarrow{h} C_3^\ell$ be the strong homomorphism determined by p (cf. Theorem 2.4.21). Then we obtain: $h(q_1) = \top$ and $h(q_2) \leq a$. Hence strong homomorphisms with codomain C_3^ℓ separate elements in $\mathcal{Q}$.
(b) Let $(\mathcal{Q}, *)$ be a left-sided and idempotent quantale in which C_3^ℓ-valued strong homomorphisms separate elements of $\mathcal{Q}$. We show that prime elements are order

generating in $\mathcal{Q}$ (i.e. $(\mathcal{Q}, *)$ is spatial). For this purpose we choose $q_1 \in \mathcal{Q} \setminus \{\top\}$ and put

$$q_2 = \bigwedge\{p \in \mathcal{Q} \mid p \text{ prime in } (\mathcal{Q}, *) \text{ and } q_1 \leq p\}.$$

Let us assume $q_1 \neq q_2$ and choose a strong homomorphism $\mathcal{Q} \xrightarrow{h} C_3^\ell$ with the property $h(q_1) \neq h(q_2)$. Since $q_1 \leq q_2$ and h is isotone, $h(q_2) \not\leq h(q_1)$ follows. Further, let $p \in \mathcal{Q}$ be the prime element corresponding to h (cf. Lemma 2.4.20) — i.e. $p = \bigvee\{q \in \mathcal{Q} \mid h(q) \leq a\}$. Now we distinguish the following cases.
Case 1: If $h(q_1) = \bot$ and $a \leq h(q_2)$, then $q_1 * \top \leq p$ and $q_2 * \top \not\leq p$ follow. Now we define an element $q_0 \in \mathcal{Q}$ by $q_0 = \bigvee\{z \in \mathcal{Q} \mid z * \top \leq p\}$. Since $q_0 * \top \leq p$, it is easily seen that q_0 is right-sided (i.e. two-sided) and prime. Now we infer from $q_1 * \top \leq p$ that $q_1 \leq q_0$ holds. Finally, we apply the definition of q_2 and obtain:

$$q_2 * \top \leq q_0 * \top \leq p,$$

which is a contradiction to $q_2 * \top \not\leq p$.
Case 2: If $h(q_1) = a$ and $h(q_2) = \top$, then the relations $q_1 \leq p$ and $q_2 \not\leq p$ hold — a property which is obviously a contradiction to the definition of q_2.
We conclude from the previous cases that the assumption is false — i.e. q_1 and q_2 coincide. Hence $(\mathcal{Q}, *)$ is spatial. □

The next proposition is an immediate corollary of Theorem 2.4.24 and is a more precise version of the main theorem in [103].

Proposition 2.4.25. *Let $(\mathcal{Q}, *)$ be a left-sided and idempotent quantale, and $\sigma(\mathcal{Q})$ be the set of all prime elements in $(\mathcal{Q}, *)$. Then $(\mathcal{Q}, *)$ is spatial if and only if $(\mathcal{Q}, *)$ is isomorphic to a subquantale of $\big(C_3^\ell\big)^{\sigma(\mathcal{Q})}$.*

Corollary 2.4.26. *Let $(\mathcal{Q}, *)$ be a left-sided and idempotent quantale, and $\varepsilon_{\mathcal{Q}}$ be the counit of $\mathsf{F} \dashv \mathsf{Pt}_{3_\ell}$ at $(\mathcal{Q}, *)$. Then $\varepsilon_{\mathcal{Q}}$ is an isomorphism if and only if $(\mathcal{Q}, *)$ is a spatial quantale.*

Proof. By definition the homomorphism $\mathcal{Q} \xrightarrow{\varepsilon_{\mathcal{Q}}} \tau_{\mathcal{Q}}$ is surjective. Hence $\varepsilon_{\mathcal{Q}}$ is an isomorphism if and only if $\varepsilon_{\mathcal{Q}}$ is injective if and only if C_3^ℓ-valued strong homomorphisms separate elements in $\mathcal{Q}$ if and only if $(\mathcal{Q}, *)$ is spatial. □

A three-valued topological space (X, τ) is called *sober* if the unit η of $\mathsf{F} \dashv \mathsf{Pt}_{3_\ell}$ at (X, τ) is a *homeomorphism* — i.e. an isomorphism in $\mathtt{Top}(3_\ell)$. Since for all $x \in X$ and all $f \in \tau$ the relation $\mathbb{A}_f(\delta_x) = f(x)$ holds, $\eta_{(X,\tau)}$ is an isomorphism if and only if $\eta_{(X,\tau)}$ is bijective.

In order to give a characterization of sobriety we recall Kolmogoroff's type of separation axiom —– the so-called T_0-axiom.

Definition 2.4.27. A three-valued topological space (X, τ) is *Kolmogoroff separated* (i.e. a T_0-space (cf. [99])) if for all $x_1, x_2 \in X$ with $x_1 \neq x_2$ there exists an $f \in \tau$ such that $f(x_1) \neq f(x_2)$ holds.

Obviously a three-valued topological space (X, τ) is a T_0-space if and only if the unit of $\mathsf{F} \dashv \mathsf{Pt}_{3_\ell}$ at (X, τ) is injective. Hence (X, τ) is a three-valued sober space if and only if (X, τ) is a T_0-space and every strong homomorphism $\tau \xrightarrow{h} C_3^\ell$ is induced by an element of X — i.e. there exists an $x \in X$ with $h = \delta_x$.

Lemma 2.4.28. *For every left-sided and idempotent quantale* $(\mathcal{Q}, *)$ *the three-valued topological space* $(\mathsf{pt}_{3_\ell}(\mathcal{Q}), \tau_{\mathcal{Q}})$ *is sober.*

Proof. It is evident that $(\mathsf{pt}_{3_\ell}(\mathcal{Q}), \tau_{\mathcal{Q}})$ is a T_0-space. Now let $\tau_{\mathcal{Q}} \xrightarrow{h} C_3^\ell$ be a strong homomorphism. Then the composition $p = h \circ \varepsilon_{\mathcal{Q}}$ is a strong homomorphism from $\mathcal{Q}$ to C_3^ℓ, and hence an element of $\mathsf{pt}_{3_\ell}(\mathcal{Q})$. Obviously the relation

$$\mathbb{A}_q(p) = p(q) = h(\varepsilon_{\mathcal{Q}}(q)) = h(\mathbb{A}_q)$$

holds for all $q \in \mathcal{Q}$ — i.e. $h = \delta_p$. Thus h is induced by p. □

To sum up we underline the fact that spatial, left-sided and idempotent quantales and three-valued sober spaces are equivalent concepts. A more categorical formulation of this situation is the statement that the adjunction formulated in Theorem 2.4.23 restricts to an equivalence of categories between the subcategory of three-valued sober spaces and the subcategory of spatial, left-sided and idempotent quantales.

As application of this result to Example 2.4.8 we obtain that the quantale $\mathbb{L}(A)$ of all closed left ideals of a unital C^*-algebra A can be identified with a three-valued sober space $\big(\mathsf{pt}_{3_\ell}(\mathbb{L}(A)), \tau_{\mathbb{L}(A)}\big)$. Since A has a unit, it follows from Neuman's series that $\big(\mathsf{pt}_{3_\ell}(\mathbb{L}(A)), \tau_{\mathbb{L}(A)}\big)$ is compact — i.e. the universal upper bound of $\tau_{\mathbb{L}(A)} \cong \mathbb{L}(A)$ is compact.

Before we proceed, let us dwell a little bit longer on the topologization problem of $\mathbb{L}(A)$.

Example 2.4.29. Let $(A, +, \cdot, ^*)$ be a non-commutative, unital C^*-algebra. Then we know from the theory of C^*-algebras that maximal left ideals can be identified with pure states. The situation is as follows (cf. [61, 10.2.10 Theorem]):

- A state ρ of A is pure if and only if its *left kernel* $L_\rho = \{a \in A \mid \rho(a^* \cdot a) = 0\}$ is a maximal left ideal.
- For every maximal left ideal I there exists a unique pure state ρ of A such that I coincides with the left kernel of ρ.

Further, let $\mathbb{L}(A)$ be the quantale of all closed left ideals of A. Since every closed left ideal is an intersection of maximal left ideals (see again [61, 10.2.10 Theorem (iii)]), it is sufficient to consider only maximal left ideals for the topologization of $\mathbb{L}(A)$. Therefore let $\mathfrak{p}(A)$ be the set of all pure states of A. For every closed left ideal I we introduce a map $\mathfrak{p}(A) \xrightarrow{\mathbb{A}_I} C_3^\ell$ by (cf. (2.51)):

$$\mathbb{A}_I(\rho) = \begin{cases} \bot, & I * A \subseteq L_\rho, \\ a, & I * A \nsubseteq L_\rho \text{ and } I \subseteq L_\rho, \\ \top, & I \nsubseteq L_\rho, \end{cases} \qquad \rho \in \mathfrak{p}(A).$$

Then $\tau_{\mathfrak{p}} = \{\mathbb{A}_I \mid I \in \mathbb{L}(A)\}$ is a three-valued topology on $\mathfrak{p}(A)$ and is isomorphic to $\mathbb{L}(A)$. In this sense we can also understand the three-valued topological space $(\mathfrak{p}(A), \tau_{\mathfrak{p}})$ as the topologization of the quantale $\mathbb{L}(A)$. Obviously, the sobrification of $(\mathfrak{p}(A), \tau_{\mathfrak{p}})$ is isomorphic to $\big(\mathsf{pt}_{3_\ell}(\mathbb{L}(A)), \tau_{\mathbb{L}(A)}\big)$.

We finish this subsection with a property which explains the relationship between sobriety and stratification of three-valued topological space. For this purpose let $\underline{a}$ be the constant map determined by $a \in C_3$. A three-valued topological space (X, τ) is said to be *stratified* if $\underline{a}$ belongs to τ.

Proposition 2.4.30. *Let X be a nonempty set. Then a three-valued sober space (X, τ) is never stratified — i.e. $\underline{a} \notin \tau$.*

Proof. Let us fix an element $x \in X$ and define a two-sided prime element of τ by $f_0 = \bigvee\{f \in \tau \mid f(x) = \bot\}$. Further, let $\tau \xrightarrow{h} C_3^\ell$ be the unique strong homomorphism corresponding to f_0 (cf. Theorem 2.4.21). Because $\underline{a} \not\leq f_0$ the property $h(\underline{a}) = \top$ follows from (2.51). Moreover, the sobriety of (X, τ) ensures the existence of an element $z \in X$ such that $h = \delta_z$ holds. If we now assume $\underline{a} \in \tau$, then we obtain $h(\underline{a}) = \underline{a}(z) = a$, which is a contradiction to $h(\underline{a}) = \top$. Hence (X, τ) cannot be stratified. □

Exercises

2.4.1. (See Exercises 2.2.1 and 2.3.1) Let C_3 be the chain with three elements. Referring to the comment after Exercise 2.3.1 there exists a unique idempotent, left-sided, but not right-sided quantale and a unique idempotent, right-sided, but not left-sided quantale on C_3. Prove:

(a) On C_3 there exists a unique idempotent, but not left-sided and not right-sided quantale which is necessarily commutative and unital.
(b) On C_3 there exists a unique two-sided and idempotent quantale which is necessarily commutative.
(c) On C_3 there exists precisely seven two-sided, but not idempotent quantales. Only two of them are non-commutative.

2.4.2. Let C_3 be the chain with three elements provided with the following multiplications (see (18) and (8) in Exercise 2.2.1):

$*_r$	$\bot$	a	$\top$
$\bot$	$\bot$	$\bot$	$\bot$
a	$\bot$	a	a
$\top$	$\bot$	$\top$	$\top$

$*$	$\bot$	a	$\top$
$\bot$	$\bot$	$\bot$	$\bot$
a	$\bot$	$\bot$	$\bot$
$\top$	$\bot$	a	$\top$

Show

(a) $(C_3, *_r)$ is a right-sided, idempotent and spatial quantale, but not left-sided.
(b) $(C_3, *)$ is a two-sided and spatial quantale, but non-idempotent.
(c) The identity 1_{C_3} is neither a strong homomorphism $(C_3, *_r) \to C_3^\ell$ nor a strong homomorphism $(C_3, *) \to C_3^\ell$.
(d) There exists only a single strong homomorphism $(C_3, *_r) \xrightarrow{h} C_3^\ell$.
(e) There exists only a single strong homomorphism $(C_3, *) \xrightarrow{h} C_3^\ell$.
(f) Conclude from (d) – (e) that in Theorems 2.4.21 and 2.4.24 neither the assumption of left-sidedness nor the assumption of idempotency imposed on $(X, *)$ can be dropped.

2.4.3. Let X and Y be complete lattices. With every arbitrary map $X \xrightarrow{f} Y$ we associate a tensor $\overline{f} \in X \otimes Y$ defined by $\overline{f} = \bigvee_{x \in X} x \otimes f(x)$. If additionally $(X, \wedge)$ and $(Y, \wedge)$ are frames, then show that $\overline{f}$ has the following representation:

$$\overline{f}(x) = \bigvee \{ f(z) \wedge \overline{f}(x) \mid x \wedge z \neq \bot \}, \qquad x \in X \setminus \{\bot\}.$$

(Hint: For $x \in X$ observe that $x \otimes \overline{f}(x) \leq \overline{f}$ and subsequently use Corollary 2.4.13 and the property that the universal bimorphism is meet-preserving.)

2.4.4. Let $\mathbb{M}_2$ be the C^*-algebra of all square matrices of order 2 with complex coefficients. Further, let A be an element of $\mathbb{M}_2$ with $\mathrm{rank}(A) = 1$. Show:

(a) If $\mathbb{L}$ is the left ideal of $\mathbb{M}_2$ generated by A — i.e. $\mathbb{L} = \{B \cdot A \mid B \in \mathbb{M}_2\}$, then there exist complex numbers $\alpha, \beta \in \mathbb{C}$ such that $|\alpha|^2 + |\beta|^2 = 1$ and the following relation holds:

$$\mathbb{L} = \left\{ \alpha \cdot \begin{pmatrix} a & 0 \\ b & 0 \end{pmatrix} + \beta \cdot \begin{pmatrix} 0 & a \\ 0 & b \end{pmatrix} \mid a, b \in \mathbb{C} \right\}.$$

(b) Every non-trivial and proper left ideal of $\mathbb{M}_2$ is maximal.

2.4.5. Let $(X, *)$ be a *factor* quantale — i.e. the subquantale of all two-sided elements of $(X, *)$ coincides with $\{\bot, \top\}$. Show: If $(X, *)$ is left-sided and semi-unital, then $(X, *)$ is idempotent and the multiplication $*$ has the following form (cf. Exercise 2.3.10):

$$x * y = \begin{cases} \bot, & x = \bot, \\ y, & x \neq \bot, \end{cases} \qquad x, y \in X.$$

In particular, $(X, *)$ is spatial (cf. Exercise 2.3.10(b)).

2.4.6. Let $\mathbb{M}_n$ be the C^*-algebra of all square matrices of order n with complex coefficients, and let $\mathbb{L}(\mathbb{M}_n)$ be the quantale of all left ideals of $\mathbb{M}_n$ provided with the ideal multiplication $*$. Show that $\big(\mathbb{L}(\mathbb{M}_n), *\big)$ is a factor and conclude from Exercise 2.4.5 that the ideal multiplication has the following form:

$$L_1 * L_2 = \begin{cases} \{0\}, & L_1 = \{0\}, \\ L_2, & L_1 \neq \{0\}, \end{cases} \qquad L_1, L_2 \in \mathbb{M}_n.$$

(Hint: Jordan's Theorem.)

2.4.7. Let c be a nucleus on a quantale $(X, *)$ and $A = c(X)$ be its range. Show that the following assertions are equivalent:

(a) The quotient $(A, \star) = (X, *)/c$ is a left-sided and idempotent quantale.
(b) The nucleus c satisfies the conditions $c(\top * x) = c(x)$ and $c(x * x) = c(x)$ for all $x \in X$.

2.4.8. Let $(C_3^\ell, *_\ell)$ be the left-sided, non-commutative and idempotent chain of three elements.

(a) Show that $(C_3^\ell, *_\ell)$ is spatial and construct the corresponding three-valued sober space $\big(\mathsf{pt}_{3_\ell}(C_3^\ell), \tau_{C_3^\ell}\big)$.
(b) Consider now the singleton $\{1_{C_3^\ell}\}$ as a three-valued topological subspace of $\big(\mathsf{pt}_{3_\ell}(C_3^\ell), \tau_{C_3^\ell}\big)$ ("open subsets" are the restrictions of those in $\tau_{C_3^\ell}$). Show that the three-valued topology on $\{1_{C_3^\ell}\}$ is stratified and isomorphic to $(C_3^\ell, *_\ell)$.

2.4.9. Let $(C_3^\ell, *_\ell)$ be the left-sided, non-commutative and idempotent chain of three elements. Further, let X be a set and $X \xrightarrow{\underline{a}} C_3$ be the constant map determined by a, where a is the unique element of C_3 which is different from the universal bounds of C_3.

(a) Show that $\underline{a}$ is a left-unit in $((C_3^\ell)^X, *)$.
(b) Let (X, τ) be a three-valued topological space and $\sigma = \{f \vee (g * \underline{a}) \mid f, g \in \tau\}$. Show that (X, σ) is a stratified three-valued topological space.

2.4.10. (Cf. [94]) Let $(X, \wedge)$ and $(Y, \wedge)$ be frames. Further, $\mathtt{Ant}(X, Y)$ denotes the complete lattice of all antitone maps $X \xrightarrow{f} Y$ ordered pointwisely — i.e.

$$f \le g \iff f(x) \le g(x) \quad \text{for all } x \in X.$$

On $\mathtt{Ant}(X, X)$ we introduce a binary operation $*$ by

$$(g * f)(x) = \bigvee\{g(y) \mid y \neq \bot,\ y \le f(x)\}, \qquad f, g \in \mathtt{Ant}(X, X). \tag{2.56}$$

Verify the following assertions:

(a) $(\mathtt{Ant}(X, X), *)$ is a quantale.
(b) The map $\mathtt{Ant}(X, X) \xrightarrow{h} X \otimes X$ defined by (cf. Exercise 2.4.3)

$$h(f) = \bigvee_{x \in X} x \otimes f(x), \qquad f \in \mathtt{Ant}(X, X)$$

is surjective and join-preserving (i.e. a regular epimorphism in Sup). If $X \otimes X$ is viewed as a subset of $\mathtt{Ant}(X, X)$, then the right adjoint map $h^{\vdash}$ of h coincides with the inclusion map. Moreover, for all $f, g \in \mathtt{Ant}(X, X)$ the associated closure operator $c_h = h^{\vdash} \circ h$ satisfies the following properties:

(i) $c_h(g) * f = g * f$.
(Hint: Exercise 2.4.3.)

(ii) $g * c_h(f) \le c_h(g * f)$.
(Hint: Fix $x_0, y \in X$ with $x_0 \neq \bot, y \neq \bot$ and $x_0 \otimes y \le \bigvee_{x \in X} x \otimes f(x)$. Since X is a frame, use again Corollary 2.4.13 and show

$$x_0 = \bigvee \{x_0 \wedge x \mid x \in X : \ x \wedge x_0 \neq \bot, \ y \wedge f(x) \neq \bot\}.$$

Further, for $x \in X$ with $x \wedge x_0 \neq \bot$ and $y \wedge f(x) \neq \bot$ observe that

$$(x \wedge x_0) \otimes g(y) \le x \otimes \big((g * f)(x)\big)$$

and then conclude that $x_0 \otimes g(y) \le \bigvee_{x \in X} x \otimes ((f * g)(x))$.)

(c) The tensor product $X \otimes X$ is a quantale w.r.t. the multiplication $\star$ defined by

$$g \star f = \bigvee_{x \in X} x \otimes ((g * f)(x)), \qquad f, g \in X \otimes X,$$

where $*$ is the multiplication of the quantale $(\mathtt{Ant}(X, X), *)$ (cf. (2.56)).
(Hint: Conclude from (b) and Exercise 2.2.3 that the closure operator c_h associated with h is a nucleus.)

2.4.11. Let $\mathbb{M}_2$ be the C^*-algebra of all square matrices of order 2 with complex coefficients and $\mathbb{L}(\mathbb{M}_2)$ be the quantale of all left ideals of $\mathbb{L}(\mathbb{M}_2)$ (cf. Example 2.4.8). Further, let C_3^{ℓ} be the non-commutative, idempotent, left-sided three-chain $C_3 = \{\bot, a, \top\}$. Show:

(a) The set $\sigma(\mathbb{L}(\mathbb{M}_2))$ of all prime elements in $(\mathbb{L}(\mathbb{M}_2), *)$ coincides with the set $\{P \in \mathbb{L}(\mathbb{M}_2) \mid P \neq \mathbb{M}_2\}$.
(Hint: Exercise 2.4.6 and see also Exercise 2.3.10 (b).)

(b) There exists a bijective map between $\{P \in \mathbb{L}(\mathbb{M}_2) \mid \{0\} \neq P \neq \mathbb{M}_2\}$ and the set $\mathfrak{p}(\mathbb{M}_2)$ of all pure states of $\mathbb{M}_2$.
(Hint: Exercise 2.4.4 (b)).

(c) If $\big(\mathsf{pt}(\mathbb{L}(\mathbb{M}_2)), \tau_{\mathbb{L}(\mathbb{M}_2)}\big)$ is the three-valued (i.e. C_3^{ℓ}-valued) sober space corresponding to $\mathbb{M}_2$, then the maps of $\tau_{\mathbb{L}(\mathbb{M}_2)}$ have the following form:

$$\mathbb{A}_{\{0\}} = \underline{\bot} \ \text{ and } \ \mathbb{A}_I(P) = \begin{cases} \top, & I \neq P, \\ a, & P = I, \end{cases} \quad \text{where } \ I \neq \{0\}, \ P \in \sigma(\mathbb{L}(\mathbb{M}_2)).$$

2.5 Balanced and Bisymmetric Quantales

The definition of the spectrum of a non-commutative C^*-algebra (see Sect. 2.5.3 infra) forces us to deal simultaneously with left- and right-symmetric quantales (cf. Lemma 2.4.3 (i)). For this purpose we introduce the following terminology.

A quantale $(X, *)$ is *bisymmetric* if the relation

$$(x_1 * x_2) * (x_3 * x_4) = (x_1 * x_3) * (x_2 * x_4)$$

holds for all $x_1, x_2, x_3, x_4 \in X$.

Since the multiplication of quantales is associative by definition, every commutative quantale is bisymmetric. In the case of unital quantales commutativity and bisymmetry are even equivalent conditions. Hence the bisymmetry axiom will play a significant rôle in the case of non-commutative *and* non-unital quantales.

With regard to Lemma 2.4.3 (i) a typical class of non-commutative and bisymmetric quantales is given by non-commutative, idempotent and left-sided (or right-sided) quantales (cf. Sect. 2.4). In this section we will encounter more examples of non-commutative and bisymmetric quantales which are in general *not* idempotent.

As a first property we show that the tensor product preserves the bisymmetry axiom in the case of balanced quantales.

Proposition 2.5.1. *Let $(X, *)$ and $(Y, *)$ be balanced quantales. Then $(X \otimes Y, \star)$ is bisymmetric if and only if $(X, *)$ and $(Y, *)$ are bisymmetric.*

Proof. By Proposition 2.3.14 (i) and (ii) the tensor product transmits the bisymmetry to its factors. On the other hand, since the relation

$$\begin{aligned}\big((x_1 \otimes y_1) \star (x_2 \otimes y_2)\big) &\star \big((x_3 \otimes y_3) \star (x_4 \otimes y_4)\big)\\ &= \big((x_1 * x_2) * (x_3 * x_4)\big) \otimes \big((y_1 * y_2) * (y_3 * y_4)\big)\end{aligned}$$

holds for each $x_1, x_2, x_3, x_4 \in X$ and $y_1, y_2, y_3, y_4 \in Y$, the bisymmetry of $(X, *)$ and $(Y, *)$ entails the bisymmetry of their tensor product. □

Example 2.5.2. Let X be a complete lattice with at least three elements. Further, let $*_\ell$ and $*_r$ be binary operations on X (in the sense of `Set`) given by (see Exercise 2.3.10)

$$x *_\ell y = \begin{cases} \bot, & x = \bot, \\ y, & x \neq \bot, \end{cases} \quad \text{and} \quad x *_r y = \begin{cases} \bot, & y = \bot, \\ x, & y \neq \bot, \end{cases} \qquad x, y \in X.$$

Then it follows from Exercise 2.3.10 (a) that $X^\ell = (X, *_\ell)$ is a left-sided and idempotent (but not right-sided) quantale and $X^r = (X, *_r)$ is a right-sided and idempotent (but not left-sided) quantale. Hence we conclude from Lemma 2.3.12, Propositions 2.3.19, 2.5.1 and Theorem 2.4.11 that their tensor product $(X^\ell \otimes X^r, \star)$ is a semi-unital, semi-integral and bisymmetric, but not idempotent, quantale. A typical example of a non-idempotent tensor of $X^\ell \otimes X^r$ is given by $(x \otimes \top) \vee (\top \otimes x)$, where x is an element of X with $\bot \neq x \neq \top$.

The next proposition gives a condition under which the semi-integral regularization of quantales leads to bisymmetric and even balanced quantales. For this purpose we introduce the following notation.

If $(X, *)$ is an arbitrary quantale, then $\mathbb{L}(X)$ and $\mathbb{R}(X)$ denote the subquantales of all left-sided elements and all right-sided elements of X.

Proposition 2.5.3. *Let $(X, *)$ be a quantale. If $\mathbb{L}(X)$ is idempotent, then the semi-integral regularization $(X, \odot)$ of $(X, *)$ is a balanced and bisymmetric quantale where the multiplication is determined by* (cf. (2.42) *in Lemma 2.3.18*):

$$x_1 \odot x_2 = x_1 * \top * x_2, \qquad x_1, x_2 \in X.$$

Proof. Since $\mathbb{L}(X)$ is idempotent, $(X, *)$, and a fortiori $(X, \odot)$, is balanced. Referring again to the idempotency of $\mathbb{L}(X)$, we use the left-symmetry of $\mathbb{L}(X)$ (cf. Lemma 2.4.3 (i)) and obtain for each $x_1, x_2, x_3, x_4 \in X$:

$$\begin{aligned}(x_1 \odot x_2) \odot (x_3 \odot x_4) &= x_1 * \top * x_2 * \top * x_3 * \top * x_4 \\ &= x_1 * \top * x_3 * \top * x_2 * \top * x_4 = (x_1 \odot x_3) \odot (x_2 \odot x_4).\end{aligned}$$

Hence $(X, \odot)$ is balanced and bisymmetric. □

Motivated by Proposition 2.5.3 and the situation given by C^*-algebras (cf. Example 2.4.8) we introduce now the following terminology due to C.J. Mulvey and J.W. Pelletier (cf. [84]).

Definition 2.5.4. An involutive quantale $(X, *, ')$ is a *Gelfand quantale* if and only if the following conditions hold[4]:

(i) $\mathbb{L}(X)$ is idempotent.
(ii) $x \leq x * x'$ for each $x \in \mathbb{L}(X)$.

In every Gelfand quantale two-sided elements are hermitian, and $x = x * x' * x$ holds for each left-sided or right-sided element $x \in X$. Moreover, if $\mathfrak{Q}_{\mathsf{RL}}(X)$ is the subquantale of a Gelfand quantale $(X, *, ')$ generated by $\mathbb{L}(X) \cup \mathbb{R}(X)$, then $(\mathfrak{Q}_{\mathsf{RL}}(X), *, ')$ is again a Gelfand quantale. It is worthwhile to note that every element of $\mathfrak{Q}_{\mathsf{RL}}(X)$ has the following simple representation:

$$x = \bigvee_{i \in I} r_i * \ell_i, \qquad r_i \in \mathbb{R}(X) \text{ and } \ell_i \in \mathbb{L}(X). \tag{2.57}$$

In this sense we are performing the "ideal multiplication" in a "wrong way". This intuitive view of the ideal multiplication will play a certain rôle when we formulate

[4]In [84] C.J. Mulvey and J.W. Pelletier require the existence of a unit — a property which seems to be superfluous in this context.

the spectrum of a non-commutative C^*-algebra in Sect. 2.5.3. Now we continue with two important examples.

Example 2.5.5. Let $(A, \cdot, +, ^*)$ be a unital C^*-algebra with unit e and $\mathbb{MAX}$ be the complete lattice of all closed linear subspaces of A ordered by set-inclusion. If M and N are closed linear subspaces, then the product $M * N$ is given by the closure of the set of all finite sums $\sum_{i \in I} a_i \cdot b_i$ with $a_i \in M$ and $b_i \in N$ — i.e. $M * N = \overline{MN}$. It is easily seen that $(\mathbb{MAX}, *)$ is a unital quantale where the unit coincides with the 1-dimensional subspace $\langle e \rangle$ generated by the unit e of A. Moreover, the involution * on A induces an involution $'$ on $\mathbb{MAX}$ as follows:

$$M' = \{a^* \mid a \in M\}, \qquad M \in \mathbb{MAX}.$$

Then $'$ coincides with the formation taking adjoint closed linear subspaces, and the quadruple $(\mathbb{MAX}, *, \langle e \rangle, ')$ is an involutive and unital quantale.
Since left-sided (resp. right-sided) elements of $\mathbb{MAX}$ are closed left (resp. right) ideals of A and the restriction of $*$ to ideals coincides with the ideal multiplication, it follows immediately from Corollary 2.4.5 and Corollary 2.4.6 that $(\mathbb{MAX}, *, \langle e \rangle, ')$ is a unital Gelfand quantale.
Finally, let $\mathfrak{Q}_{RL}(\mathbb{MAX})$ be the Gelfand quantale generated by all closed left ideals and all closed right ideals of A. Since the unit of $\mathbb{MAX}$ is a 1-dimensional subspace of A, we conclude from (2.57) that $\langle e \rangle$ is not contained in $\mathfrak{Q}_{RL}(\mathbb{MAX})$ provided A is not 1-dimensional. Hence this property is a further confirmation of an observation we made at the begining of this section — namely the significant rôle of non-unital quantales in the ideal theory of non-commutative C^*-algebras.

Now we return to Example 2.3.31 from the perspective of Gelfand quantales.

Example 2.5.6. Let X be a complete lattice with at least two different elements endowed with an order-reversing involution $x \longmapsto x^\perp$, and let $([X, X], \circ, 1_X, ')$ be the involutive and unital quantale of all join-preserving self-maps of X (cf. Example 2.3.31). Since every left-sided element f of $[X, X]$ has the form

$$f(x) = f_u(x) = \begin{cases} \top, & x \not\leq u, \\ \bot, & x \leq u, \end{cases} \qquad x \in X,$$

the subquantale $\mathbb{L}([X, X])$ is isomorphic to $X^\ell = (X, *_\ell)$ (cf. Example 2.5.2), and the isomorphism is given by $f \longmapsto u_f = \bigvee\{x \in X \mid f(x) = \bot\}$ for each $f \in \mathbb{L}([X, X])$. Consequently, $\mathbb{L}([X, X])$ is idempotent. Further, we observe that

$$(f_u)'(x) = \begin{cases} u^\perp, & x \neq \bot, \\ \bot, & x = \bot, \end{cases} \qquad x \in X.$$

Hence, if f is left-sided, then $f \leq f \circ f'$ holds if and only if $u_f = \top$ or $u_f^{\perp} \not\leq u_f$. This means that $([X, X], \circ, 1_X, ')$ is a Gelfand quantale if and only if $(X, {}^{\perp})$ is a complete *ortho-lattice* — i.e. $x \wedge x^{\perp} = \perp$ for all $x \in X$.

With regard to Proposition 2.5.3 we can summarize the previous result as follows.

Corollary 2.5.7. *The semi-integral regularization of every Gelfand quantale is a balanced and bisymmetric quantale.*

In order to describe some categorical constructions we introduce now the category `BSQuant` of balanced and bisymmetric quantales and strong homomorphisms. Obviously, `BSQuant` is located between `LQuant` (cf. Sect. 2.4) and `BQuant` (cf. Sect. 2.3.1) and can therefore be considered as a non-idempotent extension of `LQuant` (cf. Example 2.5.2). In this context the reader may notice that non-commutative and unital quantales are objects of `BQuant`, but not of `BSQuant`.

Theorem 2.5.8. *The category* `BSQuant` *is a reflective subcategory of* `BQuant`.

Proof. Let $(X, *)$ be a balanced quantale. In order to construct the balanced and bisymmetric reflection $(X^{\flat}, *^{\flat})$ of $(X, *)$ we proceed as usual and consider the set $\mathbb{F}$ of all nuclei c on $(X, *)$ satisfying the following condition:

$$c\big((x_1 * x_2) * (x_3 * x_4)\big) = c\big((x_1 * x_3) * (x_2 * x_4)\big), \qquad x_1, x_2, x_3, x_4 \in X.$$

Then the meet c_0 of $\mathbb{F}$ is contained in $\mathbb{F}$. Consequently the regular quotient $(X^{\flat}, *^{\flat}) = (c_0(X), \star)$ w.r.t. c_0 is a balanced and bisymmetric quantale, and the quotient map $X \xrightarrow{\eta_X^{\flat}} X^{\flat}$ is a strong homomorphism. If $(Y, *)$ is a further balanced and bisymmetric quantale and $X \xrightarrow{h} Y$ is a further strong homomorphism, then the nucleus $c_h = h^{\vdash} \circ h$ associated with h is obviously contained in $\mathbb{F}$. Thus h factors uniquely through $\eta_X^{\flat}$ — this means that $(X^{\flat}, *^{\flat})$ is the balanced and bisymmetric reflection of $(X, *)$. □

It is easily seen that `BSQuant` is complete, and because of Theorem 2.5.8, the cocompleteness of `BQuant` is transmitted to `BSQuant`. Hence we can summarize the previous results as follows.

FACT I. *The category* `BSQuant` *is a complete, cocomplete and monoidal category.*

In the following considerations we will investigate the question to what extent the tensor product plays a rôle in the construction of coproducts in `BSQuant`. With regard to Theorem 2.3.35 it is sufficient to require that the left factor of the tensor product has a left unit, while the right factor has a right unit. This observation motivates the following non-commutative version of Theorem 2.3.35 which also generalizes Theorem 2.4.12.

Theorem 2.5.9. *Let $(X, *)$ be a strictly left-sided, bisymmetric quantale, and let $(Y, *)$ be a strictly right-sided, bisymmetric quantale. Then the tensor product $(X \otimes Y, \star)$ with the embeddings j_X and j_Y as coproduct injections is the coproduct of $(X, *)$ and $(Y, *)$ in* `BSQuant`. *In particular, the coproduct of $(X, *)$ and $(Y, *)$ in* `BSQuant` *is semi-unital.*

Proof. By Lemma 2.3.12 (ii), Propositions 2.3.14 (i) and (ii) and 2.5.1, the tensor product $(X \otimes Y, \star)$ is a semi-unital (a fortiori balanced) and bisymmetric quantale, and the embeddings j_X and j_Y are strong homomorphisms. Let $(Z, *)$ be a further balanced and bisymmetric quantale, and let $X \xrightarrow{h_X} Z$ and $Y \xrightarrow{h_Y} Z$ be further strong homomorphisms. We show that there exists a unique strong homomorphism $X \otimes Y \xrightarrow{k} Z$ such that $h_X = k \circ j_X$ and $h_Y = k \circ j_Y$ hold.

(a) (Uniqueness) Since $(X, *)$ is strictly left-sided and $(Y, *)$ is strictly right-sided, we can base our argumentation on the following important relation

$$(\top \otimes y) \star (x \otimes \top) = (\top * x) \otimes (y * \top) = x \otimes y, \qquad x \in X,\ y \in Y.$$

If $X \otimes Y \xrightarrow{k} Z$ is a strong homomorphism satisfying the conditions $h_X = k \circ j_X$ and $h_Y = k \circ j_Y$, then we obtain:

$$k(x \otimes y) = k\big((\top \otimes y) \star (x \otimes \top)\big) = h_Y(y) * h_X(x), \qquad x \in X,\ y \in Y. \quad (2.58)$$

Since every tensor is a join of elementary tensors, k is uniquely determined.

(b) (Existence) Let $Z \otimes Z \xrightarrow{\circledast} Z$ be the binary operation corresponding to the multiplication $*$ in Z. Motivated by (2.58), we now define a join-preserving map $X \otimes Y \xrightarrow{k} Z$ by $k = \circledast \circ c_{ZZ} \circ (h_X \otimes h_Y)$, where c_{ZZ} is the Z-component of the symmetry in `Sup` (cf. Lemma 2.1.15). We show that k is a strong homomorphism. Since $(Z, *)$ is balanced, we obtain

$$k(\top \otimes \top) = h_Y(\top) * h_X(\top) = \top * \top = \top$$

— i.e. k preserves the respective universal upper bounds. Further, let $x_1 \otimes y_1$ and $x_2 \otimes y_2$ be two elementary tensors in $X \otimes Y$. Now we use the bisymmetry of $(Z, *)$ and observe:

$$\begin{aligned} k\big((x_1 \otimes y_1) \star (x_2 \otimes y_2)\big) &= k\big((x_1 * x_2) \otimes (y_1 * y_2)\big) \\ &= h_Y(y_1 * y_2) * h_X(x_1 * x_2) \\ &= (h_Y(y_1) * h_Y(y_2)) * (h_X(x_1) * h_X(x_2)) \\ &= (h_Y(y_1) * h_X(x_1)) * (h_Y(y_2) * h_X(x_2)) \\ &= k(x_1 \otimes y_1) * k(x_2 \otimes y_2). \end{aligned}$$

Since every tensor is a join of elementary tensors, the map k is a strong homomorphism. Finally, for $x \in X$ and $y \in Y$ we obtain:

$$k(j_X(x)) = k(x \otimes \top) = h_Y(\top) * h_X(x) = h_X(\top) * h_X(x) = h_X(\top * x) = h_X(x),$$
$$k(j_Y(y)) = k(\top \otimes y) = h_Y(y) * h_X(\top) = h_Y(y) * h_Y(\top) = h_Y(y * \top) = h_Y(y).$$

Thus k satisfies the desired properties. □

The next corollary is a first application of Theorem 2.5.9.

Corollary 2.5.10. *Let $(X, *)$ be a quantale such that the subquantale $\mathbb{L}(X)$ of all left-sided elements and the subquantale $\mathbb{R}(X)$ of all right-sided elements are idempotent. Further, let $\mathfrak{Q}_{\mathsf{RL}}(X)$ be the subquantale generated by $\mathbb{L}(X) \cup \mathbb{R}(X)$. Then the semi-integral regularization of $\mathfrak{Q}_{\mathsf{RL}}(X)$ is a regular quotient of the coproduct $(\mathbb{L}(X) \otimes \mathbb{R}(X), \star)$ of $\mathbb{L}(X)$ and $\mathbb{R}(X)$.*

Proof. Let $\mathbb{L}(X) \xrightarrow{q_{\mathbb{L}}} X$ and $\mathbb{R}(X) \xrightarrow{q_{\mathbb{R}}} X$ be the respective embeddings. Further, let $\odot$ be the multiplication in $\mathfrak{Q}_{\mathsf{RL}}(X)$ induced by the semi-integral regularization of $\mathfrak{Q}_{\mathsf{RL}}(X)$ — i.e. $x \odot y = x * \top * y$ for each $x, y \in \mathfrak{Q}_{\mathsf{RL}}(X)$. Then $(\mathfrak{Q}_{\mathsf{RL}}(X), \odot)$ is balanced and bisymmetric. Referring to (2.58), the unique strong homomorphism $\mathbb{L}(X) \otimes \mathbb{R}(X) \xrightarrow{h} \mathfrak{Q}_{\mathsf{RL}}(X)$ with $q_{\mathbb{L}} = h \circ j_{\mathbb{L}(X)}$ and $q_{\mathbb{R}} = h \circ j_{\mathbb{R}(X)}$ is given by:

$$h(\ell \otimes r) = r \odot \ell = r * \ell, \qquad \ell \in \mathbb{L}(X),\ r \in \mathbb{R}(X).$$

By (2.57) the homomorphism h is surjective — i.e. a regular epimorphism in `BSQuant`. □

In the following we will apply Theorem 2.5.9 to the special setting where there exists additionally a pair $X \underset{\vartheta_Y}{\overset{\vartheta_X}{\rightleftarrows}} Y$ of join-preserving anti-homomorphisms between the strictly left-sided quantale $(X, *)$ and the strictly right-sided quantale $(Y, *)$ with the properties $\vartheta_Y \circ \vartheta_X = 1_X$ and $\vartheta_X \circ \vartheta_Y = 1_Y$. By Theorem 2.3.33 there exists a unique involution $'$ on $X \otimes Y$ such that $(X \otimes Y, \star, ')$ is an involutive quantale and the relations

$$(x \otimes \top)' = \top \otimes \vartheta_X(x) \quad \text{and} \quad (\top \otimes y)' = \vartheta_Y(y) \otimes \top$$

hold for each $x \in X$ and $y \in Y$. Under these assumptions we formulate the following:

Corollary 2.5.11. *Let $(Z, *)$ be a balanced and bisymmetric quantale. Further, let $Z \xrightarrow{q_X} X$ and $Z \xrightarrow{q_Y} Y$ be strong homomorphisms with the property*

$$\vartheta_X \circ q_X = q_Y \quad \textit{and} \quad \vartheta_Y \circ q_Y = q_X.$$

The pushout of q_X and q_Y in the sense of `BSQuant` *is the coequalizer $X \otimes Y \xrightarrow{\pi} S$ of $Z \underset{j_Y \circ q_Y}{\overset{j_X \circ q_X}{\rightrightarrows}} X \otimes Y$. Moreover, there exists a unique involution $'$ on $(S, *)$ such that $(S, *, ')$ is an involutive quantale and π is an involutive homomorphism.*

Proof. Since $(X \otimes Y, \star)$ with j_X and j_Y is the coproduct of $(X, *)$ and $(Y, *)$ (cf. Theorem 2.5.9), the first assertion of the corollary is evident. In order to show the existence of an involution on $(S, *)$ transforming the quotient map π into an involutive homomorphism it is sufficient to prove that the nucleus c_π associated with π is involutive (see Sect. 2.2.2).
Referring to the construction of coequalizers, c_π is the smallest nucleus c on $X \otimes Y$ satisfying the following condition

$$c\big((j_X \circ q_X)(z)\big) = c\big((j_Y \circ q_Y)(z)\big), \qquad z \in Z. \tag{2.59}$$

Now we use the involution on $X \otimes Y$ and introduce a further nucleus $\widehat{c}$ on $X \otimes Y$ by:

$$\widehat{c}(f) = \big(c_\pi(f')\big)', \qquad f \in X \otimes Y.$$

Then the relations

$$\big(\widehat{c}((j_X \circ q_X)(z))\big)' = c_\pi\big((q_X(z) \otimes \top)'\big) = c_\pi\big(\top \otimes ((\vartheta_X \circ q_X)(z))\big) = c_\pi(\top \otimes q_Y(z))$$

and

$$\big(\widehat{c}((j_Y \circ q_Y)(z))\big)' = c_\pi\big((\top \otimes q_Y(z))'\big) = c_\pi\big(((\vartheta_Y \circ q_Y)(z)) \otimes \top\big) = c_\pi(q_X(z) \otimes \top)$$

hold for all $z \in Z$. Since c_π satisfies (2.59), we obtain:

$$\widehat{c}((j_X \circ q_X)(z)) = \big(c_\pi((j_Y \circ q_Y)(z))\big)' = \big(c_\pi((j_X \circ q_X)(z))\big)' = \widehat{c}((j_Y \circ q_Y)(z)).$$

Hence $\widehat{c}$ also satisfies (2.59). In particular, $c_\pi \le \widehat{c}$ follows, which means that c_π is involutive. □

Let $(G, *, ')$ be a Gelfand quantale. Then we can put Corollary 2.5.11 in concrete terms as follows. Let $\mathbb{I}(G)$ be the subquantale of all two-sided elements of G, $\mathbb{L}(G)$ be the subquantale of all left-sided elements, and $\mathbb{R}(G)$ be the subquantale of all right-sided elements. Further, let $\mathbb{I}(G) \xrightarrow{q_{\mathbb{L}}} \mathbb{L}(G)$ and $\mathbb{I}(G) \xrightarrow{q_{\mathbb{R}}} \mathbb{R}(G)$ be the respective embeddings. The respective restrictions of the involution $'$ on G to $\mathbb{L}(G)$ and $\mathbb{R}(G)$ give rise to a pair of join-preserving anti-homomorphisms $\mathbb{L}(G) \underset{\vartheta_{\mathbb{R}}}{\overset{\vartheta_{\mathbb{L}}}{\rightleftarrows}} \mathbb{R}(G)$, which satisfy the properties $\vartheta_{\mathbb{R}} \circ \vartheta_{\mathbb{L}} = 1_{\mathbb{L}(G)}$ and $\vartheta_{\mathbb{L}} \circ \vartheta_{\mathbb{R}} = 1_{\mathbb{R}(G)}$. Since all two-sided elements of a Gelfand quantale are hermitian, the relations

$$\vartheta_{\mathbb{L}} \circ q_{\mathbb{L}} = q_{\mathbb{R}} \quad \text{and} \quad \vartheta_{\mathbb{R}} \circ q_{\mathbb{R}} = q_{\mathbb{L}}$$

follow. Finally, let $(\rho_{\mathbb{L}}, \rho_{\mathbb{R}}, \mathbb{P}(G))$ be the pushout of $q_{\mathbb{L}}$ and $q_{\mathbb{R}}$ in the sense of `BSQuant` and

$$\mathbb{L}(G) \otimes \mathbb{R}(G) \xrightarrow{\pi_G} \mathbb{P}(G)$$

be the coequalizer of $\mathbb{I}(G) \underset{j_{\mathbb{R}(G)} \circ q_{\mathbb{R}}}{\overset{j_{\mathbb{L}(G)} \circ q_{\mathbb{L}}}{\rightrightarrows}} \mathbb{L}(G) \otimes \mathbb{R}(G)$. Hence the following diagram is commutative:

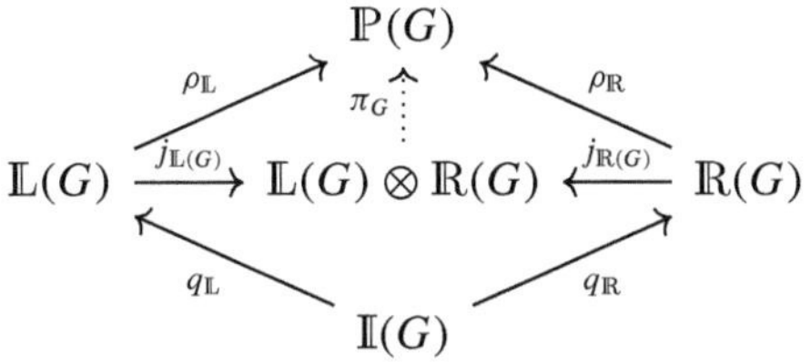

(2.60)

By Corollary 2.5.11 there exists an involution on the pushout $\mathbb{P}(G)$ transmitting the coequalizer $\mathbb{L}(G) \otimes \mathbb{R}(G) \xrightarrow{\pi_G} \mathbb{P}(G)$ to an involutive, surjective homomorphism.

Theorem 2.5.12. *Let $(G, *, ')$ be a Gelfand quantale. Then the following assertions hold*:

(i) *The strong homomorphism $\rho_{\mathbb{L}} = \pi_G \circ j_{\mathbb{L}(G)}$ is monic, and the range of $\rho_{\mathbb{L}}$ coincides with the subquantale $\mathbb{L}(\mathbb{P}(G))$ of all left-sided elements of $\mathbb{P}(G)$.*

(ii) *The strong homomorphism $\rho_{\mathbb{R}} = \pi_G \circ j_{\mathbb{R}(G)}$ is monic, and the range of $\rho_{\mathbb{R}}$ coincides with the subquantale $\mathbb{R}(\mathbb{P}(G))$ of all right-sided elements of $\mathbb{P}(G)$.*

(iii) *The strong homomorphism $\rho_{\mathbb{L}} \circ q_{\mathbb{L}} = \rho_{\mathbb{R}} \circ q_{\mathbb{R}}$ is monic, and the range of $\rho_{\mathbb{L}} \circ q_{\mathbb{L}}$ coincides with the subquantale $\mathbb{I}(\mathbb{P}(G))$ of all two-sided elements of $\mathbb{P}(G)$.*

Proof. We restrict ourselves to the proof of (i) (since the verification of assertions (ii) and (iii) is analogous, we leave it to the reader). By construction, $\mathbb{P}(G)$ is semi-unital. We consider the multiplication $\odot$ on G, which is induced by the semi-integral regularization of $(G, *, ')$. Hence $(G, \odot, ')$ is a balanced, bisymmetric and involutive quantale (cf. Proposition 2.5.3). Finally, let $\mathbb{L}(G) \xrightarrow{\iota_{\mathbb{L}}} G$ and $\mathbb{R}(G) \xrightarrow{\iota_{\mathbb{R}}} G$ be the strong homomorphisms determined by the respective inclusion maps. Since $\mathbb{L}(G) \otimes \mathbb{R}(G)$ is the coproduct of $\mathbb{L}(G)$ and $\mathbb{R}(G)$ in `BSQuant` (cf. Theorem 2.5.9), there exists a unique strong homomorphism $\mathbb{L}(G) \otimes \mathbb{R}(G) \xrightarrow{h} G$ with the properties $h \circ j_{\mathbb{L}(G)} = \iota_{\mathbb{L}}$ and $h \circ j_{\mathbb{R}(G)} = \iota_{\mathbb{R}}$ — this means that h is uniquely determined by the following condition (cf. (2.58)):

$$h(\ell \otimes r) = r \odot \ell = r * \ell, \qquad r \in \mathbb{R}(G),\ \ell \in \mathbb{L}(G).$$

Since $(\ell \otimes r)' = r' \otimes \ell'$ (cf. (2.43) in the proof of Theorem 2.3.33), it is easily seen that h is even an involutive homomorphism. Now we observe that

$$h \circ j_{\mathbb{L}(G)} \circ q_{\mathbb{L}} = h \circ j_{\mathbb{R}(G)} \circ q_{\mathbb{R}}.$$

Hence h factors through π_G — i.e. there exists a unique (involutive) strong homomorphism $\mathbb{P}(G) \xrightarrow{h_0} G$ such that $h = h_0 \circ \pi_G$. Finally, we note that:

$$h_0\big(\rho_{\mathbb{L}}(\ell)\big) = h\big(j_{\mathbb{L}(G)}(\ell)\big) = \iota_{\mathbb{L}}(\ell) = \ell, \qquad \ell \in \mathbb{L}(G).$$

Thus $\rho_{\mathbb{L}}$ is an injective strong homomorphism.
Further, since strong homomorphisms preserve left-sided elements, it is sufficient to show that for each left-sided element $x \in \mathbb{P}(G)$ there exists an $\ell \in \mathbb{L}(G)$ with $\rho_{\mathbb{L}}(\ell) = x$. For this purpose we consider a representation of a left-sided element $x \in \mathbb{P}(G)$ having the following form:

$$x = \bigvee_{i \in I} \pi_G(\ell_i \otimes r_i), \qquad \ell_i \in \mathbb{L}(G),\ r_i \in \mathbb{R}(G).$$

Since $\mathbb{P}(G)$ is semi-unital, we obtain:

$$x = \pi_G(\top \otimes \top) * x = \bigvee_{i \in I} \pi_G\big((\top * \ell_i) \otimes (\top * r_i)\big) = \bigvee_{i \in I} \pi_G\big(\ell_i \otimes (\top * r_i)\big).$$

Now we use the fact that π_G is the coequalizer of $\mathbb{I}(G) \underset{j_{\mathbb{R}(G)} \circ q_{\mathbb{R}}}{\overset{j_{\mathbb{L}(G)} \circ q_{\mathbb{L}}}{\rightrightarrows}} \mathbb{L}(G) \otimes \mathbb{R}(G)$ and observe that the relation

$$\begin{aligned}
\pi_G\big(\ell_i \otimes (\top * r_i)\big) &= \pi_G\big((\top \otimes (\top * r_i)) \star (\ell_i \otimes \top)\big)\\
&= \pi_G\big(\top \otimes (\top * r_i)\big) * \pi_G(\ell_i \otimes \top)\\
&= \pi_G\big((\top * r_i) \otimes \top\big) * \pi_G(\ell_i \otimes \top)\\
&= \pi_G\big(((\top * r_i) \otimes \top) \star (\ell_i \otimes \top)\big)\\
&= \pi_G\big((\top * r_i * \ell_i) \otimes \top\big)
\end{aligned}$$

holds for all $i \in I$. Hence

$$x = \bigvee_{i \in I} \pi_G\big((\top * r_i * \ell_i) \otimes \top\big) = \pi_G\big(j_{\mathbb{L}(G)}\big(\bigvee_{i \in I} \top * r_i * \ell_i\big)\big) = \rho_{\mathbb{L}}\big(\bigvee_{i \in I} \top * r_i * \ell_i\big)$$

follows, and assertion (i) is verified. □

We conclude from the previous theorem that for any Gelfand quantale $(G, *, ')$ the subquantale $\mathbb{L}(G)$ is isomorphic to the subquantale $\mathbb{L}(\mathbb{P}(G))$ of all left-sided elements of $\mathbb{P}(G)$. Hence $\mathbb{P}(G)$ is again a Gelfand quantale. This observation motivates us to strengthen the axioms formulated in Definition 1.8 in [55] and to fix the following terminology.

Definition 2.5.13. An involutive, bisymmetric and semi-unital quantale $(X, *, ')$ is called a *quantic frame* if and only if $(X, *, ')$ satisfies the following properties:

(a) Every left-sided (right-sided) element of $(X, *)$ is idempotent.
(b) Every two-sided element is hermitian.
(c) If $\mathbb{L}(X)$ is the subquantale of all left-sided elements, $\mathbb{R}(X)$ is the subquantale of all right-sided elements, and $\mathbb{I}(X)$ is the subquantale of all two-sided elements of $(X, *)$, then the following diagram is a pushout square in `BSQuant`

$$\begin{array}{ccccc} & & X & & \\ & \nearrow & & \nwarrow & \\ \mathbb{L}(X) & & & & \mathbb{R}(X) \\ & \nwarrow & & \nearrow & \\ & & \mathbb{I}(X) & & \end{array}$$

Since the structure of semi-unital and semi-integral quantales is inherited by the tensor product (cf. Lemma 2.3.12 and Proposition 2.3.19) as well as by regular quotients in `BSQuant`, every quantic frame $(X, *, ')$ satisfies the following properties for all $x, y, z \in X$:

- $x * x * y = x * y = x * y * y$,
- $x * y * z = x * y' * z$.

Furthermore, if $(G, *, ')$ is a Gelfand quantale, then it follows from (2.60) and Theorem 2.5.12 that $\mathbb{P}(G)$ is a quantic frame. Hence the Gelfand quantale $\mathbb{P}(G)$ is also called the *quantic frame associated with* $(G, *, ')$. Obviously, a commutative quantic frame is always a frame in the traditional sense (cf. [58]).

Further, let X be an arbitrary complete lattice. Referring to Example 2.5.2, we can construct a non-commutative, left-sided, idempotent quantale $X^\ell = (X, *_\ell)$ and a non-commutative, right-sided, idempotent quantale $X^r = (X, *_r)$ on X. Since the identity 1_X induces a pair of join-preserving anti-isomomorphisms between X^ℓ and X^r, there exists an involution $'$ on $X^\ell \otimes X^r$ such that $(X^\ell \otimes X^r, \star, ')$ is an involutive quantale (cf. Theorem 2.3.33). Obviously, the subquantale $\mathbb{I}(X^\ell \otimes X^r)$ of all two-sided elements is the initial object of `BSQuant` — namely the Boolean algebra consisting of two elements. Hence $(X^\ell \otimes X^r, \star, ')$ is a quantic frame (cf. Exercise 2.3.10(d), (e) and Theorem 2.5.9). Moreover, $(X^\ell \otimes X^r, \star, ')$ is a Gelfand quantale, because for every $x \in X$ with $x \neq \bot$ the following relation holds for all non-trivial left-sided elements of $X^\ell \otimes X^r$ (cf. Exercise 2.3.10 (d)):

$$(x \otimes \top) \star (x \otimes \top)' = (x \otimes \top) \star (\top \otimes x) = (x *_\ell \top) \otimes (\top *_r x) = \top \otimes \top \geq x \otimes \top.$$

In particular, if X is the three-chain C_3, then $(C_3^\ell \otimes C_3^r, \star, ')$ is the "smallest" non-commutative quantic frame (cf. Example 2.3.34). Hence we understand the involutive quantale $(C_3^\ell \otimes C_3^r, \star, ')$ as the *quantization of the two-chain* C_2.

Since quantic frames are by definition semi-unital and bisymmetric, the previous results motivate the investigation of the topologization of arbitrary semi-unital quantales. In this context we will show that spatial and semi-unital quantales are always bisymmetric (cf. Corollary 2.5.19 infra).

2.5.1 Semi-unital Quantales and Strong Homomorphisms

In this subsection we consider strong homomorphisms having the "smallest" non-commutative quantic frame $C_3^\ell \otimes C_3^r$ as codomain. We maintain here the notation from Example 2.3.26 and recall that the quantale $C_3^\ell \otimes C_3^r$ consists of six elements — i.e. $C_3^\ell \otimes C_3^r = (\{\top, \alpha, \lambda, \rho, \beta, \bot\}, \star)$.

Since C_3^ℓ is isomorphic to the subquantale $\mathbb{L}(C_3^\ell \otimes C_3^r)$ of all left-sided tensors of $C_3^\ell \otimes C_3^r$, the following results can be understood as a generalization of Lemma 2.4.20, Theorems 2.4.21 and 2.4.24. In this context the non-elementary tensor $\alpha = (a \otimes \top) \vee (\top \otimes a)$ of $C_3^\ell \otimes C_3^r$ will play a crucial rôle.

We begin with a refinement of the concept of prime elements. An element p of a quantale $(X, *)$ is *strongly prime* if $p \neq \top$ and the following implication holds for all $x_1, x_2 \in X$:

$$x_1 * x_2 \leq p \quad \Longrightarrow \quad (x_1 * \top) \vee x_1 \leq p \;\text{ or }\; (\top * x_2) \vee x_2 \leq p.$$

It is interesting to note that in the case of two-sided quantales the property of being strongly prime already goes back to W. Krull 1928 (cf. [69]).

Every strongly prime element is prime, but in general not vice versa (cf. Exercise 2.5.4 (c)). Further, an element p of a quantale $(X, *)$ is strongly prime if and only if p, viewed as an element of the semi-unitalization $(\overline{X}, \overline{*})$ of $(X, *)$ (cf. Sect. 2.3.1), is prime in $(\overline{X}, \overline{*})$.

In the case of semi-unital quantales, an element is prime if and only if it is strongly prime.

Lemma 2.5.14. *Let $(X, *)$ be a quantale, and $X \xrightarrow{h} C_3^\ell \otimes C_3^r$ be a strong homomorphism. Then the element $p \in X$ defined by*

$$p = \bigvee\{x \in X \mid h(x) \leq \alpha\}$$

is a strongly prime element of $(X, *)$.

Proof. The relation $h(p) \leq \alpha$ follows from the property that h is join-preserving. Since h is strong, p is necessarily different from $\top$. Now we choose $x_1, x_2 \in X$ with $x_1 * x_2 \leq p$. Because of the semi-integrality of $C_3^\ell \otimes C_3^r$ we obtain:

$$h(x_1 * \top) \star h(\top * x_2) \leq h(x_1) \star \top \star \top \star h(x_2) \leq h(x_1) \star h(x_2) = h(x_1 * x_2) \leq \alpha.$$

Since $C_3^\ell \otimes C_3^r$ is balanced, $h(x_1 * \top) \neq \top$ or $h(\top * x_2) \neq \top$. Using again the property that h is strong we observe that $h(x_1 * \top)$ is right-sided and $h(\top * x_2)$ is left-sided. Hence $h(x_1 * \top) \leq \rho$ or $h(\top * x_2) \leq \lambda$ follows — i.e. $x_1 * \top \leq p$ or $\top * x_2 \leq p$.
Finally, if $h(x_1 * \top) \leq \rho$, then $h(x_1) \neq \top$ — i.e. $h(x_1) \leq \alpha$. Analogously we verify that $h(\top * x_2) \leq \lambda$ implies $h(x_2) \leq \alpha$. To sum up, the relation $(x_1 * \top) \vee x_1 \leq p$ or $(\top * x_2) \vee x_2 \leq p$ holds — i.e. p is strongly prime. □

Theorem 2.5.15. *Let $(X, *)$ be a semi-unital quantale. Then for every prime element p in $(X, *)$ there exists a unique strong homomorphism $X \xrightarrow{h} C_3^\ell \otimes C_3^r$ satisfying the condition*

$$p = \bigvee\{x \in X \mid h(x) \leq \alpha\}. \tag{2.61}$$

Proof. (a) (Uniqueness) Let $X \xrightarrow{h} C_3^\ell \otimes C_3^r$ be a strong homomorphism satisfying (2.61). Since h is join-preserving, the equivalence $x \le p \iff h(x) \le \alpha$ holds for all $x \in X$. We distinguish the following cases where we refer frequently to the multiplication table of $\star$ (cf. Example 2.3.26):

- If $\top * x * \top \le p$, then $\top \star h(x) \star \top \le \alpha$. Hence $h(x) = \bot$.
- If $\top * x * \top \not\le p$, $x * \top \le p$ and $\top * x \le p$, then $\top \star h(x) \star \top = \top$, $h(x) \star \top \le \rho$ and $\top \star h(x) \le \lambda$. Hence $h(x) = \beta$.
- If $x * \top \not\le p$ and $\top * x \le p$, then $h(x) \star \top = \top$ and $\top \star h(x) \le \lambda$. Hence $h(x) = \lambda$.
- If $x * \top \le p$ and $\top * x \not\le p$, then $h(x) \star \top \le \rho$ and $\top \star h(x) = \top$. Hence $h(x) = \rho$.
- If $x \le p$, $x * \top \not\le p$ and $\top * x \not\le p$, then $h(x) \le \alpha$, $h(x) \star \top = \top$ and $\top \star h(x) = \top$. Hence $h(x) = \alpha$.
- If $x \not\le p$ then $h(x) = \top$.

Since $(X, *)$ is semi-unital, it follows from the previous cases that h is uniquely determined by (2.61).
(b) (Existence) The cases discussed in the uniqueness proof suggest to define a strong homomorphism $X \xrightarrow{h} C_3^\ell \otimes C_3^r$ as follows:

$$h(x) = \begin{cases} \bot, & \top * x * \top \le p, \\ \beta, & \top * x * \top \not\le p,\ x * \top \le p,\ \top * x \le p, \\ \lambda, & x * \top \not\le p,\ \top * x \le p, \\ \rho, & x * \top \le p,\ \top * x \not\le p, \\ \alpha, & x \le p,\ x * \top \not\le p,\ \top * x \not\le p, \\ \top, & x \not\le p. \end{cases} \tag{2.62}$$

Referring to the Hasse diagram in Example 2.3.26 it is evident that h is join-preserving and the relation (2.61) holds. Since $p \ne \top$ the definition of h implies $h(\top) = \top$. In order to prove the homomorphism property of h — i.e. the preservation of the respective semigroup operations, we will frequently use the prime property of p and again the fact that $(X, *)$ is semi-unital. For $x_1, x_2 \in X$ we examine the following cases:

- $\top * x_1 * x_2 * \top \le p \iff (\top * x_1 * \top \le p \text{ or } \top * x_2 * \top \le p)$.
  Hence $h(x_1 * x_2) = \bot = h(x_1) \star h(x_2)$.
- $(\top * x_1 * x_2 * \top \not\le p,\ x_1 * x_2 * \top \le p \text{ and } \top * x_1 * x_2 \le p) \iff$
  $(\top * x_1 * \top \not\le p,\ \top * x_2 * \top \not\le p,\ x_1 * \top \le p \text{ and } \top * x_2 \le p)$.
  Hence we obtain $h(x_1 * x_2) = \beta$, $h(x_1) \in \{\beta, \rho\}$ and $h(x_2) \in \{\beta, \lambda\}$. Since

  $$\beta = \beta \star \beta = \rho \star \lambda = \rho \star \beta = \beta \star \lambda,$$

  $h(x_1 * x_2) = h(x_1) \star h(x_2)$ follows.

– $(x_1 * x_2 * \top \not\leq p$ and $\top * x_1 * x_2 \leq p) \iff (\top * x_2 \leq p,\ x_1 * \top \not\leq p$ and $\top * x_2 * \top \not\leq p)$ — i.e. $h(x_1) \in \{\lambda, \alpha, \top\}$ and $h(x_2) \in \{\beta, \lambda\}$. Since

$$\lambda = \lambda \star \beta = \lambda \star \lambda = \alpha \star \beta = \alpha \star \lambda = \top \star \beta = \top \star \lambda,$$

$h(x_1 * x_2) = h(x_1) \star h(x_2)$ follows.

– $(x_1 * x_2 * \top \leq p$ and $\top * x_1 * x_2 \not\leq p) \iff (\top * x_1 * \top \not\leq p, \top * x_2 \not\leq p$ and $x_1 * \top \leq p)$ — i.e. $h(x_1) \in \{\beta, \rho\}$ and $h(x_2) \in \{\rho, \alpha, \top\}$. Since

$$\rho = \beta \star \rho = \rho \star \rho = \beta \star \alpha = \rho \star \alpha = \beta \star \top = \rho \star \top,$$

$h(x_1 * x_2) = h(x_1) \star h(x_2)$ follows.

– The case $\top * x_1 * x_2 \not\leq p$, $x_1 * x_2 * \top \not\leq p$ and $x_1 * x_2 \leq p$ is empty.

– $x_1 * x_2 \not\leq p \iff (x_1 * \top \not\leq p$ and $\top * x_2 \not\leq p)$ — i.e. $h(x_1) \in \{\lambda, \alpha, \top\}$ and $h(x_2) \in \{\rho, \alpha, \top\}$. Since $\top = \lambda \star \rho$, we obtain $h(x_1) \star h(x_2) = h(x_1 * x_2)$.

To sum up, h is a strong homomorphism satisfying the property (2.61). □

Corollary 2.5.16. *Let $(X, *, ')$ be an involutive, semi-unital quantale. Further, let $X \xrightarrow{h} C_3^\ell \otimes C_3^r$ be a strong homomorphism. Then h is an involutive homomorphism if and only if the prime element p determined by h is hermitian, where*

$$p = \bigvee\{x \in X \mid h(x) \leq \alpha\}.$$

Proof. Let $'$ be the involution on $C_3^\ell \otimes C_3^r$ constructed in Example 2.3.34. Since $\alpha' = \alpha$, $\lambda' = \rho$ and $\beta' = \beta$, the assertion follows immediately from (2.62). □

We can summarize the previous results as follows.

Fact II. *In the case of semi-unital quantales $(X, *)$, prime elements of $(X, *)$ and strong homomorphisms $X \xrightarrow{h} C_3^\ell \otimes C_3^r$ are equivalent concepts. Moreover, if $(X, *, ')$ is an involutive and semi-unital quantale, then strong homomorphisms with codomain $C_3^\ell \otimes C_3^r$ are involutive if and only if the corresponding prime element is hermitian.*

Let $(X, *)$ be a strictly left-sided quantale and $(Y, *)$ be a strictly right-sided quantale. Since prime elements of the tensor product $(X \otimes Y, \star)$ are determined by prime elements of its factors (cf. Lemma 2.3.22 and Theorem 2.3.25), it is interesting to see how this property is reflected by the corresponding strong homomorphisms. For this purpose we recall that every prime element of $(X \otimes Y, \star)$ has the form

$$(p \otimes \top) \vee (\top \otimes q)$$

where p is a prime element of $(X, *)$ and q is a prime element of $(Y, *)$.

Further, let $j_\ell = j_{C_3^\ell}$ and $j_r = j_{C_3^r}$ be the coproduct injections of the coproduct $C_3^\ell \otimes C_3^r$ (cf. Theorem 2.5.9) and $X \xrightarrow{j_X} X \otimes Y$ and $Y \xrightarrow{j_Y} X \otimes Y$ be the canonical embeddings (cf. (2.18)).

Theorem 2.5.17. *Let $(X, *)$ be a strictly left-sided quantale, $(Y, *)$ be a strictly right-sided quantale, and let $X \otimes Y \xrightarrow{h} C_3^\ell \otimes C_3^r$ be a strong homomorphism. Further, let $(p \otimes \top) \vee (\top \otimes q)$ be the prime element determined by h in the sense of Lemma 2.5.14 — i.e.*

$$(p \otimes \top) \vee (\top \otimes q) = \bigvee\{f \in X \otimes Y \mid h(f) \le \alpha\}.$$

Then the strong homomorphisms $X \xrightarrow{h_\ell} C_3^\ell$ and $Y \xrightarrow{h_r} C_3^r$ determined by p and q (in the sense of Theorem 2.4.21) are the unique strong homomorphisms making the following diagram commutative:

$$\begin{array}{ccccc} X & \xrightarrow{j_X} & X \otimes Y & \xleftarrow{j_Y} & Y \\ {\scriptstyle h_\ell}\downarrow & & \downarrow{\scriptstyle h} & & \downarrow{\scriptstyle h_r} \\ C_3^\ell & \xrightarrow{j_\ell} & C_3^\ell \otimes C_3^r & \xleftarrow{j_r} & C_3^r \end{array}$$

Proof. Obviously, C_3^ℓ is isomorphic to the subquantale $\mathbb{L}(C_3^\ell \otimes C_3^r)$ of all left-sided elements of $C_3^\ell \otimes C_3^r$, and C_3^r is isomorphic to the subquantale $\mathbb{R}(C_3^\ell \otimes C_3^r)$ of all right-sided elements of $C_3^\ell \otimes C_3^r$. Since $(X, *)$ is strictly left-sided and $(Y, *)$ is strictly right-sided, every elementary tensor $x \otimes y$ of $X \otimes Y$ has the form

$$x \otimes y = (\top \otimes y) \star (x \otimes \top). \tag{2.63}$$

Since h is a strong homomorphism, the following relation holds for all $x \in X$:

$$\begin{aligned}(h \circ j_X)(x) &= h(x \otimes \top) = h\big((\top \otimes \top) \star (x \otimes \top)\big) \\ &= h(\top \otimes \top) \star h(x \otimes \top) = \top \star \big(h \circ j_X\big)(x).\end{aligned}$$

Thus $h \circ j_X$ factors through j_ℓ. Analogously we verify that $h \circ j_Y$ factors through j_r. Hence there exist $X \xrightarrow{h_\ell} C_3^\ell$ and $Y \xrightarrow{h_r} C_3^r$ such that $h \circ j_X = j_\ell \circ h_\ell$ and $h \circ j_Y = j_r \circ h_r$. Referring again to (2.63) we obtain for all $x \in X$ and $y \in Y$:

$$h(x \otimes y) = j_r(h_r(y)) \star j_\ell(h_\ell(x)). \tag{2.64}$$

Since every tensor is a join of an appropriate family of elementary tensors, the following relation holds

$$(p \otimes \top) \vee (\top \otimes q) = \bigvee\{x \otimes y \in X \otimes Y \mid j_r(h_r(y)) \star j_\ell(h_\ell(x)) \le \alpha\},$$

where obviously p is a prime element of $(X, *)$ and q is prime element of $(Y, *)$.

Now let p_ℓ be the prime element determined by h_ℓ (in the sense of Lemma 2.4.20) — i.e.

$$p_\ell = \bigvee\{x \in X \mid h_\ell(x) \le a\}.$$

Since $h(p \otimes \top) \leq \alpha$ implies $j_\ell(h_\ell(p)) \leq \lambda = j_\ell(a)$, the relation $h_\ell(p) \leq a$ follows — i.e. $p \leq p_\ell$. On the other hand, we infer from $h(p_\ell \otimes \top) = j_\ell(h_\ell(p_\ell)) \leq j_\ell(a) = \lambda$ that

$$p_\ell \otimes \top \leq (p \otimes \top) \vee (\top \otimes q)$$

holds. Since $q \neq \top$ the relation $p_\ell \leq p$ also follows (cf. Exercise 2.1.4 (b)) — i.e. $p = p_\ell$. Analogously we verify that q is the prime element determined by h_r. □

Since every quantale $(X, *)$ is a subquantale of its semi-unitalization, we conclude from Theorem 2.5.15 that for every strongly prime element p of $(X, *)$ there exists a unique strong homomorphism $X \xrightarrow{h} C_3^\ell \otimes C_3^r$ satisfying the condition:

$$p = \bigvee\{x \in X \mid h(x) \leq \alpha\}.$$

If we now refer to Lemma 2.5.14, then we can reformulate Fact II as follows: Strongly prime elements of *arbitrary* quantales $(X, *)$ and strong homomorphisms $X \xrightarrow{h} C_3^\ell \otimes C_3^r$ are equivalent concepts.

Having made these general comments we now give a characterization of spatial and semi-unital quantales.

Theorem 2.5.18. *A semi-unital quantale $(X, *)$ is spatial if and only if strong homomorphisms $X \xrightarrow{h} C_3^\ell \otimes C_3^r$ separate elements in X — i.e. for $x_1 \neq x_2$ there exists an h with $h(x_1) \neq h(x_2)$.*

Proof. With regard to Theorem 2.5.15 the necessity is evident. On the other hand, we assume that strong homomorphisms $X \xrightarrow{h} C_3^\ell \otimes C_3^r$ separate elements in X and show that prime elements in $(X, *)$ are order generating. For this purpose, we fix an element $x_1 \in X$ and define an element $x_2 \in X$ by

$$x_2 = \bigwedge\{p \in X \mid p \text{ prime in } (X, *) \text{ and } x_1 \leq p\}. \tag{2.65}$$

Obviously, $x_1 \leq x_2$. Further, we notice that the following equivalences hold for all prime elements p in $(X, *)$:

$$x_1 \leq p \iff x_2 \leq p, \tag{2.66}$$
$$\top * x_1 \leq p \iff \top * x_2 \leq p, \tag{2.67}$$
$$x_1 * \top \leq p \iff x_2 * \top \leq p, \tag{2.68}$$
$$\top * x_1 * \top \leq p \iff \top * x_2 * \top \leq p. \tag{2.69}$$

Trivially (2.66) follows from the definition of x_2. If $\top * x_1 \leq p$, then we define a left-sided prime element $q_0 \in X$ by

$$q_0 = \bigvee\{x \in X \mid \top * x \leq p\}.$$

Hence $x_1 \le q_0$ holds, and $x_2 \le q_0$ follows from (2.66). Since $(X, *)$ is semi-unital, we obtain $\top * x_2 \le \top * q_0 = q_0 \le p$. Hence (2.67) is verified. Analogously we establish (2.68). Finally, we assume $\top * x_1 * \top \le p$. Now we construct a two-sided prime element s_0 in $(X, *)$ by

$$s_0 = \bigvee\{x \in X \mid \top * x * \top \le p\}.$$

Hence $x_1 \le s_0$ holds, and $x_2 \le s_0$ follows from (2.66). So we obtain $\top * x_2 * \top \le \top * s_0 * \top = s_0 \le p$, and (2.69) is verified.
In the following considerations we now identify prime elements p in $(X, *)$ with strong homomorphisms $X \xrightarrow{h} C_3^\ell \otimes C_3^r$ and vice versa (cf. Lemma 2.5.14 and Theorem 2.5.15). Then we infer from formula (2.62) and the equivalences (2.66)–(2.69) that $h(x_2) = h(x_1)$ holds for all strong homomorphism $X \xrightarrow{h} C_3^\ell \otimes C_3^r$. Since strong homomorphisms with codomain $(C_3^\ell \otimes C_3^r, *)$ separate elements in X, the relation $x_1 = x_2$ follows — i.e. $(X, *)$ is spatial. □

Corollary 2.5.19. *Let $(X, *)$ be a semi-unital quantale and $\sigma(X)$ be the set of all prime elements in $(X, *)$. Then $(X, *)$ is spatial if and only if $(X, *)$ is isomorphic to a subquantale of $(C_3^\ell \otimes C_3^r)^{\sigma(X)}$ in the sense of* `BSQuant` *— i.e. the embedding is a strong homomorphism. In particular, spatial and semi-unital quantales are bisymmetric.*

Proof. Since $C_3^\ell \otimes C_3^r$ is bisymmetric, the assertion follows immediately from Theorem 2.5.18. □

2.5.2 Topological Representations of Semi-Unital quantales

If in Sect. 2.4.2 we replace the idempotent and left-sided quantale $(C_3^\ell, *_\ell)$ by the "smallest" quantic frame $(C_3^\ell \otimes C_3^r, \star)$, then we can describe the concept of six-valued topologies as follows.

Let X be an arbitrary set and $(C_3^\ell \otimes C_3^r)^X$ be the set of all maps $X \xrightarrow{f} C_3^\ell \otimes C_3^r$. Then by analogy with Sect. 2.4.2 we extend the quantale structure of $C_3^\ell \otimes C_3^r$ pointwisely to $(C_3^\ell \otimes C_3^r)^X$. Hence $\big((C_3^\ell \otimes C_3^r)^X, \star\big)$ is again a semi-unital, semi-integral and bisymmetric quantale.

Now we introduce a *six-valued topology* on X as a subset τ of $(C_3^\ell \otimes C_3^r)^X$ satisfying the following conditions (see also [49, p. 150]):

(O1) τ is closed under arbitrary joins in the sense of $(C_3^\ell \otimes C_3^r)^X$ — i.e. the inclusion map $\tau \overset{\iota}{\hookrightarrow} (C_3^\ell \otimes C_3^r)^X$ is join-preserving.
(O2) The universal upper bound $\underline{\top}$ of $(C_3^\ell \otimes C_3^r)^X$ belongs to τ.
(O3) If $f, g \in \tau$, then $f \star g \in \tau$.

If τ is a six-valued topology on X, then the pair (X, τ) is called a *six-valued topological space*. Obviously, every six-valued topology is a semi-unital, semi-integral and bisymmetric subquantale of $\big((C_3^\ell \otimes C_3^r)^X, \star\big)$.

Further, let (X, τ) and (Y, σ) be six-valued topological spaces. By analogy with the situation of the three-valued case, a map $X \xrightarrow{\varphi} Y$ is *continuous* if $g \circ \varphi \in \tau$ for all $g \in \sigma$. Again six-valued topological spaces and continuous maps form a category denoted by $\texttt{Top}(6)$.

Now let $(\mathscr{Q}, *)$ be an arbitrary semi-unital quantale and $\mathsf{pt}_6(\mathscr{Q})$ be the set of all strong homomorphisms $\mathscr{Q} \xrightarrow{h} C_3^\ell \otimes C_3^r$. For every $q \in \mathscr{Q}$ we introduce a map $\mathsf{pt}_6(\mathscr{Q}) \xrightarrow{\mathbb{A}_q} C_3^\ell \otimes C_3^r$ by

$$\mathbb{A}_q(h) = h(q) \qquad h \in \mathsf{pt}_6(\mathscr{Q}).$$

Then it is easily seen that

$$\tau_{\mathscr{Q}} = \{\mathbb{A}_q \mid q \in \mathscr{Q}\}$$

is a six-valued topology on $\mathsf{pt}_6(\mathscr{Q})$, and the six-valued topological space

$$\mathsf{Pt}_6(\mathscr{Q}, *) = (\mathsf{pt}_6(\mathscr{Q}), \tau_{\mathscr{Q}})$$

is called the *six-valued topological representation* of $(\mathscr{Q}, *)$. Again we have an adjoint situation between the dual category of semi-unital quantales and the category of six-valued topological spaces, which is an extension of the adjoint situation between $\texttt{LQuant}^{op}$ and $\texttt{Top}(3_\ell)$ (cf. Theorem 2.4.23). Every semi-unital and spatial quantale is equivalent to a six-valued sober space. For details of this situation, the reader is referred to [52, Sect. 5].

Finally, by construction the six-valued topology $\tau_{\mathscr{Q}}$ is isomorphic to the spatial reflection of $\mathscr{Q}$. In fact, because of (S) in Lemma 2.3.28 (see also Corollary 2.3.29) and (2.65)–(2.69) the following relation holds for all $q \in \mathscr{Q}$:

$$\mathbb{A}_q(h) = \mathbb{A}_{c_s(q)}(h), \qquad h \in \mathsf{pt}_6(\mathscr{Q}).$$

We finish this subsection with the topological representation of the tensor product of idempotent quantales. For this purpose we fix a left-sided, idempotent quantale $(L, *)$ and a right-sided, idempotent quantale $(R, *)$. Since their tensor product is the coproduct of $(L, *)$ and $(R, *)$ in $\texttt{BSQuant}$ (cf. Theorem 2.5.9), for every strong homomorphism $L \otimes R \xrightarrow{h} C_3^\ell \otimes C_3^r$ there exist unique strong homomorphisms $L \xrightarrow{h_\ell} C_3^\ell$ and $R \xrightarrow{h_r} C_3^r$ making the following diagram commutative (cf. Theorem 2.5.17):

$$\begin{array}{ccccc} L & \xrightarrow{j_L} & L\otimes R & \xleftarrow{j_R} & R \\ {\scriptstyle h_\ell}\downarrow & & \downarrow{\scriptstyle h} & & \downarrow{\scriptstyle h_r} \\ C_3^\ell & \xrightarrow{j_\ell} & C_3^\ell\otimes C_3^r & \xleftarrow{j_r} & C_3^r \end{array}$$

Hence for every strong homomorphism $L \otimes R \xrightarrow{h} C_3^\ell \otimes C_3^r$ there exists a unique pair $(h_\ell, h_r) \in \mathsf{pt}_{3_\ell}(L) \times \mathsf{pt}_{3_r}(R)$ such that the following relation holds (cf. (2.58)):

$$h(q \otimes s) = h\big((\top \otimes s) \star (q \otimes \top)\big) = j_r(h_r(s)) \star j_\ell(h_\ell(q)), \qquad q \in L,\ s \in R. \tag{2.70}$$

Since every tensor $f \in L \otimes R$ is a join of an appropriate family of elementary tensors — i.e. $f = \bigvee_{i\in I} q_i \otimes s_i$, we obtain:

$$\mathbb{A}_f = \bigvee_{i\in I} (\mathbb{A}_{s_i} \star \mathbb{A}_{q_i}), \qquad q_i \in L,\ s_i \in R, \tag{2.71}$$

where we have identified C_3^ℓ with $\mathbb{L}(C_3^\ell \otimes C_3^r)$ and C_3^r with $\mathbb{R}(C_3^\ell \otimes C_3^r)$.

Hence the topological representation $(\mathsf{pt}_6(L \otimes R), \tau_{L\otimes R})$ of $(L \otimes R, \star)$ is the topological product in the sense of $\mathtt{Top}(6)$ of the three-valued topological spaces $\mathsf{Pt}_{3_\ell}(L, *)$ and $\mathsf{Pt}_{3_r}(R, *)$ viewed as six-valued topological spaces. In particular, the *six-valued product topology* $\tau_{L\otimes R}$ is the subquantale of $(C_3^\ell \otimes C_3^r)^{\mathsf{pt}_{3_\ell}(L)\times \mathsf{pt}_{3_r}(R)}$ generated by $\{\mathbb{A}_s \star \mathbb{A}_q \mid q \in L,\ s \in R\}$ (cf. (2.71)).

We record the following theorem.

Theorem 2.5.20. *Let $(L, *)$ be a left-sided, idempotent quantale and $(R, *)$ be a right-sided, idempotent quantale. Then the topological representation of their tensor product $(L \otimes R, \star)$ is the topological product (in the sense of $\mathtt{Top}(6)$) of the three-valued topological spaces $\mathsf{Pt}_{3_\ell}(L, *)$ and $\mathsf{Pt}_{3_r}(R, *)$ viewed as six-valued topological spaces.*

2.5.3 The Spectrum of a C*-Algebra and its Topological Representation

Let $(A, +, \cdot, e)$ be a unital C^*-algebra and $\mathbb{MAX}$ be the unital Gelfand quantale of all closed linear subspaces of A (cf. Example 2.5.5). Then the *spectrum* of A is the quantic frame $\mathbb{P}(\mathbb{MAX})$ associated with $\mathbb{MAX}$ and is denoted by $\mathrm{sp}(A)$. In particular, $\mathrm{sp}(A)$ is a bisymmetric and semi-unital Gelfand quantale. If A is commutative, then $\mathrm{sp}(A)$ coincides with the traditional spectrum of A given by the frame of all closed (two-sided) ideals of A. If A is not commutative, then we distinguish the following cases.

Case 1. Let $(A, +, \cdot, e)$ be a simple C^*-algebra — i.e. the quantale of all closed two-sided ideals coincides with the Boolean algebra $\{0, 1\}$. Then the spectrum of A is the tensor product of the quantale $\mathbb{L}(A)$ of all closed left ideals with the quantale

$\mathbb{R}(A)$ of all closed right ideals of A. Obviously, $\mathrm{sp}(A)$ is the coproduct of $\mathbb{L}(A)$ and $\mathbb{R}(A)$ in `BSQuant`.
Case 2. If $(A, +, \cdot, e)$ is not simple — e.g. the C^*-algebra of all bounded operators on an infinite-dimensional Hilbert space, then we cannot say much and can only argue that $\mathrm{sp}(A)$ is a certain regular quotient of the tensor product $\mathbb{L}(A) \otimes \mathbb{R}(A)$.

In order to investigate the topologization of $\mathrm{sp}(A)$ we first give a characterization of prime elements of quantic frames.

Lemma 2.5.21. *Let $(X, *, ')$ be a quantic frame and $\mathbb{L}(X) \otimes \mathbb{R}(X) \xrightarrow{\pi} X$ be the coequalizer of the pair of strong homomorphisms $\mathbb{I}(X) \underset{\varphi_{\mathbb{R}}}{\overset{\varphi_{\mathbb{L}}}{\rightrightarrows}} \mathbb{L}(X) \otimes \mathbb{R}(X)$ defined by:*

$$\varphi_{\mathbb{L}}(x) = x \otimes \top, \quad \varphi_{\mathbb{R}}(x) = \top \otimes x, \qquad x \in \mathbb{I}(X).$$

Then an element $p \in X$ is prime in $(X, *)$ if and only if there exist prime elements $\ell \in \mathbb{L}(X)$ and $r \in \mathbb{R}(X)$ satisfying the following properties:

(i) $\pi\big((\ell \otimes \top) \vee (\top \otimes r)\big) = p$.
(ii) *For all $z \in \mathbb{I}(X)$ the equivalence $z \leq \ell \iff z \leq r$ holds.*

Proof. (a) Let p be prime in $(X, *)$. Since π, $j_{\mathbb{L}(X)}$ and $j_{\mathbb{R}(X)}$ are strong homomorphisms, $\ell = (\pi \circ j_{\mathbb{L}(X)})^{\vdash}(p)$ is prime in $\mathbb{L}(X)$ and $r = (\pi \circ j_{\mathbb{R}(X)})^{\vdash}(p)$ is prime in $\mathbb{R}(X)$. By Theorem 2.3.25 the relation $\pi^{\vdash}(p) = (\ell \otimes \top) \vee (\top \otimes r)$ follows. Hence the surjectivity of π implies property (i). Further, for every two-sided element z we obtain the following chain of equivalences:

$$\begin{aligned} z \leq \ell = (\pi \circ j_{\mathbb{L}(X)})^{\vdash}(p) &\iff \pi(z \otimes \top) = \pi(\top \otimes z) \leq p \\ &\iff z \leq r = (\pi \circ j_{\mathbb{R}(X)})^{\vdash}(p). \end{aligned}$$

Hence the property (ii) also holds.
(b) We fix $p \in X$ with the property that there exist two prime elements $\ell \in \mathbb{L}(X)$ and $r \in \mathbb{R}(X)$ satisfying (i) and (ii). Then we consider the strong homomorphism $\mathbb{L}(X) \otimes \mathbb{R}(X) \xrightarrow{h} C_3^\ell \otimes C_3^r$ corresponding to the prime element $(\ell \otimes \top) \vee (\top \otimes r)$ (cf. Lemma 2.3.22 and Theorem 2.5.15), and we fix $z \in \mathbb{I}(X)$. Because of the following equivalences

$$z \leq \ell \iff z \otimes \top \leq (\ell \otimes \top) \vee (\top \otimes r) \quad \text{and} \quad z \leq r \iff \top \otimes z \leq (\ell \otimes \top) \vee (\top \otimes r)$$

we conclude from (ii) that $h \circ \varphi_{\mathbb{L}} = h \circ \varphi_{\mathbb{R}}$. Hence h factors through π — this means that $\pi((\ell \otimes \top) \vee (\top \otimes r))$ is prime. So, the primality of p now follows from (i). □

Corollary 2.5.22. *Let $(X, *, ')$ be a quantic frame and $\mathbb{L}(X) \otimes \mathbb{R}(X) \xrightarrow{\pi} X$ be the coequalizer of the pair $\mathbb{I}(X) \underset{\varphi_{\mathbb{R}}}{\overset{\varphi_{\mathbb{L}}}{\rightrightarrows}} \mathbb{L}(X) \otimes \mathbb{R}(X)$ of strong homomorphisms given by:*

$$\varphi_{\mathbb{L}}(x) = x \otimes \top, \quad \varphi_{\mathbb{R}}(x) = \top \otimes x, \qquad x \in \mathbb{I}(X).$$

An element $p \in X$ is a hermitian prime element if and only if there exists a hermitian prime element q of $(\mathbb{L}(X) \otimes \mathbb{R}(X), \star)$ *such that* $p = \pi(q)$.

Proof. Since the right adjoint map of an involutive homomorphism is involutive (cf. Lemma 2.2.14), the necessity of the condition is evident. On the other hand, let q be a hermitian prime element of $(\mathbb{L}(X) \otimes \mathbb{R}(X), \star)$ with $p = \pi(q)$. Then there exist prime elements $\ell \in \mathbb{L}(X)$ and $r \in \mathbb{R}(X)$ such that $q = (\ell \otimes \top) \vee (\top \otimes r)$. Since

$$(\ell \otimes \top) \vee (\top \otimes r) = \big((\ell \otimes \top) \vee (\top \otimes r)\big)' = (r' \otimes \top) \vee (\top \otimes \ell'),$$

$\ell' = r$ follows. Since every two-sided element $z \in \mathbb{I}(X)$ is hermitian, we observe that the elements ℓ and ℓ' satisfy the condition (ii) of Lemma 2.5.21. Hence we conclude from Lemma 2.5.21 that $\pi(q)$ is a hermitian prime element of $(X, *)$ where we have used the property that π is involutive. □

Having made these preparations we are now going to construct the Gelfand topology on the quantic frame $\mathbb{P}(\mathbb{MAX})$ associated with $\mathbb{MAX}$. For this purpose we fix a pure state ρ of A and consider its left kernel L_ρ. Then L_ρ is a maximal left ideal of A and consequently a prime element of $\mathbb{L}(A) \cong \mathbb{L}(\mathbb{P}(\mathbb{MAX}))$. Now, we introduce the following strong homomorphisms $\mathbb{L}(A) \xrightarrow{h_\ell} C_3^\ell$ and $\mathbb{R}(A) \xrightarrow{h_r} C_3^r$ as follows (cf. Example 2.4.29):

$$h_\ell(I) = \begin{cases} \bot, & I * A \subseteq L_\rho, \\ a, & I * A \nsubseteq L_\rho,\ I \subseteq L_\rho, \\ \top, & I \nsubseteq L_\rho, \end{cases} \qquad h_r(J) = \begin{cases} \bot, & A * J \subseteq (L_\rho)', \\ a, & A * J \nsubseteq (L_\rho)',\ J \subseteq (L_\rho)', \\ \top, & J \nsubseteq (L_\rho)' \end{cases}$$

for each $I \in \mathbb{L}(A)$ and $J \in \mathbb{R}(A)$. Further, let $\mathbb{L}(A) \otimes \mathbb{R}(A) \xrightarrow{h} C_3^\ell \otimes C_3^r$ be the involutive and strong homomorphism corresponding to the hermitian prime element $(L_\rho \otimes A) \vee (A \otimes (L_\rho)')$ (cf. Corollary 2.5.16). Since $\mathbb{L}(A) \otimes \mathbb{R}(A)$ is the coproduct of $\mathbb{L}(A)$ and $\mathbb{R}(A)$ in the sense of `BSQuant`, we conclude from Theorem 2.5.17 that h has the following form (cf. (2.64)):

$$h(I \otimes J) = j_r\big(h_r(J)\big) \star j_\ell\big(h_\ell(I)\big), \qquad I \in \mathbb{L}(A),\ J \in \mathbb{R}(A).$$

Now we invoke Corollary 2.5.22 (see also Lemma 2.5.21) and observe that h factors through $\pi_{\mathbb{MAX}}$. Hence for every pure state ρ of A there exists a unique involutive and strong homomorphism $\mathbb{P}(\mathbb{MAX}) \xrightarrow{h_\rho} C_3^\ell \otimes C_3^r$ such that the following relation holds:

$$h_\rho\big(\pi_{\mathbb{MAX}}(I \otimes J)\big) = j_r\big(h_r(J)\big) \star j_\ell\big(h_\ell(I)\big), \qquad I \in \mathbb{L}(A),\ J \in \mathbb{R}(A), \tag{2.72}$$

where h_ℓ is the three-valued strong homomorphism corresponding to the left kernel of ρ and h_r is the strong three-valued homomorphism corresponding to the adjoint of the left kernel (i.e. the right kernel) of ρ.

Finally, let $\mathfrak{p}(A)$ be the set of all pure states of A. Then every element g of

$$\mathbb{P}(\mathbb{MAX}) = \{\pi_{\mathbb{MAX}}(f) \mid f \in \mathbb{L}(A) \otimes \mathbb{R}(A)\}$$

induces a map $\mathfrak{p}(A) \xrightarrow{\mathbb{A}_g} C_3^\ell \otimes C_3^r$ by

$$\mathbb{A}_g(\rho) = h_\rho(g), \qquad \rho \in \mathfrak{p}(A).$$

Obviously, $\tau_A = \{\mathbb{A}_g \mid g \in \mathbb{P}(\mathbb{MAX})\}$ is a six-valued topology on $\mathfrak{p}(A)$ and is called the *Gelfand topology* of A. Since all strong homomorphisms h_ρ are involutive, the Gelfand topology is evidently *involutive* — i.e.

$$\mathbb{A}_g \in \tau_A \quad \Longrightarrow \quad (\mathbb{A}_g)' \in \tau_A.$$

A list of further features are the following properties of τ_A:

- If $(A, +, \cdot, ^*)$ is commutative, then the Gelfand topology τ_A coincides with the traditional Gelfand topology.
- In the case of non-commutative C^*-algebras the Gelfand topology is in general a six-valued topology.
- Every closed left ideal I of A can be identified with a $C_3^\ell \otimes C_3^r$-valued map $\mathbb{A}_{\pi_{\mathbb{MAX}}(I \otimes A)}$ and vice versa.
- Every closed right ideal J of A can be identified with a $C_3^\ell \otimes C_3^r$-valued map $\mathbb{A}_{\pi_{\mathbb{MAX}}(A \otimes J)}$ and vice versa.
- $\{\mathbb{A}_{\pi_{\mathbb{MAX}}(I \otimes J)} \mid I \in \mathbb{L}(A),\ J \in \mathbb{R}(A)\}$ is a join-basis of τ_A.

The last property of τ_A motivates us to give a more transparent representation of the map $\mathbb{A}_{\pi_{\mathbb{MAX}}(I \otimes J)}$. For this purpose we recall the construction of $\mathbb{A}_I$ in Example 2.4.29 and the dual formulation of $\mathbb{A}_I$ for a closed right ideal J:

$$\mathbb{A}_J(\rho) = \begin{cases} \bot, & A * J \subseteq (L_\rho)', \\ a, & A * J \not\subseteq (L_\rho)',\ J \subseteq (L_\rho)', \\ \top, & J \not\subseteq (L_\rho)', \end{cases} \qquad \rho \in \mathfrak{p}(A),\ J \in \mathbb{R}(A).$$

If we now identify elements of C_3^ℓ with elements of $C_3^\ell \otimes C_3^r$ in the sense of j_ℓ (e.g. a with $a \otimes \top$) and elements of C_3^r with elements of $C_3^\ell \otimes C_3^r$ in the sense of j_r (e.g. a with $\top \otimes a$), then we infer from formula (2.72) that the following relation holds:

$$\mathbb{A}_{\pi_{\mathbb{MAX}}(I \otimes J)}(\rho) = \mathbb{A}_J(\rho) \star \mathbb{A}_I(\rho), \qquad \rho \in \mathfrak{p}(A). \tag{2.73}$$

Finally, it seems that in general the Gelfand topology comprises more information about the underlying non-commutative C^*-algebra than the Jacobson topology can usually offer. For example, in the non-commutative setting primitive ideals do not distinguish maximal left (right) ideals. Hence the Jacobson topology cannot describe the mutual interrelationship between pure states. However, in the case of the Gelfand topology τ_A we have the following situation (cf. (2.73)):

$$\mathbb{A}_{\pi_{\mathbb{MAX}}(I\otimes J)}(\rho) = \begin{cases} \top, & I \not\subseteq L_\rho,\ J \not\subseteq (L_\rho)', \\ \lambda, & I \subseteq L_\rho,\ I * A \not\subseteq L_\rho,\ J \not\subseteq (L_\rho)', \\ \rho, & I \not\subseteq L_\rho,\ J \subseteq (L_\rho)',\ A * J \not\subseteq (L_\rho)', \\ \beta, & I \subseteq L_\rho,\ I * A \not\subseteq L_\rho,\ J \subseteq (L_\rho)',\ A * J \not\subseteq (L_\rho)', \\ \bot, & I * A \subseteq L_\rho \text{ or } A * J \subseteq (L_\rho)', \end{cases}$$

where L_ρ is the left kernel of ρ. For a further illustration of this situation, the reader is referred to Exercise 2.5.5.

The reconstruction of a non-commutative C^*-algebra $(A, +, \cdot, ^*)$ from its Gelfand topology τ_A is an open problem at the moment. One of the obstacles is the construction of an algebra of "non-commutative functions" defined on $\mathfrak{p}(A)$.

Exercises

2.5.1. (See Exercises 2.2.1, 2.3.1 and 2.4.1) Let C_3 be the chain with three elements. We recall from Exercise 2.3.1 (i) that there are precisely 9 balanced quantale structures defined on C_3. Show that any quantale on C_3 is bisymmetric.

Comment. Exercise 2.5.1 underlines the importance of the bisymmetry axiom when one moves from the two-valued to the many-valued setting.

2.5.2. Let $(X, *)$ be a non-commutative and two-sided quantale. Show:

(a) If $(X, *)$ bisymmetric, then $(X, *)$ is not semi-unital.
(b) The semi-unitalization (cf. Sect. 2.3.1) of $(X, *)$ is not bisymmetric.

2.5.3. Show that any non-commutative and unital quantale is balanced, but not bisymmetric.

2.5.4. Let C_3 be the three-chain provided with the following multiplication

$*$	$\bot$	a	$\top$
$\bot$	$\bot$	$\bot$	$\bot$
a	$\bot$	$\bot$	a
$\top$	$\bot$	$\bot$	$\top$

(see (7) in Exercise 2.2.1). Show

(a) $(C_3, *)$ is a non-idempotent, non-commutative, bisymmetric, balanced and two-sided quantale.
(b) $(C_3, *)$ is spatial, but not semi-unital.
(c) The universal lower bound in $(C_3, *)$ is prime, but not strongly prime.
(d) There exists a single strong homomorphism $(C_3, *) \xrightarrow{h} C_3^\ell \otimes C_3^r$, and it satisfies $\bigvee\{x \in C_3 \mid h(x) \le \alpha\} = a$. (Consequently the assumption of $(X, *)$ being semi-unital cannot be dropped in Theorem 2.5.18.)

(e) The semi-unitalization (cf. Sect. 2.3.1) $(\overline{C_3}, \overline{*})$ of $(C_3, *)$ is isomorphic to the chain C_4 consisting of four elements $\bot$, a, b and $\top$ provided with the multiplication given by the multiplication table:

$\overline{*}$	$\bot$	a	b	$\top$
$\bot$	$\bot$	$\bot$	$\bot$	$\bot$
a	$\bot$	$\bot$	a	a
b	$\bot$	$\bot$	b	b
$\top$	$\bot$	a	b	$\top$

Show that the elements a and b of C_4 are two-sided and prime in $(C_4, \overline{*})$ where b corresponds to the universal upper bound in C_3.

2.5.5. Let $\mathbb{M}_3$ be the C^*-algebra of all square matrices of order 3 with complex coefficients. It is well-known that every pure state is a vector state — i.e. for every pure state ρ of $\mathbb{M}_3$ there exists a unit vector $\mathbf{e}_\rho$ of $\mathbb{C}^3$ such that $\rho(A) = \langle A\mathbf{e}_\rho, \mathbf{e}_\rho\rangle$ holds for all $A \in \mathbb{M}_3$ where $\langle\ ,\ \rangle$ is the usual inner product in $\mathbb{C}^3$.

(a) Show that $L_\rho = \{A \in \mathbb{M}_3 \mid A\mathbf{e}_\rho = \mathbf{0}\}$ is the left kernel of ρ.
(b) Let I be a left ideal and J be a right ideal of $\mathbb{M}_3$ given by:

$$I = \left\{\begin{pmatrix} 0 & \alpha & -\alpha \\ 0 & \beta & -\beta \\ 0 & \gamma & -\gamma \end{pmatrix} \,\middle|\, \alpha, \beta, \gamma \in \mathbb{C}\right\}, \quad J = \left\{\begin{pmatrix} \alpha & \beta & \gamma \\ 0 & 0 & 0 \\ -\alpha & -\beta & -\gamma \end{pmatrix} \,\middle|\, \alpha, \beta, \gamma \in \mathbb{C}\right\}.$$

Further, let ρ_1, ρ_2, ρ_3 and ρ_4 be pure states of $\mathbb{M}_3$ determined by the following unit vectors:

$$\mathbf{e}_{\rho_1} = (1, 0, 0), \quad \mathbf{e}_{\rho_2} = (0, 1, 0), \quad \mathbf{e}_{\rho_3} = (0, 0, 1), \quad \mathbf{e}_{\rho_4} = \frac{1}{\sqrt{3}}(1, 1, 1).$$

Show that $\mathbb{A}_{\pi_{\mathbb{MAX}}(I \otimes J)}$ attains the following values:

$$\mathbb{A}_{\pi_{\mathbb{MAX}}(I \otimes J)}(\rho_1) = \lambda, \quad \mathbb{A}_{\pi_{\mathbb{MAX}}(I \otimes J)}(\rho_2) = \rho,$$
$$\mathbb{A}_{\pi_{\mathbb{MAX}}(I \otimes J)}(\rho_3) = \top, \quad \mathbb{A}_{\pi_{\mathbb{MAX}}(I \otimes J)}(\rho_4) = \beta.$$

2.5.6. Let $\mathbb{M}_2$ be the C^*-algebra of all square matrices of order 2 with complex coefficients. Further, let $\mathbb{L}(\mathbb{M}_2)$ be the quantale of all left ideals and $\mathbb{R}(\mathbb{M}_2)$ be the quantale of all right ideals. We know that every non-trivial, proper left (right) ideal is maximal (cf. Exercise 2.4.4 (b)).

(a) Show that the set $\sigma\big(\mathbb{L}(\mathbb{M}_2) \otimes \mathbb{R}(\mathbb{M}_2)\big)$ of all prime elements of the tensor product $\mathbb{L}(\mathbb{M}_2) \otimes \mathbb{R}(\mathbb{M}_2)$ can be identified with the set

$$\{(P_1, P_2) \mid P_1 \in \mathbb{L}(\mathbb{M}_2),\ P_2 \in \mathbb{R}(\mathbb{M}_2),\ P_1 \neq \mathbb{M}_2,\ P_2 \neq \mathbb{M}_2\}.$$

(b) First we recall $C_3^\ell = (C_3, *_\ell)$ and $C_3^r = (C_3, *_r)$. If L is a non-trivial left ideal and R is a non-trivial right ideal, then we consider the corresponding maps $\mathbb{L}(\mathbb{M}_2) \setminus \{\mathbb{M}_2\} \xrightarrow{\mathbb{A}_L} C_3^\ell$ and $\mathbb{R}(\mathbb{M}_2) \setminus \{\mathbb{M}_2\} \xrightarrow{\mathbb{A}_R} C_3^r$ having the following form:

$$\mathbb{A}_L(P) = \begin{cases} \top, & L \not\subseteq P, \\ a, & L = P, \\ \bot, & L = \{0\}, \end{cases} \quad \text{and} \quad \mathbb{A}_R(P) = \begin{cases} \top, & R \not\subseteq P, \\ a, & R = P, \\ \bot, & R = \{0\}. \end{cases}$$

If C_3^ℓ is identified with $\mathbb{L}(C_3^\ell \otimes C_3^r)$ and C_3^r with $\mathbb{L}(C_3^\ell \otimes C_3^r)$, then verify (cf. (2.73)):

$$\mathbb{A}_{L\otimes R}(P_1, P_2) = \mathbb{A}_R(P_2) \star \mathbb{A}_L(P_1), \qquad P_1 \neq \mathbb{M}_2,\ P_2 \neq \mathbb{M}_2.$$

(c) If L_1, L_2 and L_3 are pairwise distinct, non-trivial and proper left ideals of $\mathbb{M}_2$ and R_1, R_2 and R_3 are pairwise distinct, non-trivial and proper right ideals of $\mathbb{M}_2$. Then confirm the following relations:

(i) $\mathbb{A}_{L_1\otimes R_1} \vee \mathbb{A}_{L_2\otimes R_2} \vee \mathbb{A}_{L_3\otimes R_3} = \mathbb{A}_{\mathbb{M}_2\otimes\mathbb{M}_2}$.
(ii) $(L_1 \otimes R_1) \vee (L_2 \otimes R_2) \vee (L_3 \otimes R_3) \neq \mathbb{M}_2 \otimes \mathbb{M}_2$.

Conclude from the previous observations that $\mathbb{L}(\mathbb{M}_2) \otimes \mathbb{R}(\mathbb{M}_2)$ is not spatial.

The next Exercises 2.5.7 and 2.5.8 form a preparation for Exercise 2.5.9

2.5.7. Let A be a square matrix of order n and $\mathbb{I}$ be the left (right) ideal generated by A. If $\operatorname{rank}(A) < n$, then show that the linear dimension of $\mathbb{I}$ is given by $\dim(\mathbb{I}) = n \cdot \operatorname{rank}(A)$ and $\mathbb{I}$ has a base $\{D_{rq} \mid r = 1, \dots, n;\ q = 1, \dots, \operatorname{rank}(A)\}$ such that $\operatorname{rank}(D_{rq}) = 1$ for all r and q.
(Hint: Let $A = (\mathbf{a}_1, \dots, \mathbf{a}_n)$ with $\operatorname{rank}(A) = k < n$ and $\mathbf{a}_j = (a_{ij})_{i=1}^n$, $j = 1, \dots, n$. Without loss of generality, we can assume that $\{\mathbf{a}_1, \dots, \mathbf{a}_k\}$ is linearly independent. Then for $l = 1, \dots, k$ and $p = k+1, \dots, n$ there exist $d_{l\,p} \in \mathbb{C}$ such that the relation $\mathbf{a}_p = \sum_{l=1}^k d_{lp}\mathbf{a}_l$ holds. Further, let $B = (b_{ij})_{i,j=1}^n$ be an arbitrary square matrix of order n. We put:

$$c_{i\,m} = \sum_{j=1}^n b_{ij} \cdot a_{j\,m}, \qquad i \in \{1, \dots, n\}, \qquad m \in \{1, \dots, n\}.$$

Then $c_{ip} = \sum_{l=1}^k c_{i\,l} \cdot d_{lp}$ follows for $p = k+1, \dots, n$. Finally, for $r = 1, \dots, n$ and $q = 1, \dots, k$ we define $D_{rq} = (g_{ij}^{r,q})_{i,j=1}^n$ by:

$$g_{ij}^{r,q} = \begin{cases} 0, & i \neq r, \\ 0, & i = r,\ j \in \{1, \dots, k\} \setminus \{q\}, \\ 1, & i = r,\ j = q, \\ d_{qj}, & i = r,\ j = k+1, \dots, n. \end{cases}$$

Now $\operatorname{rank}(D_{rq}) = 1$ and $\sum_{r=1}^{n}\left(\sum_{q=1}^{k} c_{rq} \cdot D_{rq}\right) = B \cdot A$ follow. Since for every $i = 1, \ldots, n$ and for every k-tuple $(c_{iq})_{q=1}^{k}$ the system of linear equations

$$\sum_{j=1}^{n} b_{ij} \cdot a_{jq} = c_{iq}, \qquad q = 1, \ldots, k$$

has a solution $(b_{i1}, \ldots, b_{in})$, the left ideal $\mathbb{I}$ generated by A has dimension $k \cdot n$.)

2.5.8. Let $\mathbb{M}_n$ be the C^*-algebra of all square matrices of order n with complex coefficients. For every square matrix C of order n with $\operatorname{rank}(C) = 1$ there exists a left ideal $\mathbb{L}$ and a right ideal $\mathbb{R}$ such that $\mathbb{R} * \mathbb{L} = \{z \cdot C \mid z \in \mathbb{C}\}$.
(Hint: Without loss of generality we can assume that the given matrix $C = (c_{ij})_{i,j=1}^{n}$ with $\operatorname{rank}(C) = 1$ has the following form:
Let us consider two $(n-1)$-tuples $(p_1, \ldots, p_{n-1}) \in \mathbb{C}^{n-1}$ and $(q_1, \ldots, q_{n-1}) \in \mathbb{C}^{n-1}$ such that c_{ij} is given by:

$$c_{ij} = \begin{cases} 1, & i = 1,\ j = 1, \\ p_{i-1}, & i \neq 1,\ j = 1, \\ q_{j-1}, & i = 1,\ j \neq 1, \\ q_{j-1} \cdot p_{i-1}, & i \neq 1, j \neq 1. \end{cases}$$

For every vector $\mathbf{x} = (x_1, \ldots, x_n) \in \mathbb{C}^n$ and $\mathbf{y} = (y_1, \ldots, y_n) \in \mathbb{C}^n$ we introduce matrices $A_{\mathbf{x}} = (a_{ij})_{i,j=1}^{n}$ and $B_{\mathbf{y}} = (b_{ij})_{i,j=1}^{n}$ as follows:

$$a_{ij} = \begin{cases} x_i, & j = 1, \\ x_i \cdot q_{j-1}, & j \neq 1, \end{cases} \quad \text{and} \quad b_{ij} = \begin{cases} y_j, & i = 1, \\ p_{i-1} \cdot y_j, & i \neq 1. \end{cases}$$

Then $\mathbb{L}_{\mathbf{q}} = \{A_{\mathbf{x}} \mid \mathbf{x} \in \mathbb{C}^n\}$ is a left ideal and $\mathbb{R}_{\mathbf{p}} = \{B_{\mathbf{y}} \mid \mathbf{y} \in \mathbb{C}^n\}$ is a right ideal. Now we put $z = \sum_{j=1}^{n} y_j \cdot x_j$ and observe:

$$\sum_{j=1}^{n} b_{ij} \cdot a_{js} = \begin{cases} z & i = 1,\ s = 1, \\ p_{i-1} \cdot z & i \neq 1,\ s = 1, \\ z \cdot q_{s-1} & i = 1,\ s \neq 1, \\ p_{i-1} \cdot z \cdot q_{s-1} & i \neq 1,\ p \neq 1, \end{cases} = z \cdot c_{is}.$$

Hence $\mathbb{R}_{\mathbf{p}} * \mathbb{L}_{\mathbf{q}} = \{z \cdot C \mid z \in \mathbb{C}\}$ follows.)

2.5.9. Let n be a natural number with $2 \leq n$, $\mathbb{M}_n$ be the C^*-algebra of all square matrices of order n with complex coefficients, and let $(\mathbb{MAX}, *)$ be the quantale of all linear subspaces of $\mathbb{M}_n$ (cf. Example 2.5.5). Further, let $\mathbb{L}_n$ be the subquantale of all left ideals of $\mathbb{M}_n$, $\mathbb{R}_n$ be the subquantale of all right ideals of $\mathbb{M}_n$, and let $\mathfrak{Q}_1$ be the set of all linear subspaces of $\mathbb{M}_n$ which have a linear base $\mathbb{B}$ such that every matrix $C \in \mathbb{B}$ has rank 1. Show:

(a) $\mathfrak{Q}_1$ is a subquantale of $(\mathbb{MAX}, *)$.
(b) $\mathfrak{Q}_1$ coincides with $\mathfrak{Q}_{\mathsf{RL}}(\mathbb{MAX})$ — i.e. the subquantale generated by $\mathbb{L}_n \cup \mathbb{R}_n$.

(Hint: In order to verify Exercise 2.5.9 (b) use Exercises 2.5.7 and 2.5.8.)

2.5.10. Let $\mathbb{M}_n$ be the C^*-algebra of all square matrices of order n and $\mathfrak{Q}_1$ be the subquantale of $\mathbb{MAX}$ which consists of all linear subspaces $\mathbb{U}$ of $\mathbb{M}_n$ satisfying the following condition

$$\mathbb{U} \text{ has a linear base } \mathbb{B} \text{ such that every } C \in \mathbb{B} \text{ has rank } 1.$$

If $\widehat{\mathfrak{Q}}_1$ is the semi-integral regularization of $\mathfrak{Q}_1$, then show that $\widehat{\mathfrak{Q}}_1$ is a quotient of the coproduct of $\mathbb{L}(\mathbb{M}_n)$ and $\mathbb{R}(\mathbb{M}_n)$ in `BSQuant`.
(Hint: Exercise 2.5.9 (b) and Corollary 2.5.10.)

2.6 Frobenius Quantales

Let $(X, *)$ be a quantale and d be an element of X viewed as join-preserving map $\mathbb{1} \xrightarrow{\delta} X^{op}$. Then the right-implication $\searrow$ and the left-implication $\swarrow$ of the given quantale induce a pair $X \underset{h_\delta^\ell}{\overset{h_\delta^r}{\rightrightarrows}} X^{op}$ of join-preserving maps by the commutativity of the following diagrams:

$$\begin{array}{ccccc} X & \xrightarrow{r_X^{-1}} & X \otimes \mathbb{1} & \xrightarrow{1_X \otimes \delta} & X \otimes X^{op} \\ & \searrow^{h_\delta^r} & & & \downarrow \searrow \\ & & & & X^{op} \end{array} \tag{2.74}$$

$$\begin{array}{ccccc} X & \xrightarrow{\ell_X^{-1}} & \mathbb{1} \otimes X & \xrightarrow{\delta \otimes 1_X} & X^{op} \otimes X \\ & \searrow^{h_\delta^\ell} & & & \downarrow \swarrow \\ & & & & X^{op} \end{array} \tag{2.75}$$

Obviously, the pair $(h_\delta^r, h_\delta^\ell)$ forms a contravariant Galois connection on X. In fact, the following relation holds for all $x, y \in X$:

$$y \le h_\delta^r(x) \iff y \le x \searrow d \iff x \le d \swarrow y \iff x \le h_\delta^\ell(y).$$

The previous relation means that h_δ^ℓ is right adjoint to h_δ^r provided we view h_δ^ℓ as isotone map from X^{op} to X. Hence we conclude from Exercise 1.3.2 that h_δ^r is an isomorphism in Sup if and only if $h_\delta^\ell \circ h_\delta^r = 1_X$ and $h_\delta^r \circ h_\delta^\ell = 1_{X^{op}}$ hold — i.e. $x = d \swarrow (x \searrow d)$ and $x = (d \swarrow x) \searrow d$ for all $x \in X$.

These observations motivate the following definition.

Definition 2.6.1. Let $(X, *)$ be a quantale, $d \in X$ and $\mathbb{1} \xrightarrow{\delta} X^{op}$ be the join-preserving map with $\delta(1) = d$. The triple $(X, *, \delta)$ is called a *Frobenius quantale* if $X \xrightarrow{h_\delta^r} X^{op}$ is an isomorphism in Sup — this means that the following relation holds for all $x \in X$:

$$x = d \swarrow (x \searrow d) = (d \swarrow x) \searrow d.$$

In this context d is called the *dualizing element* of $(X, *, \delta)$.

Proposition 2.6.2. *Every Frobenius quantale $(X, *, \delta)$ with $\delta(1) = d$ satisfies the following properties*:

(i) $x \searrow z = ((d \swarrow z) * x) \searrow d$ *for all* $x, z \in X$.
(ii) $z \swarrow x = d \swarrow (x * (z \searrow d))$ *for all* $x, z \in X$.
(iii) $X \xrightarrow{h_\delta^r} X^{op}$ *is meet-preserving — i.e.* $(\bigwedge A) \searrow d = \bigvee\{a \searrow d \mid a \in A\}$ *for all* $A \subseteq X$.
(iv) $X \xrightarrow{h_\delta^\ell} X^{op}$ *is meet-preserving — i.e.* $d \swarrow (\bigwedge A) = \bigvee\{d \swarrow a \mid a \in A\}$ *for all* $A \subseteq X$.

Proof. The properties (i) and (ii) follow from the relation (E) in the proof of Theorem 2.3.6 and the following observations:

$$x \searrow z = x \searrow ((d \swarrow z) \searrow d) \quad \text{and} \quad z \swarrow x = (d \swarrow (z \searrow d)) \swarrow x.$$

Since h_δ^r is bijective, we infer from Exercise 1.3.2 (a)(iii) that $h_\delta^r \dashv h_\delta^\ell$ implies $h_\delta^\ell \dashv h_\delta^r$. Hence (iii) follows. If we view h_δ^ℓ as an isotone map from X to X^{op}, then we can verify (iv) analogously. □

In the following considerations we will always identify the join-preserving map $\mathbb{1} \xrightarrow{\delta} X^{op}$ with its dualizing element $d = \delta(1)$ of X.

Proposition 2.6.3. *Every Frobenius quantale is unital.*

Proof. Let $(X, *, d)$ be a Frobenius quantale. By Proposition 2.6.2 (i) the relation $((d \swarrow d) * x) \searrow d = x \searrow d$ holds for all $x \in X$. Since d is dualizing, $(d \swarrow d) * x = x$ follows — i.e. $d \swarrow d$ is a left unit of $*$. Analogously we verify that $d \searrow d$ is a right unit of $*$. Now we observe:

$$d \swarrow d = (d \swarrow d) * (d \searrow d) = d \searrow d.$$

Hence $d \swarrow d$ and $d \searrow d$ coincide and constitute the unit of $(X, *)$. □

Left- and right-sided elements (cf. Sect. 2.3) play a special rôle in Frobenius quantales.

Lemma 2.6.4. *Let $(X, *, d)$ be a Frobenius quantale and $x \in X$.*

(i) *If $x \in X$ is left-sided, then $x \searrow d = x \searrow \perp$ holds.*
(ii) *If $x \in X$ is right-sided, then $d \swarrow x = \perp \swarrow x$ holds.*

Proof. In order to verify (i) we proceed as follows. Let z be a left-sided element of X with $z \leq d$. Then $d \swarrow z = \top$ holds. Since d is dualizing, $z = \perp$ follows from Proposition 2.6.2 (iii). Now we choose an arbitrary left-sided element $x \in X$. Since $*$ is associative, $x * (x \searrow d)$ is left-sided, and we deduce from $x * (x \searrow d) \leq d$ that $x * (x \searrow d) = \perp$ is valid. Hence $x \searrow d = x \searrow \perp$ follows. — Analogously we prove (ii). □

Corollary 2.6.5. *Every Frobenius quantale satisfies the Frobenius property — i.e. for each right-sided element x and for each left-sided element y the following hold:*

$$x = (\perp \swarrow x) \searrow \perp \quad \textit{and} \quad y = \perp \swarrow (y \searrow \perp). \qquad \text{(Frobenius Property)}$$

Proof. Let $(X, *, d)$ be a Frobenius quantale and $x \in X$ be right-sided. Then the relation $d \swarrow x = \perp \swarrow x$ follows from Lemma 2.6.4 (ii). Since $\perp$ is the zero element of $(X, *)$, the element $\perp \swarrow x$ is left-sided. Now we use Lemma 2.6.4 (i) and obtain:

$$x = (d \swarrow x) \searrow d = (\perp \swarrow x) \searrow d = (\perp \swarrow x) \searrow \perp.$$

If y is left-sided, then $y = \perp \swarrow (y \searrow \perp)$ can be verified analogously. □

The next theorem gives a sufficient condition under which the tensor product of Frobenius quantales is again a Frobenius quantale.

Theorem 2.6.6. *Let $(X, *, d_X)$ and $(Y, *, d_Y)$ be Frobenius quantales. If X or Y is completely distributive, then the tensor product $(X \otimes Y, \star)$ of $(X, *, d_X)$ and $(Y, *, d_Y)$ is again a Frobenius quantale with the dualizing element d_0 determined by:*

$$d_0 = (d_X \otimes \top) \vee (\top \otimes d_Y). \tag{2.76}$$

Proof. (a) Let $d_0 \in X \otimes Y$ be defined by (2.76). In a first step we show that for $x \in X$ and $y \in Y$ the following relations hold:

$$(x \otimes y) \searrow d_0 = \big((x \searrow d_X) \otimes \top\big) \vee \big(\top \otimes (y \searrow d_Y)\big) \tag{2.77}$$
$$d_0 \swarrow (x \otimes y) = \big((d_X \swarrow x) \otimes \top\big) \vee \big(\top \otimes (d_Y \swarrow y)\big). \tag{2.78}$$

In order to verify (2.77) we choose an elementary tensor $u \otimes v \in X \otimes Y$ such that

$$(x \otimes y) \star (u \otimes v) \leq d_0.$$

Hence $(x * u) \otimes (y * v) \leq (d_X \otimes \top) \vee (\top \otimes d_Y)$ follows from the definition of $\star$ and d_0. By Exercise 2.1.4 (b) we obtain $x * u \leq d_X$ or $y * v \leq d_Y$ — i.e.

$$u \otimes v \leq \big((x \searrow d_X) \otimes \top\big) \vee \big(\top \otimes (y \searrow d_Y)\big).$$

On the other hand we observe:

$$(x \otimes y) \star \big(((x \searrow d_X) \otimes \top) \vee (\top \otimes (y \searrow d_Y))\big) \leq \big(d_X \otimes (y * \top)\big) \vee \big((x * \top) \otimes d_Y\big) \leq d_0.$$

Hence (2.77) follows. The formula (2.78) can be verified analogously.

With regard to the following considerations it is also worthwhile to consider the special cases $x \otimes \top$ and $\top \otimes y$. Referring to Proposition 2.6.2 (iii) and (iv) we use the fact that every dualizing element d satisfies the properties $\top \searrow d = \bot$ and $d \swarrow \top = \bot$. Hence we obtain from (2.77) and (2.78):

$$(x \otimes \top) \searrow d_0 = (x \searrow d_X) \otimes \top,\; d_0 \swarrow (x \otimes \top) = (d_X \swarrow x) \otimes \top,$$
$$(\top \otimes y) \searrow d_0 = \top \otimes (y \searrow d_Y),\; d_0 \swarrow (\top \otimes y) = \top \otimes (d_Y \swarrow y).$$

(b) Let $f = \bigvee_{i \in I} x_i \otimes y_i$ be an arbitrary tensor of $X \otimes Y$ with a nonempty index set I, and let $\mathscr{P}(I)$ be the power set of I. Then we conclude from (2.77) and Exercise 2.1.4 (d) that the following relation holds:

$$\begin{aligned} f \searrow d_0 &= \bigwedge_{i \in I} \big((x_i \otimes y_i) \searrow d_0\big) \\ &= \bigwedge_{i \in I} \big((x_i \searrow d_X) \otimes \top\big) \vee \big(\top \otimes (y_i \searrow d_Y)\big) \\ &= \bigvee_{A \subseteq I} \Big(\big(\bigwedge_{i \in A} (x_i \searrow d_X)\big) \otimes \big(\bigwedge_{i \notin A} (y_i \searrow d_Y)\big)\Big) \\ &= \bigvee_{A \subseteq I} \Big(\big((\bigvee_{i \in A} x_i) \searrow d_X\big) \otimes \big((\bigvee_{i \notin A} y_i) \searrow d_Y\big)\Big). \end{aligned}$$

Now we use (2.78) and the fact that d_X and d_Y are dualizing elements:

$$\begin{aligned} &d_0 \swarrow (f \searrow d_0) \\ &= \big((\bigvee_{i \in I} x_i) \otimes \top\big) \wedge \big(\top \otimes (\bigvee_{i \in I} y_i)\big) \wedge \bigwedge_{\varnothing \subsetneqq A \subsetneqq I} \Big(\big((\bigvee_{i \in A} x_i) \otimes \top\big) \vee \big(\top \otimes (\bigvee_{i \notin A} y_i)\big)\Big) \\ &= \big((\bigvee_{i \in I} x_i) \otimes (\bigvee_{i \in I} y_i)\big) \wedge \bigwedge_{\varnothing \subsetneqq A \subsetneqq I} \Big(\big((\bigvee_{i \in A} x_i) \otimes \top\big) \vee \big(\top \otimes (\bigvee_{i \notin A} y_i)\big)\Big). \end{aligned}$$

The aim of the following consideration is to verify $d_0 \swarrow (f \searrow d_0) \leq f$. Without loss of generality, we can assume that I contains at least two different indices. Now we apply Exercise 2.1.4 (e) and obtain:

$$\begin{aligned} d_0 \swarrow (f \searrow d_0) = &\Big(\bigwedge_{\varnothing \subsetneqq A \subsetneqq I} (\bigvee_{i \in A} x_i) \otimes (\bigvee_{i \in I} y_i)\Big) \vee \Big((\bigvee_{i \in I} x_i) \otimes \big(\bigwedge_{\varnothing \subsetneqq A \subsetneqq I} (\bigvee_{i \notin A} y_i)\big)\Big) \vee \\ &\vee \bigvee_{\substack{\mathbb{A} \subseteq \mathscr{P}(I) \setminus \{\varnothing, I\},\\ \varnothing \neq \mathbb{A} \neq \mathscr{P}(I) \setminus \{\varnothing, I\}}} \Big(\bigwedge_{A \in \mathbb{A}} (\bigvee_{i \in A} x_i)\Big) \otimes \Big(\bigwedge_{A \notin \mathbb{A}} (\bigvee_{i \notin A} y_i)\Big). \end{aligned}$$

As a first observation we note that

$$\bigwedge_{\varnothing \subsetneqq A \subsetneqq I} \big(\bigvee_{i \in A} x_i\big) = \bigwedge_{i \in I} x_i \quad \text{and} \quad \bigwedge_{\varnothing \subsetneqq A \subsetneqq I} \big(\bigvee_{i \notin A} y_i\big) = \bigwedge_{i \in I} y_i .$$

Hence the following relations

$$\big(\bigwedge_{\varnothing \subsetneqq A \subsetneqq I} \big(\bigvee_{i \in A} x_i\big)\big) \otimes \big(\bigvee_{i \in I} y_i\big) \le \bigvee_{i \in I} (x_i \otimes y_i),$$

$$\big(\bigvee_{i \in I} x_i\big) \otimes \big(\bigwedge_{\varnothing \subsetneqq A \subsetneqq I} \big(\bigvee_{i \notin A} y_i\big)\big) \le \bigvee_{i \in I} (x_i \otimes y_i)$$

are obvious. Now we fix $\mathbb{A} \subseteq \mathscr{P}(I) \setminus \{\varnothing, I\}$ such that $\varnothing \neq \mathbb{A} \neq \mathscr{P}(I) \setminus \{\varnothing, I\}$. Without loss of generality, we can assume that X is completely distributive and choose $z \in X$ such that z is totally below $\bigwedge_{A \in \mathbb{A}} \big(\bigvee_{i \in A} x_i\big)$ — i.e. $z \lhd \bigwedge_{A \in \mathbb{A}} \big(\bigvee_{i \in A} x_i\big)$. Hence for every $A \in \mathbb{A}$ there exists an $i_A \in A$ such that $z \le x_{i_A}$. Now we distinguish the subsequent cases:
Case 1: $\{i_A \mid A \in \mathbb{A}\} = I$. Then we have the following estimate:

$$z \otimes \big(\bigwedge_{A \notin \mathbb{A}} \big(\bigvee_{i \notin A} y_i\big)\big) \le \big(\bigwedge_{i \in I} x_i\big) \otimes \big(\bigvee_{i \in I} y_i\big) \le \bigvee_{i \in I} (x_i \otimes y_i).$$

Case 2: $\{i_A \mid A \in \mathbb{A}\} \neq I$. Then $B = I \cap \complement\{i_A \mid A \in \mathbb{A}\}$ is nonempty and not contained in $\mathbb{A}$ by definition of B — i.e. $B \notin \mathbb{A}$. Now we obtain:

$$z \otimes \big(\bigwedge_{A \notin \mathbb{A}} \big(\bigvee_{i \notin A} y_i\big)\big) \le \big(\bigwedge_{A \in \mathbb{A}} x_{i_A}\big) \otimes \big(\bigvee_{i \notin B} y_i\big) \le \bigvee_{A \in \mathbb{A}} (x_{i_A} \otimes y_{i_A}).$$

Since the totally below relation $\lhd$ is approximating, we have shown:

$$\big(\bigwedge_{A \in \mathbb{A}} \big(\bigvee_{i \in A} x_i\big)\big) \otimes \big(\bigwedge_{A \notin \mathbb{A}} \big(\bigvee_{i \notin A} y_i\big)\big) \le \bigvee_{i \in I} (x_i \otimes y_i).$$

Hence the relation $d_0 \swarrow (f \searrow d_0) \le f$ is verified.
Analogously we prove $(d_0 \swarrow f) \searrow d_0 \le f$. Thus d_0 is a dualizing element of the tensor product $(X \otimes Y, \star)$. □

A residuated preordered semigroup $(X, \le, *, \searrow, \swarrow)$ (cf. Example 2.3.7) has a dualizing element d if and only if $d \swarrow (x \searrow d) = x = (d \swarrow x) \searrow d$ holds for all $x \in X$. In the next theorem we show that the MacNeille completion of a residuated preordered semigroup with a dualizing element is a Frobenius quantale.

Lemma 2.6.7. *Let $(X, \le, *, \searrow, \swarrow)$ be a residuated preordered semigroup with a dualizing element d and $(\mathsf{Dwn}(X), \boxdot)$ be the quantale of downclosed subsets induced by $(X, \le, *, \searrow, \swarrow)$. Then the relation:*

$$\downarrow d \swarrow (A \searrow \downarrow d) = \mathsf{C}(A) = (\downarrow d \swarrow A) \searrow \downarrow d \tag{2.79}$$

holds for all $A \in \mathsf{Dwn}(X)$, *where* $\mathsf{C} = \mathsf{L} \circ \mathsf{U}$ *is the nucleus on* $(\mathsf{Dwn}(X), \boxdot)$ *associated with the MacNeille completion* (*cf.* *Example 2.2.13*).

Proof. Let us consider two downclosed subsets A and B of X. Then the relation $A \boxdot B \subseteq \downarrow d$ is equivalent to the property $a * b \leq d$ for all $a \in A$ and $b \in B$. Since d is dualizing, it is easily seen that the following properties are valid:

$$A \searrow \downarrow d = \bigcup\{B \in \mathsf{Dwn}(X) \mid A \boxdot B \subseteq \downarrow d\} = \{c \searrow d \mid c \in \mathsf{U}(A)\},$$
$$\downarrow d \swarrow B = \bigcup\{A \in \mathsf{Dwn}(X) \mid A \boxdot B \subseteq \downarrow d\} = \mathsf{L}(\{d \swarrow b \mid b \in B\}).$$

Using again the dualizing property of d we infer from the previous properties:

$$\begin{aligned}\downarrow d \swarrow (A \searrow \downarrow d) &= \downarrow d \swarrow \{c \searrow d \mid c \in \mathsf{U}(A)\}\\ &= \mathsf{L}(\{d \swarrow (c \searrow d) \mid c \in \mathsf{U}(A)\})\\ &= \mathsf{L}(\mathsf{U}(A))\\ &= \mathsf{C}(A).\end{aligned}$$

Analogously we verify $(\downarrow d \swarrow B) \searrow \downarrow d = \mathsf{C}(B)$. Hence the assertion follows. □

Theorem 2.6.8. *Let* $(X, \leq, *, \searrow, \swarrow)$ *be a residuated preordered semigroup with dualizing element* d. *Then the MacNeille completion of* $(X, \leq, *, \searrow, \swarrow)$ *is a Frobenius quantale with dualizing element* $\downarrow d$.

Proof. Since in the quotient $\widehat{X}$ of $\mathsf{Dwn}(X)$ w.r.t. C (cf. Example 2.1.4 (a) and Example 2.2.13) the right-implication and left-implication are given by

$$\mathsf{C}(A) \searrow \mathsf{C}(B) = \mathsf{C}(A \searrow \mathsf{C}(B)) \quad \text{and} \quad \mathsf{C}(A) \swarrow \mathsf{C}(B) = \mathsf{C}(\mathsf{C}(A) \swarrow B)$$

for each $A, B \in \mathsf{Dwn}(X)$, we apply Lemma 2.6.7 and obtain:

$$\downarrow d \swarrow (\mathsf{C}(A) \searrow \downarrow d) = \downarrow d \swarrow \mathsf{C}(A \searrow \downarrow d) = \mathsf{C}(\mathsf{C}(A)) = \mathsf{C}(A).$$

Analogously we verify $(\downarrow d \swarrow \mathsf{C}(A)) \searrow \downarrow d = \mathsf{C}(A)$ for all $A \in \mathsf{Dwn}(X)$. □

In the following considerations we investigate the rôle of cyclic elements of quantales. We begin with a definition.

Definition 2.6.9. An element z of a quantale $(X, *)$ is *cyclic* if the equivalence

$$x * y \leq z \iff y * x \leq z$$

holds for all $x, y \in X$.

As a first trivial property of cyclic elements we have the following:

Lemma 2.6.10. *An element* z *of a quantale* $(X, *)$ *is cyclic if and only if* $x \searrow z = z \swarrow x$ *for all* $x \in X$.

Since in the case of cyclic elements z the maps $\_ \searrow z$ and $z \swarrow \_$ coincide we simply write $x \to z$ for $x \searrow z = z \swarrow x$.

Example 2.6.11. (a) Every element of a commutative quantale is cyclic.
(b) Let $(S, \cdot)$ be a semigroup with zero 0. Further, let $(\mathscr{P}(S), \odot)$ be the quantale given by the power set of S and the Minkowski multiplication induced by $\cdot$. If S does *not* contain a nilpotent element, then $\{0\}$ is cyclic element of $(\mathscr{P}(S), \odot)$.
(c) Let $(X, \leq, *)$ be an preordered semigroup and $(\mathsf{Dwn}(X), \boxdot)$ be the quantale of downclosed subsets of X. If $z \in X$ is a cyclic element of $(X, *)$, then $\downarrow z$ is a cyclic element of $(\mathsf{Dwn}(X), \boxdot)$.

A deeper property of cyclic elements is given in the next lemma.

Lemma 2.6.12. *Every cyclic element z of a quantale $(X, *)$ induces a nucleus c_z on $(X, *)$ by:*

$$c_z(x) = (x \to z) \to z, \qquad x \in X.$$

Proof. Let z be a cyclic element of $(X, *)$. It is easily seen that the map c_z defined by $c_z(x) = (x \to z) \to z$ is a closure operator on X. Further, for $x, y \in X$ we obtain:

$$\begin{aligned} x * y * ((x * y) \to z) \leq z &\implies y * ((x * y) \to z) \leq x \to z \\ &\implies y * ((x * y) \to z) * c_z(x) \leq z \\ &\implies ((x * y) \to z) * c_z(x) * c_z(y) \leq z \\ &\implies c_z(x) * c_z(y) \leq c_z(x * y). \end{aligned}$$

Hence c_z is a nucleus. □

Theorem 2.6.13. *Let z be a cyclic element of a unital quantale $(X, *)$ and c_z be the nucleus induced by z (cf. Lemma 2.6.12). Then $z \in c_z(X)$ and the regular quotient $(c_z(X), \star) = (X, *)/c_z$ is a Frobenius quantale with the dualizing element $d = z$. In particular, z is also cyclic in $(c_z(X), \star)$.*

Proof. Let z be a cyclic element of $(X, *)$ and $(c_z(X), \star)$ be the quotient of $(X, *)$ w.r.t. c_z. Since the quantale $(X, *)$ is unital (cf. Theorem 2.2.11 and Corollary 2.2.12), the relation $c_z(z) = z$ follows. Therefore it is sufficient to show that z is cyclic and dualizing in $(c_z(X), \star)$.

Let x and y be elements of X. Since

$$c_z(x) \star c_z(y) = c_z(x * y) \leq z \iff x * y \leq z,$$

z is not only cyclic in $(X, *)$, but also in $(c_z(X), \star)$. Since the right-implication and left-implication in $(c_z(X), \star)$ have the form

$$c_z(x) \searrow c_z(y) = c_z(x \searrow c_z(y)) \quad \text{and} \quad c_z(x) \swarrow c_z(y) = c_z(c_z(x) \swarrow y),$$

we obtain immediately $(c_z(x) \to z) \to z = c_z(x)$ for all $x \in X$. □

The previous theorem motivates the following definition.

Definition 2.6.14. A Frobenius quantale $(X, *, d)$ is called a *Girard quantale* if the dualizing element d is cyclic.

Theorem 2.6.13 shows that every cyclic element of a unital quantale induces a Girard quantale. On the other hand, every Girard quantale occurs in this way. In fact, if we view a Girard quantale $(X, *, d)$ as a residuated, preordered and unital semigroup with a cyclic and dualizing element d, then $\downarrow d$ is cyclic in the unital quantale $(\mathsf{Dwn}(X), \boxdot)$, and we infer from Lemma 2.6.7 that $c_{\downarrow d}$ coincides with the closure operator C on $(\mathsf{Dwn}(X), \boxdot)$. Hence $(X, *, d)$ is isomorphic to its MacNeille completion (cf. Theorem 2.6.8) and is therefore a regular quotient of $(\mathsf{Dwn}(X), \boxdot)$ w.r.t. the nucleus induced by the cyclic element $\downarrow d$.

Example 2.6.15. Let $\mathbb{M}_n$ be the C^*-algebra of all square matrices of order n with complex coefficients, and $(\mathbb{MAX}, *)$ be the unital (involutive) quantale of all linear subspaces $\mathbb{U}$ of $\mathbb{M}$. (cf. Example 2.5.5). Further, the trace $\mathbb{M}_n \xrightarrow{\text{tr}} \mathbb{C}$ is a linear form on $\mathbb{M}_n$ and its kernel

$$\mathbb{D} = \{A \in \mathbb{M}_n \mid \text{tr}(A) = 0\}$$

is a linear subspace of $\mathbb{M}_n$. Then $(\mathbb{MAX}, *, \mathbb{D})$ is a Girard quantale.
Since $\text{tr}(A \cdot B) = \text{tr}(B \cdot A)$, the subspace $\mathbb{D}$ is evidently cyclic. In order to see that $\mathbb{D}$ is dualizing we first recall that the trace of matrices induces an inner product $\langle\ ,\ \rangle$ on $\mathbb{M}_n$ by

$$\langle A, B\rangle = \text{tr}(B^* \cdot A),$$

where B^* is the adjoint matrix of B. Obviously, the following equivalence holds for all $\mathbb{U}, \mathbb{V} \in \mathbb{MAX}$:

$$\begin{aligned} \mathbb{U} * \mathbb{V} = \text{lin. hull}\,\{A \cdot B \mid A \in \mathbb{U},\ B \in \mathbb{V}\} \subseteq \mathbb{D} \\ \iff \quad \langle B, A^*\rangle = 0 \text{ for all } A \in \mathbb{U},\ B \in \mathbb{V}. \end{aligned}$$

Further, we recall that $\mathbb{W}' = \{A^* \mid A \in \mathbb{W}\}$ is the adjoint subspace of $\mathbb{W} \in \mathbb{MAX}$ (cf. Example 2.5.5), and the orthogonal complement of $\mathbb{W}$ w.r.t. $\langle\ ,\ \rangle$ is denoted by $\mathbb{W}^{\perp}$. Then for every subspace $\mathbb{U}$ of $\mathbb{M}_n$ the relation

$$\mathbb{U} \searrow \mathbb{D} = \mathbb{D} \swarrow \mathbb{U} = \mathbb{U} \to \mathbb{D} = \{A^* \mid A \in \mathbb{U}\}^{\perp} = (\mathbb{U}^*)^{\perp} = (\mathbb{U}^{\perp})^*$$

holds. Since $\big((\mathbb{U}^*)^{\perp\perp}\big)^* = \mathbb{U}$, the subspace $\mathbb{D}$ is dualizing.
Finally, as an application of the Frobenius property (cf. Corollary 2.6.5) we obtain that for each left ideal I and each right-sided ideal J of $\mathbb{M}_n$ the following relations hold (cf. [14, p. 341]):

$$I = \{0\} \swarrow (I \searrow \{0\}) \quad \text{and} \quad J = (\{0\} \swarrow J) \searrow \{0\}.$$

In the following consideration we will focus on a special class of Girard quantales supported by completely distributive lattices (see also [91]).

Example 2.6.16. Let X be a completely distributive lattice. Further, let $([X, X], \circ)$ be the unital quantale of all join-preserving self-mappings of X provided with the composition as multiplication (cf. Example 2.3.2), and let $X \xrightarrow{d} X$ be the join-preserving map defined by

$$d(z) = \bigvee\{x \in X \mid z \not\leq x\}, \qquad z \in X. \tag{2.80}$$

Then $([X, X], \circ, d)$ is a Girard quantale.

In order to see that d is dualizing and cyclic we proceed as follows. First, for $z \in X$, we define subsets A_z and B_z of X by:

$$A_z = \{y \in X \mid y \not\leq z\} \quad \text{and} \quad B_z = \{x \in X \mid z \not\leq x\}.$$

Now we fix $z \in X$ and show that

$$\bigvee_{x \in B_z} \big(\bigwedge A_x\big) = z. \tag{2.81}$$

First we notice that $B_z = \varnothing \iff z = \bot$. Therefore, without loss of generality, we can assume $z \neq \bot$. Since $z \in A_x$ for all $x \in B_z$, the relation $\bigvee_{x \in B_z} \big(\bigwedge A_x\big) \leq z$ is evident. On the other hand, $A_x \neq \varnothing$ for all $x \in B_z$. Thus for all $x \in B_z$ we can choose some $y_x \in A_x$ and observe:

$$z \leq \bigvee_{x \in B_z} y_x. \tag{2.82}$$

In fact, if we assume $z \not\leq \bigvee_{x \in B_z} y_x$, then $x_0 := \bigvee_{x \in B_z} y_x \in B_z$ and $y_{x_0} \leq x_0$, which is a contradiction to the choice of y_{x_0} — namely $y_{x_0} \in A_{x_0}$. Hence (2.82) holds. Now we invoke the complete distributivity of X and obtain:

$$z \leq \bigwedge_{(y_x)_{x \in B_z} \in \prod_{x \in B_z} A_x} \big(\bigvee_{x \in B_z} y_x\big) = \bigvee_{x \in B_z} \big(\bigwedge A_x\big).$$

Hence (2.81) is verified. The dual statement of (2.81) is the following relation:

$$\bigwedge_{y \in A_z} \big(\bigvee B_y\big) = z. \tag{2.83}$$

Further, for each $(x, y) \in X \times X$ we introduce a join-preserving map $X \xrightarrow{\vartheta_x^y} X$ by:

$$\vartheta_x^y(z) = \begin{cases} y, & z \not\leq x \\ \bot, & z \leq x, \end{cases} \qquad z \in X, \tag{2.84}$$

and conclude from (2.81) that the identity map 1_X can be represented as a join of the maps $\vartheta_x^{\alpha(x)}$:

$$1_X = \bigvee_{x \in X} \vartheta_x^{\alpha(x)} \qquad \text{where} \quad \alpha(x) = \bigwedge A_x. \tag{2.85}$$

If $X \xrightarrow{f} X$ is a join-preserving map and $f^{\vdash}$ is the right adjoint map of f, then the relation:

$$\vartheta_x^y \circ f = \vartheta_{f^{\vdash}(x)}^y$$

holds. Hence we conclude from (2.85) that every join-preserving map from X to X can be written as a join of maps of type ϑ_x^y.

(a) We show that d is cyclic. By the previous argumentation it is sufficient to show that for all $x, y, u, v \in X$ the following equivalence holds:

$$\vartheta_u^v \circ \vartheta_x^y \leq d \iff \vartheta_x^y \circ \vartheta_u^v \leq d. \tag{EE}$$

First, we establish the implication $\vartheta_u^v \circ \vartheta_x^y \leq d \implies \vartheta_x^y \circ \vartheta_u^v \leq d$ and distinguish the following cases.

Case 1. $y \leq u$ — this means that $\vartheta_u^v \circ \vartheta_x^y$ coincides with the bottom element of $[X, X]$. From (2.80) and (2.83) we obtain

$$y \leq u = \bigwedge_{z \in A_u} d(z),$$

and the relation $\vartheta_x^y \circ \vartheta_u^v \leq d$ follows.

Case 2. $y \not\leq u$. Then $\vartheta_x^v = \vartheta_u^v \circ \vartheta_x^y \leq d$ implies $v \leq \bigwedge_{z \in A_x} d(z) = x$. Hence $\vartheta_x^y \circ \vartheta_u^v$ coincides with the bottom element of $[X, X]$, and $\vartheta_x^y \circ \vartheta_u^v \leq d$ holds trivially.

If in the previous argumentation we interchange the rôle of x and u and the rôle of y and v, then we obtain a proof of the implication $\vartheta_x^y \circ \vartheta_u^v \leq d \implies \vartheta_u^v \circ \vartheta_x^y \leq d$. To sum up, we have confirmed the equivalence (EE).

(b) We show that d is dualizing. First we consider the case $f = \vartheta_x^y$. By (2.83) the relation $g \circ \vartheta_x^y \leq d$ is equivalent to $g(y) \leq \bigwedge_{z \in A_x} d(z) = x$ for all $g \in [X, X]$. Hence $d \swarrow \vartheta_x^y = \bigvee\{g \in [X, X] \mid g \circ \vartheta_x^y \leq d\}$ coincides with the elementary tensor $y \otimes x$ of the tensor product $X \otimes X^{op}$ — i.e.

$$(d \swarrow \vartheta_x^y)(z) = (y \otimes x)(z) = \begin{cases} \top, & z \not\leq y, \\ x, & z \leq y,\ z \neq \bot, \\ \bot, & z = \bot, \end{cases} \qquad z \in X. \tag{2.86}$$

In the following considerations, $y \otimes x$ will be viewed as an element of $[X, X]$. Now we consider an arbitrary map $f \in [X, X]$ and choose a subset $\{(x_i, y_i) \mid i \in I\}$ of $X \times X$ with the property

$$f = \bigvee_{i \in I} \vartheta_{x_i}^{y_i}.$$

Then we obtain $d \swarrow f = \bigwedge_{i\in I}(y_i \otimes x_i)$. Since X is completely distributive, we can describe the meet of join-preserving maps as follows:

$$\big(\bigwedge_{i\in I}(y_i \otimes x_i)\big)(u) = \bigvee_{v \lhd u}\big(\bigwedge_{i\in I}(y_i \otimes x_i)(v)\big), \qquad u \in X, \tag{2.87}$$

where $\lhd$ is the totally below relation. In order to prove $(d \swarrow f) \searrow d = f$ it is sufficient to show that for every map ϑ_x^y with the properties

$$x \neq \top \quad \text{and} \quad \big(\bigwedge_{i\in I}(y_i \otimes x_i)\big) \circ \vartheta_x^y \leq d \tag{2.88}$$

the following implication holds for all $z \in X$:

$$z \not\leq x \quad \Longrightarrow \quad y \leq \bigvee\{y_i \mid i \in I,\ z \not\leq x_i\}.$$

Since $\big(\bigwedge_{i\in I} y_i \otimes x_i\big) \circ \vartheta_x^y \leq d$, we conclude again from (2.83) that the relation

$$\big(\bigwedge_{i\in I}(y_i \otimes x_i)\big)(y) \leq \bigwedge_{z\in A_x} d(z) = x$$

holds. Now we choose $v \in X$ with $v \lhd y$. From (2.87) and (2.88) we obtain:

$$\bigwedge_{i\in I}(y_i \otimes x_i)(v) \leq x.$$

Since $x \neq \top$, the set $N = \{i \in I \mid v \leq y_i\}$ is nonempty, and the relation

$$\bigwedge_{i\in I}(y_i \otimes x_i)(v) = \bigwedge_{i\in N} x_i \leq x$$

follows. If $z \not\leq x$, then there exists an $i_0 \in N$ with $z \not\leq x_{i_0}$. Hence in the case of $z \not\leq x$ we obtain $v \leq \bigvee\{y_i \mid i \in N,\ z \not\leq x_i\}$. Since $v \lhd y$ is arbitrary and the totally below relation is approximating, we have verified the following implication

$$z \not\leq x \quad \Longrightarrow \quad y \leq \bigvee\{y_i \mid z \not\leq x_i\},$$

which means $\vartheta_x^y \leq \bigvee_{i\in I} \vartheta_{x_i}^{y_i} = f$.

Comment. Let X be a complete lattice. The map $X \times X^{op} \xrightarrow{\vartheta} [X, X]$ defined by (cf. Example 2.6.16)

$$\vartheta(y, x) = \vartheta_x^y, \qquad y \in X,\ x \in X^{op}$$

is a bimorphism in Sup. Hence there exists a unique join-preserving map $X \otimes X^{op} = [X, X]^{op} \xrightarrow{\Phi} [X, X]$ satisfying the condition

$$\Phi(y \otimes x) = \vartheta_x^y, \qquad y \in X,\ x \in X^{op}.$$

If X is completely distributive, then Example 2.6.16 shows that $([X, X], \circ, d)$ is a Girard quantale, where d is determined by (2.80). Now we consider the order isomorphism $[X, X] \xrightarrow{h_\delta^r} [X, X]^{op}$ induced by $\mathbb{1} \xrightarrow{\delta} [X, X]$ with $\delta(1) = d$. Since d is cyclic, h_δ^r is determined by (2.86)—i.e. $h_\delta^r(\vartheta_x^y) = y \otimes x$ for all $y \in X$ and $x \in X^{op}$. Since

$$\Phi \circ h_\delta^r = 1_{[X,X]} \quad \text{and} \quad h_\delta^r \circ \Phi = 1_{[X,X]^{op}},$$

we conclude that the complete lattices $[X, X]$ and $X \otimes X^{op}$ are order isomorphic (whenever X is completely distributive).

In order to underline the importance of Example 2.6.16 we discuss two special cases.

Example 2.6.17. (a) Let $X = [0, 1]$ be the real unit interval. By (2.80) the cyclic and dualizing element of the Girard quantale of all join-preserving self-maps on $[0, 1]$ coincides with the identity map of $[0, 1]$, which is obviously the unit of the quantale $([[0, 1], [0, 1]], \circ)$.
(b) Let Ω be an arbitrary set and $X = \mathscr{P}(\Omega)$ be the power set of Ω. Then the cyclic and dualizing element d of the Girard quantale $([\mathscr{P}(\Omega), \mathscr{P}(\Omega)], \circ)$ is determined by:

$$d(\{\omega\}) = \Omega \setminus \{\omega\}, \qquad \omega \in \Omega.$$

If we identify $([\mathscr{P}(\Omega), \mathscr{P}(\Omega)], \circ)$ with the Girard quantale $(\mathrm{Re}(\Omega), \circ)$ of all binary relations on Ω, then the cyclic and dualizing relation coincides with the complement of the diagonal Δ of Ω.

In the next proposition a necessary and sufficient condition is given such that the map defined by (2.80) is a dualizing element of the quantale of all join-preserving self-maps on a complete lattice X.

Proposition 2.6.18. *Let X be a complete lattice and d be the join-preserving self-mapping on X defined by*:

$$d(z) = \bigvee\{x \in X \mid z \not\leq x\}, \qquad z \in X.$$

Then the following assertions are equivalent:

(i) *d is the dualizing element of the quantale* $([X, X], \circ)$.
(ii) *X is completely distributive.*

Proof. The implication (ii) $\Longrightarrow$ (i) is shown in Example 2.6.16. Hence we only verify (i) $\Longrightarrow$ (ii). For this purpose it is sufficient to show that the totally below relation $\lhd$ is approximating in X (cf. [36, 96])—i.e. $z = \bigvee\{u \in X \mid u \lhd z\}$ for all $z \in X$. Since $[X, X]^{op}$ coincides with $X \otimes X^{op}$, it follows that every join-preserving self-map on

X can be written as a meet of elementary tensors with respect to the partial order on $[X, X]$. Hence we proceed as follows. We maintain the notation from Example 2.6.16 and choose a subset $\{(y_i, x_i) \in X \times X \mid i \in I\}$ with the property:

$$d \swarrow 1_X = \bigwedge_{i\in I} (y_i \otimes x_i).$$

Since d is dualizing we obtain:

$$1_X = \Big(\bigwedge_{i\in I} (y_i \otimes x_i)\Big) \searrow d = \bigvee_{i\in I} \vartheta_{x_i}^{y_i}.$$

Hence for all $z \in X$ the relation $z = \bigvee\{y_i \mid i \in I,\ z \not\leq x_i\}$ holds. In particular, $z \not\leq x_i$ implies $y_i \leq z$ — i.e.

$$y_i \leq \bigwedge\{z' \mid z' \not\leq x_i\}, \quad i \in I. \tag{2.89}$$

Now we refer to the notation in Example 2.6.16 and put $A_x = \{z' \in X \mid z' \not\leq x\}$ and $B_z = \{x \in X \mid z \not\leq x\}$. By (2.89) we obtain for all $z \in X$:

$$z \geq \bigvee_{x\in B_z} \big(\bigwedge A_x\big) \geq \bigvee_{z\not\leq x_i} \big(\bigwedge A_{x_i}\big) \geq \bigvee\{y_i \mid z \not\leq x_i\} = z.$$

Hence the relation (2.81) holds. Finally, we have to verify the following implication for all $z \in X$:

$$x \in B_z \quad \Longrightarrow \quad \big(\bigwedge A_x\big) \lhd z.$$

Let C be a subset of X with $z \leq \bigvee C$. If $x \in B_z$, then $z \not\leq x$. Hence there exists a $c \in C$ with $c \in A_x$, which implies $\big(\bigwedge A_x\big) \leq c$ — i.e. $\big(\bigwedge A_x\big) \lhd z$. To sum up, we conclude from the validity of (2.81) that the totally below relation is approximating. □

Referring to Example 2.6.15 and Example 2.6.16 the reader might conjecture that every dualizing element is cyclic. But this is *not* the case. A counterexample is given in Exercise 2.6.5. Hence there exist Frobenius quantales which are not Girard quantales.

Exercises

2.6.1. Let C_3 be the chain with three elements provided with the following multiplication (see (19) in Exercise 2.2.1):

$*$	$\bot$	a	$\top$
$\bot$	$\bot$	$\bot$	$\bot$
a	$\bot$	a	$\top$
$\top$	$\bot$	$\top$	$\top$

Show that $(C_3, *, a)$ is a Girard quantale.

2.6.2. Let C_3 be the chain with three elements. Show that the Girard quantale of all join-preserving self-maps of C_3 consists of six elements and describe their algebraic properties.

2.6.3. Let $(X, *)$ be a unital quantale and d be an element of X satisfying the following property for all $x \in X$:

$$d \swarrow (x \searrow d) = (d \swarrow x) \searrow d.$$

(a) Show that the closure operator c of X defined by $c(x) = d \swarrow (x \searrow d)$ for $x \in X$ is a nucleus on $(X, *)$.
(Hint: Use Exercise 2.2.3.)
(b) Show that the regular quotient of $(X, *)$ w.r.t. c is a Frobenius quantale.
(Hint: First verify $c(d) = d$ and show that for all $x \in X$ the relations

$$x \searrow d = \big(d \swarrow (x \searrow d)\big) \searrow d \quad \text{and} \quad d \swarrow x = d \swarrow \big((d \swarrow x) \searrow d\big)$$

hold. Deduce from these properties:

$$c(x) \searrow d = c(x \searrow d) = x \searrow d \quad \text{and} \quad d \swarrow c(x) = c(d \swarrow x) = d \swarrow x.\,)$$

2.6.4. Let $\mathbb{B}$ be a complete Boolean algebra with at least two different elements. Then $(\mathbb{B}, \wedge)$ is a frame with the universal lower bound $\perp$ as dualizing element. In this sense $(\mathbb{B}, \wedge, \perp)$ can be considered as a special case of a Frobenius quantale. Further, the tensor product of $(\mathbb{B}, \wedge)$ with itself is again a frame (cf. Corollary 2.4.13). In this context we recall that the universal lower bound in $\mathbb{B} \otimes \mathbb{B}$ has the form $\underline{\perp} = \perp \otimes \top = \top \otimes \perp$, and the pseudo-complement of a tensor f in the sense of $\mathbb{B} \otimes \mathbb{B}$ is given by:

$$f \to \underline{\perp} = \bigvee \{g \in \mathbb{B} \otimes \mathbb{B} \mid f \wedge g = \underline{\perp}\}.$$

On the other hand, let $f_0 \in \mathbb{B} \otimes \mathbb{B}$ be the tensor determined by the complementation in $\mathbb{B}$ — i.e.

$$f_0(x) = x \to \perp, \qquad x \in \mathbb{B}.$$

Show:

(a) If $\mathbb{B}$ is atomless, then $f_0 \to \underline{\perp} = \underline{\perp}$.
(b) Conclude from (a) that the assumption of the complete distributivity in Theorem 2.6.6 is not superfluous.
(c) If $\mathbb{B}$ is atomic and X is the set of all atoms of $\mathbb{B}$, then $f_0 \to \underline{\perp} = \bigvee_{x \in X} x \otimes x$.

2.6.5. Let $X = \{\perp, \beta, a, b, c, \alpha, \top\}$ provided with the lattice structure which can be visualized by the Hasse diagram:

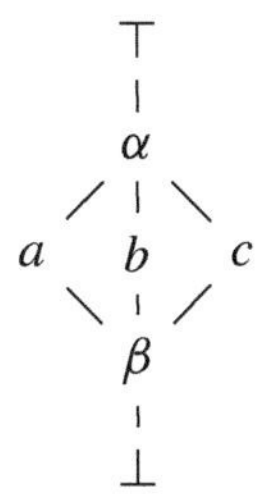

$$\alpha = a \vee b = b \vee c = c \vee a \quad \text{and} \quad \beta = a \wedge b = b \wedge c = c \wedge a.$$

Further, we consider a multiplication $*$ on X determined by the following multiplication table:

$*$	$\bot$	β	a	b	c	α	$\top$
$\bot$	$\bot$	$\bot$	$\bot$	$\bot$	$\bot$	$\bot$	$\bot$
β	$\bot$	$\bot$	$\bot$	$\bot$	$\bot$	$\bot$	β
a	$\bot$	$\bot$	β	$\bot$	β	β	a
b	$\bot$	$\bot$	β	β	$\bot$	β	b
c	$\bot$	$\bot$	$\bot$	β	β	β	c
α	$\bot$	$\bot$	β	β	β	β	α
$\top$	$\bot$	β	a	b	c	α	$\top$

Show:

(a) $(X, *)$ is an integral and non-commutative quantale.
(Hint: $\alpha * \alpha * \alpha = \bot$.)

(b) $a \searrow \bot = b = \bot \swarrow c$, $\; b \searrow \bot = c = \bot \swarrow a$, $\; c \searrow \bot = a = \bot \swarrow b$,
$\alpha \searrow \bot = \beta = \bot \swarrow \alpha$, $\; \beta \searrow \bot = \alpha = \bot \swarrow \beta$, $\; \top \searrow \bot = \bot = \bot \swarrow \top$.

(c) $X \xrightarrow{\,'\,} X$ determined by $\top' = \top, \alpha' = \alpha, a' = a, b' = c, c' = b, \beta = \beta', \bot' = \bot$ is an order-preserving involution on X such that $(X, *, \bot, ')$ is an involutive quantale.

(d) $\bot$ is a dualizing element, but *not* cyclic.

(e) $(X, *, \bot)$ is a Frobenius quantale, but *not* a Girard quantale.

2.7 Complete MV-Algebras

In this section we consider the construction of Girard quantales from the perspective of residuated, commutative lattices (cf. [35]). First we need some more terminology. A *Girard algebra* is a residuated preordered groupoid $(X, *, \searrow, \swarrow)$, which satisfies the following additional properties:

- The underlying preordered set is a bounded lattice.
- $(X, *)$ is a commutative semigroup.
- The universal lower bound is a dualizing element.

Hence in every Girard algebra the unit is necessarily the universal upper bound — i.e. every Girard algebra is integral (cf. Proposition 2.6.3). Since the left- and right-implications coincide, we simply write $(X, *, \rightarrow)$ instead of $(X, *, \searrow, \swarrow)$. Further, every Girard algebra satisfies the properties

$$x \rightarrow y = \big((x * (y \rightarrow \bot)\big) \rightarrow \bot = (y \rightarrow \bot) \rightarrow (x \rightarrow \bot), \tag{2.90}$$

$$(x \wedge y) \rightarrow \bot = (x \rightarrow \bot) \vee (y \rightarrow \bot). \tag{2.91}$$

Therefore an alternative axiomatization of Girard algebras can be given as follows:

(GA1) The partially ordered set $(X, \leq)$ is a bounded lattice with the universal bounds $\bot$ and $\top$.

(GA2) The pair $(X, *)$ is a commutative monoid (in Set) and the unit coincides with the universal upper bound $\top$ in $(X, \leq)$.

(GA3) There exists an involution $x \longmapsto x^{\perp}$ on the set X such that the following relation holds for all $x, y \in X$:

$$x \leq y \iff x * y^{\perp} = \bot.$$

It is easily seen that every Girard algebra satisfies (GA1)–(GA3), where the involution is determined by $x^{\perp} = x \rightarrow \bot$ for all $x \in X$. On the other hand, if a bounded lattice, a commutative monoid and an involution with the properties (GA1)–(GA3) are given, then we define

$$x \rightarrow y = (x * y^{\perp})^{\perp}$$

and observe that $x * y \leq z \iff y \leq x \rightarrow z$ holds for all $x, y, z \in X$. Hence $(X, *, \rightarrow)$ is a residuated commutative monoid such that $x^{\perp} = x \rightarrow \bot$ holds — i.e. the universal lower bound $\bot$ is the dualizing element of $(X, * \rightarrow)$. So (GA1)–(GA3) determine a Girard algebra, and vice versa every Girard algebra satisfies (GA1)–(GA3).

Before we present the discriminating axiom which distinguishes MV-algebras from Girard algebras we prove some important chains of equivalences in the realm of Girard algebras.

Proposition 2.7.1. *In any Girard algebra $(X, *, \rightarrow)$ the following assertions are equivalent*:

(i) *For all $x, y, z \in X$ the relation $x * (y \wedge z) = (x * y) \wedge (x * z)$ holds.*
(ii) *For all $x, y, z \in X$ the relation $(x \rightarrow y) \vee (x \rightarrow z) = x \rightarrow (y \vee z)$ holds.*
(iii) *For all $x, y \in X$ the relation $(x \rightarrow y) \vee (y \rightarrow x) = \top$ holds.*

Proof. We assume (i). Then we apply (2.90) and (2.91) and obtain for $x, y, z \in X$:

$$\begin{aligned}
x \to (y \vee z) &= x \to \big(((y \to \bot) \wedge (z \to \bot)) \to \bot\big)\\
&= \big(x * ((y \to \bot) \wedge (z \to \bot))\big) \to \bot\\
&= \big((x * (y \to \bot)) \wedge (x * (z \to \bot))\big) \to \bot\\
&= \big((x * (y \to \bot)) \to \bot\big) \vee \big((x * (z \to \bot)) \to \bot\big)\\
&= (x \to y) \vee (x \to z).
\end{aligned}$$

Hence (ii) follows. Since $(x \to y) \vee (y \to x) = ((x \vee y) \to y) \vee ((x \vee y) \to x)$, the assertion (iii) is an immediate corollary of (ii). Finally, if we assume (iii), then it is easily seen that the relation

$$\begin{aligned}
(x * y) \wedge (x * z) &= \big((x * y) \wedge (x * z)\big) * \big((y \to z) \vee (z \to y)\big)\\
&\le \big(x * (y * (y \to z))\big) \vee \big(x * (z * (z \to y))\big)\\
&\le x * (y \wedge z)
\end{aligned}$$

holds. Hence assertion (i) follows. □

It is worthwhile to note that there exist Girard algebras which do not satisfy one of the equivalent conditions of Proposition 2.7.1 (cf. Exercise 2.7.4 (c) and (d)).

If x and y are elements of a Girard algebra, then it is easily seen that

$$(x \to y) \to y \quad \text{and} \quad (y \to x) \to x$$

are upper bounds of $\{x, y\}$. On this background the following lemma has some importance.

Lemma 2.7.2. *Let $(X, *, \to)$ be a Girard algebra. Then the following assertions are equivalent*:

(i) *For all $x, y \in X$ the relation $(x \to y) \to y = x \vee y$ holds.*
(ii) *For all $x, y \in X$ the relation $(x \to y) \to y = (y \to x) \to x$ holds.*

Proof. Since the binary join $\vee$ is commutative, the implication (i) $\Longrightarrow$ (ii) is obvious. On the other hand, let us assume that (ii) holds, and let $u \in X$ be an upper bound of $\{x, y\}$ — i.e. $\top = x \to u$ and $\top = y \to u$. By (ii) the relation

$$(u \to y) \to y = (y \to u) \to u = \top \to u = u$$

holds. Hence we apply (2.90) and obtain:

$$\begin{aligned}
\top = (u \to y) \to (x \to y) &\le \big((u \to y) * (y \to \bot)\big) \to \big((x \to y) * (y \to \bot)\big)\\
&= \big((x \to y) \to y\big) \to \big((u \to y) \to y\big)\\
&= \big((x \to y) \to y\big) \to u.
\end{aligned}$$

Thus $(x \to y) \to y$ is the smallest upper bound of $\{x, y\}$, and (i) follows. □

Comment. The assertion (ii) in Lemma 2.7.2 is sometimes called Mangani's axiom and corresponds to axiom (A.3) in [100].

Theorem 2.7.3. *Let* $(X, *, \to)$ *be a Girard algebra. The following assertions are equivalent*:

(i) *For all* $x, y \in X$ *the following equivalence holds*:

$$x \le y \quad \Longleftrightarrow \quad \exists z \in X \text{ such that } x = y * z. \qquad \text{(Divisibility Law)}$$

(ii) *For all* $x, y \in X$ *the relation* $x * (x \to y) = x \wedge y$ *holds.*
(iii) *For all* $x, y \in X$ *the relation* $(x \to y) \to y = x \vee y$ *holds.* (MV-Property)

Proof. The equivalence of (i) and (ii) is evident. In order to verify the equivalence (ii) $\Longleftrightarrow$ (iii) we proceed as follows. First we fix $x, y \in X$ and assume that (ii) holds. Then we apply (2.91) and (2.90) and obtain:

$$\begin{aligned} x \vee y &= \big((x \to \bot) \wedge (y \to \bot)\big) \to \bot \\ &= \big((y \to \bot) * \big((y \to \bot) \to (x \to \bot)\big)\big) \to \bot \\ &= (x \to y) \to y. \end{aligned}$$

Hence (iii) is verified. On the other hand, if (iii) holds, then we apply (2.91) and (2.90) again and observe:

$$\begin{aligned} x \wedge y &= \big((x \to \bot) \vee (y \to \bot)\big) \to \bot \\ &= \big(\big((y \to \bot) \to (x \to \bot)\big) \to (x \to \bot)\big) \to \bot \\ &= x * (x \to y). \end{aligned}$$

Hence (ii) is verified. □

An MV-*algebra* is a Girard algebra in which one of the equivalent assertions of Theorem 2.7.3 holds. In this context we prefer the divisibility law and understand MV-algebras as divisible Girard algebras, while in [21] we find Mangani's axiom (cf. Lemma 2.7.2 and Theorem 2.7.3 (iii)) as a discriminating axiom for MV-algebras (cf. (MV6) on p. 7 in [21]), where $*$ plays the rôle of the dual binary operation corresponding to $\oplus$ in [21].

If we compare Girard algebras with MV-algebras, then it is important to note that the partial order of a Girard algebra is determined by the multiplication $*$ *and* the involution $x \longmapsto x^{\perp}$ (cf. (GA3)), while in an MV-algebra the partial order is completely determined by the multiplication (cf. Theorem 2.7.3 (i)). This observation has dramatic consequences. In Girard algebras the binary meet (and hence the binary join) is not completely determined by the given algebraic operations and forms a separate source of information, while in MV-algebras the divisibility law forces a complete determination of the lattice-theoretic structure by the corresponding algebraic operations. Depending on the point of view this observation has its advantages and disadvantages. Of course, the axiom system of an MV-algebra is simpler than that of a Girard algebra, because the lattice structure is a consequence of the algebraic operations. However, from a lattice-theoretic point of view the divisibility law can sometimes play the rôle of an obstacle — e.g. in general the divisibility law is *not* preserved under the MacNeille completion.

The aim of this section is to characterize those MV-algebras whose structure is fortunately preserved under the MacNeille completion. As a corollary we will show that every simple MV-algebra is an MV-subalgebra of the real unit interval provided with Łukasiewcz arithmetic conjunction. In this context we will deviate from the standard proof based on McNaughton functions and give a direct proof using the MacNeille completion.

We begin with some simple properties of MV-algebras expressed in the language of Girard algebras.

Corollary 2.7.4. *Every MV-algebra $(X, *, \to)$ satisfies the following properties for all $x, y, z \in X$:*

(i) $x * (y \wedge z) = (x * y) \wedge (x * z)$.
(ii) $x \to (x * y) = (x \to \bot) \vee y$.
(iii) *If e is idempotent, then $e \to \bot$ is also idempotent.*
(iv) *If e is idempotent, then $x * e = x \wedge e$ holds.*
(v) $x * y \le (x * x) \vee (y * y)$.

Proof. We apply Theorem 2.7.3 (ii) and obtain:

$$\begin{aligned}(x * y) \wedge (x * z) &= x * y * \big((x * y) \to (x * z)\big) = x * y * \big(y \to (x \to (x * z))\big)\\ &= x * \big(y \wedge (x \to (x * z))\big) = x * \big(x \to (x * z)\big) * \big((x \to (x * z)) \to y\big)\\ &\le x * z * (z \to y) \le x * (z \wedge y).\end{aligned}$$

Hence (i) holds. Because of the MV-property and (2.90) the property (ii) follows from

$$\begin{aligned}(x \to \bot) \vee y &= \big(y \to (x \to \bot)\big) \to (x \to \bot) = \big((x * y) \to \bot\big) \to (x \to \bot)\\ &= x \to (x * y).\end{aligned}$$

Let e be idempotent. Then property (ii) implies $\top = e \to e = (e \to \bot) \vee e$. Hence $e \to \bot$ is idempotent, and property (iii) is verified. On the other hand, it follows from Theorem 2.7.3 (ii) that

$$e * x \le e \wedge x = e * (e \to x) = e * e * (e \to x) \le e * x$$

holds for each $x \in X$, and property (iv) is verified. Finally, we conclude from property (i) that Proposition 2.7.1 (iii) is valid in any MV-algebra. Hence (v) follows from the following relation:

$$x * y = x * y * \big((y \to x) \vee (x \to y)\big) = \big(x * y * (y \to x)\big) \vee \big(x * y * (x \to y)\big)$$
$$\leq (x * x) \vee (y * y).$$ □

By the previous corollary the set of idempotent elements of an MV-algebra forms a Boolean algebra which is an MV-subalgebra of the given MV-algebra. In this sense an MV-algebra can be understood as a *non-idempotent* version of a Boolean algebra.

An MV-algebra is *complete* if the underlying lattice is complete. Hence complete MV-algebras are integral and commutative Girard quantales.

Theorem 2.7.5. *In any complete MV-algebra $(X, *, \to)$ the following properties hold*:

(i) *$(X, \leq)$ is a frame and a complete Brouwerian lattice (i.e. the dual lattice is a frame).*
(ii) *If $x \in X$ and A is a nonempty subset of X, then $\big(\bigwedge A\big) * x = \bigwedge_{a \in A}(a * x)$.*

Proof. In order to verify (i) we choose $x \in X$, $A \subseteq X$ and apply Theorem 2.7.3 (ii):

$$x \wedge \big(\bigvee A\big) = \big(\bigvee A\big) * \big(\big(\bigvee A\big) \to x\big) = \bigvee_{a \in A} \big(a * \big(\bigwedge_{b \in A} (b \to x)\big)\big) \leq \bigvee_{a \in A} \big(a * (a \to x)\big)$$
$$= \bigvee_{a \in A} (a \wedge x).$$

Hence $(X, \leq)$ is a frame. Since the universal lower bound is dualizing, $(X, \leq)$ is also a complete Brouwerian lattice.
Now we verify (ii). First we use Corollary 2.7.4 (ii) and subsequently apply the property that $(X, \leq)$ is a complete Brouwerian lattice:

$$x \to \big(x * \big(\bigwedge A\big)\big) = (x \to \bot) \vee \big(\bigwedge A\big) = \bigwedge_{a \in A} \big((x \to \bot) \vee a\big)$$
$$= \bigwedge_{a \in A} \big(x \to (x * a)\big) = x \to \big(\bigwedge_{a \in A} (x * a)\big).$$

Hence the relation

$$x * \big(\bigwedge A\big) = x * \big(x \to \big(x * \big(\bigwedge A\big)\big)\big) = x * \big(x \to \big(\bigwedge_{a \in A} (x * a)\big)\big) = \bigwedge_{a \in A} (x * a)$$

follows from Theorem 2.7.3 (ii). □

The next corollary shows that complete MV-algebras have sufficiently many idempotent elements.

Corollary 2.7.6. *Let $(X, *, \to)$ be a complete MV-algebra. Then for every $x \in X$ the element*

$$e_x = \bigwedge_{n \in \mathbb{N}} x^n$$

is idempotent where $x^n = x * \overset{n}{\cdots} * x$ *denotes the* n^{th} *power of* x.

Proof. By Theorem 2.7.5 (ii), the following relation holds:

$$e_x * e_x = \Big(\bigwedge_{n \in \mathbb{N}} x^n\Big) * \Big(\bigwedge_{m \in \mathbb{N}} x^m\Big) = \bigwedge_{n,m \in \mathbb{N}} x^{n+m} = e_x.$$

□

Comment. The element e_x defined in Corollary 2.7.6 is also called the *idempotent kernel* of x.

Referring to Theorem 2.6.8 it is evident that the MacNeille completion of any Girard algebra is an integral commutative Girard quantale. This observation implies that the MacNeille completion of any MV-algebra is always a commutative and integral Girard quantale.

The next theorem characterizes the property that the MacNeille completion preserves even the algebraic structure of an MV-algebra.

Theorem 2.7.7. *Let* $(X, *, \rightarrow)$ *be an* MV*-algebra. Then the following assertions are equivalent*:

(i) *The MacNeille completion of* $(X, *, \rightarrow)$ *is an* MV*-algebra.*
(ii) *For all* $x \in X$ *the following implication holds*:

$$\forall n \in \mathbb{N} : \; x^n \rightarrow \bot \le x \quad \Longrightarrow \quad x = \top.$$

(iii) *For all* $x, y \in X$ *the following implication holds*:

$$\forall n \in \mathbb{N} : \; x \le y^n \quad \Longrightarrow \quad x * y = x.$$

Proof. Let $\mathsf{Dwn}(X)$ be the complete lattice of all downclosed subsets of X. We maintain the notation from Example 2.2.13 and Lemma 2.6.7 and consider the nucleus $\mathsf{Dwn}(X) \xrightarrow{\mathsf{C}} \mathsf{Dwn}(X)$ on the unital quantale $(\mathsf{Dwn}(X), \boxdot)$ determined by

$$\mathsf{C}(A) = \mathsf{L}(\mathsf{U}(A)), \qquad A \in \mathsf{Dwn}(X)$$

where $\mathsf{U}(A)$ is the upclosed set of all upper bounds of A and $\mathsf{L}(\mathsf{U}(A))$ is the downclosed set of all lower bounds of $\mathsf{U}(A)$. The regular quotient of $(\mathsf{Dwn}(X), \boxdot)$ w.r.t. C is the MacNeille completion of $(X, *, \rightarrow)$ and is denoted by $(X^\sharp, \star)$.

(i) $\Longrightarrow$ (ii). Let us assume that $(X^\sharp, \star)$ is an MV-algebra. Further, we consider an element $x \in X$ such that $x^n \rightarrow \bot \le x$ (or, equivalently, $x \rightarrow \bot \le x^n$) holds for all

$n \in \mathbb{N}$. Since $(\downarrow x)^n = \downarrow x^n$, the meet of $\{(\downarrow x)^n \mid n \in \mathbb{N}\}$ coincides with the down-closed set I of all lower bounds of $\{x^n \mid n \in \mathbb{N}\}$. In particular, $\downarrow(x \to \bot) \subseteq I$ holds. Since $(X^\sharp, \star)$ is an MV-algebra, I is idempotent w.r.t. $\star$ (cf. Corollary 2.7.6), and we conclude from Corollary 2.7.4 (iv) that the relation

$$\downarrow(x \to \bot) = (\downarrow(x \to \bot)) \star I \subseteq (\downarrow(x \to \bot)) \star (\downarrow x) = \downarrow\bot$$

holds. Hence $x = \top$ follows.
(ii) $\Longrightarrow$ (iii). Let us consider $x, y \in X$ such that $x \le y^n$ holds for all $n \in \mathbb{N}$. Then we put $z = x \to (x * y)$ and observe $y \le z$. Further, we conclude from Corollary 2.7.4 (ii) and formula (2.91) that the following relation holds:

$$z \to \bot = x \wedge (y \to \bot) \le x \le y^n \le z^n.$$

Now we apply (ii) and obtain $z = \top$ — i.e. $x = x * y$.

(iii) $\Longrightarrow$ (i). Let us assume that the assertion (iii) holds. Referring to Theorem 2.7.3 it is sufficient to show that $(X^\sharp, \star)$ satisfies the divisibility law. For this purpose we choose $A, B \in X^\sharp$ with $A \subseteq B$. Our aim is to construct a downclosed set $E \in X^\sharp$ such that $A = B \star E$ holds. First we introduce a downclosed subset C of X as follows:

$$C = \{z \in X \mid \exists u \in \mathsf{U}(B),\ \exists x \in A : z \le u \to x\}.$$

(1) Let v be an upper bound of A. Then for $z \in C$ we choose $u \in \mathsf{U}(B)$ and $x \in A$ with $z \le u \to x$. We obtain for all $y \in B$:

$$y * z \le u * (u \to x) \le x \le v.$$

Hence v is an upper bound of $B \boxdot C$ — i.e. $\mathsf{U}(A) \subseteq \mathsf{U}(B \boxdot C)$.
(2) In order to verify $\mathsf{U}(B \boxdot C) \subseteq \mathsf{U}(A)$ we consider an element $w \in \mathsf{U}(B \boxdot C)$. Then we choose an arbitrary element x in A and show that for all $u \in \mathsf{U}(B)$ the element $u * (x \to w)$ is also an element of $\mathsf{U}(B)$. In fact, since every upper bound of B is also an upper bound of A, we infer from Theorem 2.7.3 (ii) that the relation $u * (u \to x) = u \wedge x = x$ holds for all $u \in \mathsf{U}(B)$. Hence for all $y \in B$ and $u \in \mathsf{U}(B)$ the relation

$$y = u * (u \to y) \le u * \big((u * (u \to x)) \to \big(y * (u \to x)\big)\big) \le u * (x \to w)$$

follows, where we have used the property that w is an upper bound of $B \boxdot C$. Hence $u * (x \to w)$ is an upper bound of B — i.e. $u * (x \to w) \in \mathsf{U}(B)$. By recursion we conclude from the previous observation that $u * (x \to w)^n \in \mathsf{U}(B)$ for all $n \in \mathbb{N}$. Since A is contained in B, the element $(x \to w)^n$ is a fortiori an upper bound of A for all $n \in \mathbb{N}$. Hence $x \le (x \to w)^n$ holds for all $x \in A$ and $n \in \mathbb{N}$. Now we apply assertion (iii) and obtain $x = x * (x \to w) \le w$ for all $x \in A$. Thus w is an upper bound of A and we have verified $\mathsf{U}(B \boxdot C) \subseteq \mathsf{U}(A)$.

(3) Finally, we use the property that C is a nucleus and obtain the following relation from (1) and (2):

$$A = \mathsf{L}(\mathsf{U}(A)) = \mathsf{L}(\mathsf{U}(B \boxdot C)) = \mathsf{C}(B \boxdot C) = \mathsf{C}(B \boxdot \mathsf{C}(C)) = B \star \mathsf{C}(C) = B \star E,$$

where $E = \mathsf{C}(C) \in X^\sharp$. Thus the divisibility law is verified in $(X^\sharp, \star)$, and $(X^\sharp, \star)$ is a complete MV-algebra. □

Example 2.7.8. (C.C. Chang's chain (cf. [20])) Let $\mathbb{N}_0$ be the set of natural numbers together with 0. On $\mathbb{N}_0$ we consider the usual order. Then we put $X = \mathbb{N}_0 \times \{1, 2\}$ and define a linear order $\leq$ on X by:

$$\begin{array}{ll} (n,1) \leq (m,1) \text{ if } m \leq n, & \\ (n,2) \leq (m,2) \text{ if } n \leq m, & \qquad n, m \in \mathbb{N}_0. \\ (n,2) \leq (m,1) & \end{array}$$

Then $(X, \leq)$ is a chain with universal bounds $\top = (0, 1)$ and $\bot = (0, 2)$. On X we define two binary operations $*$ and $\rightarrow$ as follows:

$$\begin{array}{ll} (n,1) * (m,1) = (n+m, 1), & \\ (n,2) * (m,2) = (0,2), & \\ (n,1) * (m,2) = (m,2) * (n,1) = (\max(m-n,0),2), & \\ (n,1) \rightarrow (m,1) = (\max(m-n,0),1), & \qquad n, m \in \mathbb{N}_0. \\ (n,1) \rightarrow (m,2) = (n+m,2), & \\ (n,2) \rightarrow (m,1) = (0,1), & \\ (n,2) \rightarrow (m,2) = (\max(n-m,0),1), & \end{array}$$

It is a matter of routine to confirm that $(X, *, \rightarrow)$ is an MV-algebra. Hence the MacNeille completion of $(X, *, \rightarrow)$ is an integral and commutative Girard quantale, which we denote by $(\widetilde{X}, \star)$. We show that $(\widetilde{X}, \star)$ is *not* an MV-algebra. For this purpose we write x_0 for $(1, 1)$. Then $x_0^n = (n, 1)$ follows from the definition of $*$, and the following relation holds

$$x_0^n \rightarrow \bot = (n,1) \rightarrow (0,2) = (n,2) \leq (1,1) = x_0,$$

where $\bot = (0, 2)$. In particular, the relation $x_0 \rightarrow \bot \leq x_0^n$ holds for all $n \in \mathbb{N}$. Since $x_0 = (1, 1) \neq (0, 1) = \top$, the assertion (ii) of Theorem 2.7.7 is violated. Hence we conclude from Theorem 2.7.7 that $(\widetilde{X}, \star)$ is not an MV-algebra.

An MV-algebra $(X, *, \rightarrow)$ is *simple* if and only if X contains at least two different elements and $\{\top\}$ is its only (proper) filter (cf. Exercise 2.7.7). Hence an MV-algebra is simple if and only if $\bot \neq \top$ and for every element $x \in X$ with $x \neq \top$ there exists an $n \in \mathbb{N}$ such that $x^n = \bot$ holds.

In the following considerations we prove that every simple MV-algebra is isomorphic to a MV-subalgebra of the real unit interval provided with Łukasiewicz arithmetic conjunction. In contrast to Theorem 3.5.1 in [21] we will not make use of free MV-algebras and the related concept of McNaughton functions, but we will give a direct proof based on the MacNeille completion of simple MV-algebras.

Lemma 2.7.9. *Every simple MV-algebra is a chain.*

Proof. Let $(X, *, \rightarrow)$ be a simple MV-algebra. We assume that there exist elements $x, y \in X$ which are not comparable — i.e. $x \not\leq y$ and $y \not\leq x$. Then we have $x \rightarrow y \neq \top$ and $y \rightarrow x \neq \top$. Since $(X, *, \rightarrow)$ is simple, there exists m, $n \in \mathbb{N}$ such that $(x \rightarrow y)^m = \bot$ and $(y \rightarrow x)^n = \bot$. Hence the relation $\big((x \rightarrow y) \vee (y \rightarrow x)\big)^{m+n} = \bot$ follows, which implies $(x \rightarrow y) \vee (y \rightarrow x) \neq \top$. Now we apply Proposition 2.7.1 and obtain a contradiction to Corollary 2.7.4 (i). Thus the assumption is false, and $(X, \leq)$ is a chain. □

Corollary 2.7.10. *The MacNeille completion of a simple MV-algebra is again a simple MV-algebra.*

Proof. Let $(X, *, \rightarrow)$ be a simple MV-algebra. Then its simplicity implies that $(X, *, \rightarrow)$ satisfies assertion (iii) in Theorem 2.7.7. Hence the MacNeille completion $(X^\sharp, \star)$ of $(X, *, \rightarrow)$ is a nontrivial MV-algebra. Moreover, if we choose $A \in X^\sharp$ with the property $A \neq X = \downarrow\top$, then $\mathsf{U}(A) \neq \{\top\}$ follows. Thus A has an upper bound u with $u \neq \top$. Since $(X, *, \rightarrow)$ is simple, there exists an $n \in \mathbb{N}$ such that $A^n \subseteq \downarrow u^n = \{\bot\}$ holds. Hence $(X^\sharp, \star)$ is also simple. □

The next proposition deals with the cardinality of simple and complete MV-algebras.

Proposition 2.7.11. *Let $(X, *, \rightarrow)$ be a simple and complete MV-algebra, and let $x_0 = \bigvee\{x \in X \mid x \neq \top\}$. If the relation $x_0 \neq \top$ is valid, then the cardinality of X is finite, and there exists a natural number $n \in \mathbb{N}$ such that X has the following form*

$$X = \{x_0^m \mid m \in \{0, 1, \ldots, n\}\}. \tag{2.92}$$

Proof. Let us assume $x_0 \neq \top$. The simplicity of $(X, *, \rightarrow)$ guarantees the existence of a natural number $n \in \mathbb{N}$ such that

$$\bot = x_0^n < x_0^{n-1} < \cdots < x_0^2 < x_0 < x_0^0 = \top.$$

Since X is a chain (cf. Lemma 2.7.9), for every $x \in X$ with $x \neq \top$ there exists an $m \in \{1, \ldots, n-1\}$ with the property $x_0^{m+1} \leq x \leq x_0^m$. Hence $x_0 = x_0^m \rightarrow x$ or $\top = x_0^m \rightarrow x$ follows. Finally, we invoke the divisibility law of MV-algebras and obtain $x = x_0^m * (x_0^m \rightarrow x)$. Thus either $x = x_0^{m+1}$ or $x = x_0^m$ holds. □

The next important property of a complete MV-algebra $(X, *, \rightarrow)$ is the fact that the formation of squares (i.e. $x \longmapsto x^2 = x * x$) is join-preserving. Indeed, because of Corollary 2.7.4 (v) the relation

$$\left(\bigvee A\right)^2 = \bigvee_{x \in A} x^2$$

holds for any subset A of X. Since $(X, *)$ is commutative, the formation of squares also preserves the multiplication. Hence the formation of squares is a strong homomorphism from $(X, *)$ to $(X, *)$, and its right adjoint map $x \longmapsto x^{1/2}$ is determined by

$$x^{1/2} = \bigvee\{z \in X \mid z * z \leq x\}, \qquad x \in X.$$

Since right adjoint maps of homomorphisms between (pre)quantales are closed maps (cf. Proposition 2.2.2), the relation

$$x^{1/2} * y^{1/2} \leq (x * y)^{1/2}$$

follows immediately for all $x, y \in X$. Moreover, if $x \in X$ is a *square* (i.e. there exists a $y \in X$ with $x = y * y$), then $x^{1/2}$ is even the *square root of* x *with respect to* $*$, which means that the additional property $x^{1/2} * x^{1/2} = x$ also holds.

The previous observations can be seen as a motivation to characterize those complete MV-algebras in which every element is a square. We shall also say that the MV-algebra *has square roots*.

Theorem 2.7.12. *Let* $(X, *, \rightarrow)$ *be a complete* MV*-algebra and* $\mathbb{B}$ *be the Boolean algebra of all idempotent elements of* $(X, *)$. *The following assertions are equivalent*:

(i) *Every element of* X *is a square w.r.t.* $*$.
(ii) *If an element* $x \in X$ *satisfies the condition*

$$\downarrow x = \{e \wedge x \mid e \in \mathbb{B}\}, \tag{I}$$

then x *is idempotent.*

The proof of Theorem 2.7.12 will be prepared by the following lemmata.

Lemma 2.7.13. *Let* $(X, *, \rightarrow)$ *be a complete* MV*-algebra such that the assertion* (ii) *in Theorem 2.7.12 holds. Then for all* $x \in X$ *with* $\bot \neq x$ *there exists a* $y \in X$ *satisfying the properties*:

$$y \neq \bot \ \textit{ and } \ (y \rightarrow \bot)^2 \rightarrow \bot \leq x.$$

Proof. Without loss of generality, we can restrict our interest to a non-idempotent element $x \in X$. Hence $x \neq \top$. Then we conclude from assertion (ii) of Theorem 2.7.12 that there exists an element $z \in X$ satisfying the following properties:

$$z \leq x \quad \text{and} \quad \forall e \in \mathbb{B} : \ z \neq e \wedge x. \tag{2.93}$$

Thus $z \neq \bot$ follows. Finally, we consider the element $y = z \wedge \big(x * (z \rightarrow \bot)\big) \in X$. *Step 1.* We claim $y \neq \bot$. Let us assume $y = \bot$. Then it follows from Proposition 2.6.2

and Corollary 2.7.4 (ii) that

$$\top = (z \to \bot) \vee (x \to z) = z \to \big(z * (x \to z)\big).$$

Hence $z = z * (x \to z)$, and since $*$ is distributive over nonempty meets (cf. Theorem 2.7.5 (ii)), we obtain:

$$z = z * \Big(\bigwedge_{n \in \mathbb{N}} (x \to z)^n\Big).$$

By Corollary 2.7.6 the element $e_0 = \bigwedge_{n \in \mathbb{N}} (x \to z)^n$ is idempotent. Now we apply Corollary 2.7.4 (iv) and obtain the relations

$$z = z * e_0 = z \wedge e_0 \le x \wedge e_0 \quad \text{and} \quad e_0 \wedge x = x * e_0 \le z.$$

Thus $z = x \wedge e_0$ holds — a statement which is obviously a contradiction to (2.93). Hence $y \neq \bot$ is verified.
Step 2. We claim $(y \to \bot)^2 \to \bot \le x$. First we simplify the notation and introduce the following abbreviation:

$$2w := (w \to \bot)^2 \to \bot, \qquad w \in X.$$

By Corollary 2.7.4 (v) the following relation holds for all $w_1, w_2 \in X$:

$$\begin{aligned} 2(w_1 \wedge w_2) &= \big((w_1 \to \bot) \vee (w_2 \to \bot)\big)^2 \to \bot \\ &= \big((w_1 \to \bot)^2 \vee (w_2 \to \bot)^2\big) \to \bot = (2w_1) \wedge (2w_2). \end{aligned}$$

After this deviation we refer to Corollary 2.7.4 (i) and Proposition 2.7.1 (iii) and obtain:

$$\begin{aligned} 2y &= (2y) * \big(\big((2z) \to x\big) \vee \big(x \to (2z)\big)\big) \\ &= \big((2z) \wedge \big(2\big(x * (z \to \bot)\big)\big)\big) * \big(\big((2z) \to x\big) \vee \big(x \to (2z)\big)\big) \\ &\le x \vee \big(\big(2\big(x * (z \to \bot)\big)\big) * \big(x \to (2z)\big)\big). \end{aligned}$$

Now we apply frequently (2.90) and establish the following relations:

$$2\big(x * (z \to \bot)\big) = (x \to z)^2 \to \bot = (x \to z) \to \big(x * (z \to \bot)\big)$$

and

$$x \to (2z) = (z \to \bot)^2 \to (x \to \bot) = (z \to \bot) \to (x \to z).$$

Finally, we observe:

$$
\begin{aligned}
2y &\le x \vee \big((2(x * (z \to \bot))) * (x \to (2z))\big) \\
&= x \vee \big(((z \to \bot) \to (x \to z)) * ((x \to z) \to (x * (z \to \bot)))\big) \\
&\le x \vee \big((z \to \bot) \to (x * (z \to \bot))\big) = z \vee x = x,
\end{aligned}
$$

where we have used Corollary 2.7.4 (ii). □

Lemma 2.7.14. *Let $(X, *, \to)$ be a complete MV-algebra such that the assertion* (ii) *in Theorem 2.7.12 holds. If x and y are elements of X with the property $y^2 < x$, then there exists an element $z \in X$ satisfying the following conditions*:

$$y < z \quad \textit{and} \quad z^2 \le x.$$

Proof. We put $w = x * (y^2 \to \bot) \neq \bot$ and apply Lemma 2.7.13 to w. Hence there exists a $v \in X$ with $v \neq \bot$ and $2v \le w$. Since $(X, *)$ is integral, the relation $2v \le y^2 \to \bot$ follows. Finally, we consider the element $z = (v \to \bot) \to y \in X$. Using again the integrality of $(X, *)$ we obtain $y \le z$.
Step 1. We claim $z \neq y$. Let us assume $z = y$. The relation

$$\top = z \to y = \big((v \to \bot) \to y\big) \to y = (v \to \bot) \vee y$$

follows immediately from the MV-property (cf. Theorem 2.7.3 (iii)). Now we apply Corollary 2.7.4 (v) and obtain

$$\top = \big((v \to \bot) \vee y\big)^2 \le (v \to \bot)^2 \vee y^2.$$

Hence the relation

$$v \le 2v = (2v) \wedge (y^2 \to \bot) = \bot$$

holds. Since the previous relation is a contradiction to $v \neq \bot$, we have verified $z \neq y$.
Step 2. We claim $z^2 \le x$. First we observe $z^2 \le (v \to \bot)^2 \to y^2$ and recall $2v \le w$. Then we apply (2.90) and the MV-property:

$$z^2 \le \big((2v) \to \bot\big) \to y^2 \le (w \to \bot) \to y^2 = (x \to y^2) \to y^2 = x \vee y^2 = x.$$

Finally, it follows from *Step 1* and *Step 2* that z satisfies the desired properties. □

Proof of Theorem 2.7.12 (i) $\Longrightarrow$ (ii). We consider an element $x \in X$ satisfying condition (I) in assertion (ii). We have to show that x is idempotent. Since every element of X is a square, there exists a $y \in X$ with $y^2 = x \to \bot$. Because of the integrality of $(X, *)$ we note that $y^2 \le y$. Hence $y \to \bot \le x$ follows. Then we apply (I) and select an idempotent element $e \in \mathbb{B}$ with the property $y \to \bot = e \wedge x$. In particular, $e \to \bot \le y$ holds. Since $e \to \bot$ is idempotent (cf. Corollary 2.7.4 (iii)), the relation $e \to \bot \le y^2$ follows — i.e. $y^2 \to \bot \le e$. Hence we obtain

$$y \to \bot = e \wedge x = e \wedge (y^2 \to \bot) = y^2 \to \bot,$$

which means that y is idempotent. Using again Corollary 2.7.4 (iii) we conclude from $x = y^2 \to \bot = y \to \bot$ that x is also idempotent.
(ii) $\Longrightarrow$ (i). Let $x \longmapsto x^{1/2}$ be the right adjoint map of the formation of squares in $(X, *)$. Then $x^{1/2} * x^{1/2} \le x$ follows. Let us assume $x \neq x^{1/2} * x^{1/2}$. By Lemma 2.7.14 there exists a $z \in X$ such that $x^{1/2} < z$ and $z^2 \le x$, which is a contradiction to the definition of $x^{1/2}$. Hence the assumption is false and $x^{1/2} * x^{1/2} = x$ is valid. In particular, every element of $(X, *)$ is a square. □

Corollary 2.7.15. *Every infinite, simple and complete MV-algebra has square roots.*

Proof. Let $(X, *, \to)$ be an infinite, simple and complete MV-algebra. In order to show that $(X, *, \to)$ has square roots, we have to verify that every element of X is a square w.r.t. $*$. For this purpose it is sufficient to establish assertion (ii) of Theorem 2.7.12. Therefore we choose $x \in X$ satisfying property (I). Since $(X, *, \to)$ is simple, the universal bounds are the only idempotent elements of X — i.e. $\mathbb{B} = \{\bot, \top\}$. Thus we obtain:

$$\downarrow x = \{e \wedge x \mid e \in \mathbb{B}\} = \{\bot, x\}.$$

Hence $\uparrow(x \to \bot) = \{\top, x \to \bot\}$ follows. Since X is infinite, we conclude from Proposition 2.7.11 that $x \to \bot = \top$ holds. Thus $x = \bot$ is idempotent. □

It is easily seen that the real unit interval $[0, 1]$ provided with Łukasiewicz arithmetic conjunction (cf. Sect. 2.3.4) is an infinite, simple and complete MV-algebra whose square roots are given by:

$$x^{1/2} = \tfrac{x+1}{2}, \qquad x \in [0, 1].$$

The aim of the following consideration is to show that every infinite, simple and complete MV-algebra is isomorphic to $([0, 1], *_Ł)$.

For this purpose we first summarize some properties which occur in the case of complete MV-algebras with square roots.

Theorem 2.7.16. *Let $(X, *, \to)$ be a complete MV-algebra such that every element $x \in X$ is a square. Then the formation of square roots satisfies the following properties:*

(i) $(x \to y)^{1/2} = x^{1/2} \to y^{1/2}$ *for all* $x, y \in X$.
(ii) $(x * x)^{1/2} = \bot^{1/2} \vee x$ *for all* $x \in X$.
(iii) *The element* $(\bot^{1/2} \to \bot) * (\bot^{1/2} \to \bot)$ *is idempotent.*
(iv) *For every nonempty subset* $\{x_i \mid i \in I\}$ *of* X *the following relation holds:*

$$\Big(\bigvee_{i\in I} x_i\Big)^{1/2} = \bigvee_{i\in I} x^{1/2}.$$

(v) *If* $\bot^{1/2} = \bot$, *then every element of* X *is idempotent — i.e.* $(X, *)$ *is a complete Boolean algebra.*

Proof. Since the formation of square roots is a closed map, the relation

$$(x \to y)^{1/2} \leq x^{1/2} \to y^{1/2}$$

is evident. On the other hand we conclude from

$$x * \big(x^{1/2} \to y^{1/2}\big)^2 = x^{1/2} * (x^{1/2} \to y^{1/2}) * x^{1/2} * (x^{1/2} \to y^{1/2}) \leq y$$

that $x^{1/2} \to y^{1/2} \leq (x \to y)^{1/2}$ also holds. Hence (i) follows.
Because of the MV-property and Corollary 2.7.4 (ii) we conclude from (i) that the following relations hold:

$$(x \vee y)^{1/2} = \big((x \to y) \to y\big)^{1/2} = (x^{1/2} \to y^{1/2}) \to y^{1/2} = x^{1/2} \vee y^{1/2},$$
$$x^{1/2} \to (x * x)^{1/2} = \big(x \to (x * x)\big)^{1/2} = \big((x \to \bot) \vee x\big)^{1/2} = (x \to \bot)^{1/2} \vee x^{1/2}.$$

Now we invoke the divisibility law and obtain:

$$(x * x)^{1/2} = x^{1/2} * \big((x \to \bot)^{1/2} \vee x^{1/2}\big) = x^{1/2} * \big((x^{1/2} \to \bot^{1/2}) \vee x^{1/2}\big) = \bot^{1/2} \vee x.$$

Hence (ii) is verified.
From (i) and Corollary 2.7.4 (ii) it is easily seen that the relation

$$\begin{aligned}(\bot^{1/2} \to \bot)^{1/2} &= (\bot^{1/2})^{1/2} \to \bot^{1/2}\\ &= \big((\bot^{1/2})^{1/2} \to \bot\big) \vee (\bot^{1/2})^{1/2}\\ &\leq (\bot^{1/2} \to \bot) \vee (\bot^{1/2})^{1/2}\end{aligned}$$

holds. Now we apply Corollary 2.7.4 (v) and take the 4th power on both sides. Hence

$$(\bot^{1/2} \to \bot)^2 \leq (\bot^{1/2} \to \bot)^4$$

follows, and the idempotency of $(\bot^{1/2} \to \bot)^2$ is verified.
In order to establish (iv) we consider a nonempty subset $\{x_i \mid i \in I\}$ of X. Then we conclude from Corollary 2.7.4 (v) and property (ii):

$$\big(\bigvee_{i\in I} x_i\big)^{1/2} = \big((\bigvee_{i\in I} x_i^{1/2}) * (\bigvee_{i\in I} x_i^{1/2})\big)^{1/2} = \bot^{1/2} \vee \big(\bigvee_{i\in I} x_i^{1/2}\big) = \bigvee_{i\in I} x_i^{1/2}.$$

Finally, for every element $x \in X$ the relation $x \wedge (x \to \bot) \leq \bot^{1/2}$ holds. If we now assume that $\bot^{1/2}$ coincides with $\bot$, then $(x \to \bot) \vee x = \top$ follows from (2.91). Thus x is idempotent, and assertion (v) is verified. □

Let $(X, *, \rightarrow)$ be a complete MV-algebra having square roots. As a preparation for the following we first define recursively an element $\bot^{1/2^n}$ of X for each $n \in \mathbb{N}_0$:

$$\bot^{1/2^0} := \bot \quad \text{and} \quad \bot^{1/2^n} := \big(\bot^{1/2^{n-1}}\big)^{1/2} \quad \text{for each } n \in \mathbb{N}.$$

Then for any dyadic rational number r of the real unit interval $[0, 1]$ we can define

$$\bot^0 := \top \quad \text{and} \quad \bot^r := (\bot^{1/2^n})^k, \qquad r := \tfrac{k}{2^n} \neq 0.$$

In particular, the expression $\bot^r$ is always well defined.

Lemma 2.7.17. *Let $(X, *, \rightarrow)$ be a complete MV-algebra such that every element is a square and the following property is satisfied:*

$$\bot^{1/2} \rightarrow \bot = \bot^{1/2}. \tag{2.94}$$

Further, let $\mathscr{D}$ be the MV-subalgebra of $[0, 1]$ consisting of all dyadic rational numbers of $[0, 1]$. Then the following properties are valid:

(i) *For all $r \in \mathscr{D}$ the relation $\bot^r \rightarrow \bot = \bot^{1-r}$ holds.*
(ii) *For all $r \in \mathscr{D}$ the relation $(\bot^r)^{1/2} * \bot^{1/2} = \bot^{\frac{r+1}{2}}$ holds.*
(iii) *If $*_L$ is the Łukasiewicz arithmetic conjunction in $\mathscr{D}$, then for all $r_1, r_2 \in \mathscr{D}$ the relation $(\bot^{r_1} \rightarrow \bot) * (\bot^{r_2} \rightarrow \bot) = \bot^{r_1 *_L r_2} \rightarrow \bot$ holds.*
(iv) *If $r_1, r_2 \in \mathscr{D}$, then $r_1 \leq r_2$ if and only if $\bot^{r_2} \leq \bot^{r_1}$.*

Proof. (a) In order to verify (i) we proceed as follows. If $r \in \{0, 1\}$, then the relation (i) is obvious. Therefore we assume $1 \neq r = \frac{k}{2^n} \neq 0$. Then it is sufficient to prove the following relation

$$\big(\bot^{1/2^n}\big)^k \rightarrow \bot = \big(\bot^{1/2^n}\big)^{2^n-k}, \qquad k = 1, \dots, 2^n - 1, \quad n \in \mathbb{N} \tag{2.95}$$

by induction over n and k. In the case of $n = k = 1$ the relation (2.95) coincides with (2.94). Now we assume that (2.95) holds for $k = 1$ and some $n \in \mathbb{N}$. Then we invoke (2.90) and (2.94) and obtain:

$$\begin{aligned}
\bot^{1/2^{n+1}} &= \big(\bot^{1/2^n}\big)^{1/2} = \big(\big(\bot^{1/2^n}\big)^{2^n-1} \rightarrow \bot\big)^{1/2} \\
&= \big(\big(\bot^{1/2^n}\big)^{2^n-1}\big)^{1/2} \rightarrow \bot^{1/2} \\
&= \big(\big(\bot^{1/2^{n+1}}\big)^{2^n-1} * \big(\bot^{1/2} \rightarrow \bot\big)\big) \rightarrow \bot \\
&= \big(\big(\bot^{1/2^{n+1}}\big)^{2^n-1} * \bot^{1/2}\big) \rightarrow \bot \\
&= \big(\big(\bot^{1/2^{n+1}}\big)^{2^n-1} * \big(\bot^{1/2^{n+1}}\big)^{2^n}\big) \rightarrow \bot.
\end{aligned}$$

Hence $\bot^{1/2^{n+1}} = \big(\bot^{1/2^{n+1}}\big)^{2^{n+1}-1} \rightarrow \bot$ follows, which means that (2.95) holds for $n + 1$ and $k = 1$.

Now we fix $n \in \mathbb{N}$ and assume that (2.95) holds for n and $k \le 2^n - 2$. Then we invoke (2.90) and Corollary 2.7.4 (ii) and obtain:

$$\begin{aligned}
\left(\bot^{1/2^n}\right)^{k+1} \to \bot &= \left(\bot^{1/2^n}\right)^k \to \left(\bot^{1/2^n} \to \bot\right)\\
&= \left(\bot^{1/2^n}\right)^k \to \left(\bot^{1/2^n}\right)^{2^n-1}\\
&= \left(\left(\bot^{1/2^n}\right)^k \to \bot\right) \vee \left(\bot^{1/2^n}\right)^{2^n-1-k}\\
&= \left(\bot^{1/2^n}\right)^{2^n-k} \vee \left(\bot^{1/2^n}\right)^{2^n-1-k}\\
&= \left(\bot^{1/2^n}\right)^{2^n-(k+1)}.
\end{aligned}$$

Hence (2.95) also holds for $k + 1$ and n.
(b) Since the relation (ii) is obvious for $r = 0$, we put $r = \frac{k}{2^n}$ with $1 \le k \le 2^n$ and verify (ii) as follows:

$$\left(\bot^r\right)^{1/2} * \bot^{1/2} = \left(\bot^{1/2^{n+1}}\right)^k * \left(\bot^{1/2^{n+1}}\right)^{2^n} = \left(\bot^{1/2^{n+1}}\right)^{k+2^n} = \bot^{\frac{r+1}{2}}.$$

(c) With regard to (iii) we choose dyadic numbers $r_1 = \frac{k}{2^n}$ and $r_2 = \frac{\ell}{2^m}$ from the unit interval and apply (i):

$$\begin{aligned}
(\bot^{r_1} \to \bot) * (\bot^{r_2} \to \bot) &= \left(\left(\bot^{1/2^n}\right)^{2^n-k}\right) * \left(\left(\bot^{1/2^m}\right)^{2^m-\ell}\right)\\
&= \left(\bot^{1/2^{n+m}}\right)^{2^{n+m}-2^m\cdot k+2^{n+m}-2^n\cdot \ell}.
\end{aligned}$$

Now we distinguish the following cases. If $2^m \cdot k + 2^n \cdot \ell \le 2^{n+m}$, then $r_1 *_{Ł} r_2 = 0$ and so

$$(\bot^{r_1} \to \bot) * (\bot^{r_2} \to \bot) = \bot = \bot^0 \to \bot = \bot^{r_1 *_{Ł} r_2} \to \bot$$

holds. If $2^{n+m} < 2^m \cdot k + 2^n \cdot \ell$, then $r_1 *_{Ł} r_2 = \frac{2^m\cdot k+2^n\cdot\ell-2^{n+m}}{2^{n+m}}$ and so we apply (i) again and obtain:

$$(\bot^{r_1} \to \bot) * (\bot^{r_2} \to \bot) = \bot^{1-(r_1 *_{Ł} r_2)} = \bot^{r_1 *_{Ł} r_2} \to \bot.$$

Thus we have established (iii).
(d) The implication $r_1 \le r_2 \implies \bot^{r_2} \le \bot^{r_1}$ is obvious. On the other hand, if we assume $\bot^{r_2} \le \bot^{r_1}$, then we apply (i) and (iii) and obtain:

$$\bot = \bot^{r_2} * \left(\bot^{r_1} \to \bot\right) = \left(\bot^{1-r_2} \to \bot\right) * \left(\bot^{r_1} \to \bot\right) = \bot^{(1-r_2)*_{Ł} r_1} \to \bot.$$

Hence $(1 - r_2) *_{Ł} r_1 = 0$ — i.e. $r_1 \le r_2$. □

Now we are ready to prove the announced characterization.

Theorem 2.7.18. *Every infinite, simple and complete MV-algebra is isomorphic to the real unit interval provided with the Łukasiewicz arithmetic conjunction.*

Proof. Let $(X, *, \to)$ be an infinite, simple and complete MV-algebra. Then every element of $(X, *, \to)$ is a square (cf. Corollary 2.7.15), and the infiniteness and simplicity of X imply the property $\bot^{1/2} \neq \bot$ (cf. Theorem 2.7.16 (v)). Moreover, we conclude from Theorem 2.7.16 (iv) that $\bigvee_{n\in\mathbb{N}} \bot^{1/2^n}$ is idempotent. Hence

$$\bigvee_{n\in\mathbb{N}} \bot^{1/2^n} = \top \tag{2.96}$$

follows from the simplicity of $(X, *, \to)$. Now we make use of the property that $(\bot^{1/2} \to \bot) * (\bot^{1/2} \to \bot)$ is idempotent (cf. Theorem 2.7.16 (iii)), apply Corollary 2.7.4 (iv) and obtain $\bot^{1/2} \wedge \big((\bot^{1/2} \to \bot) * (\bot^{1/2} \to \bot)\big) = \bot$. Since $\bot^{1/2} \neq \bot$ and $(X, \leq)$ is a chain (cf. Lemma 2.7.9), the relation

$$(\bot^{1/2} \to \bot) * (\bot^{1/2} \to \bot) = \bot$$

follows — this means

$$\bot^{1/2} = \bot^{1/2} \to \bot. \tag{2.97}$$

Thus we can apply Lemma 2.7.17 and define an order-preserving and order-reflecting map $\mathscr{D} \xrightarrow{\psi} X$ from the MV-algebra $\mathscr{D}$ of all dyadic rational numbers of $[0, 1]$ to X by $\psi(r) = \bot^r \to \bot$ for all $r \in \mathscr{D}$ (cf. Lemma 2.7.17 (iv)). Obviously, ψ is an embedding which preserves the respective multiplications and residuals (cf. Lemma 2.7.17 (i) and (iii)) and satisfies the following important property:

$$\psi(r)^{1/2} = (\bot^r \to \bot)^{1/2} = (\bot^r)^{1/2} \to \bot^{1/2} = \big((\bot^r)^{1/2} * \bot^{1/2}\big) \to \bot = \psi(\tfrac{r+1}{2}).$$

If $r_1 \leq r_2$ with $r_2 - r_1 \leq \frac{1}{2^n}$ for some $n \in \mathbb{N}$, then it is easily seen that the property

$$\big(\psi(r_1) \to \bot\big) * \psi(r_2) \leq \psi(\tfrac{1}{2^n}) \quad \text{i.e.} \quad \psi(r_2) * \bot^{1/2^n} \leq \psi(r_1)$$

holds. Because of (2.96) the previous relation implies that ψ is join-preserving (resp. meet-preserving) as far joins (resp. meets) exist in $\mathscr{D}$. Therefore we can extend ψ to an injective join- and meet-preserving map $[0, 1] \xrightarrow{\overline{\psi}} X$ by

$$\overline{\psi}(t) = \bigvee\{\psi(r) \mid r \in \mathscr{D},\ r < t\} = \bigwedge\{\psi(q) \mid q \in \mathscr{D},\ t < q\}, \qquad t \in [0, 1].$$

Again $\overline{\psi}$ preserves the algebraic structure. We prove the surjectivity of $\overline{\psi}$. For this purpose we fix $x \in X$. Since X is a chain, we infer from (2.96) that there exist dyadic numbers $u, v \in \mathscr{D}$ with $\psi(u) \leq x \leq \psi(v)$. Now we construct recursively two sequences $(r_n)_{n\in\mathbb{N}}$ and $(q_n)_{n\in\mathbb{N}}$ of dyadic rational numbers such that $(r_n)_{n\in\mathbb{N}}$ is increasing, $(q_n)_{n\in\mathbb{N}}$ is decreasing and the following property holds:

$$\psi(r_n) \leq x \leq \psi(q_n) \ \text{ and } \ 0 \leq q_n - r_n \leq \tfrac{1}{2^{n-1}}, \qquad n \in \mathbb{N}. \tag{2.98}$$

Step 1. $r_1 = u$, $q_1 = v$, $\psi(r_1) \le x$, $x \le \psi(q_1)$, $0 \le q_1 - r_1 \le 1$.
Step 2. Let r_n and q_n be defined such that (2.98) holds. Since X is a chain, we distinguish the following cases.
If $\psi(\frac{r_n+q_n}{2}) \le x$, then we put $r_{n+1} = \frac{r_n+q_n}{2} \ge r_n$ and $q_{n+1} = q_n$.
If $x \le \psi(\frac{r_n+q_n}{2})$, then we put $r_{n+1} = r_n$ and $0 \le q_{n+1} = \frac{r_n+q_n}{2} \le q_n$.
Hence we obtain $\psi(r_{n+1}) \le x \le \psi(q_{n+1})$ and $q_{n+1} - r_{n+1} = \frac{q_n-r_n}{2} \le \frac{1}{2^n}$ — this means that r_{n+1} and q_{n+1} again satisfy (2.98).
Now we return to the general argumentation. From $q_n *_Ł (1 - \frac{1}{2^{n-1}}) \le r_n$ we obtain:

$$\psi(q_n) * \psi(1 - \tfrac{1}{2^{n-1}}) = \psi(q_n) * \bot^{1/2^{n-1}} \le \psi(r_n), \qquad n \ge 2.$$

Hence we conclude from (2.96) and (2.98) that the relation

$$\bigvee_{n \in \mathbb{N}} \psi(r_n) = x = \bigwedge_{n \in \mathbb{N}} \psi(q_n)$$

holds. Finally, we put $\sup_{n \in \mathbb{N}} r_n = t_0 = \inf_{n \in \mathbb{N}} q_n$. Then the definition of $\overline{\psi}$ and the previous relation imply $\overline{\psi}(t_0) = x$. Hence the surjectivity of $\overline{\psi}$ is verified. In particular, $\overline{\psi}$ is an isomorphism between $(X, *, \to)$ and the real unit interval $[0, 1]$ provided with Łukasiewicz arithmetic conjunction. □

Corollary 2.7.19. *Every simple MV-algebra is isomorphic to an MV-subalgebra of* $[0, 1]$.

Proof. Let $(X, *, \to)$ be a simple MV-algebra. If X has n elements, then $(X, *, \to)$ is isomorphic to $\{\frac{k}{n-1} \mid k \in \{0, 1, \dots, n-1\}\}$ (cf. (2.92)). If X is infinite, then we can construct the MacNeille completion of X (cf. Corollary 2.7.10) and subsequently apply Theorem 2.7.18.

Since the only MV-algebra automorphism of $[0, 1]$ is the identity map, we can also express the previous corollary as follows: Every simple MV-algebra admits exactly one monomorphism to $[0, 1]$.

Exercises

2.7.1. Show that every dualizing element of a Girard algebra coincides with the universal lower bound.

2.7.2. Let $C_3 = \{\bot, a, \top\}$ be the chain with three elements.

(a) Show that on C_3 there exists a unique structure of a Girard algebra whose multiplication $*$ is determined by (see (9) in Exercise 2.2.1):

$*$	$\bot$	a	$\top$
$\bot$	$\bot$	$\bot$	$\bot$
a	$\bot$	$\bot$	a
$\top$	$\bot$	a	$\top$

(b) Show that the Girard algebra $(C_3, *)$ in (a) is an MV-algebra which is isomorphic to the MV-subalgebra $\{0, \frac{1}{2}, 1\}$ of $[0, 1]$ provided with the Łukasiewicz arithmetic conjunction.

2.7.3. Let $C_4 = \{\bot, a, b, \top\}$ be the chain with four elements such that $a < b$.

(a) Show that on $(C_4, *)$ there exists only two different Girard algebra structures whose multiplication tables are given as follows:

$*_1$	$\bot$	a	b	$\top$
$\bot$	$\bot$	$\bot$	$\bot$	$\bot$
a	$\bot$	$\bot$	$\bot$	a
b	$\bot$	$\bot$	b	b
$\top$	$\bot$	a	b	$\top$

and

$*_2$	$\bot$	a	b	$\top$
$\bot$	$\bot$	$\bot$	$\bot$	$\bot$
a	$\bot$	$\bot$	$\bot$	a
b	$\bot$	$\bot$	a	b
$\top$	$\bot$	a	b	$\top$

(Hint: Use the fact that there exists a unique order-reversing involution on C_4.)

(b) Show that $(C_4, *_1)$ is the semi-unitalization of the two-sided quantale (6) in Exercise 2.2.1 and $(C_4, *_1)$ is not an MV-algebra.

(c) Show that $(C_4, *_2)$ is the semi-unitalization of the two-sided quantale (2) in Exercise 2.2.1 and $(C_4, *_2)$ is an MV-algebra.

(d) Show that $(C_4, *_2)$ is isomorphic to the MV-subalgebra $\{0, \frac{1}{3}, \frac{2}{3}, 1\}$ of $[0, 1]$ provided with Łukasiewicz arithmetic conjunction.

2.7.4. Let $(C_3, *)$ be the MV-algebra considered in Exercise 2.7.2 (see also Exercise 2.2.1 (9)), and let $(C_3 \otimes C_3, \star)$ be the tensor product of $(C_3, *)$ with itself. Recall from Example 2.3.26 that $C_3 \otimes C_3$ consists of six elements:

$$\top = \top \otimes \top, \quad \alpha = (a \otimes \top) \vee (\top \otimes a), \quad \lambda = a \otimes \top, \quad \rho = \top \otimes a, \quad \beta = a \otimes a,$$
$$\bot = \bot \otimes \bot = \bot \otimes a = a \otimes \bot,$$

and its lattice structure is determined by the following Hasse diagram:

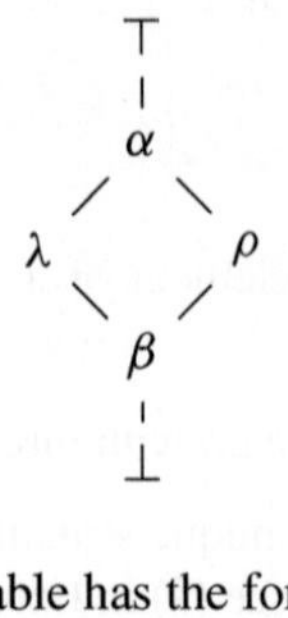

(a) Show that the multiplication table has the form

$\star$	$\bot$	β	λ	ρ	α	$\top$
$\bot$	$\bot$	$\bot$	$\bot$	$\bot$	$\bot$	$\bot$
β	$\bot$	$\bot$	$\bot$	$\bot$	$\bot$	β
λ	$\bot$	$\bot$	$\bot$	β	β	λ
ρ	$\bot$	$\bot$	β	$\bot$	β	ρ
α	$\bot$	$\bot$	β	β	β	α
$\top$	$\bot$	β	λ	ρ	α	$\top$

(b) Verify the relations $\beta \to \bot = \alpha$, $\alpha \to \bot = \beta$, $\lambda \to \bot = \lambda$ and $\rho \to \bot = \rho$.

(c) Conclude from (a) and (b) that $(C_3 \otimes C_3, \star)$ is a commutative integral Girard quantale which is the coproduct of the MV-algebra $(C_3, *)$ with itself in the category of commutative and unital quantales (cf. Theorem 2.3.35).

(d) Verify the relation $\lambda \to \rho = \alpha = \rho \to \lambda$ and conclude from this property that $(C_3 \otimes C_3, \star)$ is not an MV-algebra.

2.7.5. Deduce from Proposition 2.7.1 (iii) and Corollary 2.7.4 (i) and (v) that in any MV-algebra $(X, *, \to)$ the relation $(x \vee y)^p = x^p \vee y^p$ holds for all $x, y \in X$ and $p \in \mathbb{N}$. If in addition $(X, *, \to)$ is complete, then show that the correspondence $x \longmapsto x^p$ is join-preserving for all $p \in \mathbb{N}$.

2.7.6. Let $(X, *, \to)$ be a complete MV-algebra. Show that for every element $x \in X$ with $x \neq \top$ there exists a natural number $n \in \mathbb{N}$ such that $(x^n \to \bot) \to x \neq \top$ holds.
(Hint: If $x \neq \top$, then first verify that $\big((x \to \bot) \to \big(\bigwedge_{n \in \mathbb{N}} x^n\big)\big) \neq \top$.)

2.7.7. A filter on an MV-algebra $(X, *, \to)$ is an upclosed and nonempty subset F of X satisfying the following additional properties:

(F1) $x, y \in F \implies x * y \in F$.
(F2) $\bot \notin F$.

On the set $\mathbb{F}(X)$ of all filters on $(X, *, \to)$ we consider the set-inclusion $\subseteq$ as partial order. Verify the following properties:

(a) The partially ordered set $(\mathbb{F}(X), \subseteq)$ has maximal elements.
(Hint: Zorn's Lemma.)

(b) A filter U on $(X, *, \to)$ is maximal if and only if for every $z \notin U$ there exists a natural number $n \in \mathbb{N}$ such that $z^n \to \bot \in U$.

(c) Every maximal filter $U \in \mathbb{F}$ is prime—i.e. $x \vee y \in U \implies (x \in U$ or $y \in U)$.

(d) Let U be a maximal filter on $(X, *, \to)$. If $x \notin U$, then there exists a natural number $n \in \mathbb{N}$ such that $(x^n \to \bot) \to x \neq \top$.

2.7.8. Let x be an element of an MV-algebra $(X, *, \to)$ with the property that there exists an $n \in \mathbb{N}$ such that $(x^n \to \bot) \to x \neq \top$ holds. Show that there exists a maximal filter U on $(X, *, \to)$ with $x \notin U$.
(Hint: Deduce from Exercise 2.7.5, Proposition 2.7.1 (iii) and Corollary 2.7.4 (i) the relation

$$\big((x^n \to \bot) \to x\big)^p \vee (x^{n+1} \to \bot)^p = \top.$$

Now invoke the hypothesis $(x^n \to \bot) \to x \neq \top$ and conclude that

$$F = \{z \in X \mid \exists p \in \mathbb{N} : (x^{n+1} \to \bot)^p \leq z\}$$

is a filter on $(X, *, \to)$. Finally, apply the axiom of choice, construct a maximal filter U containing F and show that $x \notin U$.)

2.7.9. Let $(X, *, \to)$ be an MV-algebra. An equivalence relation C on X is called a *congruence* on $(X, *, \to)$ if for $(x_1, y_1) \in C$ and $(x_2, y_2) \in C$ the following additional properties are valid:

$$(x_1 * x_2, y_1 * y_2) \in C, \quad \big(x_1 \to x_2, y_1 \to y_2\big) \in C \quad \text{and} \quad (\bot, \top) \notin C.$$

Show:

(a) Every congruence C on $(X, *, \to)$ induces a filter $F_C = \{x \in X \mid (x, \top) \in C\}$ (cf. Exercise 2.7.7).
(b) Every filter $F \in \mathbb{F}(X)$ induces a congruence C_F on $(X, *, \to)$ by:

$$C_F = \{(x, y) \mid (x \to y) \wedge (y \to x) \in F\}.$$

(Hint: $y_1 * (x_1 \to x_2) * (y_1 \to x_1) * (x_2 \to y_2) \leq y_2$.)
(c) The quotient of $(X, *, \to)$ w.r.t. a filter F (i.e. w.r.t. the congruence C_F) is an MV-algebra and the corresponding quotient map q is an MV-algebra homomorphism — i.e. q preserves finite joins, the universal upper bound and the binary operations $*$ and $\to$.
(d) The quotient of an MV-algebra w.r.t. a filter U is simple if and only if U is a maximal filter. Moreover, the range of the quotient map w.r.t. a maximal filter is always isomorphic to an MV-subalgebra of the real unit interval (cf. Corollary 2.7.19).

2.7.10. Let $(X, *)$ be a complete MV-algebra. Then for every $x \in X$ with $x \neq \top$ there exists a maximal filter U on $(X, *)$ such that $x \notin U$.
(Hint: For $x \neq \top$ choose $n \in \mathbb{N}$ with $(x^n \to \bot) \to x \neq \top$ according to Exercise 2.7.6 and then apply Exercise 2.7.8.)

Comment. Exercise 2.7.10 says that every complete MV-algebra $(X, *)$ is *semi-simple* and is therefore isomorphic to an MV-subalgebra of $[0, 1]^I$, where I is the set of all maximal filters on $(X, *)$ (cf. Exercise 2.7.9 (d) and [48, Sect. 4]). In this context it is worthwhile to note that the property

$$\forall x \in X \setminus \{\top\}\ \exists n \in \mathbb{N} : (x^n \to \bot) \to x \neq \top$$

is equivalent to condition (ii) in Theorem 2.7.7, which can also be rewritten as an infinitary inference rule for the infinite-valued Łukasiewicz predicate calculus (cf. [46]).

Notes

The fact that the category Sup of complete lattices and join-preserving maps is isomorphic to the Eilenberg–Moore category of the power set monad (cf. [74, p. 57–58]) was already well-known to the Zürich school in the mid 1960s. In this sense Sup has a long history. A detailed account of the categorical properties of Sup can be found in [59], also where the importance of closure operators for quotient constructions in Sup is mentioned. The tensor product in Sup goes back to the Ph.D. thesis of D.G. Mowat (cf. [81]) and appears for the first time in a general accessible publication in 1974 (cf. [106, Lemma 1.7]). Only two years later the monoidal category Sup found its appropriate place in the literature when B. Banaschewski and E. Nelson (cf. [6]) reveal the categorical principles guiding the tensor product construction. The investigation of permanence properties of the tensor product in Sup — e.g. the preservation of complete distributivity or continuity of complete lattices — is due to Z. Shmuely and H.-J. Bandelt (cf. [7, 106]). In this context it might be interesting to add that the tensor product also exists for join semilattices (cf. [58]), but this construction is not an extension of the construction given in Theorem 2.1.8 (cf. [33]).

Prequantales/quantales already occur under the name *cl-groupoid/cl-semigroup with zero* in the third edition of G. Birkhoff's book on lattice theory (cf. [14, Corollary 1 on p. 327]). The simple fact that the multiplication of a prequantale /quantale is a bimorphism in Sup — i.e. every prequantale/quantale is a magma/semigroup in Sup — has been largely ignored in the literature. Only in [59] we do find the information that commutative and unital quantales are commutative monoids in Sup. This observation immediately leads to the insight that the tensor product of commutative and unital quantales is their coproduct (cf. [59]). In the category of semi-unital and bisymmetric quantales the proof of this fact has a non-commutative extension to the tensor product of strictly left-sided quantales with strictly right-sided quantales (cf. Theorem 2.5.9). An analysis of the tensor product of arbitrary quantales is given in [43]. Here operations defined according to Zadeh's extension principle (cf. [113]) and the compositions τ_* of probability distributive functions (cf. [104]) have been revealed as special instances of the tensor product construction.

The topological representation of left-sided, idempotent quantales and balanced, bisymmetric quantales goes back to U. Höhle, who realized that in this context the tensor product of the left-sided, non-commutative, idempotent three-chain with the right-sided, non-commutative, idempotent three-chain can be understood as the "smallest" non-commutative quantic frame and acts therefore as the *quantization* of $\mathbb{1} = \{0, 1\}$ (cf. [51, 52]).

The treatment of the law of double negation for quantales is fragmentary in more than one respect. For example, the question whether the law of double negation can be expressed in terms of the monoidal category Sup remained open in the literature, but has been solved in Sect. 2.6 of this book. Further, we have Frobenius monoids (cf. [14]), but not Frobenius quantales. L.P. Belluce was the first to investigate complete MV-algebras from the perspective of semi-simplicity (cf. [11]), but he did not show that (complete) MV-algebras are divisible and commutative Frobenius monoids (quantales). The fact that the MacNeille completion preserves the MV-algebra structure if and only if the given MV-algebra is semi-simple (see also Theorem 2.7.7) is due to U. Höhle 1995 (cf. [48]).

Chapter 3
Module Theory in Sup

In the following sections we give some applications of the theory of quantales. Traditionally, this is the field of abstract ideal theory (cf. [14]), but here we will focus on the impact of the category Sup. Since Sup is a monoidal category and quantales are semigroups in Sup, the most natural applications of quantales arise in the theory of left (right) modules on unital quantales. With regard to many-valued logics it is interesting to see that the left implication plays the rôle of a left action, while the right implication operates as a right action. In fact, the dual lattice of a unital quantale[1] $\mathfrak{Q}$ is always a $\mathfrak{Q}$-bimodule. As a second important example we refer to the fact that the algebra of "fuzzy subsets" of a set X (cf. [53, 111]) constitute the free module generated by X on the real unit interval provided with Łukasiewicz arithmetic conjunction. Hence it is not surprising that left (right) $\mathfrak{Q}$-modules are $\mathfrak{Q}$-valued complete lattices (see Sect. 3.3).

Further, in the setting of commutative and unital quantales $\mathfrak{Q}$ the category of $\mathfrak{Q}$-modules has a tensor product which is a quotient of the tensor product in Sup. This important property opens the perspective of $\mathfrak{Q}$-algebras — a theory which is not developed as yet with the exception of some well-known results related to descent theory (cf. [59]). Therefore we will draw our attention to non-commutative applications of left (right) modules.

As a typical field we choose the theory of C^*-algebras. It is remarkable to see that irreducible representations of a C^*-algebra A can be rephrased as irreducible and involutive left modules on a specific involutive and unital quantale given by all closed linear subspaces of A. The link between this observation and the spectrum of C^*-algebras as a quotient of the tensor product of the quantales of closed left ideals and closed right ideals (cf. Sect. 2.5.3) is an interesting question which has not been deeply investigated as yet. In this context Exercise 3.2.8 (b) can be considered as a first step.

---

[1]Since in this chapter elements of unital quantales are acting on complete lattices, from now on arbitrary unital quantales will always be denoted by $(\mathfrak{Q}, *, e)$.

P. Eklund et al., *Semigroups in Complete Lattices*, Developments in Mathematics 54, https://doi.org/10.1007/978-3-319-78948-4_3

This chapter finishes with a small section on automata in Sup showing that every automaton gives rise to a right module over the free unital quantale generated by its input alphabet.

3.1 Modules on Unital Quantales

In any monoidal category C the concept of modules is available — these are objects of C provided with a left or right action w.r.t. some monoid in C (cf. [73]). Here we recall the axioms of a left module, right module and bimodule in the special setting where the monoidal category is given by $(\mathsf{Sup}, \otimes, \mathbb{1}, a, c, \ell, r)$ (cf. Fact II in Sect. 2.1).

Let $(\mathfrak{Q}, *, e)$ be a unital quantale with the corresponding binary operation

$$\mathfrak{Q} \otimes \mathfrak{Q} \xrightarrow{\circledast} \mathfrak{Q}$$

determined by the multiplication $*$. A *left action* on a complete lattice X w.r.t. $(\mathfrak{Q}, *, e)$ is a join-preserving map $\mathfrak{Q} \otimes X \xrightarrow{\odot} X$ such that the following diagrams commute (cf. [73]):

$$\begin{array}{ccccc} (\mathfrak{Q} \otimes \mathfrak{Q}) \otimes X & & \xrightarrow{\quad \circledast \otimes 1_X \quad} & & \mathfrak{Q} \otimes X \\ \Big\downarrow{\scriptstyle a_{\mathfrak{Q}\mathfrak{Q}X}} & & & & \Big\downarrow{\scriptstyle \odot} \\ \mathfrak{Q} \otimes (\mathfrak{Q} \otimes X) & \xrightarrow{1_{\mathfrak{Q}} \otimes \odot} & \mathfrak{Q} \otimes X & \xrightarrow{\quad \odot \quad} & X \end{array} \tag{3.1}$$

$$\begin{array}{ccc} \mathbb{1} \otimes X & \xrightarrow{e \otimes 1_X} & \mathfrak{Q} \otimes X \\ & {\scriptstyle l_X} \searrow & \Big\downarrow{\scriptstyle \odot} \\ & & X \end{array} \tag{3.2}$$

A pair $(X, \odot)$ is a *left $\mathfrak{Q}$-module* if X is a complete lattice and $\odot$ is a left action on X w.r.t. $(\mathfrak{Q}, *, e)$. *Right $\mathfrak{Q}$-modules* are defined analogously — these are complete lattices X with *right actions* $X \otimes \mathfrak{Q} \xrightarrow{\boxdot} X$.

Because of the universal property of the tensor product in Sup we can simplify our notation and identify left (resp. right) actions with their corresponding bimorphisms. Hence instead of $\odot(\alpha \otimes x)$ (resp. $\boxdot(x \otimes \alpha)$) we also write $\alpha \odot x$ (resp. $x \boxdot \alpha$).

Since every tensor is the join of an appropriate family of elementary tensors, the commutativity of the diagrams (3.1) and (3.2) is equivalent to the following axioms:

($M1_\ell$) If $\alpha, \beta \in \mathfrak{Q}$ and $x \in X$, then $\alpha \odot (\beta \odot x) = (\alpha * \beta) \odot x$.
($M2_\ell$) If e denotes the unit of $\mathfrak{Q}$, then $e \odot x = x$ for all $x \in X$.

If we consider the transposed version of the previous axioms, then $X \otimes \mathfrak{Q} \xrightarrow{\boxdot} X$ is a right action if and only if the following axioms are satisfied:

(M1$_r$) If $\alpha, \beta \in \mathfrak{Q}$ and $x \in X$, then $(x \boxdot \alpha) \boxdot \beta = x \boxdot (\alpha * \beta)$.
(M2$_r$) If e denotes the unit of $\mathfrak{Q}$, then $x \boxdot e = x$ for all $x \in X$.

A *left $\mathfrak{Q}$-module homomorphism* is a join-preserving map $X \xrightarrow{h} Y$ such that

$$h(\alpha \odot x) = \alpha \odot h(x)$$

holds for each $\alpha \in \mathfrak{Q}$ and $x \in X$, which can be expressed by the commutativity of the following diagram:

$$\begin{array}{ccc} \mathfrak{Q} \otimes X & \xrightarrow{1_{\mathfrak{Q}} \otimes h} & \mathfrak{Q} \otimes Y \\ {\scriptstyle \odot}\big\downarrow & & \big\downarrow{\scriptstyle \odot} \\ X & \xrightarrow{\quad h \quad} & Y \end{array}$$

As usual, left $\mathfrak{Q}$-modules and left $\mathfrak{Q}$-module homomorphisms form a category denoted by $\mathtt{Mod}_\ell(\mathfrak{Q})$. *Right $\mathfrak{Q}$-module homomorphisms* and the category $\mathtt{Mod}_r(\mathfrak{Q})$. are defined analogously.

Even though we essentially restrict our interest to left $\mathfrak{Q}$-modules in this section, we begin with an interesting interrelation between left and right $\mathfrak{Q}$-modules.

First we recall that the transposed multiplication of $*$ is given by $\alpha *^\tau \beta = \beta * \alpha$ for all $\alpha, \beta \in \mathfrak{Q}$. Then $\mathfrak{Q}^\tau = (\mathfrak{Q}, *^\tau, e)$ is the transposed unital quantale of $\mathfrak{Q}$, and every left $\mathfrak{Q}$-module $(X, \odot)$ can be read as a right $\mathfrak{Q}^\tau$-module where the right action $\boxdot$ is determined by $x \boxdot \alpha = \alpha \odot x$ for each $x \in X$ and $\alpha \in \mathfrak{Q}$. Moreover, the following important properties are valid.

Proposition 3.1.1. (a) *If $(X, \boxdot)$ is a right $\mathfrak{Q}$-module, then the join-preserving map $\mathfrak{Q} \otimes X^{op} \xrightarrow{\circledcirc} X^{op}$ determined by*

$$\alpha \circledcirc x = \bigvee\{z \in X \mid z \boxdot \alpha \le x\}, \qquad \alpha \in \mathfrak{Q},\ x \in X \tag{3.3}$$

is a left action on X^{op} — i.e. $(X^{op}, \circledcirc)$ is a left $\mathfrak{Q}$-module.

(b) *If $(X, \odot)$ is a left $\mathfrak{Q}$-module, then the join-preserving map $X^{op} \otimes \mathfrak{Q} \xrightarrow{\boxcircle} X^{op}$ determined by*

$$x \boxcircle \alpha = \bigvee\{z \in X \mid \alpha \odot z \le x\}, \qquad \alpha \in \mathfrak{Q},\ x \in X \tag{3.4}$$

is a right action on X^{op} — i.e. $(X^{op}, \boxcircle)$ is a right $\mathfrak{Q}$-module.

(c) *Let $(X, \boxdot)$ be a right $\mathfrak{Q}$-module. If $\circledcirc$ is the left action on X^{op} determined by $\boxdot$, then the right action $\boxcircle$ on X determined by $\circledcirc$ coincides with $\boxdot$.*

(d) *Let $(X, \odot)$ be a left $\mathfrak{Q}$-module. If $\boxcircle$ is the right action on X^{op} determined by $\odot$, then the left action $\circledcirc$ on X determined by $\boxcircle$ coincides with $\odot$.*

Proof. Since (b) and (d) are the transposed statements of (a) and (c), respectively, we only verify (a) and (c).

(a) For this purpose we fix $x \in X$ and $\alpha, \beta \in \mathfrak{Q}$. Evidently the correspondence $(\alpha, x) \longmapsto \alpha \circledcirc x$ is a bimorphism which uniquely determines a join-preserving map $\mathfrak{Q} \otimes X^{op} \xrightarrow{\circledcirc} X^{op}$. Hence we have only to verify ($M1_\ell$) and ($M2_\ell$). By (3.3) and ($M2_r$) the condition ($M2_\ell$) is obvious. In order to prove ($M1_\ell$) we first apply ($M1_r$) and obtain:

$$\big(\alpha \circledcirc (\beta \circledcirc x)\big) \boxdot (\alpha * \beta) = \big((\alpha \circledcirc (\beta \circledcirc x)) \boxdot \alpha\big) \boxdot \beta \le (\beta \circledcirc x) \boxdot \beta \le x .$$

Hence $\alpha \circledcirc (\beta \circledcirc x) \le (\alpha * \beta) \circledcirc x$ follows. On the other hand we conclude from

$$\big(((\alpha * \beta) \circledcirc x) \boxdot \alpha\big) \boxdot \beta = \big((\alpha * \beta) \circledcirc x\big) \boxdot (\alpha * \beta) \le x$$

that $\big((\alpha * \beta) \circledcirc x\big) \boxdot \alpha \le \beta \circledcirc x$ and consequently $(\alpha * \beta) \circledcirc x \le \alpha \circledcirc (\beta \circledcirc x)$ holds. Hence ($M1_\ell$) follows.
(c) We fix now $\alpha \in \mathfrak{Q}$ and $x, z \in X$. Then the following chain of equivalences holds:

$$\begin{aligned} z \boxdot \alpha \le x &\iff z \le \alpha \circledcirc x \iff \alpha \circledcirc x \le^{op} z \\ &\iff x \le^{op} z \boxed{\circledcirc} \alpha \iff z \boxed{\circledcirc} \alpha \le x . \end{aligned}$$

Hence $\boxdot$ and $\boxed{\circledcirc}$ coincide. □

Corollary 3.1.2. *Let $(\mathfrak{Q}, *, e)$ be a unital quantale. Then the following assertions hold*:

(a) *If $(X, \boxdot)$ and $(Y, \boxdot)$ are right $\mathfrak{Q}$-modules and $X \xrightarrow{h} Y$ is a right $\mathfrak{Q}$-module homomorphism, then $(X^{op}, \circledcirc)$ and $(Y^{op}, \circledcirc)$ are left $\mathfrak{Q}$-modules and the right adjoint map $Y^{op} \xrightarrow{h^{\vdash}} X^{op}$ of h is a left $\mathfrak{Q}$-module homomorphism.*
(b) *If $(X, \odot)$ and $(Y, \odot)$ are left $\mathfrak{Q}$-modules and $X \xrightarrow{h} Y$ is a left $\mathfrak{Q}$-module homomorphism, then $(X^{op}, \boxed{\circledcirc})$ and $(Y^{op}, \boxed{\circledcirc})$ are right $\mathfrak{Q}$-modules and the right adjoint map $Y^{op} \xrightarrow{h^{\vdash}} X^{op}$ of h is a right $\mathfrak{Q}$-module homomorphism.*

Proof. Since (b) is the transposed situation of (a), we only verify (a). By Proposition 3.1.1 it is sufficient to prove that the right adjoint map $h^{\vdash}$ is a left $\mathfrak{Q}$-module homomorphism. For this purpose we choose $\alpha \in \mathfrak{Q}$, $x \in X$ and $y \in Y$. Then the following chain of equivalences hold:

$$\begin{aligned} x \le \alpha \circledcirc h^{\vdash}(y) &\iff x \boxdot \alpha \le h^{\vdash}(y) \iff h(x \boxdot \alpha) = h(x) \boxdot \alpha \le y \\ &\iff h(x) \le \alpha \circledcirc y \iff x \le h^{\vdash}(\alpha \circledcirc y) \end{aligned}$$

Hence $\alpha \circledcirc h^{\vdash}(y) = h^{\vdash}(\alpha \circledcirc y)$ — i.e. $h^{\vdash}$ is a left $\mathfrak{Q}$-module homomorphism.

We record the following general fact.

FACT I. *There exists a contravariant isomorphism* $\mathtt{Mod}_r(\mathfrak{Q}) \xrightarrow{\mathsf{F}} \mathtt{Mod}_\ell(\mathfrak{Q})$ *acting on objects and morphisms as follows*:

$$\mathsf{F}(X, \boxdot) = (X^{op}, \circledcirc), \quad X \xrightarrow{h} Y, \quad Y^{op} \xrightarrow{h^\vdash} X^{op}, \quad \mathsf{F}(h) = h^\vdash.$$

Now we continue with a characterization of $\mathfrak{Q}$-modules which also plays a significant rôle from the perspective of applications.

Proposition 3.1.3. *Let* $(\mathfrak{Q}, *, e)$ *be a unital quantale,* X *be a complete lattice, and let* $[X, X]$ *be the unital quantale of all join-preserving self-maps of* X *provided with the composition as multiplication. Further, let* $[X, X] \otimes X \xrightarrow{\mathrm{ev}_X} X$ *be the evaluation arrow — i.e.* $ev_X(f \otimes x) = f(x)$ *for all* $f \in [X, X]$ *and* $x \in X$*. Then there exists a bijective map between the set of all left actions* $\odot$ *on* X *in the sense of* $\mathtt{Sup}$ *and the set of all unital homomorphisms* $\mathfrak{Q} \xrightarrow{h} [X, X]$ *making the following diagram commutative*:

$$\begin{array}{ccc} \mathfrak{Q} \otimes X & \xrightarrow{h \otimes 1_X} & [X, X] \otimes X \\ & \searrow^{\odot} & \downarrow{\scriptstyle ev_X} \\ & & X \end{array}$$

Proof. Since $\mathtt{Sup}$ is monoidal closed (cf. Theorem 2.1.12) and $\mathrm{ev} = (\mathrm{ev}_X)_X$ is the counit of the adjoint situation $\_ \otimes X \dashv \mathrm{hom}_X$ (cf. Sect. 2.1.2), the homomorphism h coincides with the monoidal adjoint of the left-action $\odot$. The fact that h is a unital homomorphism if and only if $\odot$ is a left action can be easily verified by a diagram chase or by a simple calculation using the axioms ($\mathrm{M1}_\ell$) and ($\mathrm{M2}_\ell$) directly. □

The previous proposition says that left $\mathfrak{Q}$-modules X can be characterized simply by unital homomorphisms from $\mathfrak{Q}$ to $[X, X]$. This situation also opens a new perspective, linking an important and application-oriented interpretation to left $\mathfrak{Q}$-modules.

Remark 3.1.4. Let us consider elements of a complete lattice X as *states* of properties of some given physical system. The partial order on X reflects a certain hierarchy among states, and the join operation expresses the *superposition* of states. The *transition* of states is *coherent* with respect to the given hierarchy and is represented by join-preserving self-maps of X — so-called *transition maps*. On this basis Proposition 3.1.3 says that a *left* $\mathfrak{Q}$*-module* describes a specific *dynamic* of some physical system. Each element of the unital quantale $\mathfrak{Q}$ refers to a specific transition map, and the composition of transition maps is expressed by the multiplication in $\mathfrak{Q}$. In this context the unit of $\mathfrak{Q}$ corresponds to the identity map 1_X changing nothing.

For the sake of completeness we also present a version of Proposition 3.1.3 for right $\mathfrak{Q}$-modules.

Proposition 3.1.5. *Let* $(\mathfrak{Q}, *, e)$ *be a unital quantale,* $\mathfrak{Q}^\tau$ *be its transposed unital quantale, and let* X *be a complete lattice. Further, let* $X \otimes [X, X] \xrightarrow{\varepsilon_X} X$ *be* *the evaluation arrow (cf. the Comment after Theorem 2.1.12). Then there exists a bijective map between the set of all right actions $\boxdot$ on X (w.r.t. $\mathfrak{Q}$) in the sense of* Sup *and the set of all unital homomorphisms $\mathfrak{Q}^\tau \xrightarrow{h} [X, X]$ making the following diagram commutative:*

$$\begin{array}{ccc} X \otimes \mathfrak{Q}^\tau & \xrightarrow{1_X \otimes h} & X \otimes [X, X] \\ & \searrow^{\boxdot} & \downarrow \varepsilon_X \\ & & X \end{array}$$

Before we investigate the general properties of the category $\mathrm{Mod}_\ell(\mathfrak{Q})$, we first give some interesting examples of left $\mathfrak{Q}$-modules. For this purpose we need some further terminology.

A triple $(X, \odot, \boxdot)$ is called a $\mathfrak{Q}$-*bimodule* if $(X, \odot)$ is a left $\mathfrak{Q}$-module, $(X, \boxdot)$ is a right $\mathfrak{Q}$-module, and the associative law holds, which can be expressed by the commutativity of the following diagram:

$$\begin{array}{ccccc} (\mathfrak{Q} \otimes X) \otimes \mathfrak{Q} & & \xrightarrow{\quad \mathfrak{a}_{\mathfrak{Q}X\mathfrak{Q}} \quad} & & \mathfrak{Q} \otimes (X \otimes \mathfrak{Q}) \\ {\scriptstyle \odot \otimes 1_{\mathfrak{Q}}} \downarrow & & & & \downarrow {\scriptstyle 1_{\mathfrak{Q}} \otimes \boxdot} \\ X \otimes \mathfrak{Q} & \xrightarrow{\quad \boxdot \quad} & X & \xleftarrow{\quad \odot \quad} & \mathfrak{Q} \otimes X \end{array} \tag{3.5}$$

Due to the associativity of the tensor product in Sup (cf. Lemma 2.1.13), a triple $(X, \odot, \boxdot)$ is a $\mathfrak{Q}$-bimodule if and only if the following property holds

$$(\alpha \odot x) \boxdot \beta = \alpha \odot (x \boxdot \beta), \qquad \alpha, \beta \in \mathfrak{Q},\ x \in X. \tag{3.6}$$

Example 3.1.6. Let $(\mathfrak{Q}, *, e)$ be a unital quantale. Due to the associativity of its multiplication, $\mathfrak{Q}$ can always be considered as a $\mathfrak{Q}$-bimodule with the left action $\odot$ and the right action $\boxdot$ on $\mathfrak{Q}$ determined by the following relations for all $\alpha, \beta \in \mathfrak{Q}$:

$$\alpha \odot \beta = \alpha * \beta \quad \text{and} \quad \beta \boxdot \alpha = \beta * \alpha.$$

If we apply Proposition 3.1.1 to $(\mathfrak{Q}, \odot, \boxdot)$, then it is easily seen that $(\mathfrak{Q}^{op}, \circledcirc, \boxcircle)$ is also a $\mathfrak{Q}$-bimodule. In this context the left action $\circledcirc$ can be characterized by the left-implication, while the right action $\boxcircle$ plays the rôle of the right-implication (cf. Sect. 2.2.1) — i.e.

$$\circledcirc = \Lleftarrow \circ\, c_{\mathfrak{Q}\mathfrak{Q}^{op}} \quad \text{and} \quad \boxcircle = \Rrightarrow \circ\, c_{\mathfrak{Q}^{op}\mathfrak{Q}},$$

where $c_{\mathfrak{Q}\mathfrak{Q}^{op}}$ and $c_{\mathfrak{Q}^{op}\mathfrak{Q}}$ are the respective components of the symmetry in Sup.

A small, but interesting modification of the previous example is the following.

Example 3.1.7. Let $(\mathfrak{Q}, *)$ be an arbitrary non-unital quantale and $(\widehat{\mathfrak{Q}}, \widehat{*})$ be its unitalization (cf. Exercise 2.2.2). Then the underlying complete lattice $\widehat{\mathfrak{Q}}$ has the form

$\widehat{\mathfrak{Q}} = \mathfrak{Q} \times \{0, 1\}$ and $(\mathfrak{Q}, *)$ can be identified with a subquantale of $(\widehat{\mathfrak{Q}}, \widehat{*})$. In this context $\mathfrak{Q}$ can be provided with the structure of a $\widehat{\mathfrak{Q}}$-bimodule $(\mathfrak{Q}, \odot, \boxdot)$ as follows (see again Exercise 2.2.2):

$$\begin{aligned} (\alpha, 0) \odot x &= (\alpha, 0) \,\widehat{*}\, (x, 0) = \alpha * x, & (\alpha, 1) \odot x &= (\alpha, 1) \,\widehat{*}\, (x, 0) = (\alpha * x) \vee x, \\ x \boxdot (\alpha, 0) &= (x, 0) \,\widehat{*}\, (\alpha, 0) = x * \alpha, & x \boxdot (\alpha, 1) &= (x, 0) \,\widehat{*}\, (\alpha, 1) = (x * \alpha) \vee x. \end{aligned}$$

The next example shows that the structure of a left $\mathfrak{Q}$-module can be extended pointwisely.

Example 3.1.8. Let $(X, \leq)$ be a preordered set and $(Y, \odot)$ be a left $\mathfrak{Q}$-module. On the set $\mathbb{P}(X, Y)$ of all antitone maps $X \xrightarrow{f} Y$ we introduce a partial order by

$$f \leq g \iff f(x) \leq g(x) \quad \text{for all } x \in X.$$

Then $\mathbb{P}(X, Y)$ is a complete lattice, and the left action on $\mathbb{P}(X, Y)$ (which is also denoted by $\odot$) is given by:

$$(\alpha \odot f)(x) = \alpha \odot f(x), \qquad \alpha \in \mathfrak{Q},\ f \in \mathbb{P}(X, Y),\ x \in X. \tag{3.7}$$

Hence $(\mathbb{P}(X, Y), \odot)$ is again a left $\mathfrak{Q}$-module.

If the left $\mathfrak{Q}$-module $(Y, \odot)$ coincides with $(\mathfrak{Q}, \odot)$ (i.e. the left action is given by $\alpha \odot \beta = \alpha * \beta$ for each $\alpha, \beta \in \mathfrak{Q}$ (see Example 3.1.6)), then we simply write $\mathbb{P}(X)$ for $\mathbb{P}(X, \mathfrak{Q})$.

For each unital quantale $(\mathfrak{Q}, *, e)$ it is well-known that the forgetful functor $\mathsf{U}_1 \colon \mathtt{Mod}_\ell(\mathfrak{Q}) \to \mathtt{Sup}$ has a left adjoint functor $\mathsf{F}_1 \colon \mathtt{Sup} \to \mathtt{Mod}_\ell(\mathfrak{Q})$ sending a complete lattice X to $\mathfrak{Q} \otimes X$ where the left action $\odot$ on $\mathfrak{Q} \otimes X$ is defined for each $\alpha \in \mathfrak{Q}$ and for each elementary tensor $\beta \otimes x \in \mathfrak{Q} \otimes X$ by

$$\alpha \odot (\beta \otimes x) = (\alpha \circledast \beta) \otimes x = (\alpha * \beta) \otimes x,$$

which can also be expressed by the commutativity of the diagram (cf. [73]):

$$\begin{array}{ccc} \mathfrak{Q} \otimes (\mathfrak{Q} \otimes X) & \xrightarrow{a^{-1}_{\mathfrak{Q}\mathfrak{Q}X}} & (\mathfrak{Q} \otimes \mathfrak{Q}) \otimes X \\ & \searrow^{\odot} & \Big\downarrow {\scriptstyle \circledast \otimes 1_X} \\ & & \mathfrak{Q} \otimes X \end{array} \tag{3.8}$$

For the sake of completeness we recall the proof of this property.

Lemma 3.1.9. *Let $(\mathfrak{Q}, *, e)$ be a unital quantale and X be a complete lattice. Then the join-preserving map $\mathfrak{Q} \otimes (\mathfrak{Q} \otimes X) \xrightarrow{\odot} \mathfrak{Q} \otimes X$ determined by (3.8) is a left action on $\mathfrak{Q} \otimes X$.*

Proof. Since tensors in $\mathfrak{Q} \otimes X$ are joins of elementary tensors, we restrict our attention to elementary tensors and choose $x \in X$ and $\alpha, \beta, \gamma \in \mathfrak{Q}$. Since $*$ is associative, we obtain:

$$\begin{aligned}\alpha \odot (\beta \odot (\gamma \otimes x)) &= \alpha \odot ((\beta * \gamma) \otimes x) = (\alpha * (\beta * \gamma)) \otimes x\\ &= ((\alpha * \beta) * \gamma) \otimes x \ = (\alpha * \beta) \odot (\gamma \otimes x).\end{aligned}$$

Hence ($M1_\ell$) is verified. The axiom ($M2_\ell$) is evident. □

Addition. For $\alpha \in \mathfrak{Q}$ and $f \in \mathfrak{Q} \otimes X$ the expression $\alpha \odot f$ is explicitly given by $\alpha \odot f = \bigvee_{\beta \in \mathfrak{Q}} (\alpha * \beta) \otimes f(\beta)$ where we have made use of (2.17).

A combination of Lemma 3.1.9 with Example 3.1.8 leads to the following theorem.

Theorem 3.1.10. *Let* $(X, \leq)$ *be a preordered set,* $\mathsf{Dwn}(X)$ *be the complete lattice of all downclosed subsets of* X, $(\mathfrak{Q}, *, e)$ *be a unital quantale, and let* $\mathbb{P}(X)$ *be the left* $\mathfrak{Q}$*-module constructed in Example 3.1.8. Then there exists an isomorphism* $\mathbb{P}(X) \xrightarrow{\Phi} \mathfrak{Q} \otimes \mathsf{Dwn}(X)$ *in the sense of* $\mathtt{Mod}_\ell(\mathfrak{Q})$ *satisfying the condition*

$$\Phi(f) = \bigvee_{x \in X} f(x) \otimes {\downarrow} x, \qquad f \in \mathbb{P}(X). \tag{3.9}$$

Proof. We define a map $\mathbb{P}(X) \xrightarrow{\Phi} \mathfrak{Q} \otimes \mathsf{Dwn}(X)$ by:

$$\Phi(f)(\alpha) = \{x \in X \mid \alpha \leq f(x)\}, \qquad f \in \mathbb{P}(X),\ \alpha \in \mathfrak{Q}.$$

Since $X \xrightarrow{f} \mathfrak{Q}$ is antitone, $\Phi(f)(\alpha)$ is a downclosed subset of X for all $\alpha \in \mathfrak{Q}$. Obviously $\Phi(f)$ is join-reversing — i.e. $\Phi(f) \in \mathfrak{Q} \otimes \mathsf{Dwn}(X)$. Further, let us consider a map $\mathfrak{Q} \otimes \mathsf{Dwn}(X) \xrightarrow{\Psi} \mathbb{P}(X)$ defined by:

$$\Psi(g)(x) = \bigvee\{\alpha \in \mathfrak{Q} \mid x \in g(\alpha)\}, \qquad g \in \mathfrak{Q} \otimes \mathsf{Dwn}(X),\ x \in X. \tag{3.10}$$

The following relations hold:

$$\begin{aligned}&\bigvee\{\alpha \in \mathfrak{Q} \mid x \in \Phi(f)(\alpha)\} = \bigvee\{\alpha \in \mathfrak{Q} \mid \alpha \leq f(x)\} = f(x),\\ &\{x \in X \mid \alpha \leq \Psi(g)(x)\} = \{x \in X \mid x \in g(\alpha)\} = g(\alpha),\end{aligned}$$

where we have used the property that g is join-reversing. Hence $\Psi \circ \Phi = \mathrm{id}_{\mathbb{P}(X)}$ and $\Phi \circ \Psi = \mathrm{id}_{\mathfrak{Q} \otimes \mathsf{Dwn}(X)}$ follow. Since Φ is isotone, Φ is an order isomorphism.

In order to verify (3.9) we have to show that $\Phi(f)$ is the join of

$$\mathbb{B}_f = \{f(x) \otimes {\downarrow} x \mid x \in X\}$$

in $\mathfrak{Q} \otimes \mathsf{Dwn}(X)$. Since $x \in \Phi(f)(f(x))$, we obtain $\downarrow x \subseteq \Phi(f)(f(x))$ for all $x \in X$ — this means that $\Phi(f)$ is an upper bound of $\mathbb{B}_f$. If $g \in \mathfrak{Q} \otimes \mathsf{Dwn}(X)$ is an arbitrary upper bound of $\mathbb{B}_f$ and $\alpha \leq f(x)$, then we obtain $x \in \downarrow x \subseteq g(f(x)) \subseteq g(\alpha)$. Hence $\Phi(f)(\alpha) \subseteq g(\alpha)$ follows for all $\alpha \in \mathfrak{Q}$, and (3.9) is verified.

Finally, we show that Φ is a left $\mathfrak{Q}$-module homomorphism. In fact, the following relation holds (cf. Example 3.1.8):

$$\begin{aligned}\alpha \odot \Phi(f) &= \alpha \odot \Big(\bigvee_{x\in X} f(x) \otimes \downarrow x\Big) = \bigvee_{x\in X} \alpha \odot (f(x) \otimes \downarrow x) = \bigvee_{x\in X} (\alpha * f(x)) \otimes \downarrow x \\ &= \bigvee_{x\in X} (\alpha \odot f)(x) \otimes \downarrow x = \Phi(\alpha \odot f).\end{aligned}$$

□

Theorem 3.1.11. *Let X be a complete lattice, $(\mathfrak{Q}, *, e)$ be a unital quantale and $X \xrightarrow{\eta_X^{\mathsf{M}}} \mathfrak{Q} \otimes X$ be the join-preserving map determined by $\eta_X^{\mathsf{M}}(x) = e \otimes x$ for all $x \in X$.*[2] *Then for every left $\mathfrak{Q}$-module $(Y, \odot)$ and for every join-preserving map $X \xrightarrow{h} Y$ there exists a unique left $\mathfrak{Q}$-module homomorphism $\mathfrak{Q} \otimes X \xrightarrow{h^\sharp} Y$ making the following diagram commutative*:

$$\begin{array}{ccc} X & \xrightarrow{\eta_X^{\mathsf{M}}} & \mathfrak{Q} \otimes X \\ & {\scriptstyle h}\searrow & \big\downarrow {\scriptstyle h^\sharp} \\ & & Y \end{array} \tag{3.11}$$

Proof. (a) (Uniqueness). Let $h^\sharp$ be a left $\mathfrak{Q}$-module homomorphism making the diagram (3.11) commutative. Then for all $x \in X$ the relation $h^\sharp(e \otimes x) = h(x)$ follows. By (3.8) (cf. Lemma 3.1.9) we obtain for $\alpha \in \mathfrak{Q}$ and $x \in X$:

$$h^\sharp(\alpha \otimes x) = h^\sharp((\alpha * e) \otimes x) = h^\sharp(\alpha \odot (e \otimes x)) = \alpha \odot h^\sharp(e \otimes x) = \alpha \odot h(x).$$

Since every tensor in $\mathfrak{Q} \otimes X$ is a join of elementary tensors, $h^\sharp$ is uniquely determined by the commutativity of the diagram (3.11).

(b) (Existence). We define a join-preserving map $\mathfrak{Q} \otimes X \xrightarrow{h^\sharp} Y$ by

$$h^\sharp = \odot \circ (1_{\mathfrak{Q}} \otimes h). \tag{3.12}$$

Then for every $\alpha \in \mathfrak{Q}$ and for every elementary tensor $\beta \otimes x$ in $\mathfrak{Q} \otimes X$ the relation

$$\begin{aligned} h^\sharp(\alpha \odot (\beta \otimes x)) &= h^\sharp((\alpha * \beta) \otimes x) = (\alpha * \beta) \odot h(x) \\ &= \alpha \odot (\beta \odot h(x)) = \alpha \odot h^\sharp(\beta \otimes x) \end{aligned}$$

[2] The superscript M refers to the monad defined below Corollary 3.1.12.

holds. Since $h^\sharp$ is join-preserving and every tensor is a join of elementary tensors, $h^\sharp$ is obviously a left $\mathfrak{Q}$-module homomorphism.

Finally, since e is the unit of $\mathfrak{Q}$, we conclude from (3.12) that the relation

$$h^\sharp(e \otimes x) = e \odot h(x) = h(x)$$

holds for all $x \in X$. Hence $h^\sharp$ makes the diagram (3.11) commutative. □

Referring to Lemma 3.1.9 and Theorem 3.1.11, there exists a functor

$$\mathsf{F}_1 \colon \mathtt{Sup} \to \mathtt{Mod}_\ell(\mathfrak{Q})$$

acting on objects and morphisms as follows:

$$\mathsf{F}_1(X) = (\mathfrak{Q} \otimes X, \odot), \quad X \xrightarrow{h} Y, \qquad \mathsf{F}_1(h) = (\eta_Y^{\mathsf{M}} \circ h)^\sharp = 1_{\mathfrak{Q}} \otimes h. \tag{3.13}$$

Corollary 3.1.12. *If* $\mathsf{U}_1 \colon \mathtt{Mod}_\ell(\mathfrak{Q}) \to \mathtt{Sup}$ *is the forgetful functor, then* F_1 *is left adjoint to* U_1 *— i.e.* $\mathsf{F}_1 \dashv \mathsf{U}_1$.

Proof. The assertion follows immediately from Theorem 3.1.11. □

Let $\mathbf{M} = (\mathsf{M}, \eta^{\mathsf{M}}, \mu^{\mathsf{M}})$ be the monad on $\mathtt{Sup}$ induced by the adjoint situation $\mathsf{F}_1 \dashv \mathsf{U}_1$ (cf. Sect. 1.2). Obviously, M and μ^{M} have the following explicit form:

$$\mathsf{M}(X) = \mathfrak{Q} \otimes X, \quad X \xrightarrow{h} Y, \quad \mathsf{M}(h) = 1_{\mathfrak{Q}} \otimes h, \quad \mu_X^{\mathsf{M}} = \odot = (\circledast \otimes 1_X) \circ a_{\mathfrak{Q}\mathfrak{Q}X}^{-1}.$$

From the diagrams (3.1) and (3.2), it is not difficult to see that the Eilenberg–Moore category $\mathtt{Sup}^{\mathbf{M}}$ coincides with $\mathtt{Mod}_\ell(\mathfrak{Q})$. Hence the completeness of $\mathtt{Mod}_\ell(\mathfrak{Q})$ is inherited by the completeness of $\mathtt{Sup}$. An explicit description of colimits in $\mathtt{Mod}_\ell(\mathfrak{Q})$ is given in the next subsection.

Since the forgetful functor $\mathtt{Sup} \xrightarrow{\mathsf{U}} \mathtt{Preord}$ also has a left adjoint functor, namely Dwn (cf. Sect. 1.3.1), we can compose the adjoint situation in Corollary 3.1.12 with $\mathsf{Dwn} \dashv \mathsf{U}$ and immediately obtain the following result.

Corollary 3.1.13. *The forgetful functor from* $\mathsf{U}_0 \colon \mathtt{Mod}_\ell(\mathfrak{Q}) \to \mathtt{Preord}$ *has a left adjoint and is given by* $\mathsf{F}_0 = \mathsf{F}_1 \circ \mathsf{Dwn}$.

The monad $\mathbf{P} = \big(\mathsf{P}, \eta^{\mathsf{P}}, \mu^{\mathsf{P}}\big)$ induced by the adjoint situation $\mathsf{F}_0 \dashv \mathsf{U}_0$ is called the *monad of* $\mathfrak{Q}$*-enriched presheaves*[3] on $\mathtt{Preord}$. If we make use of the isomorphism $\mathbb{P}(X) \xrightarrow{\Phi} \mathfrak{Q} \otimes \mathsf{Dwn}(X)$ satisfying (3.9) (cf. Theorem 3.1.10), then the details of $\mathbf{P} = \big(\mathsf{P}, \eta^{\mathsf{P}}, \mu^{\mathsf{P}}\big)$ are given explicitly as follows:

---

[3]Anticipating Sect. 3.3.2 we view antitone maps $X \xrightarrow{f} \mathfrak{Q}$ as $\mathfrak{Q}$-enriched covariant presheaves $X^{op} \xrightarrow{f} \mathfrak{Q}$.

$$\mathsf{P}(X) = \mathbb{P}(X), \qquad X \xrightarrow{h} Y, \qquad \mathbb{P}(X) \xrightarrow{\mathsf{P}(h)} \mathbb{P}(Y), \qquad X \xrightarrow{\eta_X^{\mathsf{P}}} \mathbb{P}(X),$$
$$\big(\mathsf{P}(h)(f)\big)(y) = \bigvee\{f(x) \mid y \le h(x)\}, \qquad f \in \mathbb{P}(X),\ y \in Y,$$
$$\eta_X^{\mathsf{P}}(x) = \Phi^{-1}(e \otimes \downarrow x) =: 1_x, \qquad 1_x(z) = \begin{cases} e, & z \le x, \\ \bot, & z \not\le x, \end{cases} \qquad z, x \in X,$$
$$\big(\mu_X^{\mathsf{P}}(F)\big)(x) = \bigvee_{f \in \mathbb{P}(X)} F(f) * f(x), \qquad F \in \mathbb{P}(\mathbb{P}(X)), \quad x \in X.$$

Before we proceed, we confirm the previous formulas. If $X \xrightarrow{h} Y$ is an isotone map, then for $f \in \mathbb{P}(X)$ and $y_0 \in Y$ we observe:

$$\begin{aligned} \big(\mathsf{P}(h)(f)\big)(y_0) &= \Phi^{-1}\big(1_{\mathfrak{Q}} \otimes \mathsf{Dwn}(h)(\Phi(f))\big)(y_0) \\ &= \Phi^{-1}\big(\bigvee_{x \in X} f(x) \otimes \downarrow h(x)\big)(y_0) \\ &= \Phi^{-1}\big(\bigvee_{y \in Y} \big(\bigvee\{f(x) \mid y \le h(x)\}\big) \otimes \downarrow y\big)(y_0) \\ &= \bigvee\{f(x) \mid y_0 \le h(x)\}. \end{aligned}$$

The description of the unit η_X^{P} follows immediately from (3.9) and (3.10). In order to confirm the formula for the multiplication μ^{P}, we recall that

$$\mathbb{P}(\mathbb{P}(X)) \xrightarrow{\mu_X^{\mathsf{P}}} \mathbb{P}(X)$$

is the extension of the identity $1_{\mathbb{P}(X)}$ of $\mathbb{P}(X)$. In fact, the extension

$$\mathsf{Dwn}(\mathbb{P}(X)) \xrightarrow{1^\sharp_{\mathbb{P}(X)}} \mathbb{P}(X)$$

of $1_{\mathbb{P}(X)}$ to $\mathsf{Dwn}(\mathbb{P}(X))$ is given by the formation of arbitrary joins — i.e.

$$1^\sharp_{\mathbb{P}(X)}(A) = \bigvee A, \qquad A \in \mathsf{Dwn}(\mathbb{P}(X)).$$

Then the extension of the join-preserving map $1^\sharp_{\mathbb{P}(X)}$ to a left $\mathfrak{Q}$-module homomorphism $\Phi(\mathbb{P}(\mathbb{P}(X))) = \mathfrak{Q} \otimes \mathsf{Dwn}(\mathbb{P}(X)) \xrightarrow{(1^\sharp_{\mathbb{P}(X)})^\sharp} \mathbb{P}(X)$ is given in part (b) of the proof of Theorem 3.1.11, where we use the left action on $\mathbb{P}(X)$ (cf. Example 3.1.8). Hence we obtain the following relation for $F \in \mathbb{P}(\mathbb{P}(X))$ and $x \in X$:

$$\begin{aligned} \big(\mu_X^{\mathsf{P}}(F)\big)(x) &= \big((1^\sharp_{\mathbb{P}(X)})^\sharp(\Phi(F))\big)(x) = \big((1^\sharp_{\mathbb{P}(X)})^\sharp\big(\bigvee_{f \in \mathbb{P}(X)} F(f) \otimes \downarrow f\big)\big)(x) \\ &= \bigvee_{f \in \mathbb{P}(X)} \big(F(f) \odot 1^\sharp_{\mathbb{P}(X)}(\downarrow f)\big)(x) = \bigvee_{f \in \mathbb{P}(X)} F(f) * f(x). \end{aligned}$$

Theorem 3.1.14. *The Eilenberg–Moore category* $\mathtt{Preord}^{\mathbf{P}}$ *of the monad* $\mathbf{P}$ *of* $\mathfrak{Q}$*-enriched presheaves on* $\mathtt{Preord}$ *is isomorphic to* $\mathtt{Mod}_\ell(\mathfrak{Q})$.

Proof. (a) Let $(X, \odot)$ be a left $\mathfrak{Q}$-module. Then we define a join-preserving map $\mathbb{P}(X) \xrightarrow{\xi} X$ by

$$\xi(f) = \bigvee_{x \in X} f(x) \odot x, \qquad f \in \mathbb{P}(X) \tag{3.14}$$

and show that (X, ξ) is a $\mathbf{P}$-algebra. The unit axiom $\xi \circ \eta_X^{\mathsf{P}} = 1_X$ follows immediately from $(\mathrm{M2}_\ell)$. Now we apply $(\mathrm{M1}_\ell)$ and notice that for all $F \in \mathbb{P}(\mathbb{P}(X))$ the following relation holds:

$$\begin{aligned}
\big(\xi \circ \mathsf{P}(\xi)\big)(F) &= \bigvee_{x \in X} \big(\mathsf{P}(\xi)(F)\big)(x) \odot x \\
&= \bigvee_{x \in X} \big(\bigvee\{F(f) \mid f \in \mathbb{P}(X),\ x \le \xi(f)\}\big) \odot x \\
&= \bigvee_{f \in \mathbb{P}(X)} F(f) \odot \xi(f) \\
&= \bigvee_{x \in X,\, f \in \mathbb{P}(X)} F(f) \odot (f(x) \odot x) \\
&= \bigvee_{x \in X,\, f \in \mathbb{P}(X)} (F(f) * f(x)) \odot x \\
&= (\xi \circ \mu_X^{\mathsf{P}})(f).
\end{aligned}$$

Hence the associativity axiom also holds. So (X, ξ) is a $\mathbf{P}$-algebra.

Further, let $(X, \odot)$ and $(Y, \odot)$ be left $\mathfrak{Q}$-modules with the corresponding $\mathbf{P}$-algebras (X, ξ) and (Y, ζ), and let $X \xrightarrow{h} Y$ be a left $\mathfrak{Q}$-module homomorphism. Then

$$(h \circ \xi)(f) = \bigvee_{x \in X} f(x) \odot h(x) = \bigvee_{y \in Y} \big(\bigvee\{f(x) \mid x \in X,\ y \le h(x)\}\big) \odot y = (\zeta \circ \mathsf{P}(h))(f)$$

for each $f \in \mathbb{P}(X)$. Thus h is a $\mathbf{P}$-homomorphism.

(b) Let (X, ξ) be a $\mathbf{P}$-algebra. Then $(X, \le)$ is a preordered set, and $\mathbb{P}(X) \xrightarrow{\xi} X$ is an isotone map. Since $\xi \circ \eta_X^{\mathsf{P}} = 1_X$, the preordered set $(X, \le)$ is antisymmetric and hence a partially ordered set. Further, let $\mathsf{Dwn}(X) \xrightarrow{\zeta} X$ be the restriction of ξ to $\mathsf{Dwn}(X)$ — i.e. $\zeta(A) = \xi(1_A)$, where $1_A = \Phi^{-1}(e \otimes A)$. Explicitly 1_A is given by:

$$1_A(x) = \begin{cases} e, & x \in A, \\ \bot, & x \notin A, \end{cases} \qquad x \in X,\ A \in \mathsf{Dwn}(X).$$

Since $\zeta(\downarrow x) = x$ for all $x \in X$, we conclude from Lemma 1.3.3 and Theorem 1.3.4 that the partially ordered set $(X, \le)$ is a complete lattice. In particular, the join of a subset A of X is given by $\zeta(\downarrow A)$. We show that ξ is join-preserving. For this purpose we choose a subset $\mathbb{B}$ of $\mathbb{P}(X)$ and define an element F of $\mathbb{P}(\mathbb{P}(X))$ by:

$$F(f) = \begin{cases} e, & f \in \downarrow\mathbb{B}, \\ \bot, & f \notin \downarrow\mathbb{B}, \end{cases} \qquad f \in \mathbb{P}(X).$$

Then we observe:

$$\xi(\bigvee \mathbb{B}) = \big(\xi \circ \mu_X^{\mathsf{P}}\big)(F) = \big(\xi \circ \mathsf{P}(\xi)\big)(F) = \xi(1_{\downarrow\{\xi(f) \mid f \in \mathbb{B}\}}) = \bigvee_{f \in \mathbb{B}} \xi(f).$$

Hence ξ is join-preserving.

After these preparations we define the structure of a left $\mathfrak{Q}$-module on X as follows:

$$\alpha \odot x = \xi(\alpha * 1_x), \qquad \alpha \in \mathfrak{Q},\ x \in X.$$

Since ξ is join-preserving, the operation $\odot$ is evidently join-preserving in its first variable. In order to show that $\odot$ is also join-preserving in its second variable, we choose $\alpha \in \mathfrak{Q}$ and $A \subseteq X$ and introduce the following antitone maps $\mathbb{P}(X) \xrightarrow{F_1, F_2} \mathfrak{Q}$:

$$F_1 = \alpha * 1_{1_{\downarrow A}} \quad \text{and} \quad F_2 = 1_{\downarrow\{\alpha * 1_x \mid x \in A\}}.$$

Explicitly F_1 and F_2 have the form:

$$F_1(f) = \begin{cases} \alpha, & f \leq 1_{\downarrow A}, \\ \bot, & f \nleq 1_{\downarrow A}, \end{cases} \quad \text{and} \quad F_2(f) = \begin{cases} e, & \exists x \in A : f \leq \alpha * 1_x, \\ \bot, & \forall x \in A : f \nleq \alpha * 1_x. \end{cases}$$

Then we observe $\mu_X^{\mathsf{P}}(F_1) = \alpha * 1_{\downarrow A} = \mu_X^{\mathsf{P}}(F_2)$. Further, we note:

$$\bigvee_{x \in A} \alpha \odot x = \xi\big(1_{\downarrow\{\alpha \odot x \mid x \in A\}}\big) = \big(\xi \circ \mathsf{P}(\xi)\big)(F_2),$$
$$\alpha \odot \big(\bigvee A\big) = \xi(\alpha * 1_{\bigvee A}) = \big(\xi \circ \mathsf{P}(\xi)\big)(F_1).$$

Hence $\bigvee_{x \in A} \alpha \odot x = \alpha \odot \big(\bigvee A\big)$ follows from the associativity axiom. To sum up, we have shown that $\odot$ is a bimorphism.

The axiom ($M2_\ell$) is evident. In order to verify ($M1_\ell$) we choose $\alpha, \beta \in \mathfrak{Q}$ and $x \in X$. Now we consider an antitone map $\mathbb{P}(X) \xrightarrow{F} \mathfrak{Q}$ defined by:

$$F(f) = \begin{cases} \beta, & f \leq \alpha * 1_x \\ \bot, & f \nleq \alpha * 1_x, \end{cases} \qquad f \in \mathbb{P}(X).$$

Since $\mu_X^{\mathsf{P}}(F) = (\beta * \alpha) * 1_x$, we conclude again from the associativity axiom that the relation

$$(\beta * \alpha) \odot x = \big(\xi \circ \mu_X^{\mathsf{P}}\big)(F) = \big(\xi \circ \mathsf{P}(\xi)\big)(F) = \xi(\beta * 1_{\alpha \odot x}) = \beta \odot (\alpha \odot x)$$

holds. Hence $(X, \odot)$ is a left $\mathfrak{Q}$-module.

Finally, let (X, ξ) and (Y, ζ) be **P**-algebras and $X \xrightarrow{h} Y$ be a **P**-homomorphism. Then it is easily seen that the following relations hold for all $\alpha \in \mathfrak{Q}$, $x \in X$ and $A \subseteq X$:

$$h(\bigvee A) = h(\xi(1_{\downarrow A})) = \big(\zeta \circ \mathsf{P}(h)\big)(1_{\downarrow A}) = \zeta(1_{\downarrow h(A)}) = \bigvee h(A),$$
$$h(\alpha \odot x) = (h \circ \xi)(\alpha * 1_x) = (\zeta \circ \mathsf{P}(h))(\alpha * 1_x) = \zeta(\alpha * 1_{h(x)}) = \alpha \odot h(x).$$

Hence h is a left $\mathfrak{Q}$-module homomorphism. □

Remark 3.1.15. By Theorem 3.1.14 the left $\mathfrak{Q}$-module $(\mathbb{P}(X), \odot)$ (cf. Example 3.1.8) is the free left $\mathfrak{Q}$-module generated by the preordered set X. If we identify arbitrary sets X with discrete ordered sets, then $\mathbb{P}(X)$ and $\mathfrak{Q}^X$ coincide. Therefore $\mathfrak{Q}^X$ is the free left $\mathfrak{Q}$-module generated by the underlying set X (cf. p. 10 in [59]). From the point of view of fuzzy set theory the free left $\mathfrak{Q}$-module $\mathfrak{Q}^X$ is understood as the algebra of $\mathfrak{Q}$-*fuzzy subsets* of X (cf. [39]). Hence, in general, fuzzy set theory is module theory in Sup. As a confirmation of this insight we point out the well-known fact that every left $\mathfrak{Q}$-module $(X, \odot)$ is a quotient (see Fact II infra) of the free left $\mathfrak{Q}$-module $\mathfrak{Q}^X$ (cf. [59] and Exercise 3.1.9).

3.1.1 *Linear Closure Operators and Colimits in* $\mathtt{Mod}_\ell(\mathfrak{Q})$

The construction of coproducts in $\mathtt{Mod}_\ell(\mathfrak{Q})$ is straightforward. Indeed, let $\mathbb{F} = \{(X_i, \odot_i) \mid i \in I\}$ be a family of left $\mathfrak{Q}$-modules. Then we provide $\prod_{i\in I} X_i$ with the following left action:

$$\alpha \odot ((x_i)_{i\in I}) = (\alpha \odot_i x_i)_{i\in I}, \qquad \alpha \in \mathfrak{Q},\ (x_i)_{i\in I} \in \prod_{i\in I} X_i,$$

and consider the coproduct injections $X_j \xrightarrow{q_j} \prod_{i\in I} X_i$ determined by:

$$q_j(x_j) = (z_i)_{i\in I}, \quad z_i = \begin{cases} x_j, & i = j, \\ \bot, & i \neq j, \end{cases} \qquad x_j \in X_j$$

where $\bot$ is the universal lower bound in X_i. Evidently, $(\prod_{i\in I} X_i, (q_i)_{i\in I})$ is the coproduct of $\mathbb{F}$ in $\mathtt{Mod}_\ell(\mathfrak{Q})$.

Since surjective left $\mathfrak{Q}$-module homomorphisms are always epimorphisms, the next proposition can be seen as a characterization of epimorphisms in $\mathtt{Mod}_\ell(\mathfrak{Q})$.

Proposition 3.1.16. *Let $X \xrightarrow{h} Y$ be an epimorphism in the sense of* $\mathtt{Mod}_\ell(\mathfrak{Q})$*. Then h, viewed as map, is surjective.*

Proof. (a) By Example 3.1.6 the left-implication on $\mathfrak{Q}$ can be considered as a left action $\odot$ on $\mathfrak{Q}^{op}$ determined by $\alpha \odot \beta = \beta \swarrow \alpha$. Further, each element $y \in Y$

induces a left $\mathfrak{Q}$-module homomorphism $Y \xrightarrow{h_y} \mathfrak{Q}^{op}$ as follows:

$$h_y(z) = \bigvee\{\alpha \in \mathfrak{Q} \mid \alpha \odot z \le y\}, \qquad z \in Y.$$

In fact, h_y is join-reversing, and the relation $h_y(z) \odot z \le y$ holds. Further, we observe:

$$(h_y(\alpha \odot z) * \alpha) \odot z = h_y(\alpha \odot z) \odot (\alpha \odot z) \le y.$$

Hence $h_y(\alpha \odot z) * \alpha \le h_y(z)$ follows — i.e. $h_y(\alpha \odot z) \le h_y(z) \swarrow \alpha$. On the other hand, we obtain:

$$(h_y(z) \swarrow \alpha) \odot (\alpha \odot z) = ((h_y(z) \swarrow \alpha) * \alpha) \odot z \le h_y(z) \odot z \le y.$$

This means $h_y(z) \swarrow \alpha \le h_y(\alpha \odot z)$. So, we have verified that h_y is a left $\mathfrak{Q}$-module homomorphism.

(b) Let $X \xrightarrow{h} Y$ be an epimorphism in $\mathtt{Mod}_\ell(\mathfrak{Q})$ and $h^\vdash$ be its right adjoint map. Then for all $\alpha \in \mathfrak{Q}$, $x \in X$ and $y \in Y$ we obtain the following chain of equivalences:

$$\begin{aligned}
\alpha \le (h_y \circ h)(x) &\iff h(\alpha \odot x) = \alpha \odot h(x) \le y \\
&\iff \alpha \odot x \le h^\vdash(y) \\
&\iff \alpha \odot x \le (h^\vdash \circ h \circ h^\vdash)(y) \\
&\iff \alpha \odot h(x) = h(\alpha \odot x) \le (h \circ h^\vdash)(y) \\
&\iff \alpha \le (h_{(h \circ h^\vdash)(y)} \circ h)(x).
\end{aligned}$$

Hence $h_y \circ h = h_{(h \circ h^\vdash)(y)} \circ h$ follows for all $y \in Y$. If h is an epimorphism, then h_y and $h_{(h \circ h^\vdash)(y)}$ necessarily coincide. In particular, we obtain $e \le h_y(y) = h_{(h \circ h^\vdash)(y)}(y)$ and $e \le h_{(h \circ h^\vdash)(y)}\big((h \circ h^\vdash)(y)\big) = h_y((h \circ h^\vdash)(y))$, which means $(h \circ h^\vdash)(y) \le y$ and $y \le (h \circ h^\vdash)(y)$. Hence a left $\mathfrak{Q}$-module homomorphism h is an epimorphism if and only if $h \circ h^\vdash = 1_Y$ if and only if h is surjective (cf. Exercise 1.3.2 (a)). □

The next property can be derived from general principles of category theory (cf. [47, Proposition 21.16]), but we give here a direct proof.

Proposition 3.1.17. *Let $(X, \odot) \xrightarrow{h} (Y, \odot)$ be an epimorphism in $\mathtt{Mod}_\ell(\mathfrak{Q})$. Further, let $(P, \odot) \overset{p}{\underset{q}{\rightrightarrows}} (X, \odot)$ be the kernel pair of h. Then the diagram*

$$\begin{array}{ccc}
P & \xrightarrow{q} & X \\
{\scriptstyle p}\downarrow & & \downarrow{\scriptstyle h} \\
X & \xrightarrow{h} & Y
\end{array}$$

is a pullback square and a pushout square — i.e. a pulation square.

Proof. By definition, the diagram is a pullback square. In order to verify that it is also a pushout square we choose a pair $(X, \odot) \overset{k}{\underset{l}{\rightrightarrows}} (Z, \odot)$ of left $\mathfrak{Q}$-module homomorphisms with $k \circ p = l \circ q$. Since the forgetful functor from $\mathtt{Mod}_\ell(\mathfrak{Q})$ to $\mathtt{Sup}$ preserves pullbacks, we conclude from Theorem 2.1.2 that there is a unique join-preserving map $Y \xrightarrow{\pi} Z$ such that $k = \pi \circ h = l$ holds. We have to show that π preserves the respective left actions. For this purpose we fix $\alpha \in \mathfrak{Q}$ and $y \in Y$. Since h is surjective (cf. Proposition 3.1.16), there exists an $x \in X$ with $h(x) = y$. Now we obtain:

$$\pi(\alpha \odot y) = \pi(\alpha \odot h(x)) = \pi\big(h(\alpha \odot x)\big) = k(\alpha \odot x) = \alpha \odot k(x) = \alpha \odot \pi(y).$$

Hence π is a left $\mathfrak{Q}$-module homomorphism. □

As an immediate corollary we obtain from Proposition 3.1.17 that every epimorphism in $\mathtt{Mod}_\ell$ is the coequalizer of its kernel pair. Hence we can summarize the previous results as follows.

FACT II. *Surjective left $\mathfrak{Q}$-module homomorphism, epimorphism and regular epimorphism are equivalent concepts in* $\mathtt{Mod}_\ell(\mathfrak{Q})$.

A closure operator c on a left $\mathfrak{Q}$-module $(X, \odot)$ is *linear* if c is compatible with the left action in the following sense:

$$\alpha \odot c(x) \le c(\alpha \odot x), \qquad \alpha \in \mathfrak{Q},\ x \in X. \tag{3.15}$$

Obviously, every linear closure operator c on a left $\mathfrak{Q}$-module $(X, \odot)$ determines a left action $\circledcirc$ on the quotient $(c(X), \star)$ of X w.r.t. c by

$$\alpha \circledcirc c(x) = c(\alpha \odot c(x)) = c(\alpha \odot x), \qquad \alpha \in \mathfrak{Q},\ x \in X$$

such that the quotient map $X \xrightarrow{\pi} c(X)$ is transmitted to a left $\mathfrak{Q}$-module homomorphism.

Proposition 3.1.18. *The coequalizer of every pair of parallel left $\mathfrak{Q}$-module homomorphisms exists in* $\mathtt{Mod}_\ell(\mathfrak{Q})$.

Proof. If we replace nuclei by linear closure operators, the proof of Theorem 2.2.6 can be repeated verbatim. □

Further, the associated closure operator $c_h = h^\vdash \circ h$ of every left $\mathfrak{Q}$-module homomorphism $X \xrightarrow{h} Y$ is linear. In fact, the following relation holds:

$$\begin{aligned}\alpha \odot c_h(x) \le c_h(\alpha \odot c_h(x)) &= h^\vdash(h(\alpha \odot (h^\vdash \circ h)(x))) = h^\vdash(\alpha \odot (h \circ h^\vdash \circ h)(x)) \\ &= h^\vdash(\alpha \odot h(x)) = (h^\vdash \circ h)(\alpha \odot x).\end{aligned}$$

Hence every left $\mathfrak{Q}$-module homomorphism $X \xrightarrow{h} Y$ has a decomposition

$$\begin{array}{ccc} X & \xrightarrow{\pi_h} & c_h(X) \\ & \searrow^{h} & \downarrow m_h \\ & & Y \end{array}$$

into an epimorphism π_h of $\mathtt{Mod}_\ell(\mathfrak{Q})$ determined by c_h and followed by a monomorphism m_h of $\mathtt{Mod}_\ell(\mathfrak{Q})$ defined on $c_h(X)$ by $m_h(x) = h(x)$ for all $x \in c_h(X)$. Since this decomposition is unique up to isomorphism, $\mathtt{Mod}_\ell(\mathfrak{Q})$ is an (epi, mono)-category. Hence the class of quotient objects of a left $\mathfrak{Q}$-module $(X, \odot)$ is a partially ordered set and is order isomorphic to the complete lattice of all linear closure operators on $(X, \odot)$.

To sum up, we record the following important fact.

FACT III. *The category* $\mathtt{Mod}_\ell(\mathfrak{Q})$ *of left* $\mathfrak{Q}$*-modules is a complete and cocomplete (epi, mono)-category. The same results hold for right* $\mathfrak{Q}$*-modules.*

Since the forgetful functor from $\mathtt{Sup}$ to $\mathtt{Set}$ has a left adjoint functor given by the power set functor, we derive the following important result from the previous Facts II and III and Proposition 3.1.18.

Corollary 3.1.19. *The category* $\mathtt{Mod}_\ell(\mathfrak{Q})$ *(resp.* $\mathtt{Mod}_r(\mathfrak{Q})$*) is algebraic.*

3.1.2 Annihilators and Generators in Left $\mathfrak{Q}$-Modules

Let $(X, \odot)$ be a left $\mathfrak{Q}$-module and x be an element of X. The *annihilator* $\operatorname{ann}(x)$ of x is an element of $\mathfrak{Q}$ defined by

$$\operatorname{ann}(x) = \bigvee\{\alpha \in \mathfrak{Q} \mid \alpha \odot x = \bot\}.$$

Obviously, any annihilator of some element of a left $\mathfrak{Q}$-module is left-sided.

An element $x \in X$ is said to be *invariant* if $\top \odot x \le x$ holds. Obviously the universal bounds of a left $\mathfrak{Q}$-module are invariant. Moreover, if c is a linear closure operator on $(X, \odot)$, then $c(\bot)$ is always invariant.

A left $\mathfrak{Q}$-module $(X, \odot)$ is called *irreducible* if the subset of all invariant elements of X coincides with $\{\bot, \top\}$. Every linear closure operator c on an irreducible left $\mathfrak{Q}$-module is either trivial (i.e. $c(\bot) = \top$) or satisfies the property $c(\bot) = \bot$.

Finally, an element x of a left $\mathfrak{Q}$-module $(X, \odot)$ is a *generator* of $(X, \odot)$ if the following relation holds

$$X = \{\alpha \odot x \mid \alpha \in \mathfrak{Q}\}.$$

Theorem 3.1.20. *Let* $(\mathfrak{Q}, *, e)$ *be a unital quantale and* x *be a generator of a left* $\mathfrak{Q}$*-module* $(X, \odot)$*. If* $\operatorname{ann}(x)$ *is a maximal left-sided element of* $\mathfrak{Q}$*, then* $(X, \odot)$ *is irreducible.*

Proof. Let us choose an invariant element $z \in X$ with $z \neq \bot$. Since x is a generator of $(X, \odot)$ there exists an $\alpha \in \mathfrak{Q}$ such that $z = \alpha \odot x$. Then the property $z \neq \bot$ implies $\top * \alpha \not\leq \mathrm{ann}\,(x)$. In fact, we have $(\top * \alpha) \odot x = \top \odot z \geq e \odot z = z \neq \bot$.

Since the annihilator of x is a maximal left-sided element of $\mathfrak{Q}$, it follows that $\top = (\top * \alpha) \vee \mathrm{ann}\,(x)$. Now, we use again the property that x is a generator of $(X, \odot)$ and obtain:

$$\top = \top \odot x = \big((\top * \alpha) \vee \mathrm{ann}\,(x)\big) \odot x = (\top * \alpha) \odot x = \top \odot z.$$

Hence an invariant element $z \in X$ with $z \neq \bot$ necessarily coincides with $\top$. $\square$

The aim of the following considerations is to explain the rôle of annihilators in some more detail. For this purpose we first need an example.

Example 3.1.21. Let $(X, \odot)$ be a left $\mathfrak{Q}$-module and x be an invariant element of X. Then x induces a linear closure operator c_x with $c_x(z) = z \vee x$ for all $z \in X$, and the quotient of $(X, \odot)$ w.r.t. c_x coincides with the upsegment $\uparrow x$ equipped with the following left action:

$$\alpha \odot_x z = (\alpha \odot z) \vee x, \qquad \alpha \in \mathfrak{Q},\ z \in \uparrow x. \tag{3.16}$$

Obviously, the corresponding quotient map $X \xrightarrow{\pi_x} \uparrow x$ has the form $\pi_x(z) = z \vee x$ for all $z \in X$.

Further, we recall that the underlying unital quantale can always be considered as a left $\mathfrak{Q}$-module where the left action is determined by the multiplication in $\mathfrak{Q}$ (cf. Example 3.1.6). Then we can make the simple observation that every element x of a left $\mathfrak{Q}$-module $(X, \odot)$ induces a left $\mathfrak{Q}$-module homomorphisms $\mathfrak{Q} \xrightarrow{h_x} X$ by $h_x(\alpha) = \alpha \odot x$. Further, let $\mathrm{ann}\,(x)$ be the annihilator of x and $\mathfrak{Q} \xrightarrow{\pi_x} \uparrow \mathrm{ann}\,(x)$ be the quotient map constructed in Example 3.1.21 — i.e. by abuse of notation π_x has the form $\pi_x(\alpha) = \alpha \vee \mathrm{ann}\,(x)$. Since $h_x(\alpha \vee \mathrm{ann}\,(x)) = \alpha \odot x = h_x(\alpha)$, it is easily seen that h_x factors though π_x — i.e. there exists a unique left $\mathfrak{Q}$-module homomorphism $\uparrow \mathrm{ann}\,(x) \xrightarrow{k_x} X$ making the following diagram commutative:

$$\begin{array}{ccc} \mathfrak{Q} & \xrightarrow{\ \pi_x\ } & \uparrow \mathrm{ann}\,(x) \\ & {}_{h_x}\searrow & \Big\downarrow{\scriptstyle k_x} \\ & & X \end{array} \tag{3.17}$$

Finally, we determine the image of h_x. For this purpose we construct the closure operator c_{h_x} corresponding to h_x. Since the right adjoint map $h_x^{\vdash}$ of h_x has the form

$$h_x^{\vdash}(y) = y \swarrow x = \bigvee\{\alpha \in \mathfrak{Q} \mid \alpha \odot x \leq y\}, \qquad y \in X,$$

the image of h_x (i.e. the quotient of $\mathfrak{Q}$ w.r.t. c_{h_x}) is the left $\mathfrak{Q}$-module $(Z, \widehat{\odot})$, where the underlying set

$$Z = \{(\alpha \odot x) \swarrow x \mid \alpha \in \Omega\}$$

is provided with the partial order inherited by Ω and the left action $\widehat{\odot}$ is given by:

$$\beta \,\widehat{\odot}\, \big((\alpha \odot x) \swarrow x\big) = \big((\beta * \alpha) \odot x\big) \swarrow x, \qquad \beta \in \Omega.$$

By diagram (3.17) the left Ω-module $(Z, \widehat{\odot})$ is a quotient of the upsegment $\uparrow \text{ann}(x)$. If in addition x is a generator of $(X, \odot)$ — this means that h_x is an epimorphism (cf. Proposition 3.1.16), then $(Z, \widehat{\odot})$ is isomorphic to $(X, \odot)$.

3.1.3 Modules on Involutive and Unital Quantales

In this subsection we always assume that Ω is an involutive and unital quantale $(\Omega, *, e, ')$. If X is a complete lattice, then every left action $\odot$ on X induces a right action $\boxdot$ on X by:

$$x \boxdot \alpha = \alpha' \odot x, \quad \alpha \in \Omega,\ x \in X. \tag{3.18}$$

Analogously, every right action $\boxdot$ on X also induces a left action on X by:

$$\alpha \odot x = x \boxdot \alpha', \quad \alpha \in \Omega,\ x \in X. \tag{3.19}$$

Obviously, these observations give rise to isomorphisms $\mathtt{Mod}_\ell(\Omega) \xrightarrow{\mathsf{T}_\ell} \mathtt{Mod}_r(\Omega)$ and $\mathtt{Mod}_r(\Omega) \xrightarrow{\mathsf{T}_r} \mathtt{Mod}_\ell(\Omega)$ which leave morphisms invariant, but act on objects as follows:

$$\mathsf{T}_\ell(X, \odot) = (X, \boxdot) \quad \text{where } \boxdot \text{ is determined by (3.18)},$$
$$\mathsf{T}_r(X, \boxdot) = (X, \odot) \quad \text{where } \odot \text{ is determined by (3.19)}.$$

Hence we conclude from the previous Fact I that $\mathtt{Mod}_\ell(\Omega)$ and $\mathtt{Mod}_r(\Omega)$ have self-dualities given by $\mathsf{S}_\ell = \mathsf{F}\mathsf{T}_\ell$ and $\mathsf{S}_r = \mathsf{F}^{-1}\mathsf{T}_r$. In particular, S_ℓ and S_r have the following explicit form:

$$\mathsf{S}_\ell(X, \odot) = (X^{op}, \odot^{op}) \quad \text{where} \quad \alpha \odot^{op} x = \bigvee\{z \in X \mid \alpha' \odot z \le x\}, \tag{3.20}$$
$$\mathsf{S}_r(X, \boxdot) = (X^{op}, \boxdot^{op}) \quad \text{where} \quad x \boxdot^{op} \alpha = \bigvee\{z \in X \mid z \boxdot \alpha' \le x\}, \tag{3.21}$$

$$X \xrightarrow{h} Y, \quad Y^{op} \xrightarrow{h^{\vdash}} X^{op}, \quad \mathsf{S}_\ell(h) = h^{\vdash}, \quad \mathsf{S}_r(h) = h^{\vdash}.$$

The next example is a first simple application of the formulas (3.20) and (3.21).

Example 3.1.22. Let $(\Omega, *, e, ')$ be an involutive and unital quantale. Then the involution and the left- and right-implications of Ω induce the structure of a Ω-bimodule on Ω^{op} by:

$$\alpha \odot^{op} \beta = \alpha' \searrow \beta \quad \text{and} \quad \beta \boxdot^{op} \alpha = \beta \swarrow \alpha', \qquad \alpha \in \mathfrak{Q},\ \beta \in \mathfrak{Q}^{op}.$$

Referring to Example 3.1.6 it is easily seen that the involution $'$ is a $\mathfrak{Q}$-bimodule isomorphism $(\mathfrak{Q}^{op}, \circledcirc, \boxed{\boxdot}) \xrightarrow{\,'\,} (\mathfrak{Q}^{op}, \odot^{op}, \boxdot^{op})$. In fact, because of (2.35) the following relations hold:

$$(\alpha \circledcirc \beta)' = (\beta \swarrow \alpha)' = \alpha' \searrow \beta' = \alpha \odot^{op} \beta',$$
$$(\beta \boxed{\boxdot} \alpha)' = (\alpha \searrow \beta)' = \beta' \swarrow \alpha' = \beta' \boxdot^{op} \alpha.$$

The next example reveals that $\mathfrak{Q}$-bimodules determine contravariant endofunctors of $\mathtt{Mod}_\ell(\mathfrak{Q})$ (resp. $\mathtt{Mod}_r(\mathfrak{Q})$).

Example 3.1.23. Let $(\mathfrak{Q}, *, e, ')$ be an involutive and unital quantale. Further, let $(X, \odot)$ be a left $\mathfrak{Q}$-module, and let $(Y, \odot, \boxdot)$ be a $\mathfrak{Q}$-bimodule. Then the set $[X, Y]$ of all left $\mathfrak{Q}$-module homomorphisms $X \xrightarrow{f} Y$ is a complete lattice with respect to the pointwisely defined order. Since Y is a $\mathfrak{Q}$-bimodule, we can introduce the left action $\odot$ on $[X, Y]$ as follows:

$$(\alpha \odot f)(x) = f(x) \boxdot \alpha', \qquad \alpha \in \mathfrak{Q},\ f \in [X, Y],\ x \in X. \tag{3.22}$$

In fact, if f is a left $\mathfrak{Q}$-module homomorphism, the associativity axiom (3.6) implies:

$$(\alpha \odot f)(\beta \odot x) = \big(f(\beta \odot x)\big) \boxdot \alpha' = \big(\beta \odot f(x)\big) \boxdot \alpha'$$
$$= \beta \odot \big(f(x) \boxdot \alpha'\big) = \beta \odot (\alpha \odot f)(x)$$

for each $\beta \in \mathfrak{Q}$ and $x \in X$. Hence $\alpha \odot f$ is again a left $\mathfrak{Q}$-module homomorphism. The proof of the left action axioms ($M1_\ell$) and ($M2_\ell$) is obvious. Thus $([X, Y], \odot)$ is a left $\mathfrak{Q}$-module.

Finally, on this basis it is easily seen that every $\mathfrak{Q}$-bimodule $(Y, \odot, \boxdot)$ induces a contravariant endofunctor G_Y of $\mathtt{Mod}_\ell(\mathfrak{Q})$ as follows:

$$\mathsf{G}_Y(X) = [X, Y],$$
$$(X_2, \odot) \xrightarrow{h} (X_1, \odot), \quad [X_1, Y] \xrightarrow{\mathsf{G}_Y(h)} [X_2, Y], \quad \mathsf{G}_Y(h)(f) = f \circ h, \quad f \in [X_1, Y].$$

In the following considerations we will show that the $\mathfrak{Q}$-bimodule $\mathfrak{Q}^{op}$ constructed in Example 3.1.22 is a dualizer in $\mathtt{Mod}_\ell(\mathfrak{Q})$.

Lemma 3.1.24. *Let $(X, \odot)$ be a left $\mathfrak{Q}$-module. Then for every left $\mathfrak{Q}$-module homomorphism $X \xrightarrow{h} \mathfrak{Q}^{op}$ there exists a unique element $z_h \in X$ such that the following relation holds:*

$$h(x)' = \bigvee\{\alpha \in \mathfrak{Q} \mid \alpha \odot x \le z_h\}, \qquad x \in X, \tag{3.23}$$

where $'$ is the involution in $\mathfrak{Q}$.

Proof. (a) (Uniqueness) If z_1 and z_2 are elements of X such that the relation (3.23) holds, then we obtain $e \le h(z_1)'$ and $e \le h(z_2)'$, where e is the unit of $\mathfrak{Q}$. Hence the relation $z_1 = e \odot z_1 \le h(z_1)' \odot z_1 \le z_2$ follows. Interchanging the rôle of z_1 and z_2 we also have $z_2 \le z_1$. Thus z_1 and z_2 coincide.

(b) (Existence) For every left $\mathfrak{Q}$-module homomorphism $X \xrightarrow{h} \mathfrak{Q}^{op}$ we define an element $z_h \in X$ by

$$z_h = \bigvee \{y \in X \mid e \le h(y)\}.$$

Since h, viewed as map from X to $\mathfrak{Q}$, is join-reversing, $e \le h(z_h)$ holds. Now we choose $x \in X$ and put $\beta_x = \bigvee\big\{\alpha \in \mathfrak{Q} \mid \alpha \odot x \le z_h\big\}$. Since $\beta_x \odot x \le z_h$, we obtain $e \le h(z_h) \le h(\beta_x \odot x)$. Since $X \xrightarrow{h} \mathfrak{Q}^{op}$ is a left $\mathfrak{Q}$-module homomorphism, $e \le \beta_x \odot^{op} h(x) = \beta_x' \searrow h(x)$ follows — this means $\beta_x \le h(x)'$.

On the other hand, we observe $e \le h(x) \searrow h(x) = h(x)' \odot^{op} h(x) = h\big(h(x)' \odot x\big)$. Then the definition of z_h implies $h(x)' \odot x \le z_h$. Hence $h(x)' \le \beta_x$ holds. To sum up, we have shown $h(x)' = \beta_x$ for all $x \in X$. This means that z_h satisfies the condition (3.23). □

Lemma 3.1.25. *Let $(X, \odot)$ be a left $\mathfrak{Q}$-module. Then every element $z \in X$ induces a left $\mathfrak{Q}$-module homomorphism $X \xrightarrow{h_z} \mathfrak{Q}^{op}$ by*

$$h_z(x) = \bigvee \{\alpha \in \mathfrak{Q} \mid \alpha' \odot x \le z\}, \qquad x \in X. \tag{3.24}$$

Proof. Evidently, $X \xrightarrow{h_z} \mathfrak{Q}$ is join-reversing — i.e. $X \xrightarrow{h_z} \mathfrak{Q}^{op}$ is join-preserving. In particular, $h_z(x)' \odot x \le z$ for all $x \in X$. Now we fix $\alpha \in \mathfrak{Q}$ and observe:

$$\begin{aligned}(\alpha \odot^{op} h_z(x))' \odot (\alpha \odot x) &= (\alpha' \searrow h_z(x))' \odot (\alpha \odot x) = (h_z(x)' \swarrow \alpha) \odot (\alpha \odot x)\\ &= ((h_z(x)' \swarrow \alpha) * \alpha) \odot x \le h_z(x)' \odot x \le z.\end{aligned}$$

Hence $\alpha \odot^{op} h_z(x) \le h_z(\alpha \odot x)$ follows. On the other hand, the relation

$$(h_z(\alpha \odot x)' * \alpha) \odot x = h_z(\alpha \odot x)' \odot (\alpha \odot x) \le z$$

implies $(h_z(\alpha \odot x)' * \alpha)' \le h_z(x)$ — i.e. $h_z(\alpha \odot x)' * \alpha \le h_z(x)'$. Therefore we obtain $h_z(\alpha \odot x)' \le h_z(x)' \swarrow \alpha$ — i.e.

$$h_z(\alpha \odot x) \le (h_z(x)' \swarrow \alpha)' = \alpha' \searrow h_z(x) = \alpha \odot^{op} h_z(x).$$

Hence h_z is a left $\mathfrak{Q}$-module homomorphism. □

Theorem 3.1.26. *Let S_ℓ be the self-duality of $\mathtt{Mod}_\ell(\mathfrak{Q})$ and $\mathsf{G}_{\mathfrak{Q}^{op}}$ be the contravariant endofunctor induced by the $\mathfrak{Q}$-bimodule $\mathfrak{Q}^{op}$. The correspondence $z \longmapsto h_z$ determined by (3.24) is a natural isomorphism $\Gamma = (\Gamma_X)_X \colon \mathsf{S}_\ell \to \mathsf{G}_{\mathfrak{Q}^{op}}$.*

Proof. (a) Let $(X, \odot)$ be a left $\mathfrak{Q}$-module. With regard to Lemma 3.1.24 and Lemma 3.1.25 we know already that the map $X^{op} \xrightarrow{\Theta_X} [X, \mathfrak{Q}^{op}]$ defined by

$$\Theta_X(z) = h_z, \quad h_z(x) = \bigvee\{\alpha \in \mathfrak{Q} \mid \alpha' \odot x \leq z\}, \qquad x, z \in X$$

is bijective. Now we show that Θ_X is a left $\mathfrak{Q}$-module homomorphism. Obviously, Θ_X is join-preserving, which means in this context:

$$h_{\bigwedge Z}(x) = \bigvee\{\alpha \in \mathfrak{Q} \mid \alpha \odot x \leq \bigwedge Z\} = \bigwedge_{z \in Z} h_z(x), \qquad Z \subseteq X.$$

Now let $\odot^{op}$ be the left action on X^{op} (cf. (3.20)). Then for $\alpha \in \mathfrak{Q}$ we obtain:

$$\alpha' \odot (h_{\alpha \odot^{op} z}(x)' \odot x) \leq \alpha' \odot (\alpha \odot^{op} z) \leq z, \qquad x \in X.$$

Hence $\alpha' * h_{\alpha \odot^{op} z}(x)' \leq h_z(x)'$ follows — i.e.

$$h_{\alpha \odot^{op} z}(x) \leq (\alpha' \searrow h_z(x)')' = h_z(x) \swarrow \alpha = (\alpha \odot h_z)(x), \qquad x \in X$$

(cf. Example 3.1.22 and (3.22) in Example 3.1.23). On the other hand we observe:

$$\alpha' \odot ((\alpha' \searrow h_z(x)') \odot x) = (\alpha' * (\alpha' \searrow h_z(x)')) \odot x \leq h_z(x)' \odot x \leq z.$$

Then $(\alpha' \searrow h_z(x)') \odot x \leq \alpha \odot^{op} z$ holds, and the following relation follows:

$$(\alpha \odot h_z)(x) = h_z(x) \swarrow \alpha = (\alpha' \searrow h_z(x)')' \leq h_{\alpha \odot^{op} z}(x).$$

Thus we have shown $\alpha \odot h_z = h_{\alpha \odot^{op} z}$ — i.e. Θ_X is a left $\mathfrak{Q}$-module homomorphism.

(b) In order to verify the naturality of $(\Gamma_X)_X$ we consider a further left $\mathfrak{Q}$-module $(Y, \odot)$ and a left $\mathfrak{Q}$-module homomorphism $X \xrightarrow{k} Y$. If $k^{\vdash}$ is the right adjoint map of k, then for all $\alpha \in \mathfrak{Q}$, $x \in X$ and $y \in Y$ the equivalence

$$\alpha' \odot k(x) = k(\alpha' \odot x) \leq y \iff \alpha' \odot x \leq k^{\vdash}(y)$$

holds. Hence we obtain $h_y(k(x)) = h_{k^{\vdash}(y)}(x)$ for all $x \in X$ and $y \in Y$ — this means the commutativity of the following diagram:

$$\begin{array}{ccc}
X^{op} & \xleftarrow{\quad k^{\vdash} \quad} & Y^{op} \\
{\scriptstyle \Theta_X}\downarrow & & \downarrow{\scriptstyle \Theta_Y} \\
[X, \mathfrak{Q}^{op}] & \xleftarrow{\mathsf{G}_{\mathfrak{Q}^{op}}(k)} & [Y, \mathfrak{Q}^{op}]
\end{array}$$

□

As an immediate corollary of Theorem 3.1.26 we record the following:

FACT IV. *In the case of involutive and unital quantales the $\mathfrak{Q}$-bimodule $\mathfrak{Q}^{op}$ defined in Example 3.1.22 is the dualizer in the category* $\mathtt{Mod}_\ell(\mathfrak{Q})$ *of left $\mathfrak{Q}$-modules. An analogous result holds for* $\mathtt{Mod}_r(\mathfrak{Q})$.

3.1.4 The Tensor Product in $\mathtt{Mod}(\mathfrak{Q})$

In this subsection $(\mathfrak{Q}, *, e)$ always denotes a *commutative* and *unital* quantale. Due to the commutativity of $\mathfrak{Q}$ left $\mathfrak{Q}$-modules can always be viewed as $\mathfrak{Q}$-bimodules. Thus we will only speak of $\mathfrak{Q}$-modules and $\mathfrak{Q}$-module homomorphisms and the corresponding category is denoted by $\mathtt{Mod}(\mathfrak{Q})$. In this context we will always apply actions as left actions. Moreover, the left- and right-implication of $\mathfrak{Q}$ coincide and are denoted by $\rightarrow$.

The aim of the following considerations is to show that, in the case of commutative and unital quantales, the tensor product of $\mathfrak{Q}$-modules $(X, \odot)$ and $(Y, \odot)$ exists and is given by the $\mathfrak{Q}$-module $[X, Y^{op}]^{op}$ of join-reversing $\mathfrak{Q}$-module homomorphisms up to an isomorphism. Referring to Example 3.1.23, we first recall the action $\odot$ on $[X, Y^{op}]$:

$$(\alpha \odot f)(x) = \alpha \odot^{op} f(x), \qquad \alpha \in \mathfrak{Q},\ f \in [X, Y^{op}],\ x \in X,$$

where $\odot^{op}$ denotes the action on Y^{op} and is given by (3.20). Now we apply (3.20) again and note that the action on $[X, Y^{op}]^{op}$ attains the following form:

$$\alpha \odot^{op} f = \bigwedge\{g \in [X, Y^{op}]^{op} \mid f \leq \alpha \odot g\}, \qquad \alpha \in \mathfrak{Q},\ f \in [X, Y^{op}]^{op}. \quad (3.25)$$

Further $\otimes$ always denotes the tensor product in $\mathtt{Sup}$. Since meets in $[X, Y^{op}]^{op}$ are computed pointwisely, $[X, Y^{op}]^{op}$ is obviously the quotient of $X \otimes Y$ w.r.t. the closure operator c_0 on $X \otimes Y$ defined by:

$$c_0(f) = \bigwedge\{g \in [X, Y^{op}]^{op} \mid f \leq g\}, \qquad f \in X \otimes Y. \quad (3.26)$$

We begin with a characterization of $\mathfrak{Q}$-module homomorphisms in $X \otimes Y$.

Lemma 3.1.27. *Let $(X, \odot)$ and $(Y, \odot)$ be $\mathfrak{Q}$-modules. Further, let f be a tensor of $X \otimes Y$. Then the following assertions are equivalent:*

(i) *$X \xrightarrow{f} Y^{op}$ is a $\mathfrak{Q}$-module homomorphism — i.e. $f \in [X, Y^{op}]^{op}$.*
(ii) *For all $x \in X$, $y \in Y$ and $\alpha \in \mathfrak{Q}$ the following equivalence holds:*

$$(\alpha \odot x) \otimes y \leq f \iff x \otimes (\alpha \odot y) \leq f.$$

Proof. (i) $\Longrightarrow$ (ii). Since $X \xrightarrow{f} Y^{op}$ is a $\mathfrak{Q}$-module homomorphism, the following chain of equivalences hold for all $x \in X$ and $y \in Y$:

$$\begin{aligned}(\alpha \odot x) \otimes y \leq f &\iff y \leq f(\alpha \odot x) = \alpha \odot^{op} f(x)\\ &\iff \alpha \odot y \leq f(x)\\ &\iff x \otimes (\alpha \odot y) \leq f.\end{aligned}$$

(ii) $\Longrightarrow$ (i). Let us fix $x \in X$ and $\alpha \in \mathfrak{Q}$. Because of (ii) we conclude from $(\alpha \odot x) \otimes f(\alpha \odot x) \leq f$ that $x \otimes \big(\alpha \odot f(\alpha \odot x)\big) \leq f$ holds. Consequently we obtain $\alpha \odot f(\alpha \odot x) \leq f(x)$. In particular, the relation $f(\alpha \odot x) \leq \alpha \odot^{op} f(x)$ follows.

On the other hand, it is easily seen that $x \otimes \big(\alpha \odot \big(\alpha \odot^{op} f(x)\big)\big) \leq x \otimes f(x) \leq f$ holds. Now we apply (ii) again and obtain $(\alpha \odot x) \otimes (\alpha \odot^{op} f(x)) \leq f$. In particular, $\alpha \odot^{op} f(x) \leq f(\alpha \odot x)$ follows. Hence we have verified $f(\alpha \odot x) = \alpha \odot^{op} f(x)$. □

Definition 3.1.28. Let $(X, \odot)$, $(Y, \odot)$ and $(Z, \odot)$ be $\mathfrak{Q}$-modules. A bimorphism $X \times Y \xrightarrow{b} Z$ in Sup is a *bimorphism* in $\mathtt{Mod}(\mathfrak{Q})$ if b satisfies the following property

$$b(\alpha \odot x, y) = \alpha \odot b(x, y) = b(x, \alpha \odot y), \qquad \alpha \in \mathfrak{Q},\ x \in X,\ y \in Y$$

— this means that b is a $\mathfrak{Q}$-module homomorphism in each variable separately.

A pair $\big(b_0, (M, \odot)\big)$ is called a *tensor product* of $(X, \odot)$ and $(Y, \odot)$ if $(M, \odot)$ is a $\mathfrak{Q}$-module and $X \times Y \xrightarrow{b_0} M$ is a bimorphism in $\mathtt{Mod}(\mathfrak{Q})$ such that the following universal property holds:

For every $\mathfrak{Q}$-module $(Z, \odot)$ and every bimorphism $X \times Y \xrightarrow{b} Z$ in $\mathtt{Mod}(\mathfrak{Q})$ there exists a unique $\mathfrak{Q}$-module homomorphism $M \xrightarrow{h_b} Z$ making the following diagram commutative:

$$\begin{array}{ccc} X \times Y & \xrightarrow{b_0} & M \\ {\scriptstyle b}\downarrow & \swarrow{\scriptstyle h_b} & \\ Z & & \end{array}$$

Comment. The tensor product of $\mathfrak{Q}$-modules is unique up to isomorphism. In this context b_0 is called the *universal bimorphism in* $\mathtt{Mod}(\mathfrak{Q})$.

In order to prove the existence of the tensor product of $\mathfrak{Q}$-modules we begin with an example showing that universal bimorphisms in Sup induce specific bimorphisms in $\mathtt{Mod}(\mathfrak{Q})$.

Example 3.1.29. Let $(X, \odot)$ and $(Y, \odot)$ be $\mathfrak{Q}$-modules. Since $\top = \alpha \odot^{op} \top$, the universal upper bound $\top \otimes \top$ is always a $\mathfrak{Q}$-module homomorphism. Further, let c_0 be the closure operator on $X \otimes Y$ determined by (3.26). Then for all $(x, y) \in X \times Y$ we define an element $x \otimes_{\mathfrak{Q}} y$ of $[X, Y^{op}]^{op}$ by:

$$x \otimes_{\mathfrak{Q}} y = c_0(x \otimes y) = \bigwedge\{f \in [X, Y^{op}]^{op} \mid x \otimes y \leq f\}. \tag{3.27}$$

Since $\otimes$ is a bimorphism in Sup, the correspondence $(x, y) \longmapsto x \otimes_{\mathfrak{Q}} y$ is also a bimorphism in Sup. We show that $(x, y) \longmapsto x \otimes_{\mathfrak{Q}} y$ is even a bimorphism in

$\texttt{Mod}(\mathfrak{Q})$. For this purpose we fix $(x, y) \in X \times Y$ and $\alpha \in \mathfrak{Q}$ and conclude from Lemma 3.1.27 and (3.25) that the following relation holds:

$$\begin{aligned}(\alpha \odot x) \otimes_{\mathfrak{Q}} y &= \bigwedge\{f \in [X, Y^{op}]^{op} \mid (\alpha \odot x) \otimes y \leq f\}\\ &= \bigwedge\{f \in [X, Y^{op}]^{op} \mid x \otimes (\alpha \odot y) \leq f\}\\ &= x \otimes_{\mathfrak{Q}} (\alpha \odot y)\\ &= \bigwedge\{f \in [X, Y^{op}]^{op} \mid \alpha \odot y \leq f(x)\}\\ &= \bigwedge\{f \in [X, Y^{op}]^{op} \mid y \leq \alpha \odot^{op} f(x)\}\\ &= \bigwedge\{f \in [X, Y^{op}]^{op} \mid x \otimes y \leq \alpha \odot f\}\\ &= \bigwedge\{f \in [X, Y^{op}]^{op} \mid x \otimes_{\mathfrak{Q}} y \leq \alpha \odot f\}\\ &= \alpha \odot^{op} (x \otimes_{\mathfrak{Q}} y).\end{aligned}$$

Hence the correspondence $(x, y) \longmapsto x \otimes_{\mathfrak{Q}} y$ is a bimorphism in $\texttt{Mod}(\mathfrak{Q})$. By abuse of notation we also denote this correspondence by $\otimes_{\mathfrak{Q}}$.

Theorem 3.1.30. *If $(X, \odot)$ and $(Y, \odot)$ are $\mathfrak{Q}$-modules, then $([X, Y^{op}]^{op}, \otimes_{\mathfrak{Q}})$ is the tensor product of $(X, \odot)$ and $(Y, \odot)$.*

Proof. Let $(Z, \odot)$ be a further $\mathfrak{Q}$-module and $X \times Y \xrightarrow{b} Z$ be a bimorphism in $\texttt{Mod}(\mathfrak{Q})$. Since the quotient map from $X \otimes Y$ to $[X, Y^{op}]^{op}$ is join-preserving, every element $f \in [X, Y^{op}]^{op}$ is a join of an appropriate family $\{x_i \otimes_{\mathfrak{Q}} y_i \mid i \in I\}$. Hence the uniqueness of the $\mathfrak{Q}$-module homomorphism $[X, Y^{op}]^{op} \xrightarrow{h_b} Z$ with the property $b(x, y) = h_b(x \otimes_{\mathfrak{Q}} y)$ is evident. In order to verify the existence of h_b we proceed as follows. Since b is also a bimorphism in $\texttt{Sup}$, there exists a unique join-preserving map $X \otimes Y \xrightarrow{k_b} Z$ such that $b(x, y) = k_b(x \otimes y)$ for all $(x, y) \in X \times Y$. Let c_{k_b} be the closure operator associated with k_b — i.e. $c_{k_b} = k_b^{\vdash} \circ k_b$. Since every tensor of $X \otimes Y$ is a join of elementary tensors, the closure operator c_{k_b} has the following form:

$$c_{k_b}(f) = \bigvee\{x \otimes y \mid b(x, y) \leq k_b(f)\}, \qquad f \in X \otimes Y.$$

We show that $c_{k_b}(f)$ is a $\mathfrak{Q}$-module homomorphism from $(X, \odot)$ to $(Y^{op}, \odot^{op})$ for all $f \in X \otimes Y$. In fact, the following chain of equivalences holds for $\alpha \in \mathfrak{Q}$, $x \in X$ and $y \in Y$:

$$\begin{aligned}(\alpha \odot x) \otimes y \leq c_{k_b}(f) &\iff k_b\big((\alpha \odot x) \otimes y\big) \leq k_b(f)\\ &\iff b(\alpha \odot x, y) = b(x, \alpha \odot y) \leq k_b(f)\\ &\iff x \otimes (\alpha \odot y) \leq c_{k_b}(f).\end{aligned}$$

Thus we conclude from Lemma 3.1.27 that $c_{k_b}(f) \in [X, Y^{op}]^{op}$. Hence $c_0 \leq c_{k_b}$ follows. This means that k_b factors through the quotient map $X \otimes Y \to [X, Y^{op}]^{op}$ — i.e. there exists a (unique) join-preserving map $[X, Y^{op}]^{op} \xrightarrow{h_b} Z$ such that $k_b(f) = (h_b \circ c_0)(f)$ holds. In particular, the subsequent relation follows:

$$b(x,y) = k_b(x \otimes y) = h_b(c_{\mathfrak{Q}}(x \otimes y)) = h_b(x \otimes_{\mathfrak{Q}} y), \qquad x \in X,\ y \in Y.$$

In completing the proof we show that h_b is a $\mathfrak{Q}$-module homomorphism. Therefore we choose $\alpha \in \mathfrak{Q}$ and $f \in [X, Y^{op}]^{op}$ and consider a family $\{x_i \otimes_{\mathfrak{Q}} y_i \mid i \in I\}$ with the property $f = \bigvee_{i\in I} x_i \otimes_{\mathfrak{Q}} y_i$. Then we obtain:

$$\begin{aligned} h_b(\alpha \odot^{op} f) &= h_b\big(\bigvee_{i\in I} \alpha \odot^{op} (x_i \otimes_{\mathfrak{Q}} y_i)\big) \\ &= \bigvee_{i\in I} h_b\big((\alpha \odot x_i) \otimes_{\mathfrak{Q}} y_i\big) = \bigvee_{i\in I} b(\alpha \odot x_i, y_i) = \bigvee_{i\in I} \alpha \odot b(x_i, y_i) \\ &= \alpha \odot \big(\bigvee_{i\in I} b(x_i, y_i)\big) = \alpha \odot \big(\bigvee_{i\in I} h_b(x_i \otimes_{\mathfrak{Q}} y_i)\big) = \alpha \odot h_b(f). \end{aligned}$$

Hence h_b is in fact a $\mathfrak{Q}$-module homomorphism. □

Motivated by the previous theorem we introduce the following notation and terminology.

Notation. The $\mathfrak{Q}$-module $[X, Y^{op}]^{op}$ is denoted by $X \otimes_{\mathfrak{Q}} Y$ and is called the *tensor product* of the $\mathfrak{Q}$-modules $(X, \odot)$ and $(Y, \odot)$. In this context $X \times Y \xrightarrow{\otimes_{\mathfrak{Q}}} X \otimes_{\mathfrak{Q}} Y$ is called the universal bimorphism in $\mathsf{Mod}(\mathfrak{Q})$. Further, every element of $X \otimes_{\mathfrak{Q}} Y$ is said to be a *tensor*, and tensors of the form $x \otimes_{\mathfrak{Q}} y$ with $x \in X$ and $y \in Y$ are called *elementary tensors*. In this context we note the important fact that every tensor is a join of an appropriate family of elementary tensors.

In the following considerations we briefly describe algebraic properties of the tensor product in $\mathsf{Mod}(\mathfrak{Q})$.

By the universal property of the tensor product the object function $(X, Y) \longmapsto X \otimes_{\mathfrak{Q}} Y$ can be completed to a bifunctor, also denoted by $\otimes_{\mathfrak{Q}}$. In fact, for every pair of $\mathfrak{Q}$-module homomorphisms $X_1 \xrightarrow{h_1} Y_1$ and $X_2 \xrightarrow{h_2} Y_2$ there exists a unique $\mathfrak{Q}$-module homomorphism

$$X_1 \otimes_{\mathfrak{Q}} X_2 \xrightarrow{h_1 \otimes_{\mathfrak{Q}} h_2} Y_1 \otimes_{\mathfrak{Q}} Y_2$$

making the following diagram commutative:

$$\begin{array}{ccc} X_1 \times X_2 & \xrightarrow{\otimes_{\mathfrak{Q}}} & X_1 \otimes_{\mathfrak{Q}} X_2 \\ {\scriptstyle h_1 \times h_2}\big\downarrow & & \vdots\,{\scriptstyle h_1 \otimes_{\mathfrak{Q}} h_2} \\ Y_1 \times Y_2 & \xrightarrow{\otimes_{\mathfrak{Q}}} & Y_1 \otimes_{\mathfrak{Q}} Y_2 \end{array}$$

We show that for every $\mathfrak{Q}$-module $(X, \odot)$ the endofunctor $X \otimes_{\mathfrak{Q}} \_$ of $\mathsf{Mod}(\mathfrak{Q})$ has a right adjoint functor. For this purpose, we fix $(X, \odot)$ and introduce an endofunctor $\hom_X$ of $\mathsf{Mod}(\mathfrak{Q})$ as follows:

$$\hom_X(Y) = [X, Y],$$

$$Y_1 \xrightarrow{h} Y_2, \quad [X, Y_1] \xrightarrow{\hom_X(h)} [X, Y_2], \quad \hom_X(h)(f) = h \circ f, \quad f \in [X, Y_1].$$

Corollary 3.1.31. *For every $\mathfrak{Q}$-module $(X, \odot)$ the functor* $\hom_X$ *is right adjoint to the functor $X \otimes_{\mathfrak{Q}} \_$.*

Proof. Let $(X, \odot)$ be a fixed $\mathfrak{Q}$-module. Then for every further $\mathfrak{Q}$-module $(Y, \odot)$ there exist a $\mathfrak{Q}$-module homomorphism $Y \xrightarrow{\eta_Y} [X, X \otimes_{\mathfrak{Q}} Y]$ defined by:

$$\eta_Y(y)(x) = x \otimes_{\mathfrak{Q}} y$$

for all $x \in X$ and $y \in Y$. It is easily seen that $\eta = (\eta_Y)_Y$ is a natural transformation from $\mathrm{id}_{\mathtt{Mod}(\mathfrak{Q})}$ to $\hom_X \circ (X \otimes_{\mathfrak{Q}} \_)$. We show that η is the unit of the adjoint situation $(X \otimes_{\mathfrak{Q}} \_) \dashv \hom_X$. For this purpose we first notice that for every $\mathfrak{Q}$-module $(Z, \odot)$ every $\mathfrak{Q}$-module homomorphism $Y \xrightarrow{h} [X, Z]$ can be identified with a bimorphism $X \times Y \xrightarrow{b_h} Z$ in $\mathtt{Mod}(\mathfrak{Q})$ as follows:

$$b_h(x, y) = h(y)(x), \qquad x \in X, \ y \in Y.$$

By the universal property of the tensor product in $\mathtt{Mod}(\mathfrak{Q})$ (cf. Definition 3.1.28 and Theorem 3.1.30) the proof of Theorem 2.1.12 can be transferred to the setting of $\mathfrak{Q}$-modules. Hence the unique $\mathfrak{Q}$ homomorphism $X \otimes_{\mathfrak{Q}} Y \xrightarrow{\ulcorner h \urcorner} Z$ making the diagram

$$\begin{array}{ccc} Y & \xrightarrow{\eta_Y} & [X, X \otimes_{\mathfrak{Q}} Y] \\ {\scriptstyle h}\downarrow & \swarrow {\scriptstyle \hom_X(\ulcorner h \urcorner)} & \\ [X, Z] & & \end{array}$$

commutative is the unique extension of b_h to a $\mathfrak{Q}$-module homomorphism. □

Let $(X, \odot)$, $(Y, \odot)$ and $(Z, \odot)$ be $\mathfrak{Q}$-modules. For $x \in X, y \in Y, z \in Z$ and $\alpha \in \mathfrak{Q}$ it is easily seen that the following relation holds:

$$x \otimes_{\mathfrak{Q}} \big(y \otimes_{\mathfrak{Q}} (\alpha \odot z)\big) = x \otimes_{\mathfrak{Q}} \big(\alpha \odot^{op} (y \otimes_{\mathfrak{Q}} z)\big) = \alpha \odot^{op} \big(x \otimes_{\mathfrak{Q}} (y \otimes_{\mathfrak{Q}} z)\big).$$

Hence by analogy with Lemmas 2.1.13 and 2.1.15 we conclude from the universal property of the tensor product $\otimes_{\mathfrak{Q}}$ (cf. Definition 3.1.28 and Theorem 3.1.30) that the bifunctor $\otimes_{\mathfrak{Q}}$ is associative and symmetric. Finally, by Theorem 3.1.26 the $\mathfrak{Q}$-module $\mathfrak{Q}$ is obviously the unit object of $\otimes_{\mathfrak{Q}}$.

To sum up, we have established the following important

FACT V. *If $(\mathfrak{Q}, *, e)$ is a commutative and unital quantale, then the category $\mathtt{Mod}(\mathfrak{Q})$ is a symmetric and monoidal closed category provided with a self-duality turning $\mathtt{Mod}(\mathfrak{Q})$ into a star-autonomous category.*

By Fact V we can apply the constructions in Sect. 1.1 to $\mathtt{Mod}(\mathfrak{Q})$. Hence in the case of commutative and unital quantales we have (free) magmas in $\mathtt{Mod}(\mathfrak{Q})$ — these are (*free*) *algebras* — and semigroups in $\mathtt{Mod}(\mathfrak{Q})$ — these are *associative algebras*. Even though the study of these structures is beyond the scope of this book, we finish this subsection with an example related to the ideal theory of C^*-algebras.

Example 3.1.32. Let $(A, +, \cdot)$ be a unital C^*-algebra, $\mathbb{R}(A)$ be the quantale of all closed right ideals of A, and $\mathbb{I}(A)$ be the subquantale of all closed two-sided ideals of A (cf. Sect. 2.5.3). Then $\mathbb{R}(A)$ is obviously an $\mathbb{I}(A)$-module. In particular, the corresponding left action $\odot$ and right action $\boxdot$ are given by $I \odot R = R \boxdot I = R * I$, where $*$ denotes the ideal multiplication. Since the ideal multiplication of closed right (left) ideals is idempotent (cf. Corollary 2.4.5), $\mathbb{R}(A)$ is right-symmetric and $\mathbb{I}(A)$ is a frame (cf. Lemma 2.4.3). Thus the following relation holds:

$$(R_1 * R_2) \boxdot I = R_1 * (R_2 \boxdot I) = (R_1 \boxdot I) * R_2, \qquad I \in \mathbb{I}(A),\ R_1, R_2 \in \mathbb{R}(A).$$

This means that the ideal multiplication in $\mathbb{R}(A)$ is a bimorphism in the sense of the category $\mathtt{Mod}(\mathbb{I}(A))$. Hence there exists a unique extension of $*$ to the tensor product of $\mathbb{I}(A)$-modules making the following diagram commutative:

$$\begin{array}{ccc} \mathbb{R}(A) \times \mathbb{R}(A) & \xrightarrow{\ \otimes_{\mathbb{I}(A)}\ } & \mathbb{R}(A) \otimes_{\mathbb{I}(A)} \mathbb{R}(A) \\ & \searrow^{*} & \vdots\ \circledast \\ & & \mathbb{R}(A) \end{array}$$

We can summarize the previous observations in the following statement: The pair $(\mathbb{R}(A), \circledast)$ is an idempotent and in general not commutative semigroup in $\mathtt{Mod}(\mathbb{I}(A))$ — i.e. a non-commutative $\mathbb{I}(A)$-algebra in $\mathtt{Sup}$. For more details on right (left) $\mathfrak{Q}$-modules, we refer to Sects. 3.3.3 and 3.3.4.

Exercises

3.1.1. Let X be a complete lattice and $[X, X]$ be the unital quantale of all join-preserving self-maps of X provided with the multiplication given by the composition of maps. Show that the evaluation arrow induces the structure of a left $[X, X]$-module on X.
(Hint: Proposition 3.1.3.)

3.1.2. Let $C_3 = \{\bot, a, \top\}$ be the chain with three elements and $C_3^\ell = (C_3, *_l)$ (resp. $C_3^r = (C_3, *_r)$) be the non-commutative, idempotent and left-sided (resp. right-sided) quantale on C_3 (cf. Example 2.3.26). Then 1_{C_3} is an anti-isomorphism between C_3^ℓ and C_3^r — i.e. C_3^ℓ is the transposed quantale of C_3^r. Further, let $\widehat{C_3^\ell}$ be the unitalization of C_3^ℓ, $\widehat{C_3^r}$ be the unitalization of C_3^r (cf. Exercise 2.2.2), and let $([C_3, C_3], \circ, 1_{C_3})$ be the unital quantale of all join-preserving self-maps of C_3 (cf. Example 2.3.31).

(a) Show that there exist exactly six unital homomorphisms $\widehat{C_3^\ell} \xrightarrow{h} [C_3, C_3]$.

(b) Conclude from (a) that there exist exactly six right $\widehat{C_3^r}$-modules on C_3 (cf. Proposition 3.1.5). Describe explicitly the corresponding right actions on C_3 with respect to $\widehat{C_3^r}$.

3.1.3. Let $(X, \odot)$ be a left $\mathfrak{Q}$-module, and $X \xrightarrow{\iota} X$ be an order-preserving involution on X. Show that $\mathfrak{Q} \otimes X \xrightarrow{\tilde{\odot}} X$ given by

$$\alpha \mathbin{\tilde{\odot}} x = \iota(\alpha \odot \iota(x)), \quad \alpha \in \mathfrak{Q},\ x \in X$$

is again a left action on X.

3.1.4. Let $\mathfrak{Q}$ be a frame (i.e. $* = \wedge$) and $\alpha \longmapsto \alpha^{\perp}$ be an order-reversing involution on $\mathfrak{Q}$. Further, let $\mathfrak{Q} \otimes \mathfrak{Q}^{op} \xrightarrow{\odot} \mathfrak{Q}^{op}$ be a join-preserving map given by $\alpha \odot \beta = \alpha^{\perp} \vee \beta$ for all $\alpha, \beta \in \mathfrak{Q}$. Show that $(\mathfrak{Q}^{op}, \odot)$ is a left $\mathfrak{Q}$-module.

Comment. If $\mathfrak{Q} = [0, 1]$ and $\alpha^{\perp} = 1 - \alpha$, then the left action $\odot$ is also known as *Kleene–Dienes implication*.

3.1.5. Let $(\mathfrak{Q}, *, d, ')$ be an involutive Frobenius quantale. Further, let $(\mathfrak{Q}, \odot)$ and $(\mathfrak{Q}^{op}, \odot^{op})$ be the left $\mathfrak{Q}$-modules constructed in Examples 3.1.6 and 3.1.22, respectively. Show that the join-preserving map $\mathfrak{Q} \xrightarrow{\delta} \mathfrak{Q}^{op}$ given by $\delta(\beta) = \beta' \searrow d$ for each $\beta \in \mathfrak{Q}$ is a left $\mathfrak{Q}$-module isomorphism.

3.1.6. Let $(\mathfrak{Q}, *, e)$ be a unital quantale viewed as a left $\mathfrak{Q}$-module (cf. Example 3.1.6). Further, we choose a left-sided (i.e. invariant) element $z \in \mathfrak{Q}$ and consider the left $\mathfrak{Q}$-module $\uparrow z$ constructed in Example 3.1.21. Show that $x = z \vee e$ is a generator of $\uparrow z$ and its annihilator coincides with z.

3.1.7. Let $\mathbb{M}_2$ be the C^*-algebra of all square matrices of order 2 with real coefficients, and let $(\mathbb{MAX}, *, \langle e \rangle, ')$ be the involutive and unital quantale of all linear subspaces of $\mathbb{M}_2$ (cf. Example 2.5.5). We view $\mathbb{MAX}$ as a left $\mathbb{MAX}$-module (cf. Example 3.1.6) and consider a left ideal L and a linear subspace U of $\mathbb{M}_2$ defined as follows:

$$L = \left\{ \left(\begin{smallmatrix} a & 0 \\ b & 0 \end{smallmatrix}\right) \mid a, b \in \mathbb{R} \right\}, \qquad U = \left\{ \left(\begin{smallmatrix} a & 0 \\ b & c \end{smallmatrix}\right) \mid a, b, c \in \mathbb{R} \right\}.$$

Show that the following properties hold:

(a) If $W \in \uparrow L$ with $L \neq W \neq \mathbb{M}_2$, then there exists a unique real number $\alpha \in [0, \pi)$ such that $W = L + V_{\alpha}$, where the 1-dimensional linear subspace V_{α} of $\mathbb{M}_2$ is given by:

$$V_{\alpha} = \left\{ \lambda \cdot \left(\begin{smallmatrix} 0 & \cos\alpha \\ 0 & \sin\alpha \end{smallmatrix}\right) \mid \lambda \in \mathbb{R} \right\}.$$

In particular, $L \vee L^* = L + L^* = L + V_0$ and $U = L + V_{\frac{\pi}{2}}$.

(b) Let $\uparrow L$ be the left $\mathbb{MAX}$-module with the left action $\odot_L$ defined by (3.16) — i.e.

$$M \odot_L W = (M * W) \vee L, \qquad M \in \mathbb{MAX},\ W \in \uparrow L$$

(cf. Example 3.1.21). Show that the linear subspace U is a generator of $(\uparrow L, \odot_L)$ and its annihilator ann (U) coincides with L (cf. Exercise 3.1.6).

(c) Deduce from (b) that the left $\mathbb{MAX}$-module $(\uparrow L, \odot_L)$ is irreducible. (Hint: Exercise 2.4.4 (b).)

3.1.8. Let $(\mathfrak{Q}, *, e)$ be a commutative and unital quantale. Further, let $(X, \odot)$ and $(Y, \odot)$ be $\mathfrak{Q}$-modules. Show that every tensor $f \in X \otimes_{\mathfrak{Q}} Y$ satisfies the following property:

$$f(x) = \bigvee\{y \in Y \mid x \otimes_{\mathfrak{Q}} y \le f\}, \qquad x \in X.$$

(Hint: If $X \xrightarrow{f} Y^{op}$ is a left $\mathfrak{Q}$-module homomorphism, then by definition of the tensor product $\otimes_{\mathfrak{Q}}$ the equivalence $x \otimes y \le f \iff x \otimes_{\mathfrak{Q}} y \le f$ holds for all $x \in X$ and $y \in Y$.)

3.1.9. Let $(\mathfrak{Q}, *, e)$ be a unital quantale, $(X, \odot)$ be a left $\mathfrak{Q}$-module, and let $(X, \le)$ be the underlying partially ordered set of X. Show that $(X, \odot)$ is a quotient of the free left $\mathfrak{Q}$-module $(\mathbb{P}(X), \odot)$ generated by $(X, \le)$ in the sense of $\mathsf{Mod}_\ell(\mathfrak{Q})$ (cf. Remark 3.1.15). More precisely, show that the following diagram is commutative:

$$\begin{array}{ccccc} \mathfrak{Q}^X & \xrightarrow{\ \pi\ } & \mathbb{P}(X) & \xrightarrow{\ \xi\ } & X \\ & {\scriptstyle \eta_X} \nwarrow & \uparrow {\scriptstyle \eta_X^{\mathbb{P}}} & \nearrow {\scriptstyle 1_X} & \\ & & X & & \end{array}$$

where η_X and the epimorphisms π and ξ in $\mathsf{Mod}_\ell(\mathfrak{Q})$ are determined as follows:

$$\big(\eta_X(x)\big)(z) = \begin{cases} e, & z = x, \\ \bot, & z \ne x, \end{cases} \qquad x, z \in X,$$

$$\big(\pi(f)\big)(x) = \bigvee\{f(z) \mid x \le z\}, \qquad x \in X,\ f \in \mathfrak{Q}^X,$$

$$\xi(f) = \bigvee_{x \in X} f(x) \odot x, \qquad f \in \mathbb{P}(X).$$

3.1.10. (Cf. [86]) Show that in $\mathsf{Mod}_\ell(\mathfrak{Q})$ the amalgamation property holds — this means that for any monomorphism $(Y, \odot) \xrightarrow{m} (X, \odot)$ in $\mathsf{Mod}(\mathfrak{Q})$ the pushout square

$$\begin{array}{ccc} Y & \stackrel{m}{\rightarrowtail} & X \\ {\scriptstyle m} \downarrow & & \downarrow {\scriptstyle \varphi_2} \\ X & \xrightarrow{\ \varphi_1\ } & P \end{array}$$

is also a pullback square.

(Hint: First notice that Proposition 3.1.17 also holds for right $\mathfrak{Q}$-modules and subsequently apply Fact I in Sect. 3.1.)

3.2 Involutive Left $\mathfrak{Q}$-modules

3.2.1 2-Forms in Sup

Let $\otimes$ be the tensor product in Sup and $\mathbb{1}$ be the unit object in Sup (cf. Sect. 2.1.2). A *2-form* on a complete lattice X is a join-preserving map $X \otimes X \xrightarrow{\varphi} \mathbb{1}$. A 2-form φ on X is *symmetric* if

$$\varphi(x_1 \otimes x_2) = \varphi(x_2 \otimes x_1)$$

holds for each $x_1, x_2 \in X$, which can be expressed by the commutativity of the following diagram:

$$\begin{array}{ccc} X \otimes X & \xrightarrow{c_{XX}} & X \otimes X \\ & \searrow^{\varphi} & \downarrow \varphi \\ & & \mathbb{1} \end{array}$$

where c_{XX} is the XX-component of the symmetry in Sup (cf. Fact II in Sect. 2.1.2).

In order to give a characterization of 2-forms we recall that $\mathbb{1}^{op}$ is the dualizing object of Sup (cf. Lemma 2.1.14). Since $\mathbb{1}$ and $\mathbb{1}^{op}$ are isomorphic in Sup, for every complete lattice Z there exists a join-preserving bijective map $Z^{op} \xrightarrow{\vartheta_Z} [Z, \mathbb{1}]$ determined by:

$$\vartheta_Z(z) = h \iff h(u) = \begin{cases} 0, & u \le z, \\ 1, & u \nleq z, \end{cases} \qquad z \in Z.$$

In the case of $X \otimes X$ this means that for every 2-form $X \otimes X \xrightarrow{\varphi} \mathbb{1}$ there exists a unique join-reversing map $X \xrightarrow{f_\varphi} X$ (i.e. $f_\varphi \in [X, X^{op}]$) such that the following relation holds for all $x_1, x_2 \in X$

$$x_2 \le f_\varphi(x_1) \iff x_1 \otimes x_2 \le f_\varphi \iff \varphi(x_1 \otimes x_2) = 0. \tag{3.28}$$

A 2-form φ is symmetric if and only if the pair (f_φ, f_φ) is a contravariant Galois connection — i.e. $x_1 \le f_\varphi(x_2) \iff x_2 \le f_\varphi(x_1)$ for all $x_1, x_2 \in X$.

Definition 3.2.1. A pair (X, φ_X) is called a *symmetric lattice* if X is a complete lattice and φ_X is a symmetric 2-form on X. Now let (X, φ_X) and (Y, φ_Y) be symmetric lattices. A join-preserving map $X \xrightarrow{h} Y$ is said to be *orthogonal* if

$$\varphi_Y(h(x_1) \otimes h(x_2)) = \varphi_X(x_1 \otimes x_2)$$

holds for each $x_1, x_2 \in X$, which can be expressed by the commutativity of the following diagram:

$$\begin{array}{ccc} X \otimes X & \xrightarrow{h \otimes h} & Y \otimes Y \\ & \searrow^{\varphi_X} & \downarrow \varphi_Y \\ & & \mathbb{1} \end{array}$$

Symmetric lattices and orthogonal and join-preserving maps constitute a category which we denote by $\mathbb{O}(\texttt{Sup})$.

Before we proceed, we first give a characterization of orthogonal maps in terms of the category Sup.

Lemma 3.2.2. *Let (X, φ_X) and (Y, φ_Y) be symmetric lattices with the corresponding join-reversing maps $X \xrightarrow{f_{\varphi_X}} X$ and $Y \xrightarrow{f_{\varphi_Y}} Y$ in the sense of* (3.28). *Further, let $X \xrightarrow{h} Y$ be a join-preserving map and $h^{\vdash}$ be its right adjoint map. Then the following assertions are equivalent*:

(i) *The map h is orthogonal.*
(ii) *The following diagram is commutative*:

$$\begin{array}{ccc} X & \xrightarrow{h} & Y \\ f_{\varphi_X} \downarrow & & \downarrow f_{\varphi_Y} \\ X^{op} & \xleftarrow{h^{\vdash}} & Y^{op} \end{array}$$

Proof. It follows from (3.28) that a join-preserving map $X \xrightarrow{h} Y$ is orthogonal if and only if the equivalence

$$x_2 \leq f_{\varphi_X}(x_1) \iff h(x_2) \leq f_{\varphi_Y}(h(x_1))$$

holds for all $x_1, x_2 \in X$. If we now apply the right adjoint map $h^{\vdash}$ of h to the previous relation, then we obtain for all $x_1, x_2 \in X$:

$$x_2 \leq f_{\varphi_X}(x_1) \iff x_2 \leq (h^{\vdash} \circ f_{\varphi_Y} \circ h)(x_1).$$

Hence the assertion follows. □

In order to construct the tensor product of 2-forms, we need the isomorphism $\mathbb{1} \otimes \mathbb{1} \xrightarrow{\ell_{\mathbb{1}}} \mathbb{1}$ and the isomorphism $(X \otimes Y) \otimes (X \otimes Y) \xrightarrow{\Theta_{XYXY}} (X \otimes X) \otimes (Y \otimes Y)$ interchanging the second and third factor in the tensor product (cf. Appendix A.2). Hence we proceed as follows. If $X \otimes X \xrightarrow{\varphi_X} \mathbb{1}$ and $Y \otimes Y \xrightarrow{\varphi_Y} \mathbb{1}$ are 2-forms, then $\varphi_{X \otimes Y} = \ell_{\mathbb{1}} \circ (\varphi_X \otimes \varphi_Y) \circ \Theta_{XYXY}$ is a 2-form on $X \otimes Y$ and is called the *tensor product* of the 2-forms φ_X and φ_Y.

Lemma 3.2.3. *Let (X, φ_X) and (Y, φ_Y) be symmetric lattices. Further, let $X \xrightarrow{f_{\varphi_X}} X$ and $Y \xrightarrow{f_{\varphi_Y}} Y$ be the join-reversing maps corresponding to φ_X and φ_Y in the sense of* (3.28). *Then $(X \otimes Y, \varphi_{X \otimes Y})$ is again a symmetric lattice and the join-reversing* *map* $X \otimes Y \xrightarrow{f_{\varphi_{X\otimes Y}}} X \otimes Y$ *corresponding to* $\varphi_{X\otimes Y}$ *attains the following values on elementary tensors*:

$$f_{\varphi_{X\otimes Y}}(x \otimes y) = \big(f_{\varphi_X}(x) \otimes \top\big) \vee \big(\top \otimes f_{\varphi_Y}(y)\big), \qquad x \in X,\ y \in Y. \tag{3.29}$$

Proof. Since every tensor is a join of elementary tensors, the symmetry of $\varphi_{X\otimes Y}$ follows immediately from that of φ_X and φ_Y. Further, the join-reversing map corresponding to the tensor product $\varphi_{X\otimes Y}$ of φ_X and φ_Y is given on elementary tensors $x \otimes y$ as follows:

$$f_{\varphi_{X\otimes Y}}(x \otimes y) = \bigvee\{u \otimes v \in X \otimes Y \mid \ell_{\mathbb{1}}\big(\varphi_X(x \otimes u) \otimes \varphi_Y(y \otimes v)\big) = 0\}.$$

Since

$$\ell_{\mathbb{1}}\big(\varphi_X(x \otimes u) \otimes \varphi_Y(y \otimes v)\big) = 0 \iff \big(\varphi_X(x \otimes u) = 0 \quad \text{or} \quad \varphi_Y(y \otimes v) = 0\big),$$

the relation (3.29) follows. □

The next proposition shows that the tensor product of orthogonal maps is again orthogonal.

Proposition 3.2.4. *Let* (X_i, φ_{X_i}) *and* (Y_i, φ_{Y_i}) $(i = 1, 2)$ *be symmetric lattices. If* $X_1 \xrightarrow{h} X_2$ *and* $Y_1 \xrightarrow{k} Y_2$ *are orthogonal maps, then* $h \otimes k$ *is orthogonal with respect to the tensor product of the respective 2-forms — i.e. with respect to* $\varphi_{X_1\otimes Y_1}$ *and* $\varphi_{X_2\otimes Y_2}$.

Proof. Let $X_1 \xrightarrow{h} X_2$ and $Y_1 \xrightarrow{k} Y_2$ be orthogonal maps. Since the exchange of the second and third factor of the tensor product is a natural transformation (see (1.13)), the following diagram is commutative:

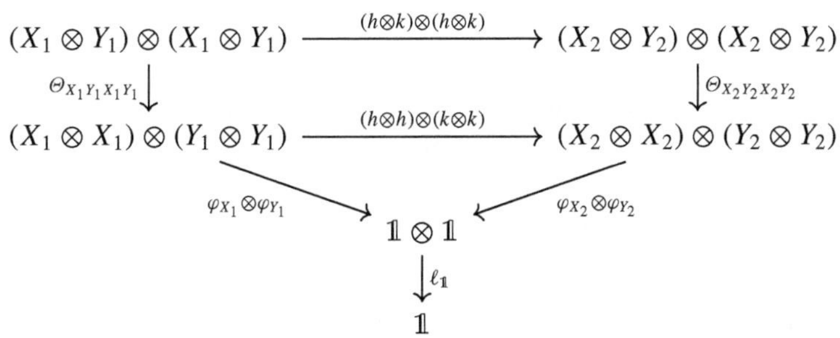

Hence $h \otimes k$ is orthogonal. □

The previous results can be summarized as follows. The tensor product of symmetric lattices exists always. Moreover, in the presence of symmetric lattices the components of the natural isomorphisms occurring in the coherence conditions of

Sup are orthogonal (cf. Exercise 3.2.1). Hence $\mathbb{O}(\mathsf{Sup})$ is a symmetric and monoidal category.

Since Sup is monoidal closed (cf. Theorem 2.1.12), every 2-form φ on X can be identified with its monoidal adjoint map $X \xrightarrow{\ulcorner\varphi\urcorner} [X, \mathbb{1}]$, which is obviously determined by $(\ulcorner\varphi\urcorner(x_1))(x_2) = \varphi(x_1 \otimes x_2)$ for each $x_1, x_2 \in X$. This way of looking at 2-forms suggests to introduce the following terminology.

A symmetric 2-form φ on X is

- *dense* if its monoidal adjoint map preserves the respective universal bounds — i.e. $\ulcorner\varphi\urcorner(\top) = \underline{\top}$.
- *faithful* if its monoidal adjoint $\ulcorner\varphi\urcorner$ is a monomorphism in Sup — this means an injective and join-preserving map.

Proposition 3.2.5. *Let X be a complete lattice, φ be a symmetric 2-form on X and $X \xrightarrow{f_\varphi} X$ be its corresponding join-reversing map. Then the following hold:*

(i) *φ is dense if and only if $f_\varphi(\top) = \bot$.*
(ii) *φ is faithful if and only if f_φ is an involution on X — i.e. $f_\varphi \circ f_\varphi = \mathrm{id}_X$.*

Proof. Let $X \xrightarrow{\ulcorner\varphi\urcorner} [X, \mathbb{1}]$ be the monoidal adjoint map of φ. Since the chain of equivalences

$$(\ulcorner\varphi\urcorner(x_1))(x_2) = 0 \iff \varphi(x_1 \otimes x_2) = 0 \iff x_1 \otimes x_2 \leq f_\varphi \iff x_2 \leq f_\varphi(x_1)$$

holds for all $x_1, x_2 \in X$, we can express $\ulcorner\varphi\urcorner$ in terms of f_φ as follows:

$$(\ulcorner\varphi\urcorner(x_1))(x_2) = \begin{cases} 0, & x_2 \leq f_\varphi(x_1), \\ 1, & x_2 \not\leq f_\varphi(x_1), \end{cases} \qquad x_1, x_2 \in X. \tag{3.30}$$

Obviously φ is dense if and only if $f_\varphi(\top) = \bot$. Since (f_φ, f_φ) is a contravariant Galois connection, we obtain $f_\varphi \circ f_\varphi \circ f_\varphi = f_\varphi$. Hence $\ulcorner\varphi\urcorner$ is injective (i.e. φ is faithful) if and only if f_φ is an involution. □

By Proposition 3.2.5 (i), a symmetric 2-form φ on a complete lattice X is dense if and only if for all $x \in X$ the implication

$$\varphi(\top \otimes x) = 0 \implies x = \bot$$

holds. Obviously every faithful 2-form is dense.

Referring again to Proposition 3.2.5 (i) it follows immediately from formula (3.29) that the density of symmetric 2-forms is preserved under their tensor product. However, in the case of faithful and symmetric 2-forms the situation changes, and we refer to the following:

Warning. In general the tensor product does not preserve faithful and symmetric 2-forms (see: Exercise 3.2.5).

The next result goes back to P. Resende [97] and is an immediate corollary of Proposition 3.2.5 (ii).

Corollary 3.2.6. *Let X be a complete lattice. There exists a bijective map between the set of all symmetric and faithful 2-forms φ on X and the set of all order-reversing involutions $x \longmapsto x^{\perp}$ on X such that the following condition is satisfied*

$$\varphi(x_1 \otimes x_2) = 0 \iff x_2 \leq x_1^{\perp}, \qquad x_1, x_2 \in X.$$

From the previous corollary we gain the important understanding that order-reversing involutions on complete lattices are not only related to non-classical negations, but they play also an important *geometric* rôle expressed by the so-called *orthogonality relation*:

$$x_1 \perp x_2 \iff x_2 \leq x_1^{\perp}, \qquad x_1, x_2 \in X. \tag{3.31}$$

Further, let (X, φ_X) be a symmetric lattice, and $X \xrightarrow{f_{\varphi_X}} X$ be the join-reversing map corresponding to the 2-form φ_X in the sense of (3.28). Using again the fact that $(f_{\varphi_X}, f_{\varphi_X})$ is a contravariant Galois connection it is easily seen that φ_X induces a closure operator c_{φ_X} on X by $c_{\varphi_X}(x) = f_{\varphi_X}(f_{\varphi_X}(x))$ for each $x \in X$. Since $f_{\varphi_X} \circ f_{\varphi_X} \circ f_{\varphi_X} = f_{\varphi_X}$, the map f_{φ_X} restricts to an order-reversing involution on the quotient $c_{\varphi_X}(X)$ of X w.r.t. c_{φ_X}. Hence the quotient map is obviously orthogonal, and the quotient of (X, φ_X) w.r.t. c_{φ_X} is a symmetric lattice with a faithful 2-form.

In the next proposition we present some properties of orthogonal and join-preserving maps.

Proposition 3.2.7. *Let (X, φ_X) and (Y, φ_Y) be symmetric lattices and let $X \xrightarrow{f_{\varphi_X}} X$ and $Y \xrightarrow{f_{\varphi_Y}} Y$ be the join-reversing maps corresponding to φ_X and φ_Y in the sense of* (3.28). *If $X \xrightarrow{h} Y$ is a join-preserving map, then the following assertions are valid*:

(i) *If h is orthogonal and φ_X is faithful, then h is injective* (*i.e. a monomorphism in* `Sup`).
(ii) *If h is injective and preserves the respective join-reversing maps — i.e. $h \circ f_{\varphi_X} = f_{\varphi_Y} \circ h$, then h is orthogonal.*
(iii) *If h is surjective* (*i.e. an epimorphism in* `Sup`) *and orthogonal, then h preserves the respective join-reversing maps f_{φ_X} and f_{φ_Y}.*
(iv) *If φ_X is faithful and h is surjective and orthogonal, then h is an order isomorphism.*

Proof. Let $h^{\vdash}$ be the right adjoint map of h. If h is orthogonal and φ_X is faithful (i.e. f_{φ_X} is an order-reversing involution), then we infer from Lemma 3.2.2 that the relation $1_X = f_{\varphi_X} \circ f_{\varphi_X} = f_{\varphi_X} \circ h^{\vdash} \circ f_{\varphi_Y} \circ h$ holds. Hence h has a left inverse map and assertion (i) is verified. Further, since the injectivity of h is equivalent $h^{\vdash} \circ h = 1_X$ (cf. Exercise 1.3.2 (a)) and h preserves the respective join-reversing maps f_{φ_X} and f_{φ_Y}, we apply Lemma 3.2.2 and conclude from

$$f_{\varphi_X} = h^{\vdash} \circ h \circ f_{\varphi_X} = h^{\vdash} \circ f_{\varphi_Y} \circ h$$

that assertion (ii) holds. Finally, since the surjectivity of h is equivalent to $h \circ h^{\vdash} = 1_Y$ (cf. Exercise 1.3.2 (a)) and h is orthogonal, we apply Lemma 3.2.2 again and obtain the following relation:

$$h \circ f_{\varphi_X} = h \circ h^{\vdash} \circ f_{\varphi_Y} \circ h = f_{\varphi_Y} \circ h.$$

Hence assertion (iii) follows. Since every injective and join-preserving map reflects the order — i.e. $h(x_1) \le h(x_2)$ implies $x_1 \le x_2$, the assertion (iv) is an immediate corollary of assertion (i). □

Example 3.2.8. Let $\mathscr{H}$ be a Hilbert space and $\mathscr{P}(\mathscr{H})$ be the complete lattice of all closed linear subspaces of $\mathscr{H}$. Then the orthogonal complement determines an order-reversing involution $^{\perp}$ and consequently a faithful symmetric 2-form on $\mathscr{P}(\mathscr{H})$. It is not difficult to verify that every surjective and unitary transformation $\mathscr{H} \xrightarrow{U} \mathscr{H}$ in $\mathscr{H}$ induces an orthogonal order isomorphism $\mathscr{P}(\mathscr{H}) \xrightarrow{h_U} \mathscr{P}(\mathscr{H})$ by the formation of (direct) images — i.e.

$$h_U(W) = \{U(w) \mid w \in W\}, \qquad W \in \mathscr{P}(\mathscr{H}).$$

The next example shows that there exist orthogonal order isomorphisms between the ortho-lattices of closed linear subspaces of Hilbert spaces which are *not* induced by unitary/orthogonal transformations.

Example 3.2.9. Let $\mathbb{R}^2$ be the 2-dimensional real Hilbert space and $\mathscr{P}(\mathbb{R}^2)$ be the complete lattice of all linear subspaces of $\mathbb{R}^2$. Then every 1-dimensional linear subspace W can be identified with a unit vector $(\cos\alpha, \sin\alpha)$ such that $\alpha \in [0, \pi)$. More precisely, for each $\alpha \in [0, \pi)$ we define

$$W_\alpha = \{r(\cos\alpha, \sin\alpha) \mid r \in \mathbb{R}\}.$$

Now let $[0, \frac{\pi}{2}) \xrightarrow{f} [0, \frac{\pi}{2})$ be a bijective self-map of $[0, \frac{\pi}{2})$. Then f determines a bijective self-map g of $[0, \pi)$ as follows:

$$g(\alpha) = \begin{cases} f(\alpha), & \alpha \in [0, \frac{\pi}{2}) \\ \frac{\pi}{2} + f(\alpha - \frac{\pi}{2}), & \alpha \in [\frac{\pi}{2}, \pi), \end{cases} \qquad \alpha \in [0, \pi).$$

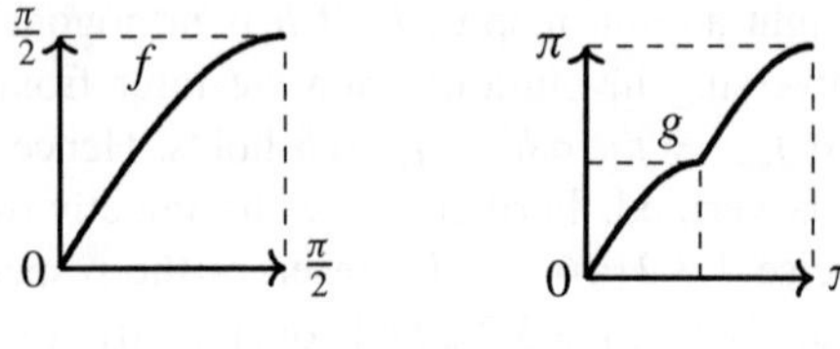

We introduce an order isomorphism $\mathscr{P}(\mathbb{R}^2) \xrightarrow{h_f} \mathscr{P}(\mathbb{R}^2)$ in the following way:

$$h_f(\{0\}) = \{0\}, \quad h_f(\mathbb{R}^2) = \mathbb{R}^2 \quad \text{and} \quad h_f(W_\alpha) = W_{g(\alpha)} \quad \text{for each } \alpha \in [0, \pi).$$

Obviously, h_f is isotone and surjective. We verify that h_f is orthogonal. Since $h_f(\{0\}) = \{0\}$ and $h_f(\mathbb{R}^2) = \mathbb{R}^2$, it is sufficient to consider 1-dimensional subspaces W_α and W_β with $\alpha, \beta \in [0, \pi)$. First we make the trivial observation:

$$W_\alpha \perp W_\beta \iff |\beta - \alpha| = \tfrac{\pi}{2}. \tag{3.32}$$

If $W_\alpha \perp W_\beta$, then without loss of generality we can assume $\alpha < \frac{\pi}{2}$ and $\beta = \alpha + \frac{\pi}{2}$. Hence the relation $g(\beta) = \frac{\pi}{2} + f(\alpha) = \frac{\pi}{2} + g(\alpha)$ holds — i.e. $h_f(W_\alpha) \perp h_f(W_\beta)$.

On the other hand, if $h_f(W_\alpha) = W_{g(\alpha)} \perp W_{g(\beta)} = h_f(W_\beta)$, then we conclude from (3.32) that $|g(\beta) - g(\alpha)| = \frac{\pi}{2}$. Without loss of generality we can assume $g(\alpha) < \frac{\pi}{2}$ and $g(\beta) = \frac{\pi}{2} + g(\alpha)$. Then $\alpha < \frac{\pi}{2} \leq \beta$ and so

$$f(\alpha) = g(\alpha) = g(\beta) - \tfrac{\pi}{2} = f(\beta - \tfrac{\pi}{2}).$$

Since f is injective, we obtain $\alpha = \beta - \frac{\pi}{2}$, and $W_\alpha \perp W_\beta$ follows from (3.32).

To sum up, we have verified that $\mathscr{P}(\mathscr{H}) \xrightarrow{h_f} \mathscr{P}(\mathscr{H})$ is an orthogonal order isomorphism. If $f \neq 1_{[0,\pi/2)}$, then h_f is *not induced* by some orthogonal transformation in $\mathbb{R}^2$ (cf. Example 3.2.8).

3.2.2 Involutive Left $\mathfrak{Q}$-Modules

Let $(\mathfrak{Q}, *, e, ')$ be an involutive and unital quantale. We enrich left $\mathfrak{Q}$-modules by symmetric 2-forms (cf. [97]).

Definition 3.2.10. A triple $(X, \odot, \varphi_X)$ is called an *involutive left $\mathfrak{Q}$-module* if (X, φ_X) is a symmetric lattice, $(X, \odot)$ is a left $\mathfrak{Q}$-module and the following relation holds for all $\alpha \in \mathfrak{Q}$ and $x_1, x_2 \in X$:

$$\varphi_X\big((\alpha \odot x_1) \otimes x_2\big) = \varphi_X\big(x_1 \otimes (\alpha' \odot x_2)\big). \tag{3.33}$$

Before we proceed, we first give a characterization of property (3.33) in terms of left $\mathfrak{Q}$-module homomorphisms.

Lemma 3.2.11. *Let $(X, \odot)$ be a left $\mathfrak{Q}$-module and φ_X be a symmetric 2-form on X. Further, let f_{φ_X} be the join-reversing map corresponding to φ_X in the sense of* (3.28). *Then the following assertions are equivalent*:

(i) *The relation $\varphi_X\big((\alpha \odot x_1) \otimes x_2\big) = \varphi_X\big(x_1 \otimes (\alpha' \odot x_2)\big)$ holds for all $x_1, x_2 \in X$ and $\alpha \in \mathfrak{Q}$.*

(ii) *The join-reversing map f_{φ_X} is a left $\mathfrak{Q}$-module homomorphism $X \xrightarrow{f_{\varphi_X}} X^{op}$ where the left action $\odot^{op}$ on X^{op} is defined in (3.20).*

Proof. Let x_1 and x_2 be arbitrary elements of X. Since (i) is equivalent to the following chain of equivalences

$$x_2 \leq f_{\varphi_X}(\alpha \odot x_1) \iff \alpha' \odot x_2 \leq f_{\varphi_X}(x_1) \iff x_2 \leq \alpha \odot^{op} f_{\varphi_X}(x_1), \quad (3.34)$$

the assertion follows. □

Example 3.2.12. Let $(\mathfrak{Q}, *, e, ')$ be an involutive and unital quantale and δ be a hermitian element of $\mathfrak{Q}$ — e.g. $\delta = \bot$. Then we introduce a join-reversing map $\mathfrak{Q} \xrightarrow{f} \mathfrak{Q}$ by $f(\beta) = \beta' \searrow \delta$ for all $\beta \in \mathfrak{Q}$. Since the relation

$$\beta \leq (\delta \swarrow \beta) \searrow \delta = (\delta' \swarrow \beta'') \searrow \delta = (\beta' \searrow \delta)' \searrow \delta = f(f(\beta))$$

holds, f determines a symmetric 2-form $\varphi_{\mathfrak{Q}}$ on $\mathfrak{Q}$ with the following explicit form:

$$\varphi_{\mathfrak{Q}}(\beta \otimes \gamma) = \begin{cases} 0, & \beta' * \gamma \leq \delta, \\ 1, & \beta' * \gamma \nleq \delta, \end{cases} \qquad \beta, \gamma \in \mathfrak{Q}. \quad (3.35)$$

Hence $(\mathfrak{Q}, \varphi_{\mathfrak{Q}})$ is a symmetric lattice. Further, we consider $\mathfrak{Q}$ as a left $\mathfrak{Q}$-module with the left action $\alpha \odot \beta = \alpha * \beta$ (cf. Example 3.1.6). Then it follows immediately from (3.35) that $(\mathfrak{Q}, \odot, \varphi_{\mathfrak{Q}})$ is an involutive left $\mathfrak{Q}$-module.

If $(\mathfrak{Q}, *, e, ', d)$ is an involutive Frobenius quantale with a hermitian dualizing element d, then d induces even the structure of an involutive left $\mathfrak{Q}$-module on $\mathfrak{Q}$ with a faithful 2-form (cf. Examples 2.6.17, 2.6.5 and Exercise 2.6.5 (c) and (e)).

The next proposition explains that Example 3.2.12 describes all possible involutive left $\mathfrak{Q}$-module structures on $(\mathfrak{Q}, \odot)$.

Proposition 3.2.13. *Let $(\mathfrak{Q}, *, e, ')$ be an involutive and unital quantale viewed as a left $\mathfrak{Q}$-module $(\mathfrak{Q}, \odot)$ in the sense of Example 3.1.6. Further, let $\varphi_{\mathfrak{Q}}$ be a symmetric 2-form on $\mathfrak{Q}$ and $f_{\varphi_{\mathfrak{Q}}}$ be the join-reversing map corresponding to $\varphi_{\mathfrak{Q}}$. Then the following assertions are equivalent:*

(i) *$(\mathfrak{Q}, \odot, \varphi_{\mathfrak{Q}})$ is an involutive left $\mathfrak{Q}$-module.*
(ii) *$f_{\varphi_{\mathfrak{Q}}}(e)$ is hermitian and $f_{\varphi_{\mathfrak{Q}}}$ has the following representation:*

$$f_{\varphi_{\mathfrak{Q}}}(\alpha) = \alpha' \searrow f_{\varphi_{\mathfrak{Q}}}(e), \qquad \alpha \in \mathfrak{Q}. \quad (3.36)$$

Proof. Let $\odot^{op}$ be the left action on $\mathfrak{Q}^{op}$ according to Example 3.1.22 (see also (3.20)). With regard to Lemma 3.2.11, the triple $(\mathfrak{Q}, \odot, \varphi_{\mathfrak{Q}})$ is an involutive left $\mathfrak{Q}$-module if and only if $\mathfrak{Q} \xrightarrow{f_{\varphi_{\mathfrak{Q}}}} \mathfrak{Q}^{op}$ is a left $\mathfrak{Q}$-module homomorphism. In particular, this means that the following relation holds:

$$f_{\varphi_{\mathfrak{Q}}}(\alpha * \beta) = \alpha' \searrow f_{\varphi_{\mathfrak{Q}}}(\beta) \qquad \alpha, \beta \in \mathfrak{Q}. \tag{3.37}$$

Now let us assume that (i) holds. Then (3.37) obviously implies (3.36). In order to verify that $f_{\varphi_{\mathfrak{Q}}}(e)$ is hermitian we refer to (3.36) and observe:

$$e \le f_{\varphi_{\mathfrak{Q}}}\big(f_{\varphi_{\mathfrak{Q}}}(e)\big) = \big(f_{\varphi_{\mathfrak{Q}}}(e)\big)' \searrow f_{\varphi_{\mathfrak{Q}}}(e).$$

Hence $\big(f_{\varphi_{\mathfrak{Q}}}(e)\big)' \le f_{\varphi_{\mathfrak{Q}}}(e)$ follows.

On the other hand, if (ii) holds, then for all $\alpha, \beta \in \mathfrak{Q}$ we obtain:

$$f_{\varphi_{\mathfrak{Q}}}(\alpha * \beta) = \alpha' \searrow (\beta' \searrow f_{\varphi_{\mathfrak{Q}}}(e)) = \alpha \odot^{op} f_{\varphi_{\mathfrak{Q}}}(\beta).$$

Hence $f_{\varphi_{\mathfrak{Q}}}$ is a left $\mathfrak{Q}$-module homomorphism. □

Example 3.2.14. Let (X, φ_X) be a symmetric lattice, $(\mathfrak{Q}, *, e, ')$ be an involutive and unital quantale provided with the left action $\alpha \odot \beta = \alpha * \beta$, and let $(\mathfrak{Q}, \odot, \varphi_{\mathfrak{Q}})$ be the involutive left $\mathfrak{Q}$-module introduced in Example 3.2.12.

Now we consider the tensor product $(\mathfrak{Q} \otimes X, \varphi_{\mathfrak{Q}\otimes X})$ of the symmetric lattices (X, φ_X) and $(\mathfrak{Q}, \varphi_{\mathfrak{Q}})$ (cf. Lemma 3.2.3) and observe that the underlying complete lattice is the free left $\mathfrak{Q}$-module generated by X (cf. (3.8)). In this context we recall that the left action on $\mathfrak{Q} \otimes X$ has the form

$$\alpha \odot (\beta \otimes x) = (\alpha * \beta) \otimes x, \qquad \alpha, \beta \in \mathfrak{Q},\ x \in X.$$

Since $(\mathfrak{Q}, \odot, \varphi_{\mathfrak{Q}})$ is an involutive left $\mathfrak{Q}$-module, $(\mathfrak{Q} \otimes X, \odot, \varphi_{\mathfrak{Q}\otimes X})$ is also an involutive left $\mathfrak{Q}$-module. In fact, for all $\alpha,\ \beta_1,\ \beta_2 \in \mathfrak{Q}$ and for all $x_1, x_2 \in X$ the following relation holds:

$$\begin{aligned}
\varphi_{\mathfrak{Q}\otimes X}\big((\alpha \odot (\beta_1 \otimes x_1)) \otimes (\beta_2 \otimes x_2)\big) &= \varphi_{\mathfrak{Q}\otimes X}\big(((\alpha * \beta_1) \otimes x_1) \otimes (\beta_2 \otimes x_2)\big)\\
&= \ell_{\mathbb{1}} \circ \big(\varphi_{\mathfrak{Q}}\big((\alpha * \beta_1) \otimes \beta_2\big) \otimes \varphi_X(x_1 \otimes x_2)\big)\\
&= \ell_{\mathbb{1}} \circ \big(\varphi_{\mathfrak{Q}}\big(\beta_1 \otimes (\alpha' * \beta_2)\big) \otimes \varphi_X(x_1 \otimes x_2)\big)\\
&= \varphi_{\mathfrak{Q}\otimes X}\big((\beta_1 \otimes x_1) \otimes ((\alpha' * \beta_2) \otimes x_2)\big)\\
&= \varphi_{\mathfrak{Q}\otimes X}\big((\beta_1 \otimes x_1) \otimes (\alpha' \odot (\beta_2 \otimes x_2))\big).
\end{aligned}$$

Morphisms between involutive left $\mathfrak{Q}$-modules are orthogonal left $\mathfrak{Q}$-module homomorphisms, and the associated category of involutive left $\mathfrak{Q}$-modules is denoted by $\mathbb{I}(\mathtt{Mod}_\ell(\mathfrak{Q}))$.

Referring to the self-duality S_ℓ of $\mathtt{Mod}_\ell(\mathfrak{Q})$ (cf. Sect. 3.1.3) we infer from Lemmas 3.2.2 and 3.2.11 that morphisms h in $\mathbb{I}(\mathtt{Mod}(\mathfrak{Q}))$ are characterized by the commutativity of the following diagram in $\mathtt{Mod}_\ell(\mathfrak{Q})$:

$$\begin{array}{ccc} X & \xrightarrow{h} & Y \\ {\scriptstyle f_{\varphi_X}}\downarrow & & \downarrow{\scriptstyle f_{\varphi_Y}} \\ X^{op} & \xleftarrow{h^{\vdash}} & Y^{op} \end{array}$$

where f_{φ_X} and f_{φ_Y} denote the join-reversing maps corresponding to the respective symmetric 2-forms in the sense of (3.28).

It follows immediately from Example 3.2.14, Proposition 3.2.4 and (3.13) that there exists a functor $\mathsf{F}\colon \mathbb{O}(\mathtt{Sup}) \to \mathbb{I}(\mathtt{Mod}_\ell(\mathfrak{Q}))$ defined by:

$$\mathsf{F}(X) = \mathfrak{Q} \otimes X, \quad X \xrightarrow{h} Y, \quad \mathsf{F}(h) = 1_{\mathfrak{Q}} \otimes h.$$

In this context the symmetric 2-form $\varphi_{\mathfrak{Q}}$ on $\mathfrak{Q}$ is determined in Example 3.2.12 where the hermitian element δ is chosen as the universal lower bound — i.e. $\delta = \bot$.

If $\mathsf{U}\colon \mathbb{I}(\mathtt{Mod}_\ell(\mathfrak{Q})) \to \mathbb{O}(\mathtt{Sup})$ is the forgetful functor, then there exists a natural transformation $id_{\mathbb{O}(\mathtt{Sup})} \to \mathsf{UF}$ (cf. Exercise 3.2.2), but in general U is *not* right adjoint to F (cf. Exercise 3.2.3). We leave the details of this situation to the reader and return to general properties of involutive left $\mathfrak{Q}$-modules.

Let f_φ be the join-reversing map corresponding to a symmetric 2-form φ in the sense of (3.28). If $(X, \odot, \varphi)$ is an involutive left $\mathfrak{Q}$-module, then we infer from Lemma 3.2.11 that the closure operator c_φ (i.e. $c_\varphi = f_\varphi \circ f_\varphi$) is linear (cf. Sect. 3.1.1). Hence the quotient of $(X, \odot, \varphi)$ w.r.t. c_φ is again a left $\mathfrak{Q}$-module and a symmetric lattice (cf. Sect. 3.2.1). Obviously, the quotient map is a morphism of $\mathbb{I}(\mathtt{Mod}_\ell(\mathfrak{Q}))$. Since the 2-form on the quotient of $(X, \odot, \varphi)$ w.r.t. c_φ is faithful, this approach always leads to involutive left $\mathfrak{Q}$-modules with a faithful 2-form. A characterization of this type of module is given in the next theorem, which is a refinement of Proposition 3.1.3.

Theorem 3.2.15. *Let $x \longmapsto x^\perp$ be an order-reversing involution on a complete lattice X and $([X, X], \circ, 1_X, ')$ be the involutive and unital quantale of all join-preserving self-maps of X constructed in Example 2.3.31. Further, let φ be the 2-form on X determined by the same order-reversing involution (cf. Corollary 3.2.6).*

(i) *If $\odot$ is a left action on X such that $(X, \odot, \varphi)$ is an involutive left $\mathfrak{Q}$-module, then there exists a unique involutive and unital homomorphism $\mathfrak{Q} \xrightarrow{h} [X, X]$ making the following diagram commutative:*

$$\begin{array}{ccc} \mathfrak{Q} \otimes X & \xrightarrow{h \otimes 1_X} & [X, X] \otimes X \\ & {\scriptstyle \odot}\searrow & \downarrow{\scriptstyle ev_X} \\ & & X \end{array}$$

where ev_X is the evaluation arrow.

(ii) *If $\mathfrak{Q} \xrightarrow{h} [X, X]$ is an involutive and unital homomorphism, then h induces a left action $\odot$ on X by the following diagram*

$$\begin{array}{ccc} \mathfrak{Q} \otimes X & \xrightarrow{h \otimes 1_X} & [X, X] \otimes X \\ & \searrow^{\odot} & \downarrow{\scriptstyle ev_X} \\ & & X \end{array}$$

such that $(X, \odot, \varphi)$ *is an involutive left* $\mathfrak{Q}$*-module.*

Proof. Let $\odot$ be a left action on X. Referring to the proof of Proposition 3.1.3 it is sufficient to show that $\odot$ satisfies (3.33) if and only if the monoidal adjoint map of $\odot$ is an involutive (unital) homomorphism.

(i) Let us assume that $\odot$ satisfies (3.33) and h is the monoidal adjoint of $\odot$. We maintain the notation of Example 2.3.31. Then we obtain the following chain of equivalences for all $\alpha \in \mathfrak{Q}$ and $x_1, x_2 \in X$:

$$\begin{aligned} x_1 \le \big(h(\alpha)^{\vdash}\big)(x_2^{\perp}) &\iff \alpha \odot x_1 = h(\alpha)(x_1) \le x_2^{\perp} \\ &\iff \varphi(x_1 \otimes (\alpha' \odot x_2)) = \varphi((\alpha \odot x_1) \otimes x_2) = 0 \\ &\iff h(\alpha')(x_2) = \alpha' \odot x_2 \le x_1^{\perp} \\ &\iff x_1 \le \big(h(\alpha')(x_2)\big)^{\perp}. \end{aligned}$$

Hence $\big(h(\alpha)^{\vdash}\big)(x_2^{\perp}) = \big(h(\alpha')(x_2)\big)^{\perp}$ follows for all $x_2 \in X$. Since $x \longmapsto x^{\perp}$ is an involution, h is involutive.

(ii) Let us assume that $\mathfrak{Q} \xrightarrow{h} [X, X]$ is an involutive and unital homomorphism. Since the left action $\odot$ on X is determined by (cf. Proposition 3.1.3):

$$\alpha \odot x = h(\alpha)(x), \qquad \alpha \in \mathfrak{Q},\ x \in X, \tag{3.38}$$

we obtain for all $\alpha \in \mathfrak{Q}$ and $x_1, x_2 \in X$:

$$\begin{aligned} \varphi\big((\alpha \odot x_1) \otimes x_2\big) = 0 &\iff h(\alpha)(x_1) = \alpha \odot x_1 \le x_2^{\perp} \\ &\iff x_1 \le \big(h(\alpha)^{\vdash}\big)(x_2^{\perp}) = \big(h(\alpha')(x_2)\big)^{\perp} \\ &\iff \alpha' \odot x_2 = h(\alpha')(x_2) \le x_1^{\perp} \\ &\iff \varphi(x_1 \otimes (\alpha' \odot x_2)) = 0. \end{aligned}$$

Hence $(X, \odot, \varphi)$ is an involutive left $\mathfrak{Q}$-module. □

We finish this subsection with a special construction of involutive left $\mathfrak{Q}$-modules which will play a prominent rôle in the theory of irreducible representations of C^*-algebras.

Theorem 3.2.16. *Let* $(\mathfrak{Q}, *, e, ')$ *be an involutive unital quantale and* q *be a maximal left-sided element of* $\mathfrak{Q}$ *with the property* $q \vee q' \neq \top$. *Further, let* $\odot_q$ *be the left action on the upsegment* $\uparrow q$ *constructed in Example 3.1.21. Then the 2-form* $(\uparrow q) \otimes (\uparrow q) \xrightarrow{\varphi} \mathbb{1}$ *determined on elementary tensors by*

$$\varphi(x_1 \otimes x_2) = \begin{cases} 0, & x_2' * x_1 \leq q \vee q', \\ 1, & x_2' * x_1 \not\leq q \vee q', \end{cases} \qquad x_1, x_2 \in \uparrow q$$

is symmetric and dense. Moreover, $(\uparrow q, \odot_q, \varphi)$ *is an involutive left* $\mathfrak{Q}$*-module.*

Proof. It is evident that φ is a symmetric 2-form and $(\uparrow q, \odot_q, \varphi)$ is an involutive left $\mathfrak{Q}$-module. Therefore we only show that φ is dense. For this purpose we choose $x \in \uparrow q$ such that $\varphi(\top \otimes x) = 0$ holds — i.e. $x' * \top \leq q \vee q'$ which is equivalent to $\top * x \leq q \vee q'$. Let us assume $x \not\leq q$. Then $\top * x$ is left-sided, and $\top * x \not\leq q$ holds, because $\mathfrak{Q}$ is semi-unital. Now we invoke the maximality of q and obtain

$$\top = (\top * x) \vee q \vee q' \leq q \vee q',$$

which is a contradiction to $q \vee q' \neq \top$. Hence the assumption is false and $x \leq q$ follows — this means that x coincides with the universal lower bound in $\uparrow q$. □

3.2.3 Representations of C*-Algebras and Involutive Left Modules

Let $\mathscr{H}$ be a Hilbert space and $\mathscr{P}(\mathscr{H})$ be the complete lattice of all closed linear subspaces of $\mathscr{H}$. On $\mathscr{P}(\mathscr{H})$ we always consider the 2-form corresponding to the order-reversing involution induced by the orthogonal complement. Further, let $L(\mathscr{H})$ be the C^*-algebra of all bounded linear operators on $\mathscr{H}$ in which the involution * is determined by the formation of adjoint operators.

A pair $(\pi, \mathscr{H})$ is a *representation* of a unital C^*-algebra $(A, +, \cdot, e)$ if $\mathscr{H}$ is a Hilbert space and $A \xrightarrow{\pi} L(\mathscr{H})$ is a $*$-homomorphism (cf. Sect. 2.4). Further, let $(\mathbb{MAX}, *, \langle e \rangle, ')$ be the involutive and unital quantale of all closed linear subspaces of A (cf. Example 2.5.5), and let $([\mathscr{P}(\mathscr{H}), \mathscr{P}(\mathscr{H})], \circ, 1_{\mathscr{P}(\mathscr{H})}, ')$ be the involutive and unital quantale of all join-preserving self-maps of $\mathscr{P}(\mathscr{H})$ constructed in Example 2.3.31. We show that every representation $(\pi, \mathscr{H})$ of a unital C^*-algebra induces an involutive and unital homomorphism $\mathbb{MAX} \xrightarrow{h_\pi} [\mathscr{P}(\mathscr{H}), \mathscr{P}(\mathscr{H})]$.

Since $A \xrightarrow{\pi} L(\mathscr{H})$ is continuous and linear, every closed linear subspace U of A determines a join-preserving map $\mathscr{P}(\mathscr{H}) \xrightarrow{f_U} \mathscr{P}(\mathscr{H})$ as follows:

$$f_U(P) = \text{top. hull}\,(\text{lin. hull}\,\{\pi(a)(x) \mid a \in U,\ x \in P\}), \qquad P \in \mathscr{P}(\mathscr{H}). \quad (3.39)$$

Further, π is a continuous algebra homomorphism. Hence the following relations hold:

$$f_{\langle e \rangle} = 1_{\mathscr{P}(\mathscr{H})}, \quad f_{V*U} = f_V \circ f_U, \qquad V, U \in \mathbb{MAX}. \quad (3.40)$$

Finally, we prove that for any closed linear subspace U of A the relation $(f_U)' = f_{U'}$ holds. First we compute the right adjoint map $f_U^{\vdash}$ of f_U. For this purpose we denote the inner product of $\mathscr{H}$ by $\langle\ ,\ \rangle$ and recall $U' = \{a^* \mid a \in U\}$. Since π is a $*$-homomorphism, we obtain for $P \in \mathscr{P}(\mathscr{H})$:

$$\begin{aligned} f_U^{\vdash}(P^{\perp}) &= \bigvee\{S \in \mathscr{P}(\mathscr{H}) \mid f_U(S) \subseteq P^{\perp}\} \\ &= \{y \in \mathscr{H} \mid \forall a \in U,\ \forall x \in P:\ \langle \pi(a)(y), x\rangle = 0\} \\ &= \{y \in \mathscr{H} \mid \forall a \in U,\ \forall x \in P:\ \langle y, \pi(a^*)(x)\rangle = 0\} \\ &= \big(f_{U'}(P)\big)^{\perp}. \end{aligned}$$

Hence $(f_U)'(P) = f_{U'}(P)$ follows. We summarize the previous results in the following lemma, which goes back to C.J. Mulvey and J.W. Pelletier (cf. [84]).

Lemma 3.2.17. *Let $(\pi, \mathscr{H})$ be a representation of a unital C^*-algebra $(A, +, \cdot, e)$. Then the map $\mathbb{MAX} \xrightarrow{h_\pi} [\mathscr{P}(\mathscr{H}), \mathscr{P}(\mathscr{H})]$ defined by*

$$h_\pi(U) = f_U, \qquad U \in \mathbb{MAX} \tag{3.41}$$

is an involutive and unital homomorphism.

Now we apply Theorem 3.2.15 and construct a left action $\odot_\pi$ on $\mathscr{P}(\mathscr{H})$, which is determined by the following commutative diagram:

$$\begin{array}{ccc} \mathbb{MAX} \otimes \mathscr{P}(\mathscr{H}) & \xrightarrow{h_\pi \otimes 1_{\mathscr{P}(\mathscr{H})}} & [\mathscr{P}(\mathscr{H}), \mathscr{P}(\mathscr{H})] \otimes \mathscr{P}(\mathscr{H}) \\ & \overset{\odot_\pi}{\searrow} & \big\downarrow \mathrm{ev}_{\mathscr{P}(\mathscr{H})} \\ & & \mathscr{P}(\mathscr{H}) \end{array}$$

Hence $(\mathscr{P}(\mathscr{H}), \odot_\pi, \varphi)$ is an involutive left $\mathbb{MAX}$-module. Since in the following considerations we need a special name for this module, we call $(\mathscr{P}(\mathscr{H}), \odot_\pi, \varphi)$ the *canonical left* $\mathbb{MAX}$ *-module* associated with the representation $(\pi, \mathscr{H})$. In this context the 2-form φ is always determined by the orthogonal complement in $\mathscr{P}(\mathscr{H})$.

It is easily seen that for all closed, linear subspaces U of A and all closed, linear subspaces P of $\mathscr{H}$ the left action $\odot_\pi$ is explicitly given by

$$U \odot_\pi P = f_U(P) \quad \text{where } f_U \text{ is defined in (3.39)}\,. \tag{3.42}$$

In order to explain the rôle of involutive left modules in the representation theory of C^*-algebras we need some more terminology.

Definition 3.2.18. Let $(\pi, \mathscr{H})$ be a representation of a C^*-algebra $(A, +, \cdot)$ with unit e. A vector $x \in \mathscr{H}$ is called *cyclic* if the topological closure of $\{\pi(a)(x) \mid a \in A\}$ coincides with $\mathscr{H}$. A representation is called *cyclic* if there exists a cyclic vector. A representation is called *irreducible* if every non-zero vector is cyclic.

Theorem 3.2.19. *Let $(\pi, \mathscr{H})$ be a representation of a unital C^*-algebra. Further, let $(\mathscr{P}(\mathscr{H}), \odot_\pi, \varphi)$ be the canonical left $\mathbb{MAX}$-module associated with $(\pi, \mathscr{H})$. Then the following assertions hold:*

(i) *The representation* $(\pi, \mathscr{H})$ *is cyclic if and only if there exists an atom* P *of* $\mathscr{P}(\mathscr{H})$ *such that* $\top \odot_\pi P = \top$, *where* $\top$ *denotes the respective universal upper bound in* $\mathbb{MAX}$ *and* $\mathscr{P}(\mathscr{H})$.
(ii) *The representation* $(\pi, \mathscr{H})$ *is irreducible if and only if the canonical left* $\mathbb{MAX}$*-module* $(\mathscr{P}(\mathscr{H}), \odot_\pi)$ *is irreducible.*

Proof. First we note that $\mathscr{P}(\mathscr{H})$ is a complete atomic lattice. In particular, atoms of $\mathscr{P}(\mathscr{H})$ coincide with 1-dimensional linear subspaces of $\mathscr{H}$, which we denote by $\langle x \rangle$. In this context x plays the rôle of some unit vector of $\mathscr{H}$ generating the corresponding 1-dimensional subspace.

Further, it follows immediately from (3.42) that the equivalence

$$\top \odot_\pi \langle x \rangle = P \iff f_A(\langle x \rangle) = P \tag{3.43}$$

holds for all closed linear subspaces P of $\mathscr{H}$. Hence with regard to (i) and (ii) it is sufficient to establish the following properties:

(1) A non-zero vector $x \in \mathscr{H}$ is cyclic if and only if $\top \odot_\pi \langle x \rangle = \top$.
(2) The canonical left $\mathbb{MAX}$-module is irreducible if and only if every non-zero vector $x \in \mathscr{H}$ is cyclic.

In fact, (1) is a direct corollary of (3.43). With regard to (2) we can argue as follows. If every non-zero vector is cyclic, then we infer from (1) that $(\mathscr{P}(\mathscr{H}), \odot_\pi)$ is irreducible. On the other hand, let us assume that $(\mathscr{P}(\mathscr{H}), \odot_\pi)$ is irreducible. Then we choose a non-zero vector $x \in \mathscr{H}$ and consider the following closed, linear subspace $P = \top \odot_\pi \langle x \rangle$. Since $\top \odot_\pi P = (\top * \top) \odot_\pi \langle x \rangle = P$, the closed subspace P is invariant. Since $(\mathscr{P}(\mathscr{H}), \odot_\pi)$ is irreducible, $P = \top$ follows. Hence x is cyclic. □

In the next subsection we recall some fundamental constructions and properties of C^*-algebras.

3.2.4 Sketch of the Gelfand–Naimark–Segal Construction

Let $(A, +, \cdot)$ always be a C^*-algebra with unit e. A linear functional $A \xrightarrow{\rho} \mathbb{C}$ is said to be *positive* if $0 \le \rho(a)$ holds for all positive elements a of A (cf. [60, p. 44 and p. 255]). A positive linear functional is a *state* of A if $\rho(e) = 1$. In the case of $A = L(\mathscr{H})$ every unit vector $x \in \mathscr{H}$ induces a state $L(\mathscr{H}) \xrightarrow{\omega_x} \mathbb{C}$ by:

$$\omega_x(T) = \langle T(x), x \rangle, \qquad T \in L(\mathscr{H}).$$

Moreover, since $*$-homomorphisms preserve positive elements of C^*-algebras, every representation $(\pi, \mathscr{H})$ of a unital C^*-algebra A and every unit vector of $\mathscr{H}$ induces a state of A by $\rho = \omega_x \circ \pi$ — i.e. $\rho(a) = \langle \pi(a)(x), x \rangle$, $a \in A$.

In what follows we show that the converse situation also holds — i.e. for every state ρ of A there exists a cyclic representation $(\pi_\rho, \mathscr{H}_\rho)$ and a cyclic vector x_ρ for π_ρ such that $\rho = \omega_{x_\rho} \circ \pi_\rho$ holds. This construction is known as the Gelfand–Naimark–Segal construction or GNS construction for short. For details the reader is referred to [60]. In what follows we give only a sketch of this method and begin with the observation that the self-adjoint element $a^* \cdot a$ is positive for every $a \in A$ (cf. [60, Theorem 4.2.6]). Hence every state ρ of A induces a semi-scalar product $\langle\ ,\ \rangle$ by

$$\langle a, b\rangle = \rho(b^* \cdot a), \qquad a, b \in A$$

on A — this means that $\langle\ ,\ \rangle$ is a hermitian bilinear form satisfying the condition $0 \le \langle a, a\rangle$ for all $a \in A$. It is well-known that for every semi-scalar product the Cauchy–Schwarz inequality holds. Hence every semi-scalar product induces a semi-norm $A \xrightarrow{p} \mathbb{R}$ by

$$p(a) = \langle a, a\rangle^{1/2}, \qquad a \in A.$$

Lemma 3.2.20. *Let ρ be a state of a unital C^*-algebra $(A, +, \cdot, e)$. Then the left kernel $L_\rho = \{a \in A \mid \rho(a^* \cdot a) = 0\}$ of ρ is a closed left ideal of A.*

Proof. Let $\langle\ ,\ \rangle$ be the semi-scalar product on A induced by the given state ρ. Then L_ρ coincides with the kernel of the associated semi-norm p — i.e. $L_\rho = p^{-1}(\{0\})$. Hence L_ρ is a linear subspace of A. In order to show that L_ρ is a left ideal, we conclude from the Cauchy–Schwarz inequality that for all $b \in A$ and $a \in L_\rho$ the relation $\rho(b^* \cdot a) = 0$ holds. If we now replace b by the expression $b^* \cdot b \cdot a$ in $\rho(b^* \cdot a) = 0$, then we obtain $0 = \rho((b^* \cdot b \cdot a)^* \cdot a) = \rho((b \cdot a)^* \cdot (b \cdot a))$. Hence $b \cdot a \in L_\rho$ follows, and L_ρ is a left ideal. Since ρ is continuous, L_ρ is obviously closed. □

Corollary 3.2.21. *Let $(\pi, \mathscr{H})$ be a representation of a unital C^*-algebra and x be a unit vector of $\mathscr{H}$. Further, let $\langle x\rangle$ be the 1-dimensional subspace of $\mathscr{H}$ generated by x and $(\mathscr{P}(\mathscr{H}), \odot_\pi)$ be the canonical left $\mathbb{MAX}$-module associated with $(\pi, \mathscr{H})$. Then the left kernel of the state $\rho = \omega_x \circ \pi$ coincides with the annihilator* ann $(\langle x\rangle)$.

Proof. Since π is a $*$-homomorphism, we obtain the following relation for an element a of the given C^*-algebra A:

$$\begin{aligned}\rho(a^* \cdot a) = \langle \pi(a^* \cdot a)(x), x\rangle &= \langle (\pi(a)^* \circ \pi(a))(x), x\rangle \\ &= \langle \pi(a)(x), \pi(a)(x)\rangle \\ &= \|\pi(a)(x)\|_2^2.\end{aligned}$$

Hence $\{a \in A \mid \rho(a^* \cdot a) = 0\} = \{a \in A \mid \pi(a)(x) = 0\}$ follows. □

Now we fix a state ρ of the unital C^*-algebra and consider its left kernel L_ρ and the associated semi-scalar product $\langle\ ,\ \rangle$. Then the quotient A/L_ρ of A w.r.t. L_p is a pre-Hilbert space with the norm

$$\|q(b)\|_2 = \rho(b^* \cdot b)^{1/2}, \qquad b \in A,$$

where $A \xrightarrow{q} A/L_\rho$ denotes the quotient map. Further, every element $a \in A$ determines a linear operator $A/L_\rho \xrightarrow{T_a} A/L_\rho$ by

$$T_a(q(b)) = q(a \cdot b), \qquad b \in A. \tag{3.44}$$

As a next step we compute the completion of A/L_ρ. This approach leads to a Hilbert space $\mathscr{H}_\rho$. Because of

$$\|T_a(q(b))\|_2 \leq \|a\| \cdot \|b\|_2, \qquad a, b \in A,$$

every operator T_a has a unique extension to a bounded operator $\widehat{T}_a$ on $\mathscr{H}_\rho$. As a result of this construction we obtain a map sending an $a \in A$ to the operator $\widehat{T}_a$ — i.e.

$$\pi_\rho(a) = \widehat{T}_a, \tag{3.45}$$

which is in fact a $*$-homomorphism $A \xrightarrow{\pi_\rho} L(\mathscr{H}_\rho)$. Hence $(\pi_\rho, \mathscr{H}_\rho)$ is a representation of $(A, +, \cdot)$. In particular, if e is the unit of A, then the vector x_ρ, viewed as the vector $q(e)$ in the completion $\mathscr{H}_\rho$, is a cyclic vector. To sum up, $(\pi_\rho, \mathscr{H}_\rho)$ is a cyclic representation of A, and the following relation holds

$$(\omega_{x_\rho} \circ \pi_\rho)(a) = \langle \pi_\rho(a)(x_\rho), x_\rho \rangle = \langle T_a(e), e \rangle = \langle a \cdot e, e \rangle = \rho(a), \qquad a \in A.$$

The representation $(\pi_\rho, \mathscr{H}_\rho)$ with the cyclic vector x_ρ of π_ρ is also called the *cyclic representation produced by the GNS construction* and is one of the basic tools in the theory of C^*-algebras.

The next theorem explains, in what sense the GNS construction is essentially unique.

Theorem 3.2.22. (cf. [60, 4.5.3]) *Let ρ be a state of a unital C^*-algebra $(A, +, \cdot, e)$ and $(\pi, \mathscr{H})$ be a cyclic representation with some cyclic vector $x \in \mathscr{H}$ such that $\rho = \omega_x \circ \pi$ holds. Further, let $(\pi_\rho, \mathscr{H}_\rho)$ be the cyclic representation of A produced by the GNS construction with the cyclic vector x_ρ. Then there exists an isomorphism — i.e. a surjective and unitary transformation $\mathscr{H}_\rho \xrightarrow{U} \mathscr{H}$ such that the following properties hold*:

$$x = U(x_\rho) \quad \textit{and} \quad \pi(a) = U \circ \pi_\rho(a) \circ U^*, \qquad a \in A.$$

Motivated by the previous theorem we introduce the following terminology. Two cyclic representations $(\pi_1, \mathscr{H}_1)$ and $(\pi_2, \mathscr{H}_2)$ of a unital C^*-algebra $(A, +, \cdot, e)$ are *equivalent* if there exists an isomorphism $\mathscr{H}_1 \xrightarrow{U} \mathscr{H}_2$ such that

$$\pi_2(a) = U \circ \pi_1(a) \circ U^*$$

holds for all $a \in A$. Then the previous theorem means the following property.

> Every cyclic representation $(\pi, \mathscr{H})$ with a cyclic vector $x \in \mathscr{H}$ is equivalent to the cyclic representation produced by the GNS construction from the state $\rho = \omega_x \circ \pi$.

If we consider a C^*-algebra A as a Banach space, then the set $\mathscr{S}(A)$ of all states of A is a convex and weak* compact subset of the topological dual space of A. By the Krein–Milman theorem $\mathscr{S}(A)$ is the closed and convex closure of its extreme points. In this context extreme points of $\mathscr{S}(A)$ are called *pure states*.

In what follows we summarize some fundamental properties of pure states and their various relationships to irreducible representations. The proofs of the following results can be found in [61, Sect. 10.2]

- The representation produced by the GNS construction from a state ρ is irreducible if and only if ρ is pure.
- If $(\pi, \mathscr{H})$ is an irreducible representation, then for all unit vectors $x \in \mathscr{H}$ the state $\rho = \omega_x \circ \pi$ is pure.
- The left kernel of a state ρ is a maximal left ideal if and only if ρ is pure.
- For every maximal left ideal L there exists a unique pure state such that L coincides with the left kernel L_ρ of ρ.
- Let L_ρ be the left kernel and N_ρ be the kernel of ρ — i.e. $N_\rho = \rho^{-1}(\{0\})$. If ρ is a pure state, then the following properties hold:

(a) $N_\rho = L_\rho \vee L'_\rho$.
(b) The quotient A/L_ρ is complete — i.e. the pre-Hilbert space A/L_ρ is already a Hilbert space.

The previous property (b) motivates us to study the topology on the (pre-)Hilbert space A/L_ρ in some more detail.

Lemma 3.2.23. *Let ρ be a pure state of a unital C^*-algebra A, and let L_ρ be its left kernel. Then the topology on the Hilbert space A/L_ρ coincides with the quotient topology induced by the topology on A.*

Proof. Let $\langle \ , \ \rangle$ be the semi-scalar product on A induced by ρ — i.e. $\langle a, b\rangle = \rho(b^* \cdot a)$ for all $a, b \in A$, and let p be the corresponding seminorm — i.e. $p(a) = \rho(a^* \cdot a)^{1/2}$ for all $a \in A$. Further, let $\| \ \|$ be the norm on A and $A \xrightarrow{q} A/L_\rho$ be the quotient map. We fix $a \in A$. Then the quotient norm $\widehat{\| \ \|}$ and the norm induced by p on A/L_ρ are given by:

$$\widehat{\|q(a)\|} = \inf_{u \in q(a)} \|u\| \quad \text{and} \quad \|q(a)\|_2 = p(u) \quad \text{for some } u \in \text{q(a)}.$$

Further, we recall the rôle of the C^*-property and the property that the left kernel L_ρ coincides with the kernel of the seminorm p. In particular, the values $p(u)$ coincide for all $u \in q(a)$, and the following relation holds:

$$\widehat{\|q(a)\|} = \inf_{u \in q(a)} \|u\| = \inf_{u \in q(a)} \|u^* \cdot u\|^{1/2}.$$

Hence we obtain

$$\|q(a)\|_2 = \inf_{u \in q(a)} p(u) = \inf_{u \in q(a)} (\rho(u^* \cdot u))^{1/2} \leq \inf_{u \in q(a)} \|u^* \cdot u\|^{1/2} = \widehat{\|} q(a) \widehat{\|}.$$

Since the quotient topology induced by a Banach space is always complete and the completeness of the topology induced by $\| \; \|_2$ follows from the fact that ρ is a pure state, we conclude from the Open Mapping Theorem (cf. [77]) in connection with the previous relation that the Hilbert space topology on A/L_ρ coincides with the quotient topology induced by the topology of A. □

Theorem 3.2.24. *Let ρ be a pure state of a unital C^*-algebra $(A, +, \cdot, e)$ and L_ρ be its left kernel. If $(\pi_\rho, \mathscr{H}_\rho)$ is the cyclic representation of A with the cyclic vector x_ρ induced by the GNS construction, then the following assertions hold:*

(i) *In the sense of $\mathbb{I}(\mathtt{Mod}_\ell(\mathbb{MAX}))$ the canonical left $\mathbb{MAX}$-module associated with $(\pi_\rho, \mathscr{H}_\rho)$ is isomorphic to the involutive left $\mathbb{MAX}$-module given by the upsegment $\uparrow L_\rho$ in $\mathbb{MAX}$ (cf. Theorem 3.2.16).*

(ii) *The 1-dimensional linear subspace $\langle x_\rho \rangle$ generated by x_ρ is a generator of the canonical left $\mathbb{MAX}$-module associated with $(\pi_\rho, \mathscr{H}_\rho)$. Further, the annihilator* ann $(\langle x_\rho \rangle)$ *of $\langle x_\rho \rangle$ coincides with the left kernel L_ρ of ρ.*

Proof. (a) Let ρ be a pure state of A. By Lemma 3.2.23 we identify A/L_ρ with $\mathscr{H}_\rho$. In particular, the quotient map q can be viewed as a surjective and continuous map from A to $\mathscr{H}_\rho$. Since for every closed linear subspace U of the upsegment $\uparrow L_\rho$ the image of U under q is closed, q induces an order isomorphism $\uparrow L_\rho \xrightarrow{\Phi} \mathscr{P}(\mathscr{H}_\rho)$ as follows:

$$\Phi(U) = q(U) := \{q(a) \mid a \in U\}, \qquad U \in \uparrow L_\rho.$$

We show that Φ is an orthogonal left $\mathbb{MAX}$-module isomorphism. For this purpose we first choose $U, V \in \uparrow L_\rho$. Since ρ is pure, the kernel of ρ coincides with $L_\rho \vee (L_\rho)'$. Hence we obtain the following chain of equivalences:

$$\begin{aligned} q(U) \perp q(V) &\iff \forall u \in U, \forall v \in V : \langle q(u), q(v) \rangle = \rho(v^* \cdot u) = 0 \\ &\iff V' * U \subseteq L_q \vee (L_q)' \iff U \perp V. \end{aligned}$$

Thus Φ is orthogonal. In order to verify that Φ is also a left $\mathbb{MAX}$-module homomorphism we proceed as follows.

First we choose an atom of $\mathbb{MAX}$ and an atom of $\uparrow L_\rho$ — i.e. a 1-dimensional, linear subspace $\langle a \rangle$ generated by some $a \in A$ and an element $b \in A$ with $b \notin L_\rho$. Further, let us consider the 1-dimensional linear subspaces $\langle a \cdot b \rangle$, $\langle q(b) \rangle$ and $\langle q(a \cdot b) \rangle$ generated by $a \cdot b$, $q(b)$ and $q(a \cdot b)$, respectively. We recall $T_a(q(b)) = q(a \cdot b)$ (cf. (3.44)) and conclude from (3.39), (3.42) and (3.45) that the following relation holds:

$$\langle a \rangle \odot_{\pi_\rho} \langle q(b) \rangle = f_{\langle a \rangle}(\langle q(b) \rangle) = \langle q(a \cdot b) \rangle = q(\langle a \cdot b \rangle). \tag{3.46}$$

Since the left action on $\uparrow L_\rho$ has the form (cf. Example 3.1.21),

$$\langle a\rangle \odot_{L_\rho} \big(\langle b\rangle \vee L_\rho\big) = (\langle a\rangle * \langle b\rangle) \vee L_\rho = \langle a \cdot b\rangle \vee L_\rho,$$

we obtain from (3.46):

$$q\big(\langle a\rangle \odot_{L_\rho} \big(\langle b\rangle \vee L_\rho\big)\big) = q(\langle a \cdot b\rangle) = \langle a\rangle \odot_{\pi_\rho} \langle q(b)\rangle = \langle a\rangle \odot_{\pi_\rho} q\big(\langle b\rangle \vee L_\rho\big).$$

Finally, we use the fact that both complete lattices $\mathbb{MAX}$ and $\uparrow L_\rho$ are atomic. In particular, every atom of $\uparrow L_\rho$ has the form $\langle b\rangle \vee L_\rho$ where $b \notin L_\rho$. Since left actions can be considered as bimorphisms, the previous relation implies that Φ is a left $\mathbb{MAX}$-module homomorphism.
(b) In order to show that the 1-dimensional, linear subspace $\langle x_\rho\rangle$ is a generator of $(\mathscr{P}(\mathscr{H}), \odot_{\pi_\rho})$ we argue as follows. Since $\mathscr{P}(\mathscr{H})$ is atomic, it is sufficient to consider a 1-dimensional, linear subspace $\langle q(a)\rangle$ generated by $q(a)$ for some $a \in A$. If we now replace the element b by the unit e of A in (3.46) and recall $q(e) = x_\rho$, then (3.46) implies:

$$\langle a\rangle \odot_{\pi_\rho} \langle x_\rho\rangle = \langle a\rangle \odot_{\pi_\rho} \langle q(e)\rangle = \langle q(a \cdot e)\rangle = \langle q(a)\rangle.$$

Hence $\langle x_\rho\rangle$ is necessarily a generator and its annihilator ann $(\langle x_\rho\rangle)$ obviously coincides with L_ρ. So assertion (ii) is verified.

As an immediate corollary of Theorem 3.2.24 (i) we obtain the following improvement of Theorem 3.2.16 in the case of the Gelfand quantale $(\mathbb{MAX}, *, \langle e\rangle, ')$.

Corollary 3.2.25. *Let L be a maximal left ideal of a unital C^*-algebra. Then the symmetric 2-form $(\uparrow L) \otimes (\uparrow L) \xrightarrow{\varphi} \mathbb{1}$ determined on elementary tensors by*

$$\varphi(U_1 \otimes U_2) = \begin{cases} 0, & U_2' * U_1 \subseteq L \vee L', \\ 1, & U_2' * U_1 \nsubseteq L \vee L', \end{cases} \qquad U_1, U_2 \in \uparrow L$$

is faithful.

3.2.5 Irreducible Representations and Irreducible Involutive Left Modules

In this subsection we consider involutive left $\mathbb{MAX}$-modules having a generator x such that its annihilator ann (x) is a maximal left ideal. Hence these left modules are always irreducible (cf. Theorem 3.1.20).

Theorem 3.2.26. *Let $(X, \odot, \varphi)$ be an involutive left $\mathbb{MAX}$-module where φ is a dense 2-form. Further, let x be a generator of $(X, \odot, \varphi)$ such that its annihilator* ann (x) *is a maximal left ideal. Then $(X, \odot, \varphi)$ is isomorphic to the involutive left $\mathbb{MAX}$-module given by the upsegment* $\uparrow$ann (x) *in the sense of* $\mathbb{I}(\mathtt{Mod}_\ell(\mathbb{MAX}))$.

Proof. Let x be a generator of $(X, \odot)$ and $\operatorname{ann}(x)$ be its annihilator. Since $\operatorname{ann}(x)$ is a maximal left ideal, x cannot be the universal lower bound of X — i.e. $x \neq \bot$. Now we define a surjective and join-preserving map $\uparrow\operatorname{ann}(x) \xrightarrow{\zeta} X$ by:

$$\zeta(B) = B \odot x, \qquad B \in \uparrow\operatorname{ann}(x).$$

If $\odot_{\operatorname{ann}(x)}$ is the left action on $\uparrow\operatorname{ann}(x)$ (cf. Example 3.1.21), then the relation

$$\zeta\big(A \odot_{\operatorname{ann}(x)} B\big) = ((A * B) \vee \operatorname{ann}(x)) \odot x = (A * B) \odot x = A \odot (B \odot x) = A \odot \zeta(B)$$

holds for all $A \in \mathbb{MAX}$ and $B \in \uparrow\operatorname{ann}(x)$. Hence ζ is a left $\mathbb{MAX}$-module homomorphism. In order to show that ζ is an isomorphism in the sense of $\mathbb{I}(\mathtt{Mod}_\ell(\mathbb{MAX}))$ we proceed as follows. Since ζ is surjective, we conclude from Corollary 3.2.25 and Proposition 3.2.7 (iv) that it is sufficient to show that ζ is orthogonal.

(a) We choose $A, B \in \uparrow\operatorname{ann}(x)$ with $A \perp B$ — i.e. $B' * A \subseteq \operatorname{ann}(x) \vee \operatorname{ann}(x)'$ (cf. (3.31) and Theorem 3.2.16). Since $(X, \odot, \varphi)$ is an involutive left $\mathbb{MAX}$-module, we apply (3.33) and obtain:

$$\begin{aligned} 0 \leq \varphi(\zeta(A) \otimes \zeta(B)) &= \varphi((B' \odot (A \odot x)) \otimes x) = \varphi(((B' * A) \odot x) \otimes x) \\ &\leq \varphi\big((\operatorname{ann}(x) \vee \operatorname{ann}(x)') \odot x) \otimes x\big) = \varphi((\operatorname{ann}(x)' \odot x) \otimes x) \\ &= \varphi(x \otimes (\operatorname{ann}(x) \odot x)) = \varphi(x \otimes \bot) = 0. \end{aligned}$$

Hence $\zeta(A) \perp \zeta(B)$ follows.

(b) First we put $x^\perp = \bigvee\{z \in X \mid z \perp x\}$ and define an element D of $\uparrow\operatorname{ann}(x)$ by

$$D = \bigvee\{C \in \mathbb{MAX} \mid C \odot x \leq x^\perp\}.$$

Since x is a generator of $(X, \odot)$, the relation $D \odot x = x^\perp$ follows. Now we show

$$D \subseteq \operatorname{ann}(x) \vee \operatorname{ann}(x)'.$$

Since $\operatorname{ann}(x)$ is a maximal left ideal, we first note that there exists a unique pure state ρ of A such that the left kernel of ρ coincides with $\operatorname{ann}(x)$. Hence

$$\operatorname{ann}(x) \vee \operatorname{ann}(x)'$$

is the kernel of ρ and has codimension 1. If we now assume $D \nsubseteq \operatorname{ann}(x) \vee \operatorname{ann}(x)'$, then $\top = A = D \vee \operatorname{ann}(x) \vee \operatorname{ann}(x)'$ follows. Hence we obtain:

$$\varphi((\top \odot x) \otimes x) = \varphi(((D \odot x) \vee ((\operatorname{ann}(x) \vee \operatorname{ann}(x)') \odot x)) \otimes x) = \varphi(x^\perp \otimes x) = 0.$$

This means $(\top \odot x) \perp x$. Since $(X, \odot)$ is irreducible (cf. Theorem 3.1.20), $\top \odot x$ necessarily coincides with the universal upper bound of X. Hence $\top \perp x$ holds, and the density of the 2-form φ implies $x = \bot$, which is a contradiction to $x \neq \bot$. Thus the assumption is false and the property

$$D \subseteq \text{ann}\,(x) \vee \text{ann}\,(x)' \tag{3.47}$$

holds. Finally, we choose $A, B \in \uparrow\!\text{ann}\,(x)$ with $\zeta(A) \perp \zeta(B)$ — this means

$$0 = \varphi((A \odot x) \otimes (B \odot x)) = \varphi((B' \odot (A \odot x)\big) \otimes x) = \varphi(((B' * A) \odot x) \otimes x).$$

Hence we obtain $(B' * A) \odot x \perp x$ — i.e. $(B' * A) \odot x \leq x^{\perp}$, and $B' * A \subseteq D$ follows. By (3.47) the relation $A \perp B$ holds.

To sum up, we have shown that ζ is an orthogonal homomorphism. □

Corollary 3.2.27. *Let $(X, \odot, \varphi)$ be an involutive left $\mathbb{MAX}$-module, and let x be a generator of $(X, \odot, \varphi)$ such that its annihilator* ann (x) *is a maximal left ideal. If the 2-form φ is dense, then φ is faithful.*

Proof. Since every maximal left ideal is the left-kernel of a pure state, the assertion follows from Corollary 3.2.25 and Theorem 3.2.26. □

Corollary 3.2.28. *Let $(X, \odot, \varphi)$ be an involutive left $\mathbb{MAX}$-module with a faithful 2-form φ. Further, let x be a generator of $(X, \odot, \varphi)$ such that its annihilator* ann (x) *is a maximal left ideal. There exists an irreducible representation $(\pi, \mathscr{H})$ of the given unital C^*-algebra $(A, +, \cdot, e)$ such that the canonical left $\mathbb{MAX}$-module associated with $(\pi, \mathscr{H})$ is isomorphic to $(X, \odot, \varphi)$.*

Proof. Let ρ be the pure state whose left kernel coincides with the annihilator of the generator x of $(X, \odot, \varphi)$. By Theorems 3.2.24 and 3.2.26 the irreducible representation produced by the GNS construction from ρ satisfies the desired property. □

The next theorem gives a characterization of equivalent irreducible representations in terms of involutive left modules.

Theorem 3.2.29. *Let $(A, +, \cdot, e)$ be a unital C^*-algebra. Then two irreducible representations are equivalent if and only if their associated canonical left $\mathbb{MAX}$-modules are isomorphic in the sense of $\mathbb{I}(\mathtt{Mod}_\ell(\mathbb{MAX}))$.*

Proof. Let $(\pi_1, \mathscr{H}_1)$ and $(\pi_2, \mathscr{H}_2)$ be irreducible representations of A, and let $(\mathscr{P}(\mathscr{H}_1), \odot_{\pi_1}, \varphi)$ and $(\mathscr{P}(\mathscr{H}_2), \odot_{\pi_2}, \varphi)$ be the corresponding canonical left $\mathbb{MAX}$-modules.

(a) If $(\pi_1, \mathscr{H}_1)$ and $(\pi_2, \mathscr{H}_2)$ are equivalent, then there exists a surjective unitary transformation $\mathscr{H}_1 \xrightarrow{\Phi} \mathscr{H}_2$ such that $\pi_2(a) = \Phi \circ \pi_1(a) \circ \Phi^*$ holds for all $a \in A$, where Φ^* is the adjoint operator of Φ. We define a surjective and join-preserving map $\mathscr{P}(\mathscr{H}_1) \xrightarrow{\Upsilon} \mathscr{P}(\mathscr{H}_2)$ by

$$\Upsilon(P) = \Phi(P) = \{\Phi(x) \mid x \in P\}, \qquad P \in \mathscr{P}(\mathscr{H}_1),$$

and show that Υ is an orthogonal left $\mathbb{MAX}$-module homomorphism. Since Φ is unitary, Υ is evidently orthogonal. Further, we fix $U \in \mathbb{MAX}$ and $P \in \mathscr{P}(\mathscr{H}_1)$. Then we obtain (cf. Sect. 3.2.3):

$$\begin{aligned}\Upsilon(U \odot_{\pi_1} P) &= \Upsilon(\text{top. closure (lin. hull}\, \{\pi_1(a)(x) \mid a \in U,\, x \in P\})) \\ &= \text{top. closure (lin. hull}\, \{\Phi(\pi_1(a)(x)) \mid a \in U,\, x \in P\}) \\ &= \text{top. closure (lin. hull}\, \{\Phi(\pi_1(a)(\Phi^*(\Phi(x)))) \mid a \in U,\, x \in P\}) \\ &= \text{top. closure (lin. hull}\, \{\pi_2(a)(y) \mid a \in U,\, y \in \Upsilon(P)\}) \\ &= U \odot_{\pi_2} \Upsilon(P).\end{aligned}$$

Hence Υ is a left $\mathbb{MAX}$-module homomorphism. So we have shown that Υ is an isomorphism in the sense of $\mathbb{I}(\mathtt{Mod}_\ell(\mathbb{MAX}))$.

(b) Let $\mathscr{P}(\mathscr{H}_1) \xrightarrow{\Upsilon} \mathscr{P}(\mathscr{H}_2)$ be an isomorphism in the sense of $\mathbb{I}(\mathtt{Mod}_\ell(\mathbb{MAX}))$. Then Υ is an order isomorphism and maps atoms of $\mathscr{P}(\mathscr{H}_1)$ to atoms of $\mathscr{P}(\mathscr{H}_2)$. Therefore, if we fix a unit vector $x \in \mathscr{H}_1$, then we can choose a unit vector $y \in \mathscr{H}_2$ with $y \in \Upsilon(\langle x\rangle)$, where $\langle x\rangle$ is the 1-dimensional, linear subspace generated by x. Since both representations are irreducible, we can consider the following pure states of A:

$$\rho_1 = \omega_x \circ \pi_1 \quad \text{and} \quad \rho_2 = \omega_y \circ \pi_2.$$

Then we infer from Corollary 3.2.21 that the left-kernel of ρ_1 coincides with the annihilator of $\langle x\rangle$ — i.e. $L_{\rho_1} = \text{ann}\,(\langle x\rangle)$, and the left-kernel of ρ_2 coincides with the annihilator of $\langle y\rangle = \Upsilon(\langle x\rangle)$ — i.e. $L_{\rho_2} = \text{ann}\,(\langle y\rangle)$. Since Υ is a left $\mathbb{MAX}$-module isomorphism, the relation $\text{ann}\,(\langle x\rangle) = \text{ann}\,(\langle y\rangle)$ follows — i.e. $L_{\rho_1} = L_{\rho_2}$. Hence the pure states ρ_1 and ρ_2 coincide. Now we consider the irreducible representation $(\pi_\rho, \mathscr{H}_\rho)$ produced by the GNS construction from the pure state $\rho = \rho_1 = \rho_2$. Then the essential uniqueness of the GNS construction (cf. Theorem 3.2.22) says that $(\pi_1, \mathscr{H}_1)$ and $(\pi_\rho, \mathscr{H}_\rho)$ as well as $(\pi_\rho, \mathscr{H}_\rho)$ and $(\pi_2, \mathscr{H}_2)$ are equivalent, and consequently $(\pi_1, \mathscr{H}_1)$ and $(\pi_2, \mathscr{H}_2)$ are equivalent. □

Let $(A, +, \cdot, e)$ be a unital C^*-algebra. If we read irreducible representations up to (unitary) equivalence and involutive left $\mathbb{MAX}$-modules up to isomorphism in the sense of $\mathbb{I}(\mathtt{Mod}_\ell(\mathbb{MAX}))$, then the previous results mean that irreducible representations of $(A, +, \cdot, e)$ can be identified with involutive left $\mathbb{MAX}$-modules $(X, \odot, \varphi)$ satisfying the following conditions:

(1) The 2-form φ is faithful.
(2) $(X, \odot)$ has a generator whose annihilator is a maximal left ideal of $(A, +, \cdot, e)$.

Exercises

3.2.1. Let (X, φ_X), (Y, φ_Y) and (Z, φ_Z) be symmetric lattices. In this context $\ell_{\mathbb{1}}$ is the 2-form on the unit object $\mathbb{1}$. Further, let

$$(X \otimes Y) \otimes Z \xrightarrow{a_{XYZ}} X \otimes (Y \otimes Z), \qquad X \otimes Y \xrightarrow{c_{XY}} Y \otimes X,$$
$$\mathbb{1} \otimes X \xrightarrow{\ell_X} X, \qquad X \otimes \mathbb{1} \xrightarrow{r_X} X$$

be the components of the natural isomorphisms occurring in the coherence conditions of the monoidal category Sup. We recall that the tensor product of symmetric lattices is constructed as the tensor product in Sup provided with the respective tensor product of 2-forms. Show:

(a) The isomorphism $(X \otimes Y) \otimes Z \xrightarrow{a_{XYZ}} X \otimes (Y \otimes Z)$ is orthogonal.
(Hint: Apply Proposition A.2.1 in Appendix A.2.)

(b) The isomorphism $X \otimes Y \xrightarrow{c_{XY}} Y \otimes X$ is orthogonal.
(Hint: First verify the following relation for objects X and Y in a symmetric monoidal category

$$\Theta_{YXYX} \circ (c_{XY} \otimes c_{XY}) = c_{(X\otimes X)(Y\otimes Y)} \circ \Theta_{XYXY},$$

where Θ_{YXYX} and Θ_{XYXY} are defined in Appendix A.2.)

(c) The isomorphism $\mathbb{1} \otimes X \xrightarrow{\ell_X} X$ is orthogonal.

(d) The isomorphism $X \otimes \mathbb{1} \xrightarrow{r_X} X$ is orthogonal.

3.2.2. Let $(\mathfrak{Q}, *, e, ')$ be an involutive and unital quantale and (X, φ_X) be a symmetric lattice. We choose the universal lower bound in $\mathfrak{Q}$ as a hermitian element — i.e. $\delta = \perp$, and consider the symmetric 2-form $\varphi_{\mathfrak{Q}}$ on $\mathfrak{Q}$ constructed in Example 3.2.12. Then $(\mathfrak{Q} \otimes X, \varphi_{\mathfrak{Q}\otimes X})$ is a symmetric lattice. Further, let $X \xrightarrow{\eta_X} \mathfrak{Q} \otimes X$ be the X-component of the unit of the adjoint situation between Sup and $\mathsf{Mod}_\ell(\mathfrak{Q})$ (cf. Sect. 3.1) — i.e. $\eta_X(x) = e \otimes x$ for all $x \in X$. Show that η_X is an orthogonal map from (X, φ_X) to $(\mathfrak{Q} \otimes X, \varphi_{\mathfrak{Q}\otimes X})$.

3.2.3. Let $\mathfrak{Q} = ([0, 1], *_Ł)$ be the real unit interval provided with the Łukasiewicz arithmetic conjunction. Then $([0, 1], *_Ł)$ is a complete MV-algebra (cf. Sect. 2.7) and an involutive left $\mathfrak{Q}$-module with the order-reversing involution $\alpha \longmapsto 1 - \alpha$ in the sense of Example 3.2.12 ($\delta = 0$). Further, let $([0, 1] \otimes [0, 1], \odot, \varphi_{[0,1]\otimes[0.1]})$ be the involutive left $\mathfrak{Q}$-module constructed in Example 3.2.14 and

$$[0, 1] \xrightarrow{\eta_{[0,1]}} [0, 1] \otimes [0, 1], \quad \eta_{[0,1]}(\alpha) = 1 \otimes \alpha, \qquad \alpha \in [0, 1]$$

be the orthogonal embedding (cf. Exercise 3.2.2).

Show that the identity map of $[0, 1]$ does not have an extension to an orthogonal left $\mathfrak{Q}$-module homomorphism — i.e. there does not exist an orthogonal left $\mathfrak{Q}$-module homomorphism $[0,1] \otimes [0,1] \xrightarrow{h} [0,1]$ making the following diagram commutative in $\mathbb{O}(\mathsf{Sup})$:

$$\begin{array}{ccc} [0,1] & \xrightarrow{\eta_{[0,1]}} & [0,1] \otimes [0,1] \\ & \searrow^{1_{[0,1]}} & \downarrow h \\ & & [0,1] \end{array}$$

(Hint: By Theorem 3.1.11 the identity map $1_{[0,1]}$ has a unique extension to a left $\mathfrak{Q}$-module homomorphism $[0,1] \otimes [0,1] \xrightarrow{h} [0,1]$ such that $1_{[0,1]} = h \circ \eta_{[0,1]}$ holds in Sup. In particular, h is given by the Łukasiewicz arithmetic conjunction as multiplication in $[0,1]$ — i.e. $h(\alpha \otimes \beta) = \alpha *_{Ł} \beta$. Since the right adjoint map $h^{\vdash}$ of h has the form $h^{\vdash}(\gamma)(\alpha) = \alpha \to \gamma = \min(1 - \alpha + \gamma, 1)$ (cf. Theorem 2.2.10), h cannot be orthogonal.)

3.2.4. A 2-form φ on a complete lattice is called *non-trivial* if $\varphi(\top \otimes \top) = 1$ holds. Now let (X_1, φ_{X_1}), (X_2, φ_{X_2}), (Y_1, φ_{Y_1}) and (Y_2, φ_{Y_2}) be symmetric lattices with non-trivial 2-forms and $X_1 \xrightarrow{h} Y_1$, $X_2 \xrightarrow{k} Y_2$ be join-preserving maps with the property $h(\top) = \top$ and $k(\top) = \top$. If $X_1 \otimes Y_1 \xrightarrow{h \otimes k} X_2 \otimes Y_2$ is orthogonal with respect to $\varphi_{X_1 \otimes Y_1}$ and $\varphi_{X_2 \otimes Y_2}$, then show that h and k are also orthogonal.

3.2.5. Let $\mathbb{B}$ be a complete Boolean algebra. On $\mathbb{B}$ we introduce a symmetric 2-form $\varphi_{\mathbb{B}}$ as follows:

$$\varphi_{\mathbb{B}}(x_1 \otimes x_2) = \begin{cases} 0, & x_1 \wedge x_2 = \bot, \\ 1, & x_1 \wedge x_2 \neq \bot, \end{cases} \qquad x_1, x_2 \in \mathbb{B}.$$

Further, let $\varphi_{\mathbb{B} \otimes \mathbb{B}}$ be the tensor product of the 2-form $\varphi_{\mathbb{B}}$ with itself. Show:

(a) $\varphi_{\mathbb{B}}$ is faithful and the corresponding order-reversing involution $f_{\varphi_{\mathbb{B}}}$ coincides with the formation of complements in $\mathbb{B}$.
(b) The join-reversing map $f_{\varphi_{\mathbb{B} \otimes \mathbb{B}}}$ corresponding to $\varphi_{\mathbb{B} \otimes \mathbb{B}}$ coincides with the formation of pseudo-complements in $\mathbb{B} \otimes \mathbb{B}$ viewed as a frame (cf. Corollary 2.4.13). (Hint: Lemma 3.2.3 and formula (2.77), resp. (2.78).)
(c) If $\mathbb{B}$ is atomless, then the 2-form $\varphi_{\mathbb{B} \otimes \mathbb{B}}$ is not faithful. (Hint: Use Exercise 2.6.4 (a).)

3.2.6. Let $(\mathfrak{Q}, *, e, ')$ be an involutive and unital quantale and $(\mathfrak{Q}, \odot, \varphi_{\mathfrak{Q}})$ be an involutive left $\mathfrak{Q}$-module such that $\alpha \odot \beta = \alpha * \beta$ for all $\alpha, \beta \in \mathfrak{Q}$. Further, let $f_{\varphi_{\mathfrak{Q}}}$ be the join-reversing map corresponding to $\varphi_{\mathfrak{Q}}$. Show that $\varphi_{\mathfrak{Q}}$ is faithful if and only if $(\mathfrak{Q}, *, f_{\varphi_{\mathfrak{Q}}}(e))$ is a Frobenius quantale (cf. Example 3.2.12).
(Hint: Proposition 3.2.13.)

3.2.7. Let X be a nonempty set and $(\mathfrak{Q}, *, e, ')$ be an involutive and unital quantale. We view X as a preordered set provided with the discrete order. Then $\mathsf{Dwn}(X)$ is the power set $\mathscr{P}(X)$ of X. Further, we consider the symmetric 2-form $\varphi_{\mathfrak{Q}}$ on $\mathfrak{Q}$ defined in Example 3.2.12 and the symmetric 2-form $\varphi_{\mathscr{P}(X)}$ on $\mathscr{P}(X)$ corresponding to the complementation in $\mathscr{P}(X)$. Then $(\mathfrak{Q} \otimes \mathscr{P}(X), \odot, \varphi_{\mathfrak{Q}\otimes\mathscr{P}(X)})$ is an involutive left $\mathfrak{Q}$-module (see Example 3.2.14).

Further, let δ be the hermitian element occurring in the construction of $\varphi_{\mathfrak{Q}}$ (cf. Example 3.2.12). We introduce a symmetric 2-form $\varphi_{\mathfrak{Q}^X}$ on the complete lattice $\mathfrak{Q}^X$ corresponding to the following join-reversing map

$$f \longmapsto f^\delta \quad \text{where} \quad f^\delta(x) = f(x)' \searrow \delta, \qquad x \in X,\ f \in \mathfrak{Q}^X.$$

Finally, we recall the left action $\odot$ on $\mathfrak{Q}^X$ determined by $(\alpha \odot f)(x) = \alpha * f(x)$ for all $x \in X$ (cf. Example 3.1.8).

(a) Show that $(\mathfrak{Q}^X, \odot, \varphi_{\mathfrak{Q}^X})$ is an involutive left $\mathfrak{Q}$-module.
(b) Show that $(\mathfrak{Q}^X, \odot, \varphi_{\mathfrak{Q}^X})$ and $(\mathfrak{Q} \otimes \mathscr{P}(X), \odot, \varphi_{\mathfrak{Q}\otimes\mathscr{P}(X)})$ are isomorphic in the sense of $\mathbb{I}(\mathtt{Mod}_\ell(\mathfrak{Q}))$.
(Hint: Apply Theorem 3.1.10, (3.9) and Lemma 3.2.3.)

3.2.8. Let $(A, +, \cdot, e)$ be a unital C^*-algebra and $\mathbb{MAX}$ be the involutive and unital quantale of all closed linear subspaces of A (cf. Example 2.5.5). We maintain the notation from Sect. 2.5.3 and recall that the spectrum $\mathrm{sp}(A)$ of $(A, +, \cdot)$ is given by the quantic frame $\mathbb{P}(\mathbb{MAX})$ associated with $\mathbb{MAX}$. Further, let $(\vartheta, \mathscr{H})$ be a representation of A and $\mathbb{MAX} \xrightarrow{h_\vartheta} [\mathscr{P}(\mathscr{H}), \mathscr{P}(\mathscr{H})]$ be the involutive and unital homomorphism determined by (3.41) (cf. Lemma 3.2.17). Show:

(a) The involutive and unital homomorphism h_ϑ is a strong homomorphism if and only if $(\vartheta, \mathscr{H})$ is an irreducible representation.
(b) If $(\vartheta, \mathscr{H})$ is an irreducible representation of A, then there exists a unique involutive homomorphism h from $\mathbb{P}(\mathbb{MAX})$ to the semi-integral regularization of $[\mathscr{P}(\mathscr{H}), \mathscr{P}(\mathscr{H})]$ satisfying the condition $h(\pi_{\mathbb{MAX}}(L \otimes R)) = h_\vartheta(R * L)$ for all closed left ideals L and all closed right ideals R.
(Hint: Let $\mathbb{MAX}_{\mathrm{sr}}$ and $[\mathscr{P}(\mathscr{H}), \mathscr{P}(\mathscr{H})]_{\mathrm{sr}}$ be the semi-integral regularizations of $\mathbb{MAX}$ and $[\mathscr{P}(\mathscr{H}), \mathscr{P}(\mathscr{H})]$, respectively. Then conclude from (a) that h_ϑ is also an involutive homomorphism from $\mathbb{MAX}_{\mathrm{sr}}$ to $[\mathscr{P}(\mathscr{H}), \mathscr{P}(\mathscr{H})]_{\mathrm{sr}}$. Further, let $\mathbb{I}(A) \xrightarrow{q_{\mathbb{L}}} \mathbb{L}(A) \xrightarrow{\iota_{\mathbb{L}}} \mathbb{MAX}_{\mathrm{sr}}$ and $\mathbb{I}(A) \xrightarrow{q_{\mathbb{R}}} \mathbb{R}(A) \xrightarrow{\iota_{\mathbb{R}}} \mathbb{MAX}_{\mathrm{sr}}$ be the respective embeddings. Now we infer from the universal property of the pushout square (cf. Definition 2.5.13 (c), see also the proof of Theorem 2.5.12) that there exists a unique involutive homomorphism $\mathbb{P}(\mathbb{MAX}) \xrightarrow{h_0} \mathbb{MAX}_{\mathrm{sr}}$ such that $h_0 \circ \rho_{\mathbb{L}} = \iota_{\mathbb{L}}$ and $h_0 \circ \rho_{\mathbb{R}} = \iota_{\mathbb{R}}$ hold. Moreover, the relation

$$h_0\big(\pi_{\mathbb{MAX}}(L \otimes R)\big) = R * L, \qquad L \in \mathbb{L}(A),\ R \in \mathbb{R}(A)$$

follows from the construction of h_0. Thus $h := h_\vartheta \circ h_0$ satisfies the desired properties.)

3.3 $\mathfrak{Q}$-Modules and (Co)Complete $\mathfrak{Q}$-Preordered Sets

Let $\mathfrak{Q} = (\mathfrak{Q}, *, e)$ be an arbitrary unital quantale and $\mathfrak{Q}^\tau$ be its transposed quantale. Since every right action over $\mathfrak{Q}$ can be read as a left action over $\mathfrak{Q}^\tau$, there exists a (covariant) isomorphism between the category $\mathtt{Mod}_r(\mathfrak{Q})$ of right $\mathfrak{Q}$-modules and the category $\mathtt{Mod}_\ell(\mathfrak{Q}^\tau)$ of left $\mathfrak{Q}^\tau$-modules. Hence this situation suggests to restrict our interest to one of these categories. We decide to focus first on right $\mathfrak{Q}$-modules and begin with a remarkable theorem.

Theorem 3.3.1. *Let $(X, \boxdot)$ be a right $\mathfrak{Q}$-module. Then the map $X \times X \xrightarrow{p} \mathfrak{Q}$ defined by*

$$p(x, y) = \bigvee\{\alpha \in \mathfrak{Q} \mid x \boxdot \alpha \le y\}, \qquad x, y \in X \tag{3.48}$$

satisfies the following properties:

(O1) $e \le p(x, x)$ *for each* $x \in X$, (Reflexivity)
(O2) $p(x, y) * p(y, z) \le p(x, z)$ *for each* $x, y, z \in X$. (Transitivity)

Proof. The property (O1) follows immediately from axiom (M2$_r$). In order to verify (O2) we use the right $\mathfrak{Q}$-module axiom (M1$_r$) and observe that the following relation holds for $x, y, z \in X$:

$$x \boxdot (p(x, y) * p(y, z)) = (x \boxdot p(x, y)) \boxdot p(y, z) \le y \boxdot p(y, z) \le z.$$

Hence (O2) follows from (3.48). □

A map $X \times X \xrightarrow{p} \mathfrak{Q}$ is called a *$\mathfrak{Q}$-valued preorder* on X (or *$\mathfrak{Q}$-preorder* for short) if p satisfies the conditions (O1) and (O2). A pair (X, p) is a *$\mathfrak{Q}$ -valued preordered set* (or *$\mathfrak{Q}$ -preordered set* for short) if X is a set and p is a $\mathfrak{Q}$-preorder on X.

If we read preordered sets as thin and small categories, then unital quantales $\mathfrak{Q}$ are monoidal biclosed and complete categories (cf. [63]), where the tensor product is given by the multiplication $*$ in $\mathfrak{Q}$. In this sense $\mathfrak{Q}$-preordered sets and $\mathfrak{Q}$-enriched categories (cf. [63]) are equivalent concepts. Obviously, the reflexivity axiom (O1) means the existence of $\mathfrak{Q}$-enriched identities, and the transitivity axiom (O2) is equivalent to the $\mathfrak{Q}$-enriched composition law in the following sense. With every map $X \times X \xrightarrow{p} \mathfrak{Q}$ we can associate a map $X \times X \xrightarrow{p^\tau} \mathfrak{Q}$ defined by $p^\tau(x, y) = p(y, x)$ for all $x, y \in X$. Then p is transitive (i.e. satisfies (O2)) if and only if p^τ satisfies the following condition for each $x, y, z \in X$:

(O2)$'$ $p^\tau(y, z) * p^\tau(x, y) \le p^\tau(x, z)$. ($\mathfrak{Q}$-Enriched Composition Law)

Hence, in the language of enriched category theory, the statement of Theorem 3.3.1 says that every right $\mathfrak{Q}$-module has an intrinsic structure of a $\mathfrak{Q}$-enriched category.

The aim of this section is to characterize right (left) $\mathfrak{Q}$-modules in terms of $\mathfrak{Q}$−preordered sets. For this purpose we make a digression on $\mathfrak{Q}$-preorders and develop $\mathfrak{Q}$-valued analogues of some significant properties explained in Sect. 1.3.

3.3.1 $\mathfrak{Q}$-Preorders and Cocompleteness

As a first important observation we note that every $\mathfrak{Q}$-preorder p on a set X has an underlying preorder $\leq_p$ given by

$$x \leq_p y \iff e \leq p(x, y). \tag{3.49}$$

Since this preorder will play an important rôle in the context of p, we introduce a special name and call $\leq_p$ the *intrinsic preorder* underlying p. In order to simplify the notation we will sometimes suppress the index p and will denote an intrinsic preorder by $\leq$, if a confusion is not possible. From the perspective of enriched category theory the intrinsic preorder obviously plays the rôle of the underlying ordinary category (cf. p. 26 in [63]).

A $\mathfrak{Q}$-preorder is *antisymmetric* if its intrinsic preorder is antisymmetric.

Let (X, p) and (Y, q) be $\mathfrak{Q}$-preordered sets. A map $X \xrightarrow{h} Y$ is a $\mathfrak{Q}$-*homomorphism* (also called $\mathfrak{Q}$-*functor* (cf. [63])) if h satisfies the following condition for all $x, z \in X$:

$$p(x, z) \leq q(h(x), h(z)).$$

Obviously $\mathfrak{Q}$-homomorphisms are always isotone w.r.t. the underlying intrinsic preorders. Moreover, $\mathfrak{Q}$-preordered sets and $\mathfrak{Q}$-homomorphisms form a category denoted by $\mathtt{Preord}(\mathfrak{Q})$.

It is easily seen that $\mathtt{Preord}(\mathfrak{Q})$ is topological over $\mathtt{Set}$. Hence $\mathtt{Preord}(\mathfrak{Q})$ is complete and cocomplete. But in the general setting of non-commutative and unital quantales it is an open question whether $\mathtt{Preord}(\mathfrak{Q})$ is a monoidal biclosed category (see Exercise 3.3.9 in the case of commutative quantales). Finally, we note that the category of preordered sets is isomorphic to a full and coreflective subcategory of $\mathtt{Preord}(\mathfrak{Q})$. In this sense $\mathtt{Preord}(\mathfrak{Q})$ is an extension of $\mathtt{Preord}$.

We begin with a property which forces maps to be $\mathfrak{Q}$-homomorphisms.

Lemma 3.3.2. *Let (X, p) and (Y, q) be $\mathfrak{Q}$-preordered sets, and let $X \underset{k}{\overset{h}{\rightleftarrows}} Y$ be a pair of maps. If h and k satisfy the following property for all $x \in X$ and $y \in Y$:*

$$q(h(x), y) = p(x, k(y)), \tag{$\mathfrak{Q}AD$}$$

then h and k are $\mathfrak{Q}$-homomorphisms.

Proof. We fix $x_1, x_2 \in X$ and conclude from (O1), (O2) and ($\mathfrak{Q}$AD) that the following relation holds:

$$\begin{aligned} p(x_1, x_2) &\leq p(x_1, x_2) * q\big(h(x_2), h(x_2)\big) = p(x_1, x_2) * p\big(x_2, k(h(x_2))\big) \\ &\leq p\big(x_1, k(h(x_2))\big) = q\big(h(x_1), h(x_2)\big). \end{aligned}$$

Hence h is a $\mathfrak{Q}$-homomorphism. If we interchange the rôle of p and q, then it is easily seen that k is also a $\mathfrak{Q}$-homomorphism. □

Motivated by the previous lemma we make the following definition (cf. [63]). A pair of $\mathfrak{Q}$-homomorphisms $(X, p) \underset{k}{\overset{h}{\rightleftarrows}} (Y, q)$ is called *$\mathfrak{Q}$-enriched adjoint* if (h, k) satisfies the condition ($\mathfrak{Q}$AD) in Lemma 3.3.2. In this context h is left adjoint to k, k is right adjoint to h, and we write $h \dashv k$.

A characterization of $\mathfrak{Q}$-enriched adjointness is given in the next lemma.

Lemma 3.3.3. *Let (X, p) and (Y, q) be $\mathfrak{Q}$-preordered sets. Further, let $\le_p$ and $\le_q$ denote the respective underlying intrinsic preorders. A pair $X \underset{k}{\overset{h}{\rightleftarrows}} Y$ of $\mathfrak{Q}$-homomorphisms is a $\mathfrak{Q}$-enriched adjoint pair if and only if the following relation holds for all $x \in X$ and $y \in Y$*

$$x \le_p k(h(x)) \quad \text{and} \quad h(k(y)) \le_q y. \qquad (\mathfrak{Q}AD')$$

Proof. The properties (O1), (3.49) and ($\mathfrak{Q}$AD) immediately imply ($\mathfrak{Q}$AD′). On the other hand, because of (3.49), the condition ($\mathfrak{Q}$AD′) is equivalent to the conjunction $e \le p(x, k(h(x)))$ and $e \le q(h(k(y)), y)$. Now we apply (O2) and obtain:

$$q(h(x), y) \le p(k(h(x)), k(y)) \le p(x, k(y)) \le q(h(x), h(k(y))) \le q(h(x), y).$$

Hence ($\mathfrak{Q}$AD) follows. □

By the previous lemma, $\mathfrak{Q}$-enriched adjointness and adjointness in the sense of the respective intrinsic preorders are equivalent concepts. Hence we suppress the adjective "$\mathfrak{Q}$-enriched" and speak of an *adjoint pair of $\mathfrak{Q}$-homomorphisms*. Referring to Sect. 1.3, it is evident that adjoint $\mathfrak{Q}$-homomorphisms are uniquely determined by each other up to equivalence.

Comment. The connection of Lemmas 3.3.2 and 3.3.3 goes back to a theorem due to H. Lai and D. Zhang. In this context the superfluous assumption of commutativity is present (cf. [71, Theorem 2.10]).

Before we can introduce "cocomplete" $\mathfrak{Q}$-valued preordered sets, we have to define the concept of a $\mathfrak{Q}$-valued downclosed subset. It is well-known that a downclosed subset A of a preordered set X can be viewed as a contravariant 2-functor from X to $\mathbb{1} = \{0, 1\}$. If we replace $\mathbb{1}$ by $\mathfrak{Q}$, then a "$\mathfrak{Q}$-valued downclosed subset" of a $\mathfrak{Q}$-preordered set (X, p) is a *$\mathfrak{Q}$-enriched contravariant presheaf* f on (X, p) — i.e. a map $X \xrightarrow{f} \mathfrak{Q}$ satisfying the following condition:

$$p(z, x) * f(x) \le f(z), \qquad x, z \in X. \qquad \text{(Left Extensionality)}$$

The set of all $\mathfrak{Q}$-enriched contravariant presheaves on (X, p) is denoted by $\mathbb{P}(X, p)$. Obviously, every element x of X induces a $\mathfrak{Q}$-enriched contravariant presheaf $\widetilde{x}$ by:

$$\widetilde{x}(z) = p(z, x), \qquad z \in X.$$

The next example describes the structure of a right $\mathfrak{Q}$-module on $\mathbb{P}(X, p)$.

Example 3.3.4. Let (X, p) be a $\mathfrak{Q}$-ordered set. On $\mathbb{P}(X, p)$ we define a partial order as follows:

$$f \le g \iff f(x) \le g(x) \quad \text{for all } x \in X. \tag{3.50}$$

Then $(\mathbb{P}(X, p), \le)$ is a complete lattice. On $\mathbb{P}(X, p)$ we introduce a right action $\boxdot$ determined by:

$$(f \boxdot \alpha)(x) = f(x) * \alpha, \qquad \alpha \in \mathfrak{Q},\ f \in \mathbb{P}(X, p),\ x \in X. \tag{3.51}$$

It is easily seen that the $\mathfrak{Q}$-preorder d on $\mathbb{P}(X, p)$ induced by $\boxdot$ in the sense of Theorem 3.3.1 has the following form for each $f, g \in \mathbb{P}(X, p)$:

$$d(f, g) = \bigvee\{\alpha \in \mathfrak{Q} \mid \forall x \in X:\ f(x) * \alpha \le g(x)\} = \bigwedge_{x \in X} \big(f(x) \searrow g(x)\big).$$

Obviously, the intrinsic preorder underlying $(\mathbb{P}(X, p), d)$ is antisymmetric and coincides with the partial order defined in (3.50). Moreover, the *$\mathfrak{Q}$-enriched Yoneda embedding* $(X, p) \xrightarrow{\eta_{(X,p)}} (\mathbb{P}(X, p), d)$ is the $\mathfrak{Q}$-homomorphism given by:

$$\eta_{(X,p)}(x) = \widetilde{x} \qquad x \in X.$$

In this context, the following relation holds for all $x \in X$ and $f \in \mathbb{P}(X, p)$:

$$d(\eta_{(X,p)}(x), f) = f(x). \tag{3.52}$$

The next lemma is a $\mathfrak{Q}$-valued version of Lemma 1.3.3 and explains under which condition the $\mathfrak{Q}$-enriched Yoneda embedding $\eta_{(X,p)}$ has a left adjoint $\mathfrak{Q}$-homomorphism.

Lemma 3.3.5. *Let (X, p) be a $\mathfrak{Q}$-preordered set and $\mathbb{P}(X, p) \xrightarrow{\xi} X$ be a $\mathfrak{Q}$-homomorphism. Then the following assertions are equivalent*:

(i) *ξ is left adjoint to $\eta_{(X,p)}$.*
(ii) *$\xi \circ \eta_{(X,p)}$ and 1_X are naturally equivalent in the sense of the underlying intrinsic preorders.*

Proof. Let $\le_p$ be the intrinsic preorder underlying (X, p). By Lemma 3.3.3 it is sufficient to establish the following equivalence:

$$\big(\forall x \in X,\ x \le_p \xi(\eta_{(X,p)}(x))\big) \iff \big(\forall f \in \mathbb{P}(X, p),\ f \le \eta_{(X,p)}(\xi(f))\big)$$

In order to verify the necessity we choose $f \in \mathbb{P}(X, p)$ and apply (3.52) and (O2):

$$f(x) = d(\widetilde{x}, f) \le p(\xi(\widetilde{x}), \xi(f)) \le p(x, \xi(\widetilde{x})) * p(\xi(\widetilde{x})), \xi(f)) \le p(x, \xi(f)), \quad x \in X.$$

Hence $f \le \eta_{(X,p)}(\xi(f))$ follows. In order to show that the condition is also sufficient we consider the special case $f = \widetilde{x} = \eta_{(X,p)}(x)$ and obtain:

$$e \le \widetilde{x}(x) \le (\eta_{(X,p)}(\xi(\widetilde{x})))(x) = p(x, \xi(\widetilde{x})).$$

Hence $x \le_p \xi(\eta_{(X,p)}(x))$ follows for all $x \in X$. □

Motivated by the previous lemma we introduce the following terminology and notation.

Definition 3.3.6. A $\mathfrak{Q}$-preordered set (X, p) is *cocomplete* (or *$\mathfrak{Q}$-enriched join-complete*) if the $\mathfrak{Q}$-enriched Yoneda embedding $X \xrightarrow{\eta_{(X,p)}} \mathbb{P}(X, p)$ has a left adjoint $\mathfrak{Q}$-homomorphism $\mathbb{P}(X, p) \xrightarrow{\xi} X$. Since ξ is uniquely determined up to equivalence, we also write $\mathsf{sup}_{(X,p)}$ instead of ξ and call $\mathsf{sup}_{(X,p)}$ the *formation of joins of $\mathfrak{Q}$-enriched contravariant presheaves* on (X, p).

A *join-complete $\mathfrak{Q}$-valued lattice* is a cocomplete $\mathfrak{Q}$-preordered set (X, p) with an antisymmetric $\mathfrak{Q}$-preorder p.

Before we proceed, we give a characterization of cocomplete $\mathfrak{Q}$-preordered sets. For this purpose we need some further terminology.

Definition 3.3.7. Let (X, p) be a $\mathfrak{Q}$-preordered set. An element $x_0 \in X$ is called an *upper bound of a $\mathfrak{Q}$-enriched contravaraint presheaf f* on (X, p) if $f \le \eta_{(X,p)}(x_0)$ holds. An upper bound x_0 of a $\mathfrak{Q}$-enriched contravariant presheaf f is called a *join of f* if for all $\alpha \in \mathfrak{Q}$ and for all $x \in X$ the following additional implication is true:

$$\big(\forall z \in X: \ f(z) * \alpha \le p(z, x)\big) \implies \alpha \le p(x_0, x).$$

From (O2) it is easily seen that an element $x_0 \in X$ is a join of some $\mathfrak{Q}$-enriched presheaf f on (X, p) if and only if the equivalence

$$\big(\forall z \in X: \ f(z) * \alpha \le p(z, x)\big) \iff \alpha \le p(x_0, x) \tag{3.53}$$

is valid for all $\alpha \in \mathfrak{Q}$ and all $x \in X$. Moreover, if the underlying $\mathfrak{Q}$-preorder is antisymmetric, then joins of $\mathfrak{Q}$-enriched contravariant presheaves are unique.

Theorem 3.3.8. *For every $\mathfrak{Q}$-preordered set (X, p) the following assertions are equivalent:*

(i) *(X, p) is cocomplete.*
(ii) *Every $\mathfrak{Q}$-enriched contravariant presheaf on (X, p) has a join.*

Proof. Let f be a $\mathfrak{Q}$-enriched contravariant presheaf on a $\mathfrak{Q}$-preordered set (X, p). By (3.53) an element $x_0 \in X$ is a join of f if and only if $p(x_0, x) = d\big(f, \eta_{(X,p)}(x)\big)$

holds for all $x \in X$. Hence (i) implies (ii).

On the other hand, if (ii) holds, then for every $\mathfrak{Q}$-enriched presheaf f on (X, p) the set $\mathbb{A}_f$ of all joins of f is nonempty. By the axiom of choice we now select a choice function $\mathbb{P}(X, p) \xrightarrow{\xi} X$ such that $\xi(f)$ is contained in $\mathbb{A}_f$ for all $f \in \mathbb{P}(X, p)$. Since $p\big(\xi(f), x\big) = d\big(f, \eta_{(X,p)}(x)\big)$ holds for all $x \in X$ and $f \in \mathbb{P}(X, p)$, we infer from Lemma 3.3.2 that ξ is a $\mathfrak{Q}$-homomorphism which is left adjoint to $\eta_{(X,p)}$. Thus (X, p) is cocomplete, and (i) is verified. □

The next lemma shows that cocompleteness of $\mathfrak{Q}$-preordered sets can be considered as an enrichment of the traditional concept of complete preordered sets.

Lemma 3.3.9. *Let (X, p) be a cocomplete $\mathfrak{Q}$-preordered set and $\leq_p$ be its underlying intrinsic preorder. Then $(X, \leq_p)$ is a complete preordered set.*

Proof. Let $\mathsf{sup}_{(X,p)}$ be the formation of joins of $\mathfrak{Q}$-enriched contravariant presheaves on (X, p). Since for all $x_1, x_2 \in X$ the chain of equivalences

$$\widetilde{x_1} \leq \widetilde{x_2} \iff e \leq p(x_1, x_2) \iff x_1 \leq_p x_2$$

holds, we can identify downclosed subsets A of X with $\mathfrak{Q}$-enriched contravariant presheaves $f_A = \bigvee_{a \in A} \widetilde{a}$. Hence we can restrict $\mathsf{sup}_{(X,p)}$ to $\mathsf{Dwn}(X)$. In this situation the assertion follows from Lemmas 3.3.5, 1.3.3 and Theorem 1.3.4. □

Comment. If we anticipate Exercise 3.3.8, then Lemma 3.3.9 is obviously a special case of the general property that cocompleteness is inherited under a change of base.

The next example plays a strategic rôle in the theory of join-complete $\mathfrak{Q}$-valued lattices.

Example 3.3.10. Let (X, p) be a $\mathfrak{Q}$-preordered set and $(\mathbb{P}(X, p), d)$ be the $\mathfrak{Q}$-preordered set of all $\mathfrak{Q}$-enriched, contravariant presheaves on (X, p). Then d is antisymmetric (cf. Example 3.3.4), and $\big(\mathbb{P}(X, p), d\big)$ is a join-complete $\mathfrak{Q}$-valued lattice.

In fact, let us consider the $\mathfrak{Q}$-homomorphism $\mathbb{P}\big(\mathbb{P}(X, p)\big) \xrightarrow{\mu_{(X,p)}} \mathbb{P}(X, p)$ given by:

$$\mu_{(X,p)}(F)(x) = \bigvee_{f \in \mathbb{P}(X,p)} f(x) * F(f), \qquad F \in \mathbb{P}(\mathbb{P}(X, p), d),\ x \in X. \quad (3.54)$$

It is easily seen that for every $\mathfrak{Q}$-enriched contravariant presheaf f on (X, p) the relation

$$\big((\mu_{(X,p)} \circ \eta_{(\mathbb{P}(X,p),d)})(f)\big)(x) = \bigvee_{g \in \mathbb{P}(X,p)} g(x) * d(g, f) = f(x)$$

holds for all $x \in X$. Hence $(\mathbb{P}(X, p), d)$ is a join-complete $\mathfrak{Q}$-valued lattice. In particular, we will also denote $\mu_{(X,p)}$ by $\mathsf{sup}_{(\mathbb{P}(X,p),d)}$.

As the next step we complete the object function $(X, p) \longmapsto \mathbb{P}(X, p)$ to an endofunctor of Preord($\mathfrak{Q}$) by:

$$(X, p) \xrightarrow{h} (Y, q), \quad \mathbb{P}(X, p) \xrightarrow{\mathbb{P}(h)} \mathbb{P}(Y, q), \quad \big(\mathbb{P}(h)(f)\big)(y) = \bigvee_{x \in X} q(y, h(x)) * f(x),$$

where $f \in \mathbb{P}(X, p)$ and $y \in Y$.

Having made these preparations the next lemma reveals an important property of $\mathfrak{Q}$-homomorphisms.

Lemma 3.3.11. *Let (X, p) and (Y, q) be cocomplete $\mathfrak{Q}$-preordered sets. Then every $\mathfrak{Q}$-homomorphism $X \xrightarrow{h} Y$ satisfies the condition:*

$$\big(\mathsf{sup}_{(Y,q)} \circ \mathbb{P}(h)\big)(f) \leq \big(h \circ \mathsf{sup}_{(X,p)}\big)(f), \qquad f \in \mathbb{P}(X, p),$$

where $\leq$ is the intrinsic preorder underlying q.

Proof. Let f be a $\mathfrak{Q}$-enriched, contravariant presheaf on (X, p). Referring to (3.52) and Lemma 3.3.5 we obtain for all $x \in X$:

$$f(x) \leq p(\mathsf{sup}_{(X,p)}(\widetilde{x}), \mathsf{sup}_{(X,p)}(f)) \leq p(x, \mathsf{sup}_{(X,p)}(f)) \leq q(h(x), h(\mathsf{sup}_{(X,p)}(f))).$$

Hence the relation

$$\big(\mathbb{P}(h)(f)\big)(y) \leq \bigvee_{x \in X} q(y, h(x)) * q(h(x), h(\mathsf{sup}_{(X,p)}(f))) \leq q\big(y, h\big(\mathsf{sup}_{(X,p)}(f)\big)\big)$$

follows for all $y \in Y$. Since $\mathfrak{Q}$-homomorphisms are isotone w.r.t. the underlying intrinsic preorders, we apply again Lemma 3.3.5 and obtain:

$$\big(\mathsf{sup}_{(Y,q)} \circ \mathbb{P}(h)\big)(f) \leq \big(\mathsf{sup}_{(Y,q)} \circ \eta_{(Y,q)}\big)\big(\big(h \circ \mathsf{sup}_{(X,p)}\big)(f)\big) \leq \big(h \circ \mathsf{sup}_{(X,p)}\big)(f).$$

Hence the assertion is verified. □

Motivated by the previous lemma we introduce the following terminology.

Definition 3.3.12. Let (X, p) and (Y, q) be cocomplete $\mathfrak{Q}$-preordered sets. Further, let $\leq_q$ be the intrinsic preorder underlying q. Then a $\mathfrak{Q}$-homomorphism $X \xrightarrow{h} Y$ is *cocontinuous* (or *$\mathfrak{Q}$-enriched join-preserving*) if the following relation holds:

$$\big(h \circ \mathsf{sup}_{(X,p)}\big)(f) \leq_q \big(\mathsf{sup}_{(Y,q)} \circ \mathbb{P}(h)\big)(f), \qquad f \in \mathbb{P}(X, p).$$

Lemma 3.3.13. *Let (X, p) and (Y, q) be cocomplete $\mathfrak{Q}$-preordered sets. Every cocontinuous ($\mathfrak{Q}$-enriched join-preserving) $\mathfrak{Q}$-homomorphism $X \xrightarrow{h} Y$ is join-preserving w.r.t. the underlying intrinsic preorders.*

Proof. Let $\mathsf{sup}_{(X,p)}$ and $\mathsf{sup}_{(Y,q)}$ be the respective formations of joins of $\mathfrak{Q}$-enriched contravariant presheaves. Since h is cocontinuous, for all $A \in \mathsf{Dwn}(X)$ the following relation holds:

$$h\Big(\mathsf{sup}_{(X,p)}\big(\bigvee_{a\in A} \widetilde{a}\big)\Big) \leq_q \mathsf{sup}_{(Y,q)}\Big(\mathbb{P}(h)\Big(\bigvee_{a\in A} \widetilde{a}\Big)\Big) = \mathsf{sup}_{(Y,q)}\Big(\bigvee_{b\in\downarrow h(A)} \widetilde{b}\Big).$$

Hence h is join-preserving w.r.t. the underlying intrinsic preorders. □

Comment. There exist join-preserving maps which are not $\mathfrak{Q}$-enriched join-preserving (cf. Exercise 3.3.6). Hence cocontinuity is a smoothness property, which is stronger than join preservation in the traditional sense.

The next theorem is a $\mathfrak{Q}$-valued version of Theorem 1.3.7.

Theorem 3.3.14. *Let (X, p) and (Y, q) be cocomplete $\mathfrak{Q}$-preordered sets, and let $X \xrightarrow{h} Y$ be a $\mathfrak{Q}$-homomorphism. Then h is cocontinuous if and only if h has a right adjoint $\mathfrak{Q}$-homomorphism.*

Proof. Every $\mathfrak{Q}$-homomorphism $(X, p) \xrightarrow{h} (Y, q)$ induces a further $\mathfrak{Q}$-homomorphism $(Y, q) \xrightarrow{\Phi} \mathbb{P}(X, p)$ as follows:

$$\Phi(y)(x) = q(h(x), y), \qquad y \in Y,\ x \in X.$$

Since (X, p) is cocomplete, we introduce a $\mathfrak{Q}$-homomorphism $(Y, q) \xrightarrow{k} (X, p)$ by $k = \mathsf{sup}_{(X,p)} \circ \Phi$. Since

$$e \leq d\big(\widetilde{x_0}, \Phi(h(x_0))\big) \leq p(x_0, k(h(x_0))), \qquad x_0 \in X,$$

the relation $x_0 \leq_p k(h(x_0))$ follows for each $x_0 \in X$, where $\leq_p$ is the intrinsic preorder underlying p.

Let us now assume that h is cocontinuous. It is easily seen that the following relation holds for all $y_0 \in Y$:

$$h(k(y_0)) \leq_q \mathsf{sup}_{(Y,q)}\big(\mathbb{P}(h)(\Phi(y_0))\big) \leq_q y_0.$$

Thus k is a right adjoint to h. On the other hand, if h has a right adjoint $\mathfrak{Q}$-homomorphism $h^{\vdash}$, then we observe

$$p(x_0, x) \leq q(h(x_0), h(x)) = p(x_0, h^{\vdash}(h(x))), \qquad x_0, x \in X.$$

Hence the relation $f \leq \mathbb{P}(h^{\vdash} \circ h)(f)$ follows for each $f \in \mathbb{P}(X, p)$. Now we apply Lemma 3.3.11 to $h^{\vdash}$ and obtain:

$$\mathsf{sup}_{(X,p)}(f) \leq_p \mathsf{sup}_{(X,p)}\big(\mathbb{P}(h^{\vdash} \circ h)(f)\big) \leq_p h^{\vdash}\big(\mathsf{sup}_{(Y,q)}\big(\mathbb{P}(h)(f)\big)\big).$$

Using again the right adjointness of $h^{\vdash}$, we infer from the previous relation that $h\big(\mathsf{sup}_{(X,p)}(f)\big) \leq_q \mathsf{sup}_{(Y,q)}\big(\mathbb{P}(h)(f)\big)$ holds. Hence h is cocontinuous. $\square$

If a $\mathfrak{Q}$-preordered set (X, p) is cocomplete, then we infer from Theorem 3.3.14 that $\mathsf{sup}_{(X,p)}$ is always cocontinuous. A further application of Theorem 3.3.14 is the following proposition.

Proposition 3.3.15. *Let (X, p) and (Y, q) be cocomplete $\mathfrak{Q}$-preordered sets, and let $X \xrightarrow{h} Y$ be a $\mathfrak{Q}$-homomorphism. Then $\mathbb{P}(X, p) \xrightarrow{\mathbb{P}(h)} \mathbb{P}(Y, q)$ is cocontinuous.*

Proof. If $f \in \mathbb{P}(X, p)$ and $g \in \mathbb{P}(Y, q)$, then it is easily seen that the following relation holds:

$$\begin{aligned}\bigwedge_{y\in Y}\Big(\big(\mathbb{P}(h)(f)\big)(y) \searrow g(y)\Big) &= \bigwedge_{x\in X}\bigwedge_{y\in Y}\Big(\big(q(y, h(x)) * f(x)\big) \searrow g(y)\Big)\\ &= \bigwedge_{x\in X}\bigwedge_{y\in Y}\Big(f(x) \searrow \big(q(y, h(x)) \searrow g(y)\big)\Big)\\ &= \bigwedge_{x\in X}\Big(f(x) \searrow g(h(x))\Big).\end{aligned}$$

Hence we conclude from Lemma 3.3.2 that the map $g \longmapsto g \circ h$ is a $\mathfrak{Q}$-homomorphism which is right adjoint to $\mathbb{P}(h)$. Hence $\mathbb{P}(h)$ is cocontinuous. $\square$

The composition of cocontinuous $\mathfrak{Q}$-homomorphisms is evidently cocontinuous. Moreover, in the case of antisymmetric $\mathfrak{Q}$-preorders we can characterize cocontinuity as follows.

Let (X, p) be a cocomplete $\mathfrak{Q}$-preordered set and (Y, q) be a join-complete $\mathfrak{Q}$-valued lattice. By Lemma 3.3.11 a $\mathfrak{Q}$-homomorphism $X \xrightarrow{h} Y$ is cocontinuous if and only if the following diagram is commutative:

$$\begin{array}{ccc} \mathbb{P}(X, p) & \xrightarrow{\mathbb{P}(h)} & \mathbb{P}(Y, q) \\ {\scriptstyle \mathsf{sup}_{(X,p)}}\big\downarrow & & \big\downarrow{\scriptstyle \mathsf{sup}_{(Y,q)}} \\ X & \xrightarrow{\quad h \quad} & Y \end{array} \tag{3.55}$$

In the case of $h = \mathsf{sup}_{(Y,q)}$ the previous diagram attains the following form:

$$\begin{array}{ccc} \mathbb{P}\big(\mathbb{P}(Y, q)\big) & \xrightarrow{\mathbb{P}(\mathsf{sup}_{(Y,q)})} & \mathbb{P}(Y, q) \\ {\scriptstyle \mu_{(Y,q)}}\big\downarrow & & \big\downarrow{\scriptstyle \mathsf{sup}_{(Y,q)}} \\ \mathbb{P}(Y, q) & \xrightarrow{\quad \mathsf{sup}_{(Y,q)} \quad} & Y \end{array} \tag{3.56}$$

which expresses the associativity of $\mathsf{sup}_{(Y,q)}$ in the case of join-complete $\mathfrak{Q}$-valued lattices.

The next theorem can be understood as an extension theorem and is a $\mathfrak{Q}$-valued analogue of Corollary 1.3.11.

Theorem 3.3.16. *Let (X, p) be a $\mathfrak{Q}$-preordered set and (Y, q) be a join-complete $\mathfrak{Q}$-valued lattice. Then for every $\mathfrak{Q}$-homomorphism $X \xrightarrow{h} Y$ there exists a unique cocontinuous $\mathfrak{Q}$-homomorphism $\mathbb{P}(X, p) \xrightarrow{h^\sharp} Y$ making the following diagram commutative*:

$$\begin{array}{ccc} X & \xrightarrow{\eta_{(X,p)}} & \mathbb{P}(X,p) \\ & {\scriptstyle h}\searrow & \vdots\, {\scriptstyle h^\sharp} \\ & & Y \end{array} \tag{3.57}$$

Proof. (Uniqueness) Let f be a $\mathfrak{Q}$-enriched contravariant presheaf on (X, p). Then f induces a $\mathfrak{Q}$-enriched contravariant presheaf F_f on $(\mathbb{P}(X, p), d)$ by

$$F_f(g) = \bigvee_{x \in X} d(g, \widetilde{x}) * f(x), \qquad g \in \mathbb{P}(X, p).$$

Then we conclude from the left extensionality of f that the following relation holds

$$\big(\mathsf{sup}_{(\mathbb{P}(X,p),d)}(F_f)\big)(z) = \bigvee_{\substack{x \in X \\ g \in \mathbb{P}(X,p)}} g(z) * d(g, \widetilde{x}) * f(x) = \bigvee_{x \in X} p(z, x) * f(x) = f(z)$$

for each $z \in X$. Hence $\mathsf{sup}_{(\mathbb{P}(X,p),d)}(F_f) = f$ follows.

Now we assume that $h^\sharp$ is a cocontinuous $\mathfrak{Q}$-homomorphism making the diagram (3.57) commutative. Since q is antisymmetric, we obtain:

$$\begin{aligned} h^\sharp(f) = \big(h^\sharp \circ \mathsf{sup}_{(\mathbb{P}(X,p),d)}\big)(F_f) &= \big(\mathsf{sup}_{(Y,q)} \circ \mathbb{P}(h^\sharp)\big)(F_f) \\ &= \mathsf{sup}_{(Y,q)}\Big(\bigvee_{\substack{x \in X \\ g \in \mathbb{P}(X,p)}} q(\_, h^\sharp(g)) * d(g, \widetilde{x}) * f(x)\Big) \\ &= \mathsf{sup}_{(Y,q)}\Big(\bigvee_{x \in X} q(\_, h^\sharp(\widetilde{x})) * f(x)\Big) \\ &= \mathsf{sup}_{(Y,q)}\Big(\bigvee_{x \in X} q(\_, h(x)) * f(x)\Big) \\ &= \mathsf{sup}_{(Y,q)} \mathbb{P}(h)(f). \end{aligned}$$

Hence $h^\sharp = \mathsf{sup}_{(Y,q)} \circ \mathbb{P}(h)$ follows. Since q is antisymmetric, $\mathsf{sup}_{(Y,q)}$ is the unique formation of joins of $\mathfrak{Q}$-enriched contravariant presheaves on (Y, q). This means that $h^\sharp$ is uniquely determined by h and the commutativity of (3.57).
(Existence) The uniqueness proof and Proposition 3.3.15 suggest to introduce a cocontinuous $\mathfrak{Q}$-homomorphism $\mathbb{P}(X, p) \xrightarrow{h^\sharp} Y$ by $h^\sharp = \mathsf{sup}_{(Y,q)} \circ \mathbb{P}(h)$. Since for all $x \in X$ the relation

$$\big(\mathsf{sup}_{(Y,q)} \circ \mathbb{P}(h)\big)(\widetilde{x}) = \mathsf{sup}_{(Y,q)}\big(q(\_, h(z)) * p(z, x)\big) = \big(\mathsf{sup}_{(Y,q)} \circ \eta_{(Y,q)}\big)(h(x))$$

holds, $h^\sharp \circ \eta_{(X,p)} = h$ follows from the antisymmetry of q. Hence $h^\sharp$ satisfies the desired properties. □

Let $\mathtt{Sup}(\mathfrak{Q})$ be the category of join-complete $\mathfrak{Q}$-valued lattices and cocontinuous $\mathfrak{Q}$-homomorphisms. Then Theorem 3.3.16 means that the forgetful functor from $\mathtt{Sup}(\mathfrak{Q})$ to $\mathtt{Preord}(\mathfrak{Q})$ has a left adjoint functor. The monad on $\mathtt{Preord}(\mathfrak{Q})$ induced by this adjoint situation has already appeared implicitly in the previous considerations and is given by the triple $\mathbf{P}(\mathfrak{Q}) = (\mathbb{P}, \eta, \mu)$, where the endofunctor $\mathbb{P}$ is defined before Lemma 3.3.11, the components of the unit η are given by the respective $\mathfrak{Q}$-enriched Yoneda embeddings (cf. Example 3.3.4) and the components of the multiplication μ coincide with the formation of joins of $\mathfrak{Q}$-enriched contravariant presheaves (cf. (3.54) in Example 3.3.10). The monad $\mathbf{P}(\mathfrak{Q})$ is also called the *monad of $\mathfrak{Q}$-enriched contravariant presheaves on* $\mathtt{Preord}(\mathfrak{Q})$. The monad of downclosed sets is obviously a submonad of $\mathbf{P}(\mathfrak{Q})$.

It follows immediately from Lemma 3.3.5 and the diagrams (3.55) and (3.56) that the Eilenberg–Moore category of $\mathbf{P}(\mathfrak{Q})$ coincides with $\mathtt{Sup}(\mathfrak{Q})$.

3.3.2 Complete $\mathfrak{Q}$-Preordered Sets

In order to introduce a concept of a complete (i.e. $\mathfrak{Q}$-enriched meet-complete) $\mathfrak{Q}$-preordered set, we need a $\mathfrak{Q}$-valued version of upclosed subsets (cf. Sect. 1.3). Since in the general, non-commutative setting of unital quantales $\mathfrak{Q}$ a principle of $\mathfrak{Q}$-valued duality is not available, we can now simply replace $\mathfrak{Q}$ by its transposed quantale $\mathfrak{Q}^\tau$ and transfer $\mathfrak{Q}$-preorders to $\mathfrak{Q}^\tau$-preorders by interchanging their arguments. If in Sect. 3.3.1 we proceed in this way, then 3.3.4 – 3.3.16 can obviously be rewritten. Hence this approach leads to the following terminology, definitions and results.

If (X, p) is a $\mathfrak{Q}$-preordered set, then a map $X \xrightarrow{f} \mathfrak{Q}$ is called a *$\mathfrak{Q}$-enriched covariant presheaf* on (X, p) if f satisfies the condition:

$$f(x) * p(x, z) \le f(z), \qquad x, z \in X. \quad \text{(Right Extensionality)}$$

On the set $\mathbb{P}^\dagger(X, p)$ of all $\mathfrak{Q}$-enriched covariant presheaves we introduce the pointwise order. Then $\mathbb{P}^\dagger(X, p)$ is a complete lattice, and the left action $\odot$ on $\mathbb{P}^\dagger(X, p)$ is determined as follows:

$$(\alpha \odot f)(x) = \alpha * f(x), \qquad \alpha \in \mathfrak{Q},\ f \in \mathbb{P}^\dagger(X, p),\ x \in X.$$

Hence $(\mathbb{P}^\dagger(X, p), \odot)$ is a left $\mathfrak{Q}$-module. If we now anticipate the formula (3.64), then the corresponding $\mathfrak{Q}$-preorder $d^\dagger$ on $\mathbb{P}^\dagger(X, p)$ is given by

$$d^\dagger(f, g) = \bigwedge_{x \in X} (f(x) \swarrow g(x)), \qquad f, g \in \mathbb{P}^\dagger(X, p).$$

Finally, the $\mathfrak{Q}$-enriched Yoneda embedding $(X, p) \xrightarrow{\eta^{\dagger}_{(X,p)}} (\mathbb{P}^{\dagger}(X, p), d^{\dagger})$ has the form:

$$\eta^{\dagger}_{(X,p)}(x) = \widetilde{x}^{\dagger}, \quad \widetilde{x}^{\dagger}(z) = p(x, z), \qquad z \in X.$$

In this context, the relation $d^{\dagger}(f, \widetilde{x}^{\dagger}) = f(x)$ holds for each $\mathfrak{Q}$-enriched covariant presheaf f and for each $x \in X$.

A $\mathfrak{Q}$-preordered set (X, p) is *complete* (or *$\mathfrak{Q}$-enriched meet-complete*) if the $\mathfrak{Q}$-enriched Yoneda embedding $\eta^{\dagger}_{(X,p)}$ has a right adjoint $\mathfrak{Q}$-homomorphism — i.e. there exists a $\mathfrak{Q}$-homomorphism $\mathbb{P}^{\dagger}(X, p) \xrightarrow{\xi} X$ such that the relation

$$d^{\dagger}\big(\eta^{\dagger}_{(X,p)}(x), f\big) = p(x, \xi(f))$$

holds for all $x \in X$ and $f \in \mathbb{P}^{\dagger}(X, p)$. Obviously, ξ is right adjoint to $\eta^{\dagger}_{(X,p)}$ if and only if $\xi \circ \eta^{\dagger}_{(X,p)}$ is equivalent to 1_X. Since right adjoint $\mathfrak{Q}$-homomorphisms of a given $\mathfrak{Q}$-homomorphism are unique up to equivalence, we denote ξ by $\mathsf{inf}_{(X,p)}$ and call $\mathsf{inf}_{(X,p)}$ the *formation of meets of $\mathfrak{Q}$-enriched covariant presheaves* on (X, p).

A *meet-complete* $\mathfrak{Q}$-valued lattice is a complete $\mathfrak{Q}$-preordered set with an anti-symmetric $\mathfrak{Q}$-preorder.

Definition 3.3.17. Let (X, p) be a $\mathfrak{Q}$-preordered set. An element $x_0 \in X$ is called a *lower bound of a $\mathfrak{Q}$-enriched covaraint presheaf* g on (X, p) if $g(z) \le \eta^{\dagger}_{(X,p)}(x_0)$ holds. A lower bound x_0 of a $\mathfrak{Q}$-enriched covariant presheaf g is called a *meet of* g if for all $\alpha \in \mathfrak{Q}$ and for all $x \in X$ the following additional implication is true:

$$\big(\forall z \in X : \alpha * g(z) \le p(x, z)\big) \implies \alpha \le p(x, x_0).$$

Again by (O2), an element $x_0 \in X$ is a meet of a covariant presheaf g on (X, p) if and only if the equivalence

$$\big(\forall z \in X : \alpha * g(z) \le p(x, z)\big) \iff \alpha \le p(x, x_0)$$

holds for all $x \in X$ and $\alpha \in \mathfrak{Q}$ — this means the validity of the following equation:

$$d^{\dagger}\big(\eta^{\dagger}_{(X,p)}(x), g\big) = p(x, x_0), \qquad x \in X.$$

Hence a $\mathfrak{Q}$-preordered set (X, p) is complete if and only if every $\mathfrak{Q}$-enriched covariant presheaf on (X, p) has a meet, where we apply of course the axiom of choice.

The next proposition is a $\mathfrak{Q}$-valued analogue of Proposition 1.3.5.

Proposition 3.3.18. *A $\mathfrak{Q}$-preordered set (X, p) is complete if and only if (X, p) is cocomplete.*

Proof. Let us consider the following $\mathfrak{Q}$-homomorphisms $\mathbb{P}^{\dagger}(X, p) \xrightarrow{\Delta} \mathbb{P}(X, p)$ and $\mathbb{P}(X, p) \xrightarrow{\Delta^{\dagger}} \mathbb{P}^{\dagger}(X, p)$ defined by

$$\Delta(f)(x) = \bigwedge_{z\in X} (p(x,z) \swarrow f(z)), \qquad f \in \mathbb{P}^{\dagger}(X,p),\ x \in X,$$

$$\Delta^{\dagger}(f)(x) = \bigwedge_{z\in X} (f(z) \searrow p(z,x)), \qquad f \in \mathbb{P}(X,p),\ x \in X.$$

Since $\widetilde{x} = \Delta(\widetilde{x}^{\dagger})$ and $\widetilde{x}^{\dagger} = \Delta^{\dagger}(\widetilde{x})$ for all $x \in X$, the assertion follows. □

Because of the previous proposition it is not necessary to distinguish between completeness and cocompleteness of $\mathfrak{Q}$-preordered sets. Hence a *complete $\mathfrak{Q}$-valued lattice* is either $\mathfrak{Q}$-enriched meet-complete or $\mathfrak{Q}$-enriched join-complete.

An example of a complete $\mathfrak{Q}$-valued lattice is $\big(\mathbb{P}^{\dagger}(X,p), d^{\dagger}\big)$. In this context the formation of meets of $\mathfrak{Q}$-enriched covariant presheaves on $\big(\mathbb{P}^{\dagger}(X,p), d^{\dagger}\big)$ is determined by:

$$\mu^{\dagger}_{(X,p)}(F)(x) = \inf_{(\mathbb{P}^{\dagger}(X,p),d^{\dagger})}(F)(x) = \bigvee_{f\in\mathbb{P}^{\dagger}(X,p)} F(f) * f(x),$$

where $F \in \mathbb{P}^{\dagger}(\mathbb{P}^{\dagger}(X,p), d^{\dagger})$.

At this point it is worthwhile to recall that in a *strict* sense the formation of meets of $\mathfrak{Q}$-enriched covariant presheaves is *not* a $\mathfrak{Q}$-valued dual concept of the formation of joins of $\mathfrak{Q}$-enriched contravariant presheaves. We can solve this problem if we add an involutive anti-endomorphism on the given quantale.

Remark 3.3.19. Let $(\mathfrak{Q}, *, e, ')$ be an involutive and unital quantale. Then with every $\mathfrak{Q}$-preordered set (X,p) we can associate the *dual* $\mathfrak{Q}$-preordered set (X, p^{op}), where the *dual* $\mathfrak{Q}$-preorder p^{op} is given by:

$$p^{op}(x_1,x_2) = p(x_2,x_1)', \qquad x_1, x_2 \in X.$$

Further, we consider $\mathfrak{Q}$ as a right $\mathfrak{Q}$-module $(\mathfrak{Q}, \boxdot)$ in the sense of Example 3.1.6. Then the $\mathfrak{Q}$-preorder on $\mathfrak{Q}$ coincides with the right-implication $\searrow$ of $\mathfrak{Q}$. Finally, if $X \xrightarrow{f} \mathfrak{Q}$ is a map, then its *conjugate map* $X \xrightarrow{f'} \mathfrak{Q}$ has the form $f'(x) = f(x)'$ for all $x \in X$. Now we make the following observations:

(i) A $\mathfrak{Q}$-enriched contravariant presheaf on (X,p) is equivalent to a $\mathfrak{Q}$-homomorphism from (X, p^{op}) to $(\mathfrak{Q}, \searrow)$ — i.e. f is a $\mathfrak{Q}$-enriched contravariant presheaf on (X,p) if and only if its conjugate map $(X, p^{op}) \xrightarrow{f'} (\mathfrak{Q}, \searrow)$ is a $\mathfrak{Q}$-homomorphism.
(ii) The right extensionality of a $\mathfrak{Q}$-enriched covariant presheaf f on (X,p) is equivalent to the requirement that $(X,p) \xrightarrow{f} (\mathfrak{Q}, \searrow)$ is a $\mathfrak{Q}$-homomorphism.

Hence f is a $\mathfrak{Q}$-enriched contravariant (covariant) presheaf on (X,p) if and only if its conjugate map f' is a $\mathfrak{Q}$-enriched covariant (contravariant) presheaf on (X, p^{op}). Based on this interrelationship we formulate the *Principle of $\mathfrak{Q}$-Enriched Duality* as follows:

> If P is a property (notion) in a $\mathfrak{Q}$-preordered set (X, p), then its *$\mathfrak{Q}$-enriched dual* property (notion) P^{op} in (X, p) is the property (notion) P phrased in (X, p^{op}) modulo involution — this means that $\mathfrak{Q}$-enriched contravariant (covariant) presheaves on (X, p) are expressed by their conjugate $\mathfrak{Q}$-enriched covariant (contravariant) presheaves on (X, p^{op}).

Having made these preparations we are now in a position to prove that joins and meets of $\mathfrak{Q}$-enriched presheaves are $\mathfrak{Q}$-valued dual notions.

In fact, let g be a $\mathfrak{Q}$-enriched covariant presheaf on a given $\mathfrak{Q}$-preordered set (X, p), and let x_0 be a join of its conjugate map — i.e. the $\mathfrak{Q}$-enriched contravariant presheaf g' on (X, p^{op}). Then x_0 is obviously a meet of g, because the following chain of equivalences holds for all $\alpha \in \mathfrak{Q}$ and $x \in X$:

$$\begin{aligned}\big(\forall z \in X : \alpha * g(z) \le p(x, z)\big) &\iff \big(\forall z \in X : g'(z) * \alpha' \le p^{op}(z, x)\big)\\ &\iff \alpha' \le p^{op}(x_0, x)\\ &\iff \alpha \le p(x, x_0).\end{aligned}$$

Finally we complete the object function $(X, p) \longmapsto \mathbb{P}^{\dagger}(X, p)$ to an endofunctor $\mathbb{P}^{\dagger}$ of `Preord`($\mathfrak{Q}$) as follows:

$$(X, p) \xrightarrow{h} (Y, q), \ \mathbb{P}^{\dagger}(X, p) \xrightarrow{\mathbb{P}^{\dagger}(h)} \mathbb{P}^{\dagger}(Y, q), \ \big(\mathbb{P}^{\dagger}(h)(f)\big)(y) = \bigvee_{x \in X} f(x) * q(h(x), y),$$

where $f \in \mathbb{P}^{\dagger}(X, p)$ and $y \in Y$. If (X, p) and (Y, q) are complete $\mathfrak{Q}$-preordered sets, then every $\mathfrak{Q}$-homomorphism $X \xrightarrow{h} Y$ satisfies the condition:

$$\big(h \circ \mathsf{inf}_{(X,p)}\big)(f) \le \big(\mathsf{inf}_{(Y,q)} \circ \mathbb{P}^{\dagger}(h)\big)(f), \qquad f \in \mathbb{P}^{\dagger}(X, p),$$

where $\le$ is the intrinsic preorder underlying q.

Moreover, a $\mathfrak{Q}$-homomorphism $(X, p) \xrightarrow{h} (Y, q)$ is called *continuous* (i.e. *$\mathfrak{Q}$-enriched meet-preserving*) if for all $\mathfrak{Q}$-enriched covariant presheaves f on (X, p) the following relation holds:

$$\big(\mathsf{inf}_{(Y,q)} \circ \mathbb{P}^{\dagger}(h)\big)(f) \le \big(h \circ \mathsf{inf}_{(X,p)}\big)(f), \qquad f \in \mathbb{P}^{\dagger}(X, p),$$

where $\le$ is again the intrinsic preorder underlying q.

Every continuous $\mathfrak{Q}$-homomorphism is meet-preserving with respect to the underlying intrinsic preorders. As a further important property we point out that a $\mathfrak{Q}$-homomorphism h has a left adjoint $\mathfrak{Q}$-homomorphism if and only if h is continuous. Hence $\mathsf{inf}_{(X,p)}$ and $\mathbb{P}^{\dagger}(h)$ are continuous. Finally, the following extension theorem holds.

Theorem 3.3.20. *Let (X, p) be a $\mathfrak{Q}$-preordered set and (Y, q) be a meet-complete $\mathfrak{Q}$-valued lattice. Then for every $\mathfrak{Q}$-homomorphism $X \xrightarrow{h} Y$ there exists a unique continuous $\mathfrak{Q}$-homomorphism $\mathbb{P}^{\dagger}(X, p) \xrightarrow{h^{\sharp}} Y$ making the following diagram commutative:*

$$\begin{array}{ccc} X & \xrightarrow{\eta^{\dagger}_{(X,p)}} & \mathbb{P}^{\dagger}(X,p) \\ & \searrow^{h} & \vdots\, h^{\sharp} \\ & & Y \end{array} \tag{3.58}$$

In particular, $h^{\sharp}$ *coincides with* $\mathsf{inf}_{(Y,q)} \circ \mathbb{P}^{\dagger}(h)$.

By analogy with Sect. 3.3.1 we can summarize the situation as follows. The triple $\mathbf{P}^{\dagger}(\mathfrak{Q}) = (\mathbb{P}^{\dagger}, \eta^{\dagger}, \mu^{\dagger})$ is a monad on $\mathtt{Preord}(\mathfrak{Q})$ and is called the *monad of* $\mathfrak{Q}$*-enriched covariant presheaves*. The Eilenberg–Moore category of $\mathbf{P}^{\dagger}(\mathfrak{Q})$ coincides with the category $\mathtt{Inf}(\mathfrak{Q})$ of meet-complete $\mathfrak{Q}$-valued lattices with continuous $\mathfrak{Q}$-homomorphisms.

3.3.3 Right $\mathfrak{Q}$-Modules and Join-Complete $\mathfrak{Q}$-Valued Lattices

Theorem 3.3.21. *Let* $(X, \boxdot)$ *be a right* $\mathfrak{Q}$*-module and* p *be the* $\mathfrak{Q}$*-preorder on* X *determined by* (3.48). *Then* (X, p) *is a join-complete* $\mathfrak{Q}$*-valued lattice and the intrinsic preorder underlying* p *coincides with the given partial order on* X. *Moreover, the formation of joins of* $\mathfrak{Q}$*-enriched contravariant presheaves on* (X, p) *attains the following form*:

$$\mathsf{sup}_{(X,p)}(f) = \bigvee_{x \in X} x \boxdot f(x), \qquad f \in \mathbb{P}(X, p).$$

Proof. Since $x \boxdot p(x, y) \le y$, the intrinsic preorder $\le_p$ coincides with the partial order $\le$ on X. In particular, the intrinsic preorder is antisymmetric. Further, let us define a map $\mathbb{P}(X, p) \xrightarrow{\xi} X$ by

$$\xi(f) = \bigvee_{x \in X} x \boxdot f(x), \qquad f \in \mathbb{P}(X, p), \tag{3.59}$$

and conclude from (3.48) and the relation

$$\bigvee_{x \in X} \big(x \boxdot f(x)\big) \boxdot d(f, g) = \bigvee_{x \in X} x \boxdot \big(f(x) * d(f, g)\big) \le \bigvee_{x \in X} x \boxdot g(x)$$

that ξ is a $\mathfrak{Q}$-homomorphism. Further, we observe:

$$\xi(\widetilde{x}) = \bigvee_{z \in X} z \boxdot p(z, x) = x, \qquad x \in X,$$

where we have used (M2$_r$) and (3.48). By Lemma 3.3.5 the $\mathfrak{Q}$-homomorphism ξ is left adjoint to $\eta_{(X,p)}$ — i.e. (X, p) is join-complete $\mathfrak{Q}$-valued lattice and ξ coincides with $\mathsf{sup}_{(X,p)}$. □

The next theorem shows that the converse of Theorem 3.3.21 also holds.

Theorem 3.3.22. *Let (X, p) be a join-complete $\mathfrak{Q}$-valued lattice and $\leq$ be the intrinsic partial order underlying p. Then $(X, \leq)$ is a complete lattice and there exists a unique right action $\boxdot$ on X satisfying the following properties*:

$$p(x, z) = \bigvee\{\alpha \in \mathfrak{Q} \mid x \boxdot \alpha \leq z\}, \qquad x, z \in X, \tag{3.60}$$

$$\mathsf{sup}_X(f) = \bigvee_{x \in X} x \boxdot f(x), \qquad f \in \mathbb{P}(X, p). \tag{3.61}$$

Proof. Let $\leq$ be the intrinsic partial order underlying p, and let $\mathsf{sup}_{(X,p)}$ be the formation of joins of $\mathfrak{Q}$-enriched contravariant presheaves on (X, p). By Lemma 3.3.9 the pair $(X, \leq)$ is a complete lattice, and $\mathsf{sup}_{(X,p)}$ is join-preserving w.r.t. the underlying intrinsic partial orders. Moreover, we know from Sect. 3.3.1 that $((X, p), \mathsf{sup}_{(X,p)})$ is a $\mathbf{P}(\mathfrak{Q})$-algebra of the monad of $\mathfrak{Q}$-enriched contravariant presheaves.

Based on these properties we construct the structure of a right $\mathfrak{Q}$-module on X as follows.

(a) Let us consider a map $X \times \mathfrak{Q} \xrightarrow{\boxdot} X$ defined by $x \boxdot \alpha = \mathsf{sup}_{(X,p)}(\widetilde{x} * \alpha)$ for each $x \in X$ and $\alpha \in \mathfrak{Q}$. We show that $\boxdot$ is a bimorphism in the sense of Sup. Since $\mathsf{sup}_{(X,p)}$ is join-preserving and $\mathbb{P}(X, p)$ is a right $\mathfrak{Q}$-module (cf. Example 3.3.4), it is evident that $\boxdot$ is join-preserving in its second argument. In order to show that $\boxdot$ is also join-preserving in its first argument we begin with a general property which is related to the associativity of forming $\mathfrak{Q}$-enriched joins.

For each $\alpha \in \mathfrak{Q}$ and for each $f \in \mathbb{P}(X, p)$ we introduce a $\mathfrak{Q}$-enriched contravaraint presheaf $F_{(f,\alpha)}$ on $(\mathbb{P}(X, p), d)$ by:

$$F_{(f,\alpha)}(g) = d(g, f) * \alpha, \qquad g \in \mathbb{P}(X, p).$$

Then $f * \alpha$ is the join of $F_{(f,\alpha)}$ — i.e. $\mu_{(X,p)}\big(F_{(f,\alpha)}\big) = f * \alpha$. Further, we apply the associativity of $\mathsf{sup}_{(X,p)}$ (cf. diagram (3.56)) and obtain:

$$\mathsf{sup}_{(X,p)}(f * \alpha) = \big(\mathsf{sup}_{(X,p)}(f)\big) \boxdot \alpha. \tag{3.62}$$

Now we distinguish the following cases.

Case 1. Let $\perp$ be the universal lower bound in $\mathbb{P}(X, p)$. Since $\mathsf{sup}_{(X,p)}$ is join-preserving, $\mathsf{sup}_{(X,p)}(\perp)$ is the universal lower bound in $(X, \leq)$, which we also denote by $\perp$. Then $\perp \boxdot \alpha = \perp$ follows from (3.62).

Case 2. Let A be a nonempty subset of X. Then we put $f = \bigvee_{a \in A} \widetilde{a}$ and observe $\mathsf{sup}_{(X,p)}(f) = \bigvee A$. Using again the property that $\mathsf{sup}_{(X,p)}$ is join-preserving we derive the following relation from (3.62):

$$\bigvee_{a\in A} a \boxdot \alpha = \mathsf{sup}_{(X,p)}(f * \alpha) = \big(\bigvee A\big) \boxdot \alpha.$$

Hence $\boxdot$ is also join-preserving in its first argument.
(b) We show that $\boxdot$ satisfies the axioms ($M1_r$) and ($M2_r$). Because of the unit axiom of $\mathbf{P}(\mathfrak{Q})$-algebras (see also Lemma 3.3.5), the condition ($M2_r$) is evident. In order to verify ($M1_r$) we choose $\alpha, \beta \in \mathfrak{Q}$, $x \in X$ and $f \in \mathbb{P}(X, p)$ such that $f = \widetilde{x} * \beta$ holds. Referring again to (3.62) we obtain:

$$\begin{aligned}(x \boxdot \beta) \boxdot \alpha &= \big(\mathsf{sup}_{(X,p)}(f)\big) \boxdot \alpha\\ &= \mathsf{sup}_{(X,p)}(f * \alpha)\\ &= \mathsf{sup}_{(X,p)}\big(\widetilde{x} * (\beta * \alpha)\big)\\ &= x \boxdot (\beta * \alpha).\end{aligned}$$

Hence ($M1_r$) is verified, and $(X, \boxdot)$ is a right $\mathfrak{Q}$-module.
(c) We fix $x, z \in X$ and observe that $\widetilde{z} * p(z, x) \leq \widetilde{x}$. Hence $z \boxdot p(z, x) \leq x$ follows from the isotonicity of $\mathsf{sup}_{(X,p)}$. On the other hand, if $x \boxdot \alpha \leq y$, then we obtain:

$$\widetilde{x}(z) * \alpha = d(\widetilde{z}, \widetilde{x} * \alpha) \leq p(z, x \boxdot \alpha) \leq p(z, x \boxdot \alpha) * p(x \boxdot \alpha, y) \leq p(z, y).$$

Hence the relation $\alpha \leq \bigwedge_{z\in X}\big(p(z, x) \searrow p(z, y)\big) = p(x, y)$ holds, and (3.60) is verified. Since every $\mathfrak{Q}$-enriched contravariant presheaf f on (X, p) is left extensional, we obtain $f = \bigvee_{x\in X} \widetilde{x} * f(x)$, and obviously the relation (3.61) follows. Finally, the uniqueness of $\boxdot$ is an immediate corollary of (3.60) and (3.61). □

In the following propositions we investigate the behaviour of certain morphisms between right $\mathfrak{Q}$-modules

Proposition 3.3.23. *Let X and Y be complete lattices and $X \xrightarrow{h} Y$ be an isotone map. Further, let $\boxdot_X$ and $\boxdot_Y$ be right actions on X and Y respectively, and let p_X and p_Y be the corresponding $\mathfrak{Q}$-preorders. Then h is a $\mathfrak{Q}$-homomorphism from (X, p_X) to (Y, p_Y) if and only if h satisfies the following condition for all $\alpha \in \mathfrak{Q}$ and $x \in X$:*

$$h(x) \boxdot_Y \alpha \leq h(x \boxdot_X \alpha). \tag{3.63}$$

Proof. If h is a $\mathfrak{Q}$-homomorphism, then we conclude from

$$\alpha \leq p_X(x, x \boxdot_X \alpha) \leq p_Y\big(h(x), h(x \boxdot_X \alpha)\big)$$

that the relation $h(x) \boxdot_Y \alpha \leq h(x \boxdot_X \alpha)$ holds. Hence (3.63) is necessary. In order to show that (3.63) is sufficient we use the isotonicity of h and observe:

$$h(x_1) \boxdot_Y p_X(x_1, x_2) \leq h\big(x_1 \boxdot_X p_X(x_1, x_2)\big) \leq h(x_2), \qquad x_1, x_2 \in X.$$

Hence $p_X(x_1, x_2) \leq p_Y(h(x_1), h(x_2))$ follows. □

Proposition 3.3.24. *Let X and Y be complete lattices and $X \xrightarrow{h} Y$ be a join-preserving map. Further, let $\boxdot_X$ and $\boxdot_Y$ be right actions on X and Y, respectively. If (X, p_X) and (Y, p_Y) are the corresponding join-complete $\mathfrak{Q}$-valued lattices (cf. Theorem 3.3.21), then the following assertions are equivalent:*

(i) *The relation $h\big(\mathsf{sup}_{(X,p_X)}(f)\big) \le \mathsf{sup}_{(Y,p_Y)}\big(\mathbb{P}(h)(f)\big)$ holds for all $\mathfrak{Q}$-enriched contravariant presheaves f on (X, p_X).*

(ii) *The relation $h(x \boxdot_X \alpha) \le h(x) \boxdot_Y \alpha$ holds for all $\alpha \in \mathfrak{Q}$ and $x \in X$.*

Proof. If (i) is valid, then we apply Theorem 3.3.21 and obtain:

$$h(x \boxdot_X \alpha) = h\big(\mathsf{sup}_{(X,p_X)}(\widetilde{x} * \alpha)\big) \le \mathsf{sup}_{(Y,p_Y)}\big(\mathbb{P}(h)(\widetilde{x} * \alpha)\big)$$
$$= \mathsf{sup}_{(Y,p_Y)}(\widetilde{h(x)} * \alpha) = h(x) \boxdot_Y \alpha.$$

Hence (ii) follows. On the other hand, we choose an $\mathfrak{Q}$-enriched contravariant presheaf f on (X, p_X). If (ii) holds, then we make use of the join preservation of h and apply again Theorem 3.3.21:

$$h\big(\mathsf{sup}_{(X,p_X)}(f)\big) = h\big(\bigvee_{x \in X} x \boxdot_X f(x)\big) = \bigvee_{x \in X} h(x \boxdot_X f(x)) \le \bigvee_{x \in X} h(x) \boxdot_Y f(x)$$
$$= \bigvee_{x \in X} \bigvee_{y \in Y} \big(y \boxdot_Y p_Y(y, h(x))\big) \boxdot_Y f(x)$$
$$= \bigvee_{y \in Y} y \boxdot_Y \big(\mathbb{P}(h)(f)\big)(y) = \mathsf{sup}_{(Y,p_Y)}\big(\mathbb{P}(h)(f)\big).$$

Hence (i) follows. □

As an immediate corollary of Propositions 3.3.23 and 3.3.24 we obtain that right $\mathfrak{Q}$-module homomorphisms and cocontinuous $\mathfrak{Q}$-homomorphisms are equivalent concepts. Hence we can summarize the previous results as follows:

FACT I. *The category $\mathtt{Sup}(\mathfrak{Q})$ of join-complete $\mathfrak{Q}$-valued lattices is isomorphic to the category $\mathtt{Mod}_r(\mathfrak{Q})$ of right $\mathfrak{Q}$-modules. Therefore right $\mathfrak{Q}$-modules play the rôle of join-complete $\mathfrak{Q}$-valued lattices.*

3.3.4 Left $\mathfrak{Q}$-Modules and Meet-Complete $\mathfrak{Q}$-Valued Lattices

Let $(X, \odot)$ be a left $\mathfrak{Q}$-module. Then the corresponding $\mathfrak{Q}$-preorder p on X attains the following form:

$$p(x, y) = \bigvee\{\alpha \in \mathfrak{Q} \mid \alpha \odot y \le x\}, \qquad x, y \in X. \tag{3.64}$$

Obviously, the relation $p(x, y) \odot y \le x$ holds for all $x, y \in X$. In this context it is important to note that the *intrinsic* preorder underlying p coincides with the *dual* order of X.

Having made these observations we can rewrite 3.3.21 – 3.3.24 as follows. Let $(X, \odot)$ be a left $\mathfrak{Q}$-module and p be the $\mathfrak{Q}$-preorder on X determined by (3.64). Then (X, p) is a meet-complete lattice and the formation $\mathsf{inf}_{(X,p)}$ of meets of $\mathfrak{Q}$-enriched covariant presheaves is given as follows:

$$\mathsf{inf}_{(X,p)}(g) = \bigvee_{x \in X} g(x) \odot x, \quad g \in \mathbb{P}^\dagger(X, p).$$

On the other hand, if (X, p) is a given meet-complete $\mathfrak{Q}$-valued lattice, then we introduce on X the partial order $\leq$ defined by the dual intrinsic order underlying p — i.e. $\leq = \leq_p^{op}$. Then $(X, \leq)$ is a complete lattice. Further, there exists a unique left action $\odot$ on $(X, \leq)$ such that the following relations hold:

- $p(x, y) = \bigvee \{\alpha \in \mathfrak{Q} \mid \alpha \odot y \leq x\}$,
- $\mathsf{inf}_{(X,p)}(g) = \bigvee_{x \in X} g(x) \odot x, \quad g \in \mathbb{P}^\dagger(X, p)$.

Finally, left $\mathfrak{Q}$-module homomorphisms and continuous $\mathfrak{Q}$-homomorphisms are equivalent concepts. Hence we can summarize the situation as follows.

FACT II. *The category* `Inf`$(\mathfrak{Q})$ *of meet-complete* $\mathfrak{Q}$*-valued lattices with continuous* $\mathfrak{Q}$*-homomorphisms is isomorphic to the category* `Mod`$_\ell(\mathfrak{Q})$ *of left* $\mathfrak{Q}$*-modules. Therefore left* $\mathfrak{Q}$*-modules play the rôle of meet-complete* $\mathfrak{Q}$*-valued lattices.*

We finish this subsection with a general remark related to the monadic basis of right (left) $\mathfrak{Q}$-modules.

Remark 3.3.25. The composition of the monad $\mathbf{P}(\mathfrak{Q})$ of contravariant presheaves with the monad $\mathbf{P}^\dagger(\mathfrak{Q})$ of covariant presheaves exists and coincides with the $\mathfrak{Q}$-valued double power monad. For details the reader is referred to [50, 108], where unital quantales have already been replaced by quantaloids.

Exercises

3.3.1. Let $(\mathfrak{Q}, *, e)$ be a unital quantale viewed as a left $\mathfrak{Q}$-module (cf. Example 3.1.6)). Show:

(a) The $\mathfrak{Q}$-preorder p on $\mathfrak{Q}$ in the sense of (3.64) coincides with the left-implication of $\mathfrak{Q}$. In particular, p is antisymmetric.
(b) If $\alpha, \beta \in \mathfrak{Q}$, then $\alpha * \beta$ is the meet of the $\mathfrak{Q}$-enriched covariant presheaf $\alpha * \widetilde{\beta}^\dagger$ on $(\mathfrak{Q}, p)$.

3.3.2. Let $\mathbb{M}_2$ be the C^*-algebra of all square matrices of order 2 with real coefficients, and let $\mathbb{MAX}$ be the involutive and unital quantale of all linear subspaces of $\mathbb{M}_2$ (cf. Example 2.5.5). We view $\mathbb{MAX}$ as a left $\mathbb{MAX}$-module (cf. Example 3.1.6) and recall the left ideal L introduced in Exercise 3.1.7 — i.e.

$$L = \left\{ \begin{pmatrix} a & 0 \\ b & 0 \end{pmatrix} \mid a, b \in \mathbb{R} \right\}.$$

Further, we consider the left $\mathbb{MAX}$-module $(\uparrow L, \odot_L)$ constructed in the sense of Example 3.1.21 and maintain the notation from Exercise 3.1.7. We fix real numbers $\alpha, \beta \in [0, \pi)$ and define a square matrix M_α of order 2 by:

$$M_\alpha = \begin{pmatrix} 0 & \cos\alpha \\ 0 & \sin\alpha \end{pmatrix}.$$

Finally, let V_α be the 1-dimensional linear subspace of $\mathbb{M}_2$ generated by M_α — i.e. $V_\alpha = \{\lambda \cdot M_\alpha \mid \lambda \in \mathbb{R}\}$.

(a) Show that all square matrices of order 2

$$X = \begin{pmatrix} x & y \\ z & u \end{pmatrix}$$

satisfying the condition $X \cdot M_\alpha \in L + V_\beta$ are solutions of the following system of linear equations:

$$\begin{cases} x\cos\alpha + y\sin\alpha = \lambda\cos\beta \\ z\cos\alpha + u\sin\alpha = \lambda\sin\beta \end{cases} \qquad \lambda \in \mathbb{R}.$$

If $\alpha \neq \frac{\pi}{2}$, then the solutions (x, y, z, u) of the previous linear equations have the form:

$$\begin{pmatrix} x \\ y \\ z \\ u \end{pmatrix} = \lambda \cdot \begin{pmatrix} \frac{\cos\beta}{\cos\alpha} \\ 0 \\ \frac{\sin\beta}{\cos\alpha} \\ 0 \end{pmatrix} + \mu \cdot \begin{pmatrix} -\sin\alpha \\ \cos\alpha \\ 0 \\ 0 \end{pmatrix} + \nu \cdot \begin{pmatrix} 0 \\ 0 \\ -\sin\alpha \\ \cos\alpha \end{pmatrix},$$

where $\lambda, \mu, \nu \in \mathbb{R}$. If $\alpha = \frac{\pi}{2}$, then the solutions attain this form:

$$\begin{pmatrix} x \\ y \\ z \\ u \end{pmatrix} = \lambda \cdot \begin{pmatrix} 0 \\ \cos\beta \\ 0 \\ \sin\beta \end{pmatrix} + \mu \cdot \begin{pmatrix} -1 \\ 0 \\ 0 \\ 0 \end{pmatrix} + \nu \cdot \begin{pmatrix} 0 \\ 0 \\ -1 \\ 0 \end{pmatrix}.$$

(b) Conclude from (a) that $\{X \in \mathbb{M}_2 \mid X \cdot M_\alpha \in L + V_\beta\}$ is a 3-dimensional subspace of $\mathbb{M}_2$.

(c) Let p be the $\mathbb{MAX}$-preorder on $\uparrow L$ determined by (3.64). Further, we consider $W_1, W_2 \in \uparrow L \setminus \{L, \mathbb{M}_2\}$. Show that $p(W_1, W_2)$ is always a 3-dimensional linear subspace of $\mathbb{M}_2$.
(Hint: Since L is a left ideal, show first that the relation $X \cdot (L + V_\alpha) \subseteq (L + V_\beta)$ is equivalent to $X \cdot M_\alpha \in V_\beta$. Subsequently use Exercise 3.1.7 (a) and the previous properties of (a) and (b).)

(d) We recall from Exercise 3.1.7 (b) that $U = L + V_{\frac{\pi}{2}}$ is a generator of $(\uparrow L, \odot_L)$. If $W_1, W_2 \in \uparrow L \setminus \{L, \mathbb{M}_2\}$, then show that the following properties are valid:

(i) $p(W, U) = W$ for all $W \in \uparrow L$.
(ii) $p(W_1, W_2) \in \uparrow L$ if and only if $W_2 = U$.
(Hint: First note $W_2 = L + V_\alpha \neq U \iff \alpha \neq \frac{\pi}{2}$ and then observe:

$$L \cdot V_\alpha = \left\{ \left(\begin{smallmatrix} 0 & a \\ 0 & b \end{smallmatrix} \right) \mid a, b \in \mathbb{R} \right\}, \qquad \alpha \neq \tfrac{\pi}{2}.)$$

3.3.3. Let $C_2 = \{0, 1\}$ be the two-chain and $C_3 = \{\bot, a, \top\}$ be the integral quantale on the three-chain satisfying the condition $a * a = \bot$ (i.e. $(C_3, *)$ is the three-valued MV-algebra — cf. (9) in Exercises 2.2.1 and 2.7.2). On C_2 we consider two antisymmetric C_3-preorders p_1 and p_2 defined by:

$$p_i(0, 0) = p_i(0, 1) = p_i(1, 1) = \top, \quad i \in \{1, 2\}, \quad p_1(1, 0) = a, \quad p_2(1, 0) = \bot.$$

Show:

(a) The intrinsic partial orders of p_1 and p_2 coincide with the usual partial order on C_2.
(b) The C_3-preordered set (C_2, p_1) is a C_3-enriched join-complete lattice.
(c) The C_3-preordered set (C_2, p_2) is not C_3-enriched join-complete lattice. (Hint: Consider the C_3-enriched contravariant presheaf f on (C_2, p_2) determined by $f(1) = f(0) = a$ and observe that $d(f, \widetilde{0}) = a$, where d is the antisymmetric C_3-preorder on $\mathbb{P}(C_2, p_2)$.)

3.3.4. Let $(\mathfrak{Q}, *, e, ')$ be an involutive and unital quantale. Further, let $(X, \boxdot)$ be a right $\mathfrak{Q}$-module and $(X^{op}, \boxdot^{op})$ be the dual right $\mathfrak{Q}$-module of $(X, \boxdot)$, where $\boxdot^{op}$ is defined by (3.21). If p is the $\mathfrak{Q}$-preorder corresponding to $(X, \boxdot)$ (cf. (3.48) and Theorem 3.3.21), then show that the dual $\mathfrak{Q}$-preorder p^{op} corresponds to $(X^{op}, \boxdot^{op})$.

3.3.5. Let $(\mathfrak{Q}, *, e)$ be a unital quantale. On $\mathfrak{Q}$ we consider the following antisymmetric $\mathfrak{Q}$-preorders π_1 and π_2:

$$\pi_1(\alpha, \beta) = \alpha \searrow \beta \quad \text{and} \quad \pi_2(\alpha, \beta) = \alpha \swarrow \beta, \qquad \alpha, \beta \in \mathfrak{Q}.$$

In this context it might be helpful to point out that the intrinsic partial order underlying π_2 coincides with the dual order of $\mathfrak{Q}$.

(a) Show that $(\mathfrak{Q}, \pi_1)$ and $(\mathfrak{Q}, \pi_2)$ are join-complete $\mathfrak{Q}$-valued lattices. In particular, verify that the respective formations of joins are given by:

$$\mathsf{sup}_{(\mathfrak{Q},\pi_1)}(f) = \bigvee_{\alpha \in \mathfrak{Q}} (\alpha * f(\alpha)), \qquad f \in \mathbb{P}(\mathfrak{Q}, \pi_1),$$
$$\mathsf{sup}_{(\mathfrak{Q},\pi_2)}(f) = \bigwedge_{\alpha \in \mathfrak{Q}} (f(\alpha) \searrow \alpha), \qquad f \in \mathbb{P}(\mathfrak{Q}, \pi_2).$$

(b) If $(\mathfrak{Q}, \boxdot)$ is the right $\mathfrak{Q}$-module corresponding to the join-complete lattice $(\mathfrak{Q}, \pi_1)$, then show that the right action $\boxdot$ is given by $\alpha \boxdot \varepsilon = \alpha * \varepsilon$.

(c) If $(\mathfrak{Q}, \boxdot)$ is the right $\mathfrak{Q}$-module corresponding to the join-complete lattice $(\mathfrak{Q}, \pi_2)$, then show that the right action $\boxdot$ is given by $\alpha \boxdot \varepsilon = \varepsilon \searrow \alpha$.

(d) Additionally we assume that there exists an involution $'$ on $\mathfrak{Q}$ such that $(\mathfrak{Q}, *, e, ')$ is an involutive quantale. Further, let π_1^{op} be the dual $\mathfrak{Q}$-preorder of π_1. Show that the involution is a $\mathfrak{Q}$-isomorphism $(\mathfrak{Q}, \pi_1^{op}) \xrightarrow{'} (\mathfrak{Q}, \pi_2)$.

3.3.6. Let us consider the real unit interval $[0, 1]$ as a complete MV-algebra provided with the Łukasiewicz arithmetic conjunction $*_Ł$. Since $*_Ł$ is commutative, both types of implications (i.e. left-implication and right-implication) coincide and are given by the so-called the Łukasiewicz implication — i.e.

$$\alpha \to \beta = \min(1 - \alpha + \beta, 1), \qquad \alpha, \beta \in [0, 1].$$

Further, we introduce an antisymmetric $[0, 1]$-preorder π together with its dual $[0, 1]$-preorder π^{op} on $[0, 1]$ as follows[4]:

$$\pi(\alpha, \beta) = \alpha \to \beta \quad \text{and} \quad \pi^{op}(\alpha, \beta) = \beta \to \alpha, \qquad \alpha, \beta \in [0, 1].$$

Then $([0, 1], \pi_1)$ and $([0, 1], \pi_2)$ are join-complete $[0, 1]$-valued lattices (see Exercise 3.3.5 (a)). Show that the map $[0, 1] \xrightarrow{h} [0, 1]$ defined by

$$h(\alpha) = 1 - \tfrac{1}{2} \cdot \alpha, \qquad \alpha \in [0, 1]$$

is a $[0, 1]$-homomorphism from $([0, 1], \pi)$ to $([0, 1], \pi^{op})$ which is not cocontinuous, but is join-preserving w.r.t. the underlying intrinsic partial orders.

(Hint: If $\alpha, \varepsilon, \gamma \in [0, 1]$ and $([0, 1], \pi) \xrightarrow{h} ([0, 1], \pi^{op})$ is a $[0, 1]$-homomorphism, then the relation $\mathbb{P}(h)(\widetilde{\alpha}^{\dagger} *_Ł \varepsilon)(\gamma) = (h(\alpha) \to \gamma) *_Ł \varepsilon$ holds. By Exercise 3.3.5 (a) we obtain $\mathsf{sup}_{([0,1],\pi)}(\widetilde{\alpha}^{\dagger} *_Ł \varepsilon) = \alpha *_Ł \varepsilon$ and $\mathsf{sup}_{([0,1],\pi^{op})}(h(\alpha) \to \_) *_Ł \varepsilon = \varepsilon \to h(\alpha)$. If h now has the form $h(\alpha) = 1 - \frac{1}{2} \cdot \alpha$, then

$$\tfrac{5}{6} = h\big(\mathsf{sup}_{([0,1],p)}(\widetilde{\tfrac{2}{3}}^{\dagger} *_Ł \tfrac{2}{3})\big) < 1 = \mathsf{sup}_{([0,1],p^{op})}\big(\mathbb{P}(h)(\widetilde{\tfrac{2}{3}}^{\dagger} *_Ł \tfrac{2}{3})\big)$$

follows. Finally, we use the fact that the intrinsic partial order underlying π^{op} coincides with the dual order of $[0, 1]$.)

3.3.7. Let $(X, \boxdot)$ be a right $\mathfrak{Q}$-module and (X, p) be the corresponding join-complete $\mathfrak{Q}$-valued lattice. If $X \xrightarrow{f} \mathfrak{Q}$ is an arbitrary map, then the *contravariant hull* $\downarrow f$ and the *covariant hull* $\uparrow f$ of f are determined as follows:

$$(\downarrow f)(x) = \bigvee_{z \in X} p(x, z) * f(z) \quad \text{and} \quad (\uparrow f)(x) = \bigvee_{z \in X} f(z) * p(z, x), \qquad x \in X.$$

[4]Because of the commutativity the involution coincides with the identity $1_{[0,1]}$.

Show that the following relations hold:

(a) $\mathsf{sup}_{(X,p)}(\downarrow f) = \bigvee_{x \in X} x \boxdot f(x)$.
(b) $\mathsf{inf}_{(X,p)}(\uparrow f) = \bigvee_{x \in X} x \boxdot \big(\bigwedge_{z \in X} p(x,z) \swarrow f(z)\big)$.
(c) $\bigvee_{x \in X} f(x) * f(x) \leq p\big(\mathsf{inf}_{(X,p)}(\uparrow f), \mathsf{sup}_{(X,p)}(\downarrow f)\big)$.

(Hint: Proposition 3.3.18.)

Comment. If (X, p) is join-complete $\mathfrak{Q}$-valued lattice, then it follows immediately from Exercise 3.3.7 (a) and (b) that joins and meets of arbitrary $\mathfrak{Q}$-valued maps (i.e. "$\mathfrak{Q}$-fuzzy subsets" of X) exist.

3.3.8. (Change of base) Let $(\mathfrak{Q}, *, e_{\mathfrak{Q}})$ and $(\mathfrak{R}, *, e_{\mathfrak{R}})$ be unital quantales. Further, let $\mathfrak{Q} \xrightarrow{h} \mathfrak{R}$ be a unital homomorphism and $h^{\vdash}$ be the right adjoint map of h. Verify the following statements:

(a) If (X, p) is a $\mathfrak{Q}$-preordered set, then $(X, h \circ p)$ is an $\mathfrak{R}$-preordered set.
(b) If (X, p) is an $\mathfrak{R}$-preordered set, then $(X, h^{\vdash} \circ p)$ is a $\mathfrak{Q}$-preordered set.
(c) If (X, p) is a cocomplete $\mathfrak{R}$-preordered set, then $(X, h^{\vdash} \circ p)$ is also a cocomplete $\mathfrak{Q}$-preordered set.

(Hint. If f is a $\mathfrak{Q}$-enriched contravariant presheaf on $(X, h^{\vdash} \circ p)$ and x_0 is a join of the $\mathfrak{R}$-enriched contravariant hull $\downarrow h \circ f$ of $h \circ f$ on (X, p), then use Theorem 3.3.8 and show that the following chain of equivalences holds for all $\alpha \in \mathfrak{Q}$ and all $x \in X$:

$$\begin{aligned}
(\forall z \in X:\ f(z) * \alpha \leq (h^{\vdash} \circ p)(z, x)) &\iff (\forall z \in X: (h \circ f)(z) * h(\alpha) \leq p(z, x)) \\
&\iff (\forall z \in X: (\downarrow h \circ f)(z) * h(\alpha) \leq p(z, x)) \\
&\iff h(\alpha) \leq p(x_0, x) \\
&\iff \alpha \leq (h^{\vdash} \circ p)(x_0, x).
\end{aligned}$$

3.3.9. Let $\mathfrak{Q} = (\mathfrak{Q}, *, e)$ be a commutative and unital quantale. Since $\mathfrak{Q}$ is commutative, we can consider a bifunctor $\mathtt{Preord}(\mathfrak{Q}) \times \mathtt{Preord}(\mathfrak{Q}) \xrightarrow{\otimes} \mathtt{Preord}(\mathfrak{Q})$ acting on objects and morphisms as follows:

$$\begin{aligned}
&(X, p_X) \otimes (Y, p_Y) = \big((X \times Y), p_X \otimes p_Y\big) \qquad \text{where} \\
&(p_X \otimes p_Y)\big((x_1, y_1), (x_2, y_2)\big) = p_X(x_1, x_2) * p_Y(y_1, y_2), \quad x_1, x_2 \in X,\ y_1, y_2 \in Y, \\
&(X_1, p_{X_1}) \xrightarrow{h} (X_2, p_{X_2}), \quad (Y_1, p_{Y_1}) \xrightarrow{k} (Y_2, p_{Y_2}), \\
&(h \otimes k)(x, y) = \big(h(x), k(y)\big), \qquad x \in X_1,\ y \in Y_1.
\end{aligned}$$

Show that $\otimes$ induces the structure of a symmetric and monoidal closed category on $\mathtt{Preord}(\mathfrak{Q})$.

(Hint: The unit object is a singleton $\{\cdot\}$ provided with $p(\cdot, \cdot) = e$ and the internal hom-object $[(X, p_X), (Y, p_Y)]$ consists of all $\mathfrak{Q}$-homomorphisms $(X, p_X) \xrightarrow{h} (Y, p_Y)$ provided with the $\mathfrak{Q}$-preorder q defined by $q(h, k) = \bigwedge_{x \in X} p_Y\big(h(x), k(x)\big)$.)

3.3.10. (Cf. [44, Theorem 3.11]). Let $(\mathfrak{Q}, *, e)$ be a unital quantale and (X, p) and (Y, q) be $\mathfrak{Q}$-preordered sets. Further, let $(\mathbb{P}(X, p), d)$ be the $\mathfrak{Q}$-preordered set of all $\mathfrak{Q}$-enriched contravariant presheaves on (X, p) and $(\mathbb{P}^\dagger(Y, q), d^\dagger)$ be the $\mathfrak{Q}$-preordered set of all $\mathfrak{Q}$-enriched covariant presheaves on (Y, q). Finally, a $\mathfrak{Q}$-*relation* (i.e. a $\mathfrak{Q}$-*distributor*) from (X, p) to (Y, q) is a map $X \times Y \xrightarrow{r} \mathfrak{Q}$ satisfying the following condition for all $x_1, x_2 \in X$ and $y_1, y_2 \in Y$:

$$p(x_2, x_1) * r(x_1, y_1) * q(y_1, y_2) \leq r(x_2, y_2).$$

Show:

(a) Every $\mathfrak{Q}$-relation $X \times Y \xrightarrow{r} \mathfrak{Q}$ induces a pair $(\mathbb{P}(X, p), d) \underset{\Psi}{\overset{\Phi}{\rightleftarrows}} (\mathbb{P}^\dagger(Y, q), d^\dagger)$ of adjoint $\mathfrak{Q}$ -homomorphisms by

$$\Phi(f)(y) = \bigwedge_{z \in X} f(z) \searrow r(z, y), \quad \Psi(g)(x) = \bigwedge_{u \in Y} r(x, u) \swarrow g(u),$$

where $f \in \mathbb{P}(X, p)$ and $g \in \mathbb{P}^\dagger(Y, q)$. In particular, $\Phi \dashv \Psi$ holds.

(b) For every adjoint pair $(\mathbb{P}(X, p), d) \underset{\Psi}{\overset{\Phi}{\rightleftarrows}} (\mathbb{P}^\dagger(Y, q), d^\dagger)$ such that $\Phi \dashv \Psi$ there exists a unique $\mathfrak{Q}$-relation $X \times Y \xrightarrow{r} \mathfrak{Q}$ satisfying the following property for all $f \in \mathbb{P}(X, p)$ and for all $g \in \mathbb{P}^\dagger(Y, q)$:

$$\Phi(f)(y) = \bigwedge_{z \in X} f(z) \searrow r(z, y), \quad \Psi(g)(x) = \bigwedge_{u \in Y} r(x, u) \swarrow g(u).$$

(Hint: Consider a map $X \times Y \xrightarrow{r} \mathfrak{Q}$ determined by

$$r(x, y) = \Phi(\widetilde{x})(y), \quad \widetilde{x}(z) = p(z, x), \quad z \in X.$$

The right extensionality of r is obvious, while the left extensionality follows from the property that Φ is a $\mathfrak{Q}$-homomorphism. Hence r is a $\mathfrak{Q}$-relation. Moreover, the relation $f(x) = d(\widetilde{x}, f) \leq d^\dagger(\Phi(\widetilde{x}), \Phi(f))$ implies

$$\Phi(f)(y) \leq \bigwedge_{x \in X} f(x) \searrow \Phi(\widetilde{x})(y).$$

On the other hand, using the adjunction we first observe that

$$r(x, y) = \Phi(\widetilde{x})(y) = d^\dagger(\Phi(\widetilde{x}), \widetilde{y}^\dagger) = d(\widetilde{x}, \Psi(\widetilde{y}^\dagger)) = \Psi(\widetilde{y}^\dagger)(x),$$

and then derive the following estimate:

$$\begin{aligned}\bigwedge_{x \in X} f(x) \searrow r(x, y) = d(f, \Psi(\widetilde{y}^\dagger)) &\leq d^\dagger(\Phi(f), \Phi \circ \Psi(\widetilde{y}^\dagger)) \\ &\leq d^\dagger(\Phi(f), \widetilde{y}^\dagger) = \Phi(f)(y).\ )\end{aligned}$$

Comment. (a) The previous exercise is a $\mathfrak{Q}$-valued version of the concept of polarities initiated by G. Birkhoff in the binary case $\mathbb{1} = \{0, 1\}$. In this context orders have been restricted to discrete orders (cf. [14]).
(b) Given an adjoint pair $(\mathbb{P}(X, p), d) \underset{\Psi}{\overset{\Phi}{\rightleftarrows}} (\mathbb{P}^\dagger(Y, q), d^\dagger)$ with $\Phi \dashv \Psi$, then the composition $\Psi \circ \Phi$ is a linear nucleus on the right $\mathfrak{Q}$-module $\mathbb{P}(X, p)$ (cf. Proposition 3.3.23) whose regular quotient w.r.t. $\Psi \circ \Phi$ can be identified with a join-complete $\mathfrak{Q}$-valued lattice. In the special case when the adjoint pair $\Phi \dashv \Psi$ is given by $(\mathbb{P}(X, p), d) \underset{\Delta}{\overset{\Delta^\dagger}{\rightleftarrows}} (\mathbb{P}^\dagger(X, p), d^\dagger)$ (cf. Proof of Proposition 3.3.18), the regular quotient w.r.t. the linear nucleus $\Delta \circ \Delta^\dagger$ can be identified with the *$\mathfrak{Q}$-valued MacNeille completion* of (X, p).

3.4 Theory of Automata in Sup

3.4.1 General Definition

Let $\mathsf{C} = (\mathsf{C}_0, \otimes, \mathbb{1}, a, c, \ell, r)$ be a symmetric and monoidal closed category. Further, we assume that the underlying category C_0 is cocomplete.

A quintuple (Z, X, δ, τ, F) is an *automaton in* C if Z and X are objects of C_0, $Z \otimes X \xrightarrow{\delta} Z$ and $\mathbb{1} \xrightarrow{\tau} Z$ are C_0-morphisms and F is a subobject of Z represented by a monomorphism $F \xrightarrow{m} Z$. Referring to the standard understanding in Computer Science, Z is the object of *states*, X represents the *input alphabet*, δ describes the *dynamic* (or the transition map) of the automaton, τ is the *initial state* and F comprises the object of *final states*.

If we now consider the endofunctor F of C_0 defined by $\mathsf{F} = \_ \otimes X$, then it is easily seen that the dynamic (Z, δ) of an automaton is an F-algebra in C_0. Since C is closed, F preserves colimits. Hence we conclude from Theorem A.1.1 that every C_0-object A generates a free F-algebra $\big(\eta_A, (A^\sharp, \delta_{A^\sharp})\big)$. As a corollary of this observation we note that the initial state τ always has a unique extension to an F-homomorphism $\tau^\sharp$ which can be expressed by the following commutative diagram:

$$\begin{array}{ccccc} \mathbb{1} & \xrightarrow{\eta_{\mathbb{1}}} & \mathbb{1}^\sharp & \xleftarrow{\delta_{\mathbb{1}^\sharp}} & \mathbb{1}^\sharp \otimes X \\ & \searrow^{\tau} & \Big\downarrow{\scriptstyle \tau^\sharp} & & \Big\downarrow{\scriptstyle \tau^\sharp \otimes 1_X} \\ & & Z & \xleftarrow{\delta} & Z \otimes X \end{array}$$

Finally, we conclude from Theorem 1.1.8 that $\mathbb{1}^\sharp$ is the free monoid in C generated by the object X. Therefore we also write X^* for $\mathbb{1}^\sharp$.

A language is a subobject L of X^* represented by a monomorphism $L \xrightarrow{\lambda} X^*$. Given an automaton $M = (Z, X, \delta, \tau, U)$, a language L is *M-regular* if and only if $\tau^\sharp \circ \lambda$ factors through m — i.e.

$$\begin{array}{ccc} L & \overset{\lambda}{\rightarrowtail} & \mathbb{1}^{\sharp} = X^{*} \\ \Big\downarrow \vdots & & \Big\downarrow \tau^{\sharp} \\ F & \overset{m}{\rightarrowtail} & Z \end{array}$$

If C_0 has pullback squares, then the largest M-regular language is given by the pullback of m along $\tau^{\sharp}$ and is called the *language recognized by* M.

We finish this subsection with a brief comment related to `Set`. In this context the tensor product coincides with the cartesian product and the unit object is given by a singleton $\{\cdot\}$. Hence the concept of automata in `Set`, as it has been introduced in the above sense, coincides with the standard concept of (deterministic) automata (cf. [56]).

3.4.2 The Situation in **Sup**

In this subsection we restrict our interest to the category `Sup` of complete lattices and join-preserving maps. Hence states have a hierarchy expressed by an order, and the superposition of states is computed by joins.

An interesting class of automata in `Sup` is generated by non-deterministic automata in `Set`. For this reason we recall the following terminology.

Definition 3.4.1. A quintuple (Z, X, Δ, z_0, F) is a *non-deterministic automaton* if Z and X are sets, Δ is a relation from $Z \times X$ to Z (i.e. $\Delta \subseteq (Z \times X) \times Z$), z_0 is an element of Z and F is a subset of Z. In this context we use the following terminology. The set Z is understood as the set of states, z_0 is the initial state, X is the input alphabet, Δ represents the non-deterministic transition map and F comprises all final states.

Since every relation between two sets can be identified with a join-preserving map between the respective power sets, it is evident that every non-deterministic automaton in `Set` gives rise to an automaton in `Sup` in the following way. Let $\mathscr{P}(X)$ and $\mathscr{P}(Z)$ be the respective power sets of X and Z. Then we identify the non-deterministic transition map Δ with a bimorphism $\mathscr{P}(Z) \times \mathscr{P}(X) \xrightarrow{b_\Delta} \mathscr{P}(Z)$ (i.e. a join-preserving map in each variable separately) as follows:

$$b_\Delta(A, B) = \{v \in Z \mid \exists (z, x) \in A \times B \,:\, (z, x, v) \in \Delta\}, \qquad A \in \mathscr{P}(Z),\ B \in \mathscr{P}(X).$$

Due to the universal property of the tensor product in `Sup` (cf. Theorem 2.1.8 and Fact II in Sect. 2.1.2) there exists a unique join-preserving map

$$\mathscr{P}(Z) \otimes \mathscr{P}(X) \xrightarrow{\delta} \mathscr{P}(Z)$$

making the following diagram commutative:

$$\begin{array}{ccc} \mathscr{P}(Z) \times \mathscr{P}(X) & \xrightarrow{\ \otimes\ } & \mathscr{P}(Z) \otimes \mathscr{P}(X) \\ & \searrow^{b_\Delta} & \vdots\, \delta \\ & & \mathscr{P}(Z) \end{array}$$

In this context it is worthwhile to note that the tensor product $\mathscr{P}(Z) \otimes \mathscr{P}(X)$ is isomorphic to $\mathscr{P}(X)^Z$ (cf. Example 2.1.9). Further, we identify the initial state z_0 with a join-preserving map $\mathbb{1} = \{0, 1\} \xrightarrow{\tau} \mathscr{P}(Z)$ by $\tau(0) = \varnothing$ and $\tau(1) = \{z_0\}$. Then $M = (\mathscr{P}(Z), \mathscr{P}(X), \delta, \tau, \mathscr{P}(F))$ is an automaton in Sup. Obviously, the dynamic $(\mathscr{P}(Z), \delta)$ of M is a functor algebra in Sup w.r.t. the endofunctor $\_ \otimes \mathscr{P}(X)$. Now we consider the extension $\mathbb{1}^\sharp = (\mathscr{P}(X))^* \xrightarrow{\tau^\sharp} \mathscr{P}(Z)$ of τ and make the following observations:

- The free unital quantale generated by the power set $\mathscr{P}(X)$ is the power set $\mathscr{P}(X^*)$ of the free monoid X^* generated by X (in the sense of Set) provided with the Minkowski multiplication.
- Since $\tau^\sharp$ is join-preserving, we can identify $\tau^\sharp$ with its restriction τ^* to the atoms of $\mathscr{P}(X^*)$ — i.e. $X^* \xrightarrow{\tau^*} \mathscr{P}(Z)$.

A language $\mathbb{L}$ is now a complete sublattice of $\mathscr{P}(X^*)$ — i.e. the inclusion map is join-preserving. As an example we consider the power set $\mathscr{P}(L)$ of a subset L of X^*, where L can be considered as a language in the sense of Set. Then $\mathscr{P}(L)$ is M-regular if and only if the restriction of τ^* to L factors through $\mathscr{P}(F)$. In this sense we arrive at a new concept of regularity in the framework of non-deterministic automata.

Now let us address the general situation of automata in Sup. For this purpose we fix a complete lattice X and consider the endofunctor F of Sup defined by $\mathsf{F} = \_ \otimes X$. Further, let $\mathbf{T} = (\mathsf{T}, \eta, \mu)$ be the F-term monad (cf. Appendix A.1). Then with every F-algebra (Y, δ_Y) we associate the corresponding structure morphism $\mathsf{T}(Y) = Y^\sharp \xrightarrow{\xi_Y} Y$ and identify (Y, δ_Y) with the $\mathbf{T}$-algebra (Y, ξ_Y). In particular, we recall that there exists a natural isomorphism $\psi_\infty \colon \mathsf{T} \to \_ \otimes \mathbb{1}^\sharp$ (cf. Sect. 1.1.2), where $\mathbb{1}^\sharp = X^*$ is the free unital quantale generated by X (cf. Theorem 1.1.8).

If $M = (Y, X, \delta, \tau, F)$ is an automaton in Sup, then the join-preserving map δ describing the dynamic of M has an extension $Y \otimes X^* \xrightarrow{\delta^*} Y$ which is determined by the following commutative diagram:

$$\begin{array}{ccc} Y \otimes X^* & \xleftarrow{\ \psi_\infty^{r_Y^{-1}}\ } & Y^\sharp = \mathsf{T}(Y) \\ \delta^* \vdots & \swarrow_{\xi_Y} & \\ Y & & \end{array} \tag{3.65}$$

where $\psi_\infty^{r_Y^{-1}}$ denotes the Y-component of ψ_∞.

Theorem 3.4.2. *If* (Y, X, δ, τ, F) *is an automaton in* Sup, *then* (Y, δ^*) *is a right* X^**-module, where* X^* *is the free unital quantale generated by the complete lattice* X.

Proof. Let $[Y, Y]$ be the complete lattice of all join-preserving maps $Y \xrightarrow{f} Y$ provided with the multiplication given by the composition. Then $([Y, Y], \circ, 1_Y)$ is a unital quantale (cf. Example 2.3.2). Since Sup is a symmetric and monoidal closed category, there exists a unique join-preserving map $X \xrightarrow{h} [Y, Y]$ such that the following diagram is commutative:

$$\begin{array}{ccc} Y \otimes X & \xrightarrow{1_Y \otimes h} & Y \otimes [Y, Y] \\ & \searrow^{\delta} & \downarrow{\scriptstyle \varepsilon_Y} \\ & & Y \end{array}$$

where ε_Y is the evaluation arrow defined on elementary tensors by $\varepsilon(y \otimes f) = f(y)$ for all $f \in [Y, Y]$ and $y \in Y$. Moreover, it is not difficult to see that the unique join-preserving map $[Y, Y] \otimes [Y, Y] \xrightarrow{m} [Y, Y]$ determined by

$$\begin{array}{ccc} Y \otimes ([Y, Y] \otimes [Y, Y]) & \xrightarrow{1_Y \otimes m} & Y \otimes [Y, Y] \\ {\scriptstyle a_{Y\,[Y,Y]\,[Y,Y]}} \uparrow & & \\ (Y \otimes [Y, Y]) \otimes [Y, Y] & & \downarrow {\scriptstyle \varepsilon_Y} \\ {\scriptstyle \varepsilon_Y \otimes 1_{[Y,Y]}} \downarrow & & \\ Y \otimes [Y, Y] & \xrightarrow{\quad \varepsilon_Y \quad} & Y \end{array} \tag{3.66}$$

acts on elementary tensors as follows:

$$m(f \otimes g)(y) = g(f(y)), \qquad y \in Y, \ f, g \in [Y, Y].$$

Hence m is the transposed multiplication of the usual composition — i.e. $m^\tau = \circ$. Now we maintain the notation from Sect. 1.1.2 and make the following observations. Since $\mathbb{1}^\sharp = X^*$ is the free unital quantale generated by the complete lattice X, there exists a unique unital homomorphism $(\mathbb{1}^\sharp, m_{\mathbb{1}^\sharp}) \xrightarrow{\widehat{h}} ([Y, Y], m)$ such that $\widehat{h} \circ \eta_X = h$ holds. Thus $\widehat{h}$ is also a unital homomorphism from $(\mathbb{1}^\sharp, m^\tau_{\mathbb{1}^\sharp})$ to $([Y, Y], \circ)$.

Because of diagram (3.65) and Proposition 3.1.5 it is sufficient to verify the commutativity of the following diagram:

$$\begin{array}{ccccc} Y^\sharp & \xrightarrow{\psi_\infty^{r_Y^{-1}}} & Y \otimes \mathbb{1}^\sharp & \xrightarrow{1_Y \otimes \widehat{h}} & Y \otimes [Y, Y] \\ & {\scriptstyle \xi_Y} \searrow & & \searrow^{\delta^*} & \downarrow {\scriptstyle \varepsilon_Y} \\ & & & & Y \end{array} \tag{3.67}$$

We refer to the primitive recursion of the free algebra algorithm and introduce two sequences of join-preserving maps $Z_k(Y) \xrightarrow{\delta_k} Y$ and $Z_k(\mathbb{1}) \xrightarrow{h_k} [Y, Y]$ as follows:

$$\delta_1 = \delta, \quad \delta_{k+1} = \delta \circ \big((\delta_k \sqcup 1_Y) \otimes 1_X\big), \quad h_1 = h \circ \ell_X, \quad h_{k+1} = m \circ \big((h_k \sqcup e) \otimes h\big),$$

where $k \in \mathbb{N}$ and $\mathbb{1} \xrightarrow{e} [Y, Y]$ is the unit of $([Y, Y], m)$ — i.e. $e(1) = 1_Y$. Further, we put $U_k = Z_k(\mathbb{1}) \sqcup \mathbb{1}$ and define recursively a third sequence of isomorphisms $Z_k(Y) \xrightarrow{\varphi_k} Y \otimes Z_k(\mathbb{1})$ by:

$$\varphi_1 = 1_Y \otimes \ell_X^{-1}, \quad \varphi_{k+1} = a_{Y\,U_k\,X} \circ \big((\varphi_k \oplus r_Y^{-1}) \otimes 1_X\big), \qquad k \in \mathbb{N}.$$

We show that for all $k \in \mathbb{N}$ the diagram

$$\begin{array}{ccccc} Z_k(Y) & \xrightarrow{\varphi_k} & Y \otimes Z_k(\mathbb{1}) & \xrightarrow{1_Y \otimes h_k} & Y \otimes [Y, Y] \\ & \searrow^{\delta_k} & & & \downarrow{\scriptstyle \varepsilon_Y} \\ & & & & Y \end{array} \tag{3.68}$$

is commutative. Since $(1_Y \otimes h_1) \circ \varphi_1 = 1_Y \otimes h$, the case $k = 1$ is trivial. Since the relation $r_Y = \varepsilon_Y \circ (1_Y \otimes e)$ holds, the diagram

$$\begin{array}{ccccc} Z_1(Y) \sqcup Y & \xrightarrow{\varphi_1 \oplus r_Y^{-1}} & Y \otimes (Z_1(\mathbb{1}) \sqcup \mathbb{1}) & \xrightarrow{1_Y \otimes (h_1 \sqcup e)} & Y \otimes [Y, Y] \\ & \searrow^{\delta_1 \sqcup 1_Y} & & & \downarrow{\scriptstyle \varepsilon_Y} \\ & & & & Y \end{array}$$

is commutative. Now we apply the endofunctor $\mathsf{F} = \_ \otimes X$ to the previous diagram and obtain:

$$\begin{array}{ccccc} & & Y \otimes Z_2(\mathbb{1}) & \xrightarrow{1_Y \otimes ((h_1 \sqcup e) \otimes h)} & Y \otimes ([Y, Y] \otimes [Y, Y]) \\ & \nearrow^{\varphi_2} & \uparrow{\scriptstyle a} & & \uparrow{\scriptstyle a} \\ Z_2(Y) & \xrightarrow{(\varphi_1 \oplus r_Y^{-1}) \otimes 1_X} & (Y \otimes U_1) \otimes X & \xrightarrow{(1_Y \otimes (h_1 \sqcup e)) \otimes h} & (Y \otimes [Y, Y]) \otimes [Y, Y] \\ & \searrow_{(\delta_1 \sqcup 1_Y) \otimes 1_X} & & & \downarrow{\scriptstyle \varepsilon_Y \otimes 1_{[Y,Y]}} \\ & & Y \otimes X & \xrightarrow{1_Y \otimes h} & Y \otimes [Y, Y] \\ & & & \searrow^{\delta} & \downarrow{\scriptstyle \varepsilon_Y} \\ & & & & Y \end{array}$$

Referring to the diagram (3.66) the relation $\delta_2 = \varepsilon_Y \circ (1_Y \otimes h_2) \circ \varphi_2$ follows. Hence by means of primitive recursion the commutativity of (3.68) can be verified. Finally, we pass to the respective direct limits and conclude from (3.68) that the diagram

$$\begin{array}{ccccc} Z_\infty(Y) & \xrightarrow{\varphi_\infty} & Y \otimes Z_\infty(\mathbb{1}) & \xrightarrow{1_Y \otimes h_\infty} & Y \otimes [Y, Y] \\ & \searrow^{\delta_\infty} & & & \downarrow{\scriptstyle \varepsilon_Y} \\ & & & & Y \end{array}$$

is commutative. Hence the relation

$$\xi_Y = \delta_\infty \sqcup 1_Y = \varepsilon_Y \circ \big(1_Y \otimes (h_\infty \sqcup e)\big) \circ (\varphi_\infty \oplus r_Y^{-1}) = \varepsilon_Y \circ (1_Y \otimes \widehat{h}) \circ \psi_\infty^{r_Y^{-1}}$$

holds, and the commutativity of (3.67) is verified. □

Comment. Theorem 3.4.2 holds not only in Sup, but in any symmetric and monoidal closed category with underlying cocomplete category (cf. [41, 42]). If, for example, $\mathfrak{Q}$ is a commutative and unital quantale, then we can replace Sup by Mod($\mathfrak{Q}$) (cf. Fact V in Sect. 3.1.4) and consequently arrive at the concept of automata in Mod($\mathfrak{Q}$). Referring to Remark 3.1.15, this observation might be the point of departure for a concept *deserving* the name "fuzzy automaton".

Finally, as an immediate corollary of Sect. 3.3.3 and Theorem 3.4.2 we obtain the important result that every automaton (Y, X, δ, τ, F) in Sup gives rise to a join-complete X^*-valued lattice Y in which the join of a map $Y \xrightarrow{f} X^*$ can be expressed by the following element $y_0 \in Y$ (cf. Exercise 3.3.7):

$$y_0 = \bigvee_{y \in Y} \delta^*\big(y, f(y)\big).$$

Notes

The theory of modules in the category Sup essentially goes back to A. Joyal and M. Tierney 1984 (cf. [59]), who have mainly focused on commutative and unital quantales. With regard to the algebraic foundations of quantale-valued structures the important fact that $\mathfrak{Q}^X$ is the free left $\mathfrak{Q}$-module generated by the set X appears already on p. 10 in [59].

Even though dual modules have been studied in [59, 91], our comprehensive explanation of the self-duality of the category of left modules on involutive, unital quantales and its dualizer seems to be new. The concept of involutive left modules on involutive, unital quantales was invented by P. Resende 2004 (cf. [97]). It was his observation that various results on the quantization of points obtained by C.J. Mulvey and J.W. Pelletier 2001 (cf. [84]) permit a more elegant formulation in terms of irreducible and involutive left modules.

The characterization of the category of right $\mathfrak{Q}$-modules by the $\mathfrak{Q}$-enriched version of the category Sup (i.e. the category of join-complete $\mathfrak{Q}$-valued lattices) goes back to I. Stubbe 2006, who developed this theory in a more general context given by quantaloid enriched categories (cf. [107]). In this context, from a historical perspective, it is interesting to note that in the case of the real unit interval viewed as integral quantale the axioms of a many-valued preorder go back to L.A. Zadeh 1971 and S.V. Ovchinnikov 1984 (cf. [88, 112]).

The pathway leading to automata in monoidal closed categories can be summarized as follows. The notion *automaton* was coined by S.C. Kleene 1951 (cf. [64, 65]) and emerged from the study of nerve nets by W.S. McCulloch and W. Pitts 1943 (cf. [76]). Categorical formulations of automata theory in general monoidal categories can be traced back to the work of L. Budach and H.-J. Hoehnke 1969/70 (cf. [17, 18]), H. Ehrig and M. Pfender 1972 (cf. [29, 30]) and J.A. Goguen 1971 (cf. [40–42]. A continuation of this development can be found in [3, 4, 74]).

Appendix A

The appendix explains the construction of free functor algebras and the role of coherence axioms in the construction of the tensor product of semigroups in symmetric monoidal categories. A graphical representation of the vertical and horizontal composition of natural transformations with applications to the composition of monads is also given.

A.1 Free Functor Algebras

Let F be an endofunctor of a category $\mathtt{C}$. A pair (X, δ_X) is an F-*algebra* if $X \in |\mathtt{C}|$ and $\mathsf{F}(X) \xrightarrow{\delta_X} X$ is a morphism of $\mathtt{C}$. A morphism $X \xrightarrow{h} Y$ of $\mathtt{C}$ is an F-*homomorphism* from (X, δ_X) to (Y, δ_Y) if h satisfies the additional property $h \circ \delta_X = \delta_Y \circ \mathsf{F}(h)$ — i.e. if the following diagram is commutative:

$$\begin{array}{ccc} \mathsf{F}(X) & \xrightarrow{\mathsf{F}(h)} & \mathsf{F}(Y) \\ {\scriptstyle \delta_X}\downarrow & & \downarrow{\scriptstyle \delta_Y} \\ X & \xrightarrow{\quad h \quad} & Y \end{array}$$

Evidently F-algebras and F-homomorphisms form a category denoted by $\mathtt{Alg}(\mathsf{F}, \mathtt{C})$.

Theorem A.1.1. ([1, 3]) *Let $\mathtt{C}$ be a cocomplete category and F be an endofunctor of $\mathtt{C}$. If F preserves countable direct limits, then for each $X \in |\mathtt{C}|$ there exists an F-algebra $(X^\sharp, \delta_{X^\sharp})$ and a $\mathtt{C}$-morphism $X \xrightarrow{\eta_X} X^\sharp$ such that the following property holds*:

> *For every F-algebra (Y, δ_Y) and for every $\mathtt{C}$-morphism $X \xrightarrow{h} Y$ there exists a unique F-homomorphism $(X^\sharp, \delta_{X^\sharp}) \xrightarrow{h^\sharp} (Y, \delta_Y)$ such that the following diagram is commutative*:

$$\begin{array}{ccc} X & \xrightarrow{\eta_X} & X^\sharp \\ & {\scriptstyle h}\searrow & \downarrow{\scriptstyle h^\sharp} \\ & & Y \end{array} \tag{A.1}$$

P. Eklund et al., *Semigroups in Complete Lattices*, Developments in Mathematics 54, https://doi.org/10.1007/978-3-319-78948-4

In particular, $(\eta_X, (X^\sharp, \delta_{X^\sharp}))$ is unique up to F*-homomorphism.*

Proof. The uniqueness of $(\eta_X, (X^\sharp, \delta_{X^\sharp}))$ follows immediately from the commutativity of (A.1). In order to verify the existence of $(\eta_X, (X^\sharp, \delta_{X^\sharp}))$ we proceed as follows. First we recall the construction of the *free algebra algorithm* (cf. [1, 3]). For this purpose we fix $X \in |\mathsf{C}|$ and the coproduct injection $X \xrightarrow{j_X} \mathsf{F}(X) \sqcup X$. Then we consider two countable direct systems $(Z_k(X), e_{m\,k})_{k\in\mathbb{N}}$ and $(W_k(X), \ell_{m\,k})_{k\in\mathbb{N}}$ defined by:

$$
\begin{aligned}
&Z_1(X) = \mathsf{F}(X), \qquad W_1(X) = \mathsf{F}(X) \sqcup X,\\
&Z_k(X) = \mathsf{F}(W_{k-1}(X)), \quad W_k(X) = Z_k(X) \sqcup X, \qquad k \geq 2,\\
&Z_1(X) \xrightarrow{e^X_{21}} Z_2(X), \quad e^X_{21} = \mathsf{F}(j_X), \quad W_1(X) \xrightarrow{\ell^X_{21}} W_2(X), \quad \ell^X_{21} = \mathsf{F}(j_X) \oplus 1_X,\\
&Z_k(X) \xrightarrow{e^X_{k+1\,k}} Z_{k+1}(X), \quad e^X_{k+1\,k} = \mathsf{F}(\ell^X_{k\,k-1}),\\
&W_k(X) \xrightarrow{\ell^X_{k+1\,k}} W_{k+1}(X), \quad \ell^X_{k+1\,k} = e^X_{k+1\,k} \oplus 1_X, \qquad k \geq 2,
\end{aligned}
$$

where $\sqcup$ and $\oplus$ denote the binary operations for the construction of the coproduct of objects and the construction of the coproduct of arrows in C.
Now we invoke the cocompleteness of C and conclude that the direct limit $Z_\infty(X)$ of $(Z_k, e^X_{m\,k})_{k\in\mathbb{N}}$ exists. In particular, the corresponding canonical arrows are denoted by $Z_k(X) \xrightarrow{e^X_k} Z_\infty(X)$.

As a next step we form the coproduct $Z_\infty(X) \sqcup X$ and denote the corresponding coproduct injections by $X \xrightarrow{\eta_X} Z_\infty(X) \sqcup X$ and $Z_\infty(X) \xrightarrow{j^X_\infty} Z_\infty(X) \sqcup X$. Then it is easily seen that $Z_\infty(X) \sqcup X$ is the direct limit of $(W_k(X), \ell^X_{m\,k})_{k\in\mathbb{N}}$ with the canonical arrows $W_k(X) \xrightarrow{e^X_k \oplus 1_X} Z_\infty(X) \sqcup X$. Moreover, since F preserves countable direct limits, $\mathsf{F}(Z_\infty(X) \sqcup X)$ is the direct limit of $\big(\mathsf{F}(W_k(X)), \mathsf{F}(\ell^X_{m\,k})\big)_{k\in\mathbb{N}}$. Now we refer to the construction of $(Z_k(X), e^X_{m\,k})_{k\in\mathbb{N}}$ and observe that $Z_k(X) = \mathsf{F}(W_{k-1}(X))$ and $e^X_{m\,k} = \mathsf{F}(\ell^X_{m-1\,k-1})$ hold for each $k \geq 2$. Hence $\mathsf{F}(Z_\infty(X) \sqcup X)$ and $Z_\infty(X)$ are isomorphic. In particular, there exists an isomorphism $\mathsf{F}(Z_\infty(X) \sqcup X) \xrightarrow{\vartheta_X} Z_\infty(X)$ satisfying the following important properties

$$
e^X_1 = \vartheta_X \circ \mathsf{F}(\eta_X) \quad \text{and} \quad e^X_k = \vartheta_X \circ \mathsf{F}\big(e^X_{k-1} \oplus 1_X\big), \quad k \geq 2, \tag{A.2}
$$

where we have applied the relations $(e^X_1 \oplus 1_X) \circ j_X = \eta_X$ and $\mathsf{F}(j_X) = e^X_{21}$. Finally, we put $X^\sharp = Z_\infty(X) \sqcup X$ and $\delta_{X^\sharp} = j^X_\infty \circ \vartheta_X$ and show that the F-algebra $(X^\sharp, \delta_{X^\sharp})$ satisfies the required properties.

(a) (Uniqueness). Let us consider an F-homomorphism $(X^\sharp, \delta_{X^\sharp}) \xrightarrow{h^\sharp} (Y, \delta_Y)$ such that the diagram (A.1) is commutative. Since $\delta_Y \circ \mathsf{F}(h^\sharp) = h^\sharp \circ \delta_{X^\sharp}$, the following relations hold:

$$
h^\sharp \circ j^X_\infty \circ e^X_1 = h^\sharp \circ \delta_{X^\sharp} \circ \mathsf{F}(\eta_X) = \delta_Y \circ \mathsf{F}(h^\sharp) \circ \mathsf{F}(\eta_X) = \delta_Y \circ \mathsf{F}(h), \tag{A.3}
$$

$$h^\sharp \circ j_\infty^X \circ e_k^X = \delta_Y \circ \mathsf{F}(h^\sharp) \circ \mathsf{F}(e_{k-1}^X \oplus 1_X) = \delta_Y \circ \mathsf{F}\big((h^\sharp \circ j_\infty^X \circ e_{k-1}^X) \sqcup h\big), \quad k \geq 2. \tag{A.4}$$

Now we invoke the universal property of direct limits and observe that the restriction of $h^\sharp$ to $Z_\infty(X)$ (i.e. $h^\sharp \circ j_\infty^X$) is uniquely determined by h.

(b) (Existence). With regard to (A.3) and (A.4) we are motivated to introduce a sequence $(h_k)_{k\in\mathbb{N}}$ of morphisms $Z_k(X) \xrightarrow{h_k} Y$ as follows:

$$h_1 = \delta_Y \circ \mathsf{F}(h) \quad \text{and} \quad h_k = \delta_Y \circ \mathsf{F}(h_{k-1} \sqcup h), \qquad k \geq 2. \tag{A.5}$$

By means of induction the relation $h_{k+1} \circ e_{k+1\,k}^X = h_k$ holds for each $k \in \mathbb{N}$. Then we conclude from the construction of direct limits that there exists a unique morphism $Z_\infty(X) \xrightarrow{h_\infty} Y$ with the property $h_\infty \circ e_k^X = h_k$ for all $k \in \mathbb{N}$. Now we put $h^\sharp = h_\infty \sqcup h$ and note $h^\sharp \circ j_\infty^X \circ e_k^X = h_k$. Further, we observe:

$$\begin{aligned}
\delta_Y \circ \mathsf{F}\big(h^\sharp\big) \circ \big(\vartheta_X^{-1} \circ e_{k+1}^X\big) &= \delta_Y \circ \mathsf{F}(h_\infty \sqcup h) \circ \mathsf{F}(e_k^X \oplus 1_X)\\
&= \delta_Y \circ \mathsf{F}(h_k \sqcup h)\\
&= h_{k+1}\\
&= h^\sharp \circ j_\infty^X \circ \vartheta_X \circ \big(\vartheta_X^{-1} \circ e_{k+1}^X\big).
\end{aligned}$$

Referring again to the universal property of direct limits we obtain $\delta_Y \circ \mathsf{F}(h^\sharp) = h^\sharp \circ j_\infty^X \circ \vartheta_X = h^\sharp \circ \delta_{X^\sharp}$ — i.e. $h^\sharp$ is an F-homomorphism. □

If an endofunctor F of a cocomplete category $\mathtt{C}$ preserves countable direct limits, then Theorem A.1.1 means that the forgetful functor $\mathsf{U}\colon \mathtt{Alg}(\mathsf{F}, \mathtt{C}) \to \mathtt{C}$ has a left adjoint functor $\mathsf{G}\colon \mathtt{C} \to \mathtt{Alg}(\mathsf{F}, \mathtt{C})$. In particular, G is determined by

$$\mathsf{G}(X) = (X^\sharp, \delta_{X^\sharp}), \quad \mathsf{G}(h) = (\eta_Y \circ h)^\sharp \quad \text{where} \quad X \xrightarrow{h} Y.$$

In this context $(X^\sharp, \delta_{X^\sharp})$ is said to be the *free* F-*algebra* generated by the object $X \in |\mathtt{C}|$. Moreover, the monad $\mathbf{T} = (\mathsf{T}, \eta, \mu)$ associated with the adjoint situation $\mathsf{G} \dashv \mathsf{U}$ is called the F-*term monad*. Since U is the forgetful functor, the multiplication $\mu = (\mu_X)_{X\in|\mathtt{C}|}$ has the following explicit form

$$(X^\sharp)^\sharp \xrightarrow{\mu_X} X^\sharp, \qquad \mu_X = (1_{X^\sharp})^\sharp, \qquad \mu_X \circ j_\infty^{X^\sharp} \circ e_1^{X^\sharp} = \delta_{X^\sharp}.$$

Finally, the construction of the F-term functor T can be simplified as follows. If $X \xrightarrow{h} Y$ is given, then we introduce recursively a sequence $\big(\widehat{h}_n\big)_{n\in\mathbb{N}}$ of morphisms $Z_n(X) \xrightarrow{\widehat{h}_n} Z_\infty(Y)$ as follows:

$$\widehat{h}_1 = e_1^Y \circ \mathsf{F}(h) \quad \text{and} \quad \widehat{h}_n = \vartheta_Y \circ \mathsf{F}\big(\widehat{h}_{n-1} \oplus h\big), \qquad n \geq 2.$$

Hence there exists a unique morphism $Z_\infty(X) \xrightarrow{\widehat{h}_\infty} Z_\infty(Y)$ satisfying the condition $\widehat{h}_\infty \circ e_n^X = \widehat{h}_n$ for all $n \in \mathbb{N}$. Now we put $h_k = j_\infty^Y \circ \widehat{h}_k$ and observe that $\mathsf{T}(h)$ is given by

$$\mathsf{T}(h) = \widehat{h}_\infty \oplus h.$$

For some applications we also recall the monadic properties of the F-term monad — i.e. the associativity axiom and the unit axiom (cf. Sect. 1.2):

$$\mu_X \circ \mu_{X^\sharp} = \mu_X \circ \mathsf{T}(\mu_X) \quad \text{and} \quad \mu_X \circ \eta_{X^\sharp} = 1_{X^\sharp} = \mu_X \circ \mathsf{T}(\eta_X). \tag{A.6}$$

In this context it is worthwhile to note that the Eilenberg–Moore category of T-algebras is isomorphic to the category $\mathtt{Alg}(\mathsf{F}, \mathtt{C})$ of F-algebras.

A.2 Coherence Axioms in the Construction of the Tensor Product of Monoids

Let us fix now four objects X, Y, U and V of a symmetric monoidal category $(\mathtt{C}, \otimes, \mathbb{1}, a, \ell, r, c)$ and define morphisms

$$(X \otimes (Y \otimes U)) \otimes V \xleftarrow{\Phi_{XYUV}} (X \otimes Y) \otimes (U \otimes V) \xrightarrow{\Phi^*_{XYUV}} X \otimes ((Y \otimes U) \otimes V)$$

by the following diagram (see also Sect. 1.1.3):

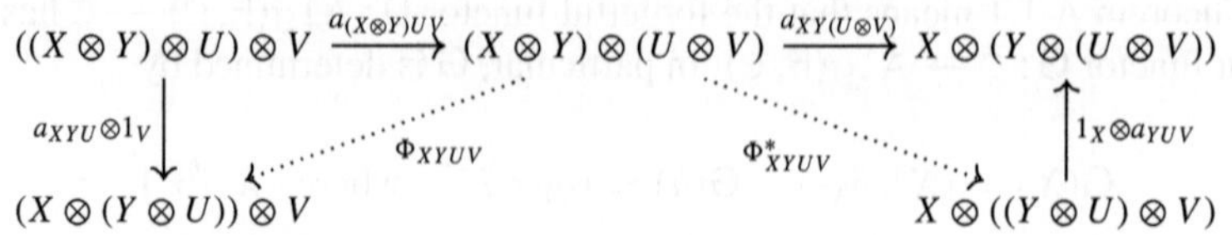

Because of the pentagonal diagram, the diagram

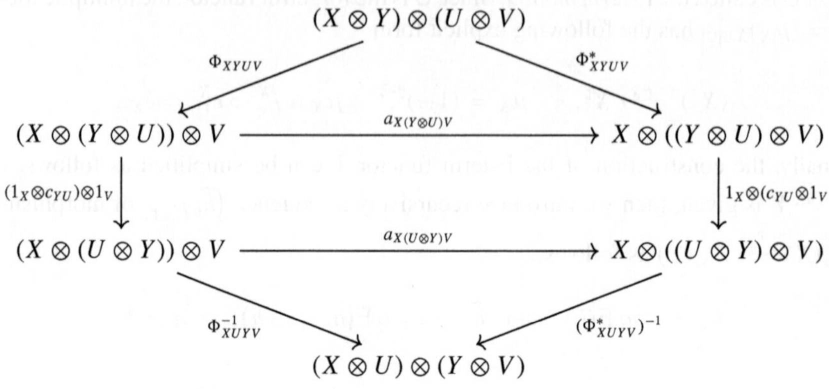

is commutative.

Hence we are in a position to introduce a morphism

$$(X \otimes Y) \otimes (U \otimes V) \xrightarrow{\Theta_{XYUV}} (X \otimes U) \otimes (Y \otimes V)$$

by

$$\begin{aligned}\Theta_{XYUV} &= \Phi^{-1}_{XUYV} \circ \big((1_X \otimes c_{YU}) \otimes 1_V\big) \circ \Phi_{XYUV} \\ &= (\Phi^*_{XUYV})^{-1} \circ \big(1_X \otimes (c_{YU} \otimes 1_V)\big) \circ \Phi^*_{XYUV}\end{aligned}$$

and verify the following proposition.

Proposition A.2.1. *Let* X, Y, U, V, K *and* L *be objects of a symmetric and monoidal category* C. *Then the following relation holds*:

$$\begin{aligned}&(a_{XUK} \otimes a_{YVL}) \circ \Theta_{(X\otimes U)(Y\otimes V)KL} \circ (\Theta_{XYUV} \otimes 1_{K\otimes L}) \\ &\quad = \Theta_{XY(U\otimes K)(V\otimes L)} \circ (1_{X\otimes Y} \otimes \Theta_{UVKL}) \circ a_{(X\otimes Y)(U\otimes V)(K\otimes L)}.\end{aligned} \tag{A.7}$$

Proof. A repeated application of the pentagonal diagram leads to the following commutative diagram:

$$\begin{array}{ccc}
\big((X \otimes Y) \otimes (U \otimes V)\big) \otimes (K \otimes L) & \xrightarrow{a} & (X \otimes Y) \otimes \big((U \otimes V) \otimes (K \otimes L)\big) \\
 & & \big\downarrow{\scriptstyle 1_{X\otimes Y}\otimes a} \\
 & & (X \otimes Y) \otimes \big(U \otimes (V \otimes (K \otimes L))\big) \\
\Big\downarrow{\scriptstyle \Phi_{XYUV}\otimes 1_{K\otimes L}} & & \big\uparrow{\scriptstyle a} \\
 & & ((X \otimes Y) \otimes U) \otimes (V \otimes (K \otimes L)) \\
 & & \big\downarrow{\scriptstyle a\otimes(1_V\otimes 1_{K\otimes L})} \\
\big((X \otimes (Y \otimes U)) \otimes V\big) \otimes (K \otimes L) & \xrightarrow{a} & (X \otimes (Y \otimes U)) \otimes (V \otimes (K \otimes L)) \\
{\scriptstyle ((a^{-1}\circ(1_X\otimes c))\otimes 1_V)\otimes 1}\big\downarrow & & \big\downarrow{\scriptstyle (a^{-1}\circ(1_X\otimes c))\otimes(1_V\otimes 1)} \\
\big(((X \otimes U) \otimes Y) \otimes V\big) \otimes (K \otimes L) & \xrightarrow{a} & ((X \otimes U) \otimes Y) \otimes (V \otimes (K \otimes L)) \\
{\scriptstyle a\otimes 1_{K\otimes L}}\big\downarrow & & \\
\big((X \otimes U) \otimes (Y \otimes V)\big) \otimes (K \otimes L) & & \Big\downarrow{\scriptstyle a} \\
{\scriptstyle a}\big\downarrow & & \\
(X \otimes U) \otimes \big((Y \otimes V) \otimes (K \otimes L)\big) & \xrightarrow{1_{X\otimes U}\otimes a} & (X \otimes U) \otimes \big(Y \otimes (V \otimes (K \otimes L))\big) \\
{\scriptstyle 1_{X\otimes U}\otimes a}\big\uparrow & & \\
(X \otimes U) \otimes \big(((Y \otimes V) \otimes K) \otimes L\big) & & \Big\uparrow{\scriptstyle 1_{X\otimes U}\otimes(1_Y\otimes a)} \\
{\scriptstyle 1_{X\otimes U}\otimes(a\otimes 1_L)}\big\downarrow & & \\
(X \otimes U) \otimes \big((Y \otimes (V \otimes K)) \otimes L\big) & \xrightarrow{1_{X\otimes U}\otimes a} & (X \otimes U) \otimes \big(Y \otimes ((V \otimes K) \otimes L)\big)
\end{array}$$

Further we observe:

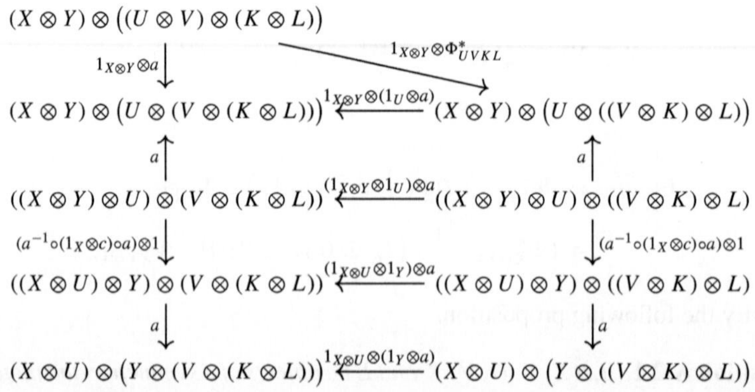

We paste the previous diagrams together and arrive at the following commutative diagram:

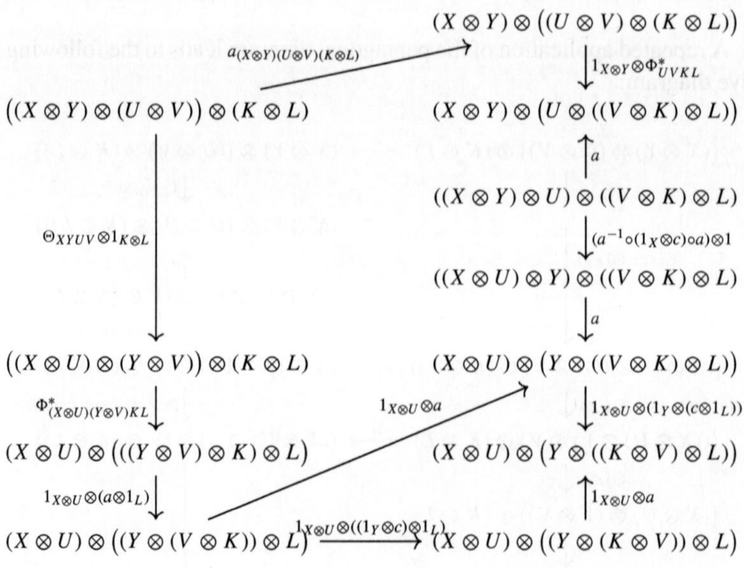

Further, we observe the commutativity of the following diagram:

$$\begin{array}{ccc}
(X\otimes Y)\otimes\big(U\otimes((V\otimes K)\otimes L)\big) & \xrightarrow{1_{X\otimes Y}\otimes(1_U\otimes(c\otimes 1_L))} & (X\otimes Y)\otimes\big(U\otimes((K\otimes V)\otimes L)\big)\\
{\scriptstyle a}\uparrow & & \uparrow{\scriptstyle a}\\
((X\otimes Y)\otimes U)\otimes((V\otimes K)\otimes L) & \xrightarrow{(1_{X\otimes Y}\otimes 1_U)\otimes(c\otimes 1_L)} & ((X\otimes Y)\otimes U)\otimes((K\otimes V)\otimes L)\\
{\scriptstyle (a^{-1}\circ(1_X\otimes c)\circ a)\otimes 1}\downarrow & & \downarrow{\scriptstyle (a^{-1}\circ(1_X\otimes c)\circ a)\otimes 1}\\
(X\otimes U)\otimes Y)\otimes((V\otimes K)\otimes L) & \xrightarrow{(1_{X\otimes U}\otimes 1_Y)\otimes(c\otimes 1_L)} & (X\otimes U)\otimes Y)\otimes((K\otimes V)\otimes L)\\
{\scriptstyle a}\downarrow & & \downarrow{\scriptstyle a}\\
(X\otimes U)\otimes\big(Y\otimes((V\otimes K)\otimes L)\big) & \xrightarrow{1_{X\otimes U}\otimes(1_Y\otimes(c\otimes 1_L))} & (X\otimes U)\otimes\big(Y\otimes((K\otimes V)\otimes L)\big)
\end{array}$$

Now we paste the previous two diagrams together and obtain the commutativity of the following diagram:

(B1)

$$\begin{array}{ccc}
 & & (X\otimes Y)\otimes\big((U\otimes V)\otimes(K\otimes L)\big)\\
 & \nearrow{\scriptstyle a_{(X\otimes Y)(U\otimes V)(K\otimes L)}} & \downarrow{\scriptstyle 1_{X\otimes Y}\otimes\Phi^*_{UVKL}}\\
\big((X\otimes Y)\otimes(U\otimes V)\big)\otimes(K\otimes L) & & (X\otimes Y)\otimes\big(U\otimes((V\otimes K)\otimes L)\big)\\
 & & \downarrow{\scriptstyle 1_{X\otimes Y}\otimes(1_U\otimes(c\otimes 1_L))}\\
 & & (X\otimes Y)\otimes(U\otimes((K\otimes V)\otimes L))\\
{\scriptstyle \Theta_{XYUV}\otimes 1_{K\otimes L}}\Bigg\downarrow & & \uparrow{\scriptstyle a}\\
 & & ((X\otimes Y)\otimes U)\otimes((K\otimes V)\otimes L)\\
 & & \downarrow{\scriptstyle (a^{-1}\circ(1_X\otimes c)\circ a)\otimes 1}\\
\big((X\otimes U)\otimes(Y\otimes V)\big)\otimes(K\otimes L) & & ((X\otimes U)\otimes Y)\otimes((K\otimes V)\otimes L)\\
{\scriptstyle \Phi^*_{(X\otimes U)(Y\otimes V)KL}}\downarrow & & \downarrow{\scriptstyle a}\\
(X\otimes U)\otimes\big(((Y\otimes V)\otimes K)\otimes L\big) & & (X\otimes U)\otimes\big(Y\otimes((K\otimes V)\otimes L)\big)\\
{\scriptstyle 1_{X\otimes U}\otimes(a\otimes 1_L)}\downarrow & & \uparrow{\scriptstyle 1_{X\otimes U}\otimes a}\\
(X\otimes U)\otimes\big((Y\otimes(V\otimes K))\otimes L\big) & \xrightarrow{1_{X\otimes U}\otimes((1_Y\otimes c)\otimes 1_L)} & (X\otimes U)\otimes\big((Y\otimes(K\otimes V))\otimes L\big)\\
 & & \uparrow{\scriptstyle 1_{X\otimes U}\otimes(a\otimes 1_L)}\\
 & & (X\otimes U)\otimes\big(((Y\otimes K)\otimes V)\otimes L\big)\\
 & & \downarrow{\scriptstyle 1_{X\otimes U}\otimes((c\otimes 1_V)\otimes 1_L)}\\
 & & (X\otimes U)\otimes\big(((K\otimes Y)\otimes V)\otimes L\big)\\
 & & \downarrow{\scriptstyle 1_{X\otimes U}\otimes a}\\
 & & (X\otimes U)\otimes\big((K\otimes Y)\otimes(V\otimes L)\big)
\end{array}$$

In the next step we use the coherence axiom (4) and again the pentagonal diagram in the instance of four objects given by K, Y, V and L. Then the following diagram is commutative:

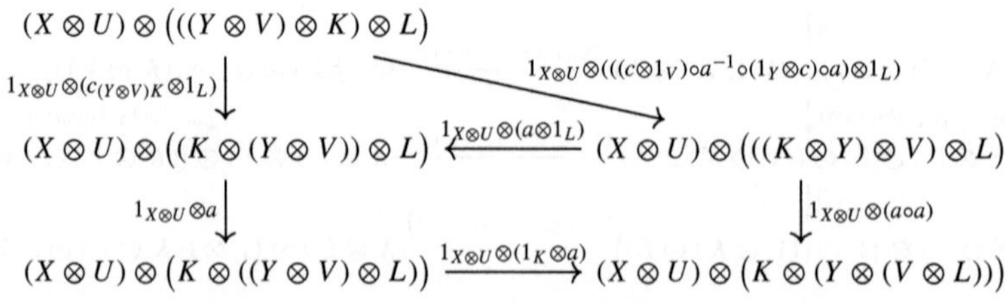

Now we paste together the previous diagram and (**B1**) and obtain:
(**B2**)

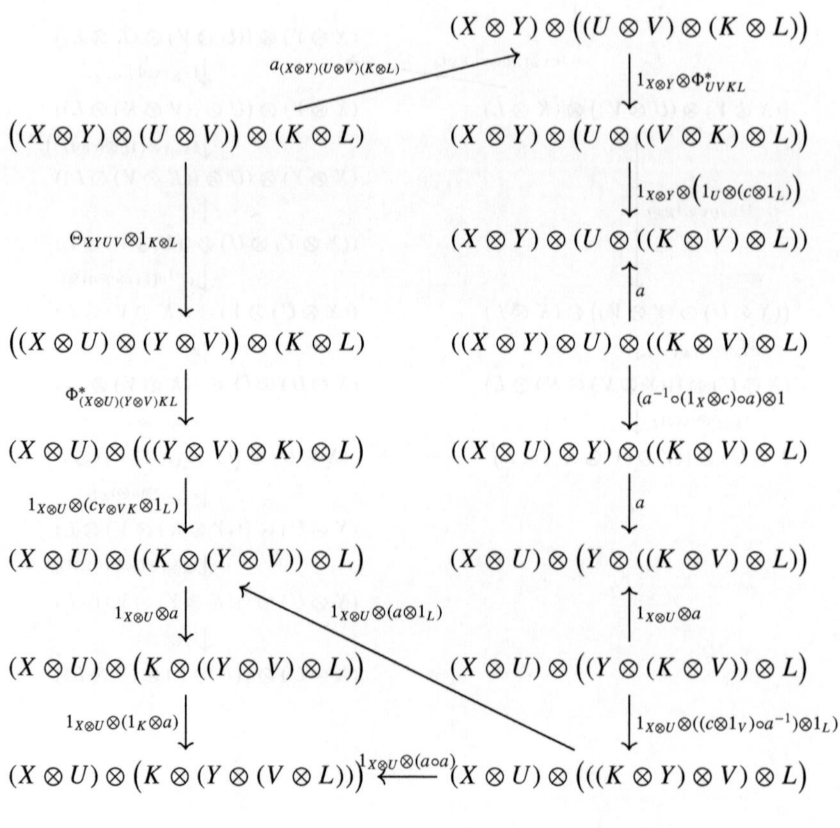

Further, we observe the commutativity of the following two diagrams:

(C)

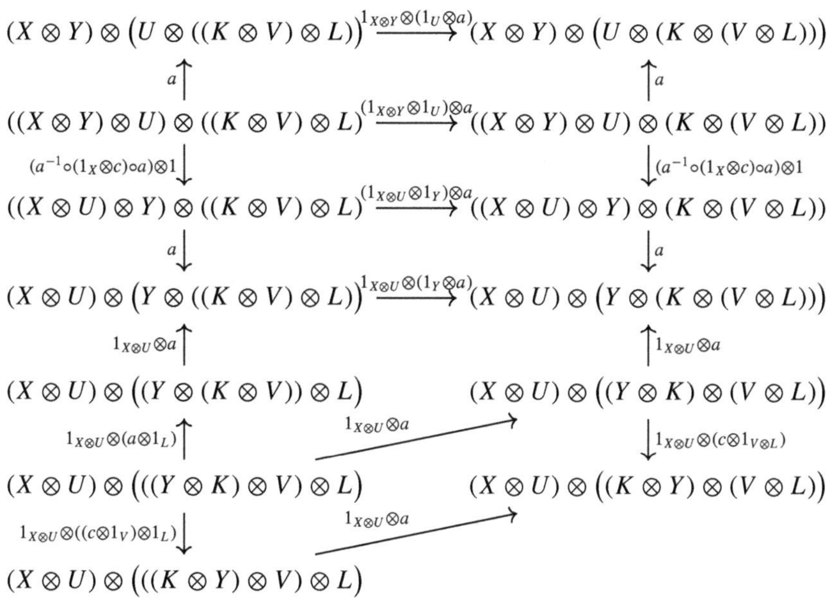

(D)

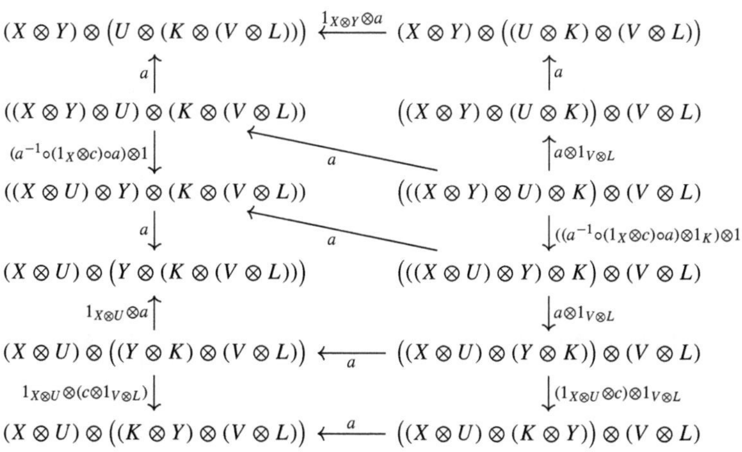

In a next step we paste together (**B2**) and (**C**) and obtain:
(**B3**)

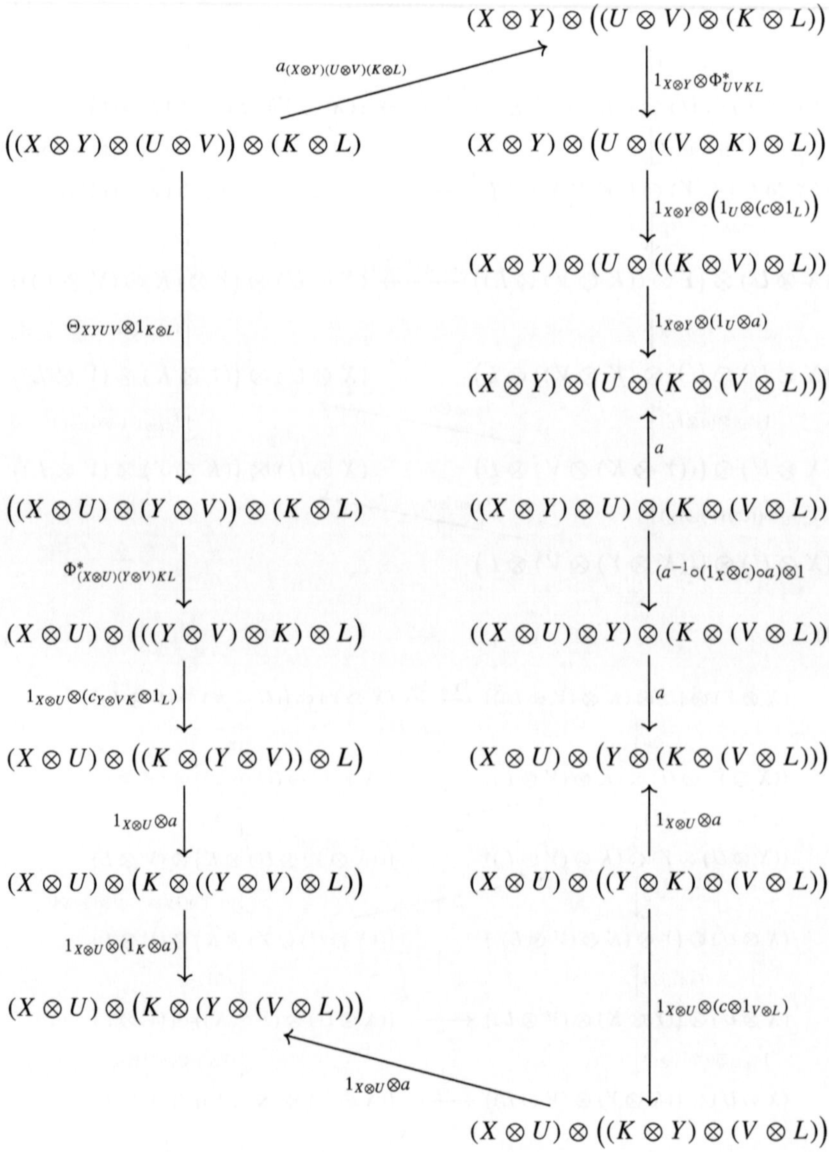

Further we paste together (**B3**) and (**D**) and obtain:
(**B4**)

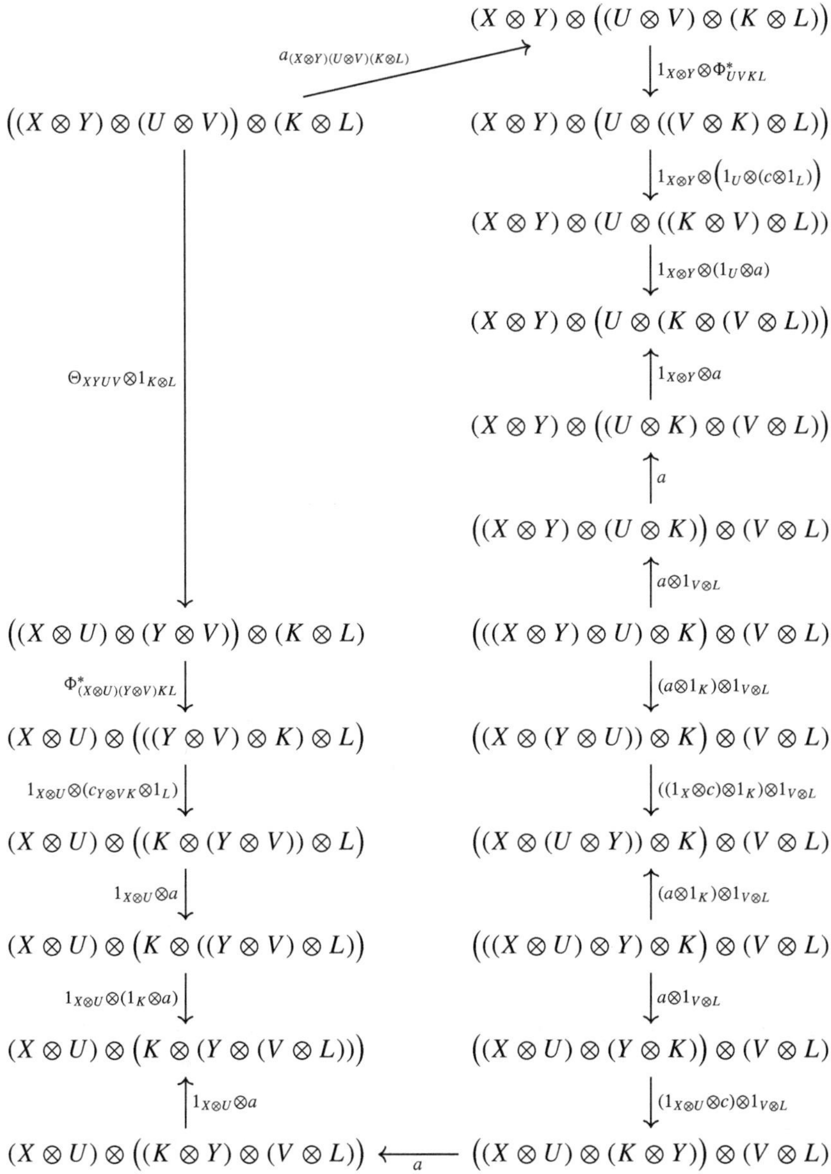

Referring to the definition of $(\Phi^*_{UKVL})^{-1}$ and Θ_{UVKL} we can simplify the previous diagram and add the pentagonal diagram in the instances of $X \otimes U$, K, Y and $V \otimes L$:

$$
\begin{array}{ccc}
 & \overset{a_{(X\otimes Y)(U\otimes V)(K\otimes L)}}{\nearrow} & (X \otimes Y) \otimes \big((U \otimes V) \otimes (K \otimes L)\big) \\
 & & \Big\downarrow{\scriptstyle 1_{X\otimes Y}\otimes\Theta_{UVKL}} \\
\big((X \otimes Y) \otimes (U \otimes V)\big) \otimes (K \otimes L) & & (X \otimes Y) \otimes \big((U \otimes K) \otimes (V \otimes L)\big) \\
\Big\downarrow{\scriptstyle \Theta_{XYUV}\otimes 1_{K\otimes L}} & & \Big\uparrow{\scriptstyle a} \\
 & & \big((X \otimes Y) \otimes (U \otimes K)\big) \otimes (V \otimes L) \\
 & & \Big\uparrow{\scriptstyle a\otimes 1_{V\otimes L}} \\
\big((X \otimes U) \otimes (Y \otimes V)\big) \otimes (K \otimes L) & & \big(((X \otimes Y) \otimes U) \otimes K\big) \otimes (V \otimes L) \\
\Big\downarrow{\scriptstyle \Phi^*_{(X\otimes U)(Y\otimes V)KL}} & & \Big\downarrow{\scriptstyle (a\otimes 1_K)\otimes 1_{V\otimes L}} \\
(X \otimes U) \otimes \big(((Y \otimes V) \otimes K) \otimes L\big) & & \big((X \otimes (Y \otimes U)) \otimes K\big) \otimes (V \otimes L) \\
\Big\downarrow{\scriptstyle 1_{X\otimes U}\otimes(c_{Y\otimes V K}\otimes 1_L)} & & \Big\downarrow{\scriptstyle ((1_X\otimes c)\otimes 1_K)\otimes 1_{V\otimes L}} \\
(X \otimes U) \otimes \big((K \otimes (Y \otimes V)) \otimes L\big) & & \big((X \otimes (U \otimes Y)) \otimes K\big) \otimes (V \otimes L) \\
\Big\downarrow{\scriptstyle 1_{X\otimes U}\otimes a} & & \Big\uparrow{\scriptstyle (a\otimes 1_K)\otimes 1_{V\otimes L}} \\
(X \otimes U) \otimes \big(K \otimes ((Y \otimes V) \otimes L)\big) & & \big(((X \otimes U) \otimes Y) \otimes K\big) \otimes (V \otimes L) \\
\Big\downarrow{\scriptstyle 1_{X\otimes U}\otimes(1_K\otimes a)} & & \Big\downarrow{\scriptstyle a\otimes 1_{V\otimes L}} \\
(X \otimes U) \otimes \big(K \otimes (Y \otimes (V \otimes L))\big) & & \big((X \otimes U) \otimes (Y \otimes K)\big) \otimes (V \otimes L) \\
\Big\uparrow{\scriptstyle 1_{X\otimes U}\otimes a} & & \Big\downarrow{\scriptstyle (1_{X\otimes U}\otimes c)\otimes 1_{V\otimes L}} \\
(X \otimes U) \otimes \big((K \otimes Y) \otimes (V \otimes L)\big) & \xleftarrow{\;a\;} & \big((X \otimes U) \otimes (K \otimes Y)\big) \otimes (V \otimes L) \\
\Big\downarrow{\scriptstyle 1_{X\otimes U}\otimes a} & & \Big\uparrow{\scriptstyle a\otimes 1_{V\otimes L}} \\
 & & \big(((X \otimes U) \otimes K) \otimes Y\big) \otimes (V \otimes L) \\
 & & \Big\downarrow{\scriptstyle a} \\
(X \otimes U) \otimes \big(K \otimes (Y \otimes (V \otimes L))\big) & \xleftarrow{\;a\;} & ((X \otimes U) \otimes K) \otimes (Y \otimes (V \otimes L)) \\
\Big\uparrow{\scriptstyle 1_{X\otimes U}\otimes(1_K\otimes a)} & & \Big\uparrow{\scriptstyle 1\otimes a} \\
(X \otimes U) \otimes \big(K \otimes ((Y \otimes V) \otimes L)\big) & \xleftarrow{\;a\;} & ((X \otimes U) \otimes K) \otimes ((Y \otimes V) \otimes L)
\end{array}
$$

Using the definition of $(\Phi^*_{(X\otimes U)K(Y\otimes V)L})^{-1}$ and $\Theta_{(X\otimes U)(Y\otimes V)KL}$ we continue to simplify the previous diagram:
(B5)

$$
\begin{array}{ccc}
 & & (X\otimes Y)\otimes\big((U\otimes V)\otimes(K\otimes L)\big) \\
 & \nearrow^{a_{(X\otimes Y)(U\otimes V)(K\otimes L)}} & \Big\downarrow{\scriptstyle 1_{X\otimes Y}\otimes\Theta_{UVKL}} \\
\big((X\otimes Y)\otimes(U\otimes V)\big)\otimes(K\otimes L) & & (X\otimes Y)\otimes\big((U\otimes K)\otimes(V\otimes L)\big) \\
 & & \Big\uparrow{\scriptstyle a} \\
\Big\downarrow{\scriptstyle \Theta_{XYUV}\otimes 1_{K\otimes L}} & & \big((X\otimes Y)\otimes(U\otimes K)\big)\otimes(V\otimes L) \\
 & & \Big\uparrow{\scriptstyle a\otimes 1_{V\otimes L}} \\
\big((X\otimes U)\otimes(Y\otimes V)\big)\otimes(K\otimes L) & & \big(((X\otimes Y)\otimes U)\otimes K\big)\otimes(V\otimes L) \\
 & & \Big\downarrow{\scriptstyle (a\otimes 1_K)\otimes 1_{V\otimes L}} \\
 & & \big((X\otimes(Y\otimes U))\otimes K\big)\otimes(V\otimes L) \\
 & & \Big\downarrow{\scriptstyle ((1_X\otimes c)\otimes 1_K)\otimes 1_{V\otimes L}} \\
 & & \big((X\otimes(U\otimes Y))\otimes K\big)\otimes(V\otimes L) \\
\Big\downarrow{\scriptstyle \Theta_{(X\otimes U)(Y\otimes V)KL}} & & \Big\uparrow{\scriptstyle (a\otimes 1_K)\otimes 1_{V\otimes L}} \\
 & & \big(((X\otimes U)\otimes Y)\otimes K\big)\otimes(V\otimes L) \\
 & & \Big\downarrow{\scriptstyle a\otimes 1_{V\otimes L}} \\
 & & \big((X\otimes U)\otimes(Y\otimes K)\big)\otimes(V\otimes L) \\
 & & \Big\downarrow{\scriptstyle (1_{X\otimes U}\otimes c)\otimes 1_{V\otimes L}} \\
((X\otimes U)\otimes K)\otimes((Y\otimes V)\otimes L) & & \big((X\otimes U)\otimes(K\otimes Y)\big)\otimes(V\otimes L) \\
\Big\downarrow{\scriptstyle 1\otimes a} & & \Big\uparrow{\scriptstyle a\otimes 1_{V\otimes L}} \\
((X\otimes U)\otimes K)\otimes(Y\otimes(V\otimes L)) & \xleftarrow{\;a\;} & \big(((X\otimes U)\otimes K)\otimes Y\big)\otimes(V\otimes L)
\end{array}
$$

In a further step we apply three times the pentagonal diagram and obtain the following commutative diagram:

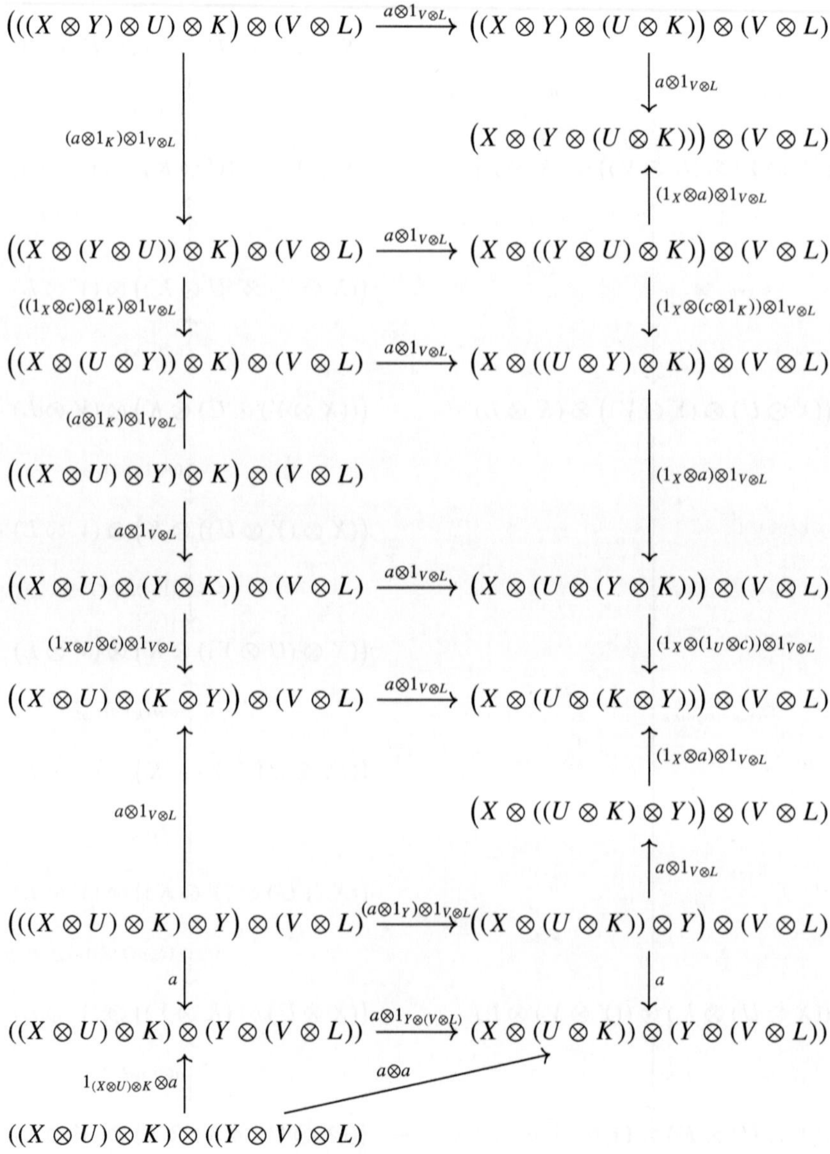

Now we apply again the coherence axiom (4) and simplify the previous diagram as follows:
(B6)

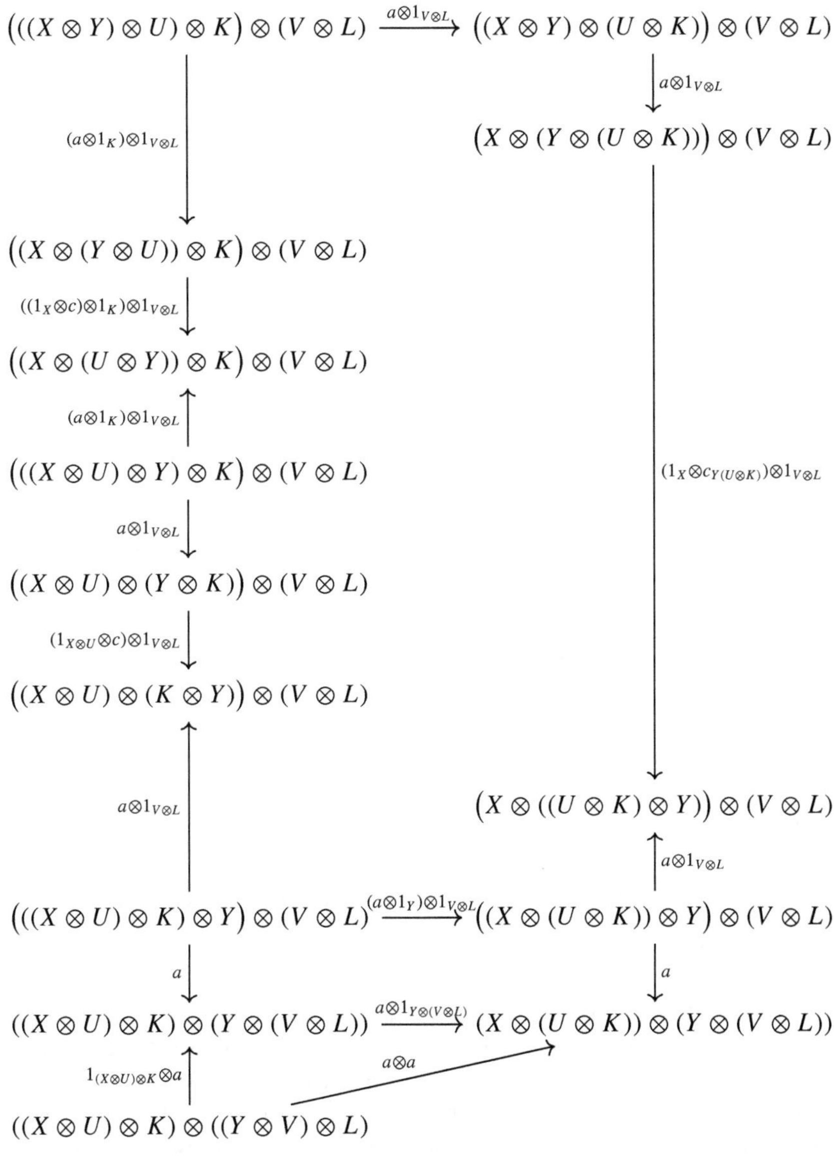

Now we paste together the diagrams (**B5**) and (**B6**) and finally arrive at the following commutative diagram:

$$
\begin{array}{ccc}
 & & (X \otimes Y) \otimes \big((U \otimes V) \otimes (K \otimes L)\big) \\
 & \nearrow^{a_{(X\otimes Y)(U\otimes V)(K\otimes L)}} & \Big\downarrow{\scriptstyle 1_{X\otimes Y}\otimes\Theta_{UVKL}} \\
\big((X \otimes Y) \otimes (U \otimes V)\big) \otimes (K \otimes L) & & ((X \otimes Y) \otimes \big((U \otimes K) \otimes (V \otimes L)\big) \\
 & & \Big\uparrow{\scriptstyle a} \\
{\scriptstyle \Theta_{XYUV}\otimes 1_{K\otimes L}}\Big\downarrow & & \big((X \otimes Y) \otimes (U \otimes K)\big) \otimes (V \otimes L) \\
 & & \Big\downarrow{\scriptstyle a\otimes 1_{V\otimes L}} \\
\big((X \otimes U) \otimes (Y \otimes V)\big) \otimes (K \otimes L) & & \big(X \otimes (Y \otimes (U \otimes K))\big) \otimes (V \otimes L) \\
 & & \Big\downarrow{\scriptstyle (1_X\otimes c_{Y(U\otimes K)})\otimes 1_{V\otimes L}} \\
 & & \big(X \otimes ((U \otimes K) \otimes Y)\big) \otimes (V \otimes L) \\
{\scriptstyle \Theta_{(X\otimes U)(Y\otimes V)KL}}\Big\downarrow & & \Big\uparrow{\scriptstyle a\otimes 1_{V\otimes L}} \\
 & & \big((X \otimes (U \otimes K)) \otimes Y\big) \otimes (V \otimes L) \\
 & & \Big\downarrow{\scriptstyle a} \\
((X \otimes U) \otimes K) \otimes ((Y \otimes V) \otimes L) & \xrightarrow{a\otimes a} & (X \otimes (U \otimes K)) \otimes (Y \otimes (V \otimes L))
\end{array}
$$

Referring once again to the definition of $\Phi_{XY(U\otimes K)(V\otimes L)}$, $\Phi^{-1}_{X(U\otimes K)Y(V\otimes L)}$ and $\Theta_{XY(U\otimes K)(V\otimes L)}$, we conclude from the previous diagram that the relation (A.7) holds. □

A.3 Graphical Representation of Proofs Involving Monad Composition

The calculus of natural transformations comprises two compositions, the ordinary composition (resp. vertical composition) and star composition (resp. horizontal composition). Since both compositions sometimes occur simultaneously, e.g., in the construction of the multiplication of composite monads, it is desirable to formulate a representation making these types of constructions more transparent. In this context the interchange law suggests to develop a graphically two-dimensional framework. In fact, the ordinary composition can pictorially be represented as a vertical stacking of respective building blocks, whereas star composition is depicted as a horizontal sequencing of such blocks. This enables us to unfold one-dimensionally written expressions into two-dimensionally written expressions, which sometimes more transparently exposes where and how to use commutativity of diagrams in proof steps.

In order to see this more precisely we may depict a natural transformation $\tau : \mathsf{F} \to \mathsf{G}$ as

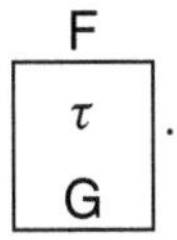

Note here that we hide the information about underlying categories. Once we use a natural transformation $\tau : \mathsf{F} \to \mathsf{G}$ it is thereby assumed that the domain category of the functors F and G is the same, and also that the codomain category of the functors F and G is the same — i.e. $\mathsf{F}, \mathsf{G} : \mathtt{C} \to \mathtt{D}$. The left and right vertical lines in the building block therefore hide the categories with which they in fact are annotated. For reasons which will become evident when we compose horizontally, we read the domain-codomain for functors from right to left — i.e. we correlate $\mathsf{F}, \mathsf{G} : \mathtt{D} \leftarrow \mathtt{C}$ with the graphical representation

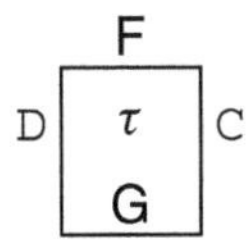

in order to be totally transparent about the notation involving underlying categories.

It is now clearly tempting to attach blocks to one another both vertically as well as horizontally in such a way that the functors of the attaching horizontal lines of the blocks match, and the categories of the attaching vertical lines match. In order to depict these attachment, let $\tau : \mathsf{F} \to \mathsf{G}$, $\sigma : \mathsf{G} \to \mathsf{H}$, $\tau' : \mathsf{F}' \to \mathsf{G}'$ and $\sigma' : \mathsf{G}' \to \mathsf{H}'$ be natural transformations with $\mathsf{F}, \mathsf{G}, \mathsf{H} : \mathtt{C} \to \mathtt{D}$ and $\mathsf{F}', \mathsf{G}', \mathsf{H}' : \mathtt{D} \to \mathtt{E}$. We then have a vertical attachment in

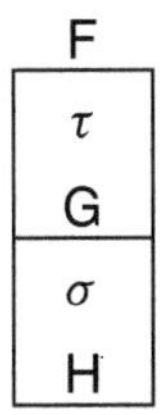

and a horizontal attachment in

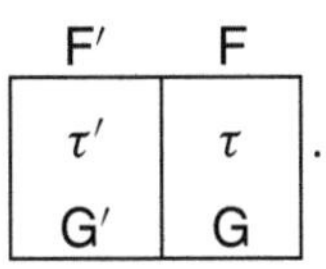

As we shall see later, the vertical attachment will be defined as the vertical composition of natural transformations. This then enables us to view the vertical two-block attachment as one block, which in turn can be attached vertically or horizontally to other blocks, as long as the functors and categories, respectively, on the horizontal and vertical lines continue to match. Similarly, the horizontal attachment will be defined as the horizontal (or star) composition of natural transformations. This similarly enables us to view the horizontal two-block attachment as one block, which can be attached to other blocks.

The construction

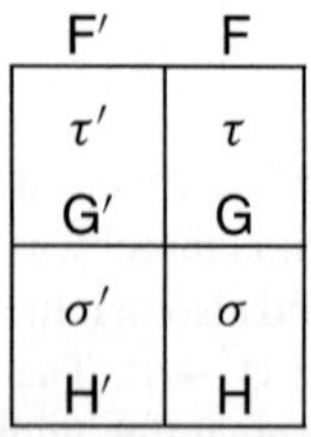

now makes sense from the viewpoint of matching domains and codomains at the vertical and horizontal lines, even if we still haven't defined what the corresponding single block natural transformation actually is. Respective functors appear here explicitly, whereas the underlying categories are hidden. Later on in this Appendix A.3, when we use blocks to facilitate proofs related to the composition of monads, we will use endofunctors, and then obviously we have one and the same category underlying the whole graphical representation.

We now define respective attachment as composition. In the case of vertical composition blocks $\tau : \mathsf{F} \rightarrow \mathsf{G}$ and $\sigma : \mathsf{G} \rightarrow \mathsf{H}$ are built, composed, and defined vertically as

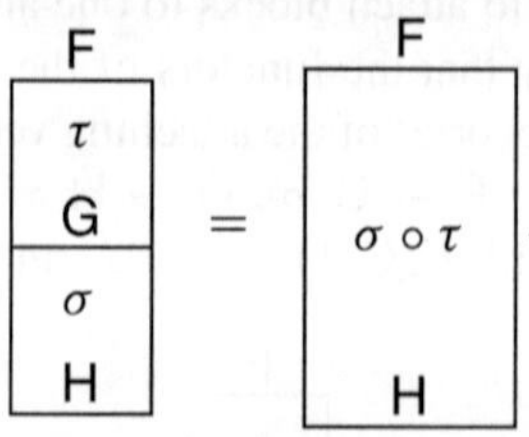

Note how the underlying categories are annotated to the vertical lines from top to bottom — i.e. we would see it as

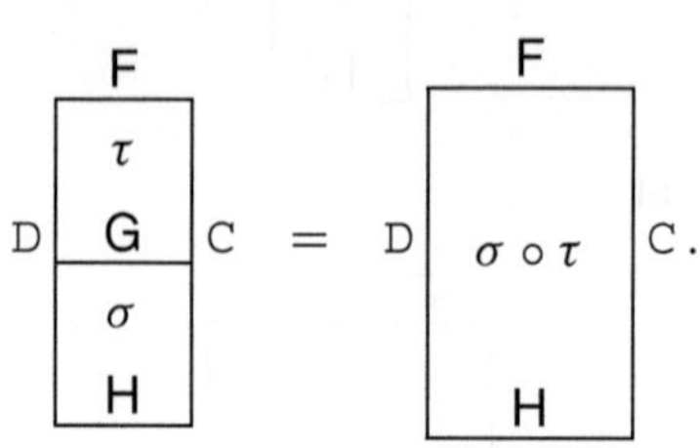

Mixing notation using expressions and graphical representation could be depicted as

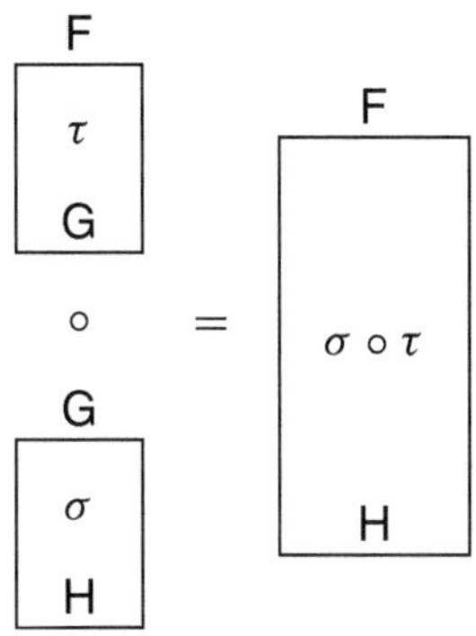

showing explicitly how the vertical line corresponds to composition. In these vertical compositions, we can indeed omit the underlying categories in the graphical representation, since there is no danger that compositions may not exist. The definition of vertical composition ensures that the respective domains and codomains of functors coincide for natural transformations used in compositions. As said before, in the case of endofunctors, it is only the information of the common underlying category that is hidden in a representation without explicitly writing out the categories as annotated to vertical lines.

In horizontal composition, there is similarly no danger of confusion in the case of using only endofunctors, but in the more general case, we obviously have to be more careful concerning what is hidden about underlying categories.

In order to define horizontal attachment of blocks expressing the horizontal composition, we first recall the following formulas (cf. [38, p. 269]):

$$\mathrm{B} \xrightarrow{\mathsf{H}} \mathrm{C} \underset{\mathsf{F'}}{\overset{\mathsf{F}}{\rightrightarrows}} \mathrm{D} \xrightarrow{\mathsf{K}} \mathrm{E}, \quad \eta\colon \mathsf{F} \to \mathsf{F}', \quad 1_{\mathsf{K}} \star \eta = \mathsf{K}\eta \quad \text{and} \quad \eta \star 1_{\mathsf{H}} = \eta\mathsf{H}. \tag{A.8}$$

Then for the horizontal composition of τ' followed by τ, denoted by the star product $\tau' \star \tau$, the following relation holds

$$\tau' \star \tau = (1_{\mathsf{G'}} \star \tau) \circ (\tau' \star 1_{\mathsf{F}}) = (\tau' \star 1_{\mathsf{G}}) \circ (1_{\mathsf{F'}} \star \tau) = \tau'\mathsf{G} \circ \mathsf{F}'\tau = \mathsf{G}'\tau \circ \tau'\mathsf{F}, \tag{A.9}$$

where the fourth equality is known as one of the five *Godement rules* (see also Sect. Notes in Chap. 1).

Note how the domain and codomain functors in $\tau' \star \tau\colon \mathsf{F}' \circ \mathsf{F} \to \mathsf{G}' \circ \mathsf{G}$ is visually denoted by the juxtaposition of two building blocks. We can then define the horizontal block attachment according to

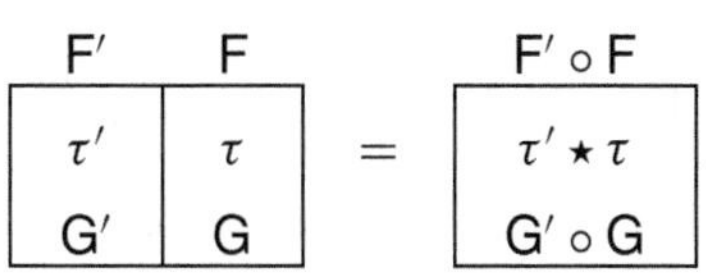

.

We may omit the sign for functorial composition, and simply write

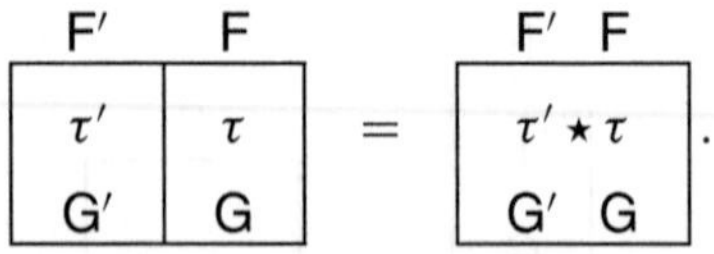

Note indeed how the juxtaposition order reflects the syntactic order of $\tau' \star \tau$, as enabled by the right-to-left reading of domain and codomain of functors. Note also how the middle vertical line shares the same category for the two boxes — i.e. mixing notation and using expressions and graphical representation we could write

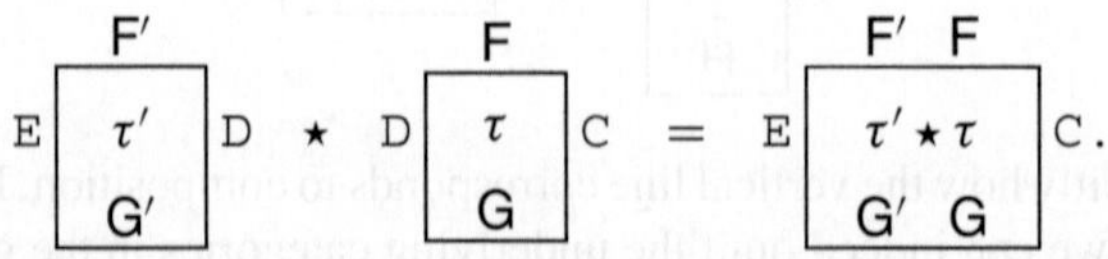

Now we proceed to represent more complex graphical representations involving vertical and horizontal composition of natural transformations. The star composition (see (A.9))

$$\tau' \star \tau = (\tau' \star 1_G) \circ (1_{F'} \star \tau) = (1_{G'} \star \tau) \circ (\tau' \star 1_F)$$

can pictorially be represented by

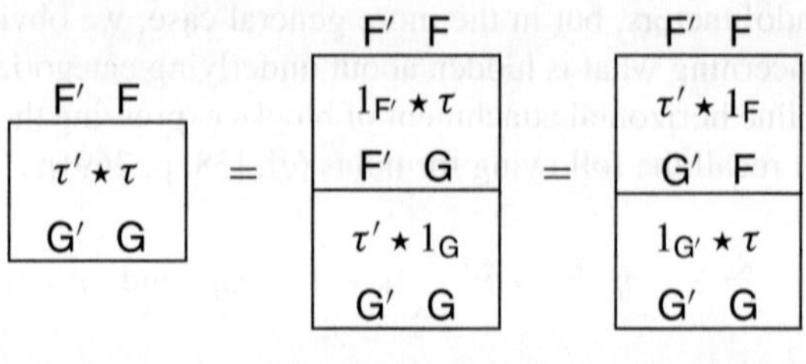

(A.10)

We now see how blocks can be attached vertically and horizontally in any order as long as the corresponding domains and codomains match. An equation like

$$1 \star (\sigma \circ \tau) = (1 \star \sigma) \circ (1 \star \tau)$$

can be written as

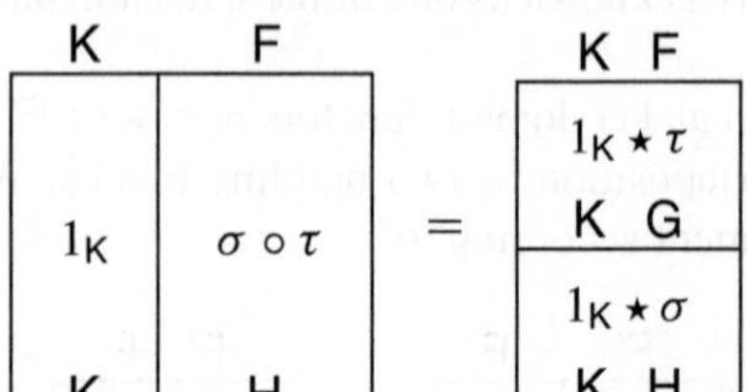

showing how a horizontality is equated with a verticality.

As a first example on how this graphical technique can be used to support proofs, we now see the following procedure

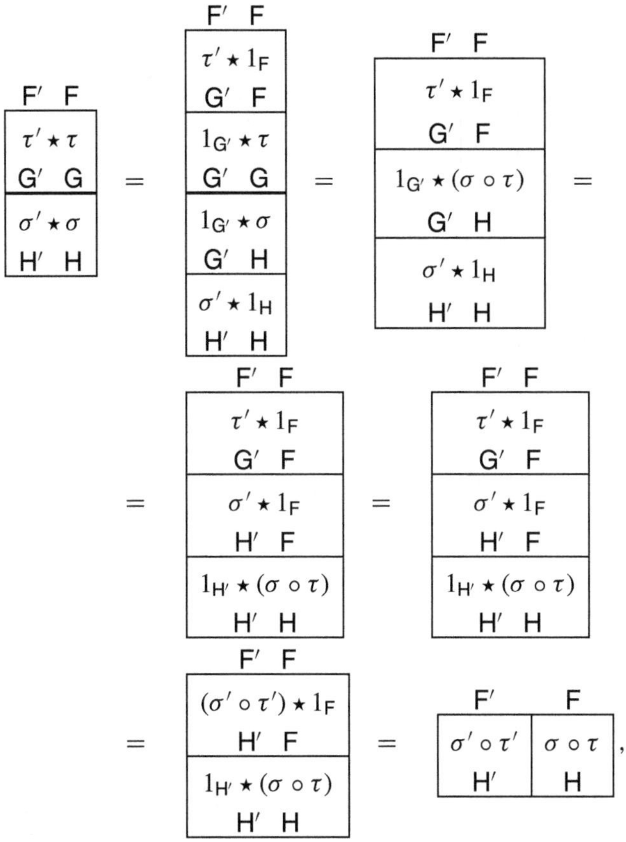

i.e. we have used the graphical representation to illuminate the proof of the *Interchange Law*

$$(\sigma' \circ \tau') \star (\sigma \circ \tau) = (\sigma' \star \sigma) \circ (\tau' \star \tau).$$

Graphically, the Interchange Law can be summarized as

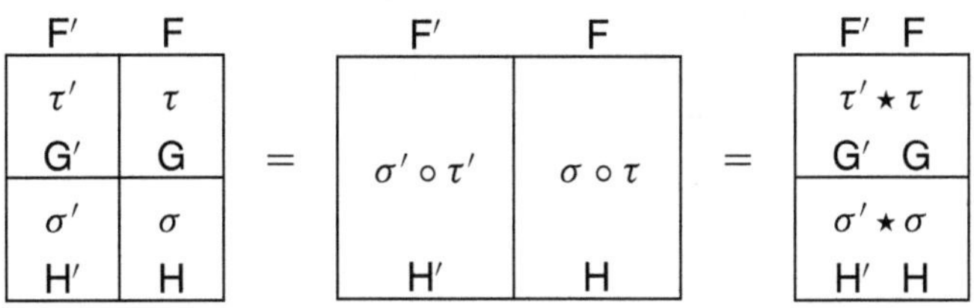

showing how blocks with particular positions can be attached vertically and horizontally in any order without altering the resulting transformation.

Monad conditions can also suitably be depicted in a similar fashion. For a monad $\mathbf{F} = (\mathsf{F}, \eta, \mu)$, the unit axiom and associativity axiom can be depicted as

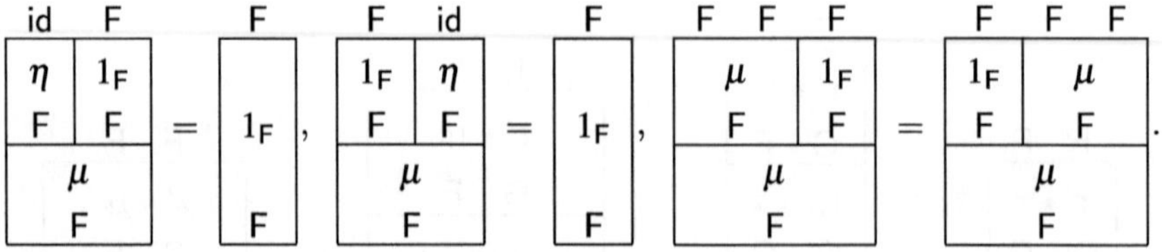

The equality

$$1_{\mathsf{F}} \circ 1_{\mathsf{F}} = 1_{\mathsf{F}}$$

can more explicitly be depicted as

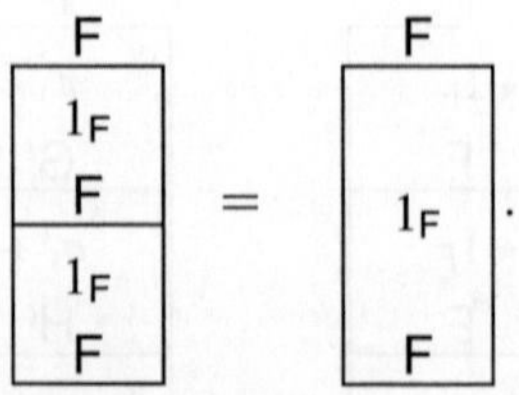

Note then how the identity transformation

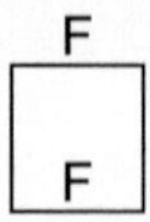

can be allowed to represent not just 1_{F}, but indeed also $1_{\mathsf{F}} \circ 1_{\mathsf{F}}$, $1_{\mathsf{F}} \circ 1_{\mathsf{F}} \circ 1_{\mathsf{F}}$, and so on. Similarly, since

$$1_{\mathsf{F}'} \star 1_{\mathsf{F}} = 1_{\mathsf{F}' \circ \mathsf{F}}$$

without explicitly using identity transformations can be depicted as

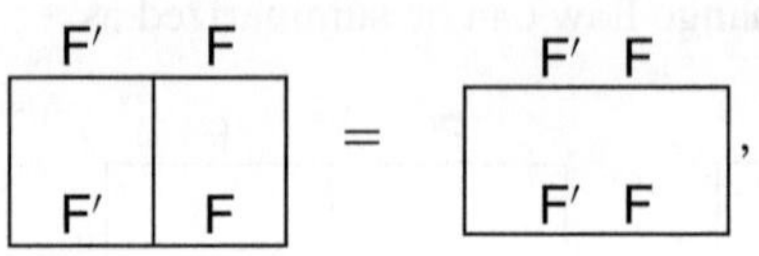

the representation

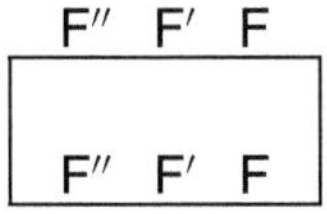

can also be depicted according to

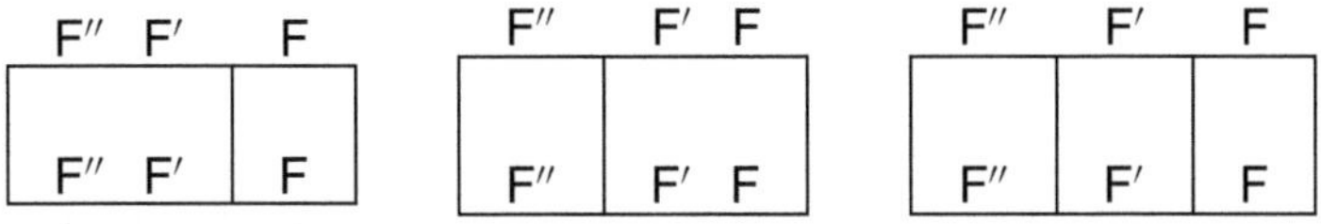

as required by the context where respective expressions of natural transformations appear in some proof steps.

In our graphical representations of proofs we apply exclusively the Interchange Law with the possible exception of given axioms or properties. For the sake of completeness we include here the depiction of the other Godement rules (cf. [38]).

Remark A.3.1. Among the other three Godement rules, for given functors

$$\mathsf{K}'\colon \mathrm{D} \to \mathrm{D}', \quad \mathsf{K}''\colon \mathrm{D}' \to \mathrm{D}'', \quad \mathsf{J}'\colon \mathrm{C}' \to \mathrm{C}, \quad \mathsf{J}''\colon \mathrm{C}'' \to \mathrm{C}',$$

and since the horizontal composition is associative, we have (cf. (A.8))

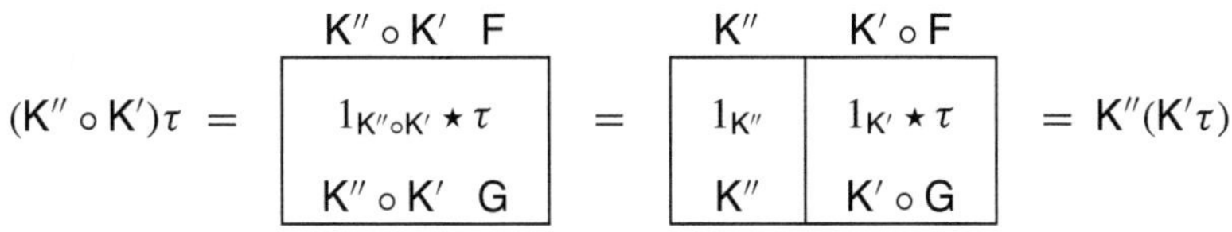

and

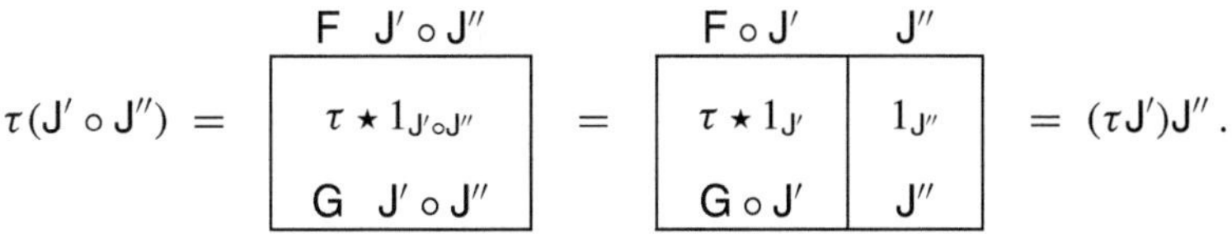

The final Godement rule is

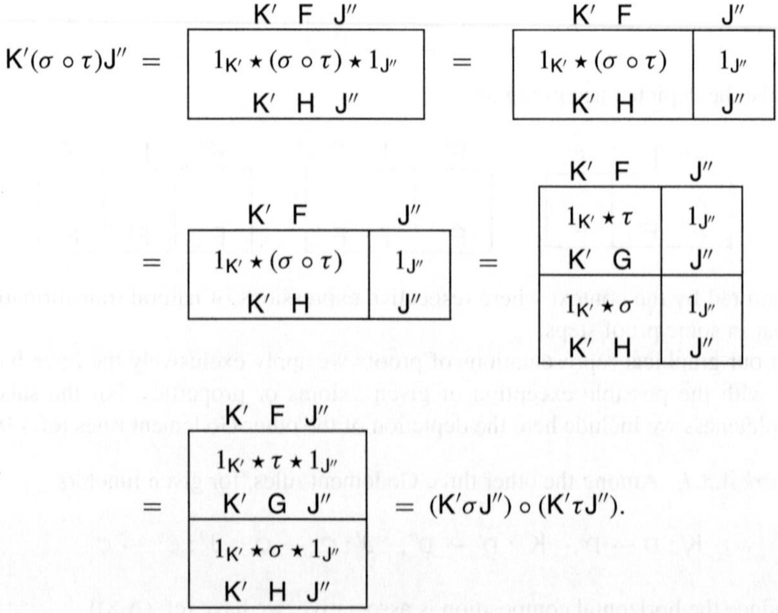

Now we finish Appendix A.3 by depicting the proof of Beck's Theorem (cf. Theorem 1.2.12) using graphical representations. First we establish the unit axiom. For this purpose we begin with the graphical representation of the diagrams (A) and (C) in Sect. 1.2:

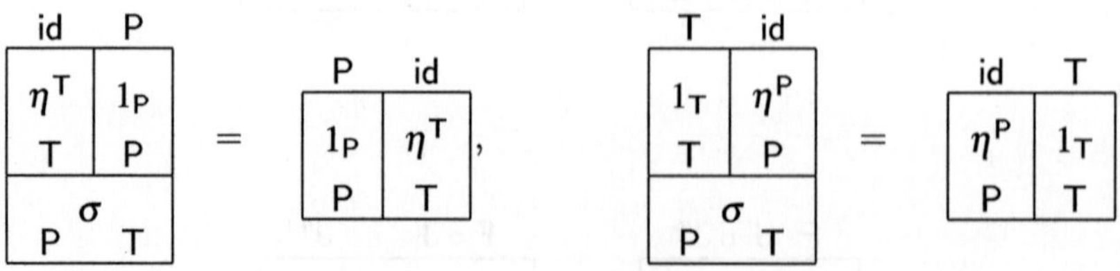

$$\mu\circ(\eta\star 1_{\mathsf{PT}}) =$$

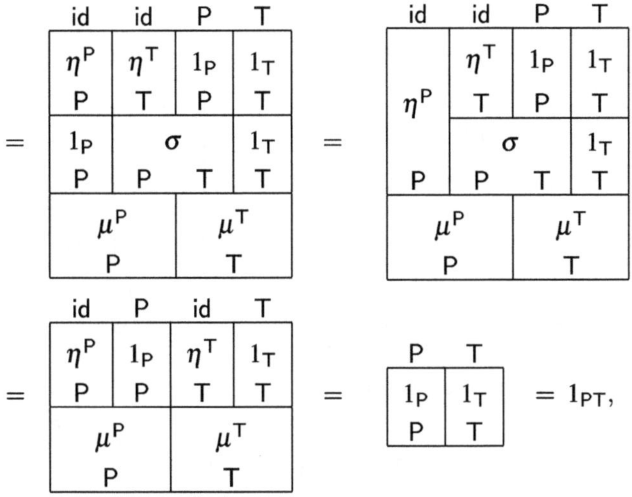

$$\mu\circ(1_{\mathsf{PT}}\star\eta) =$$

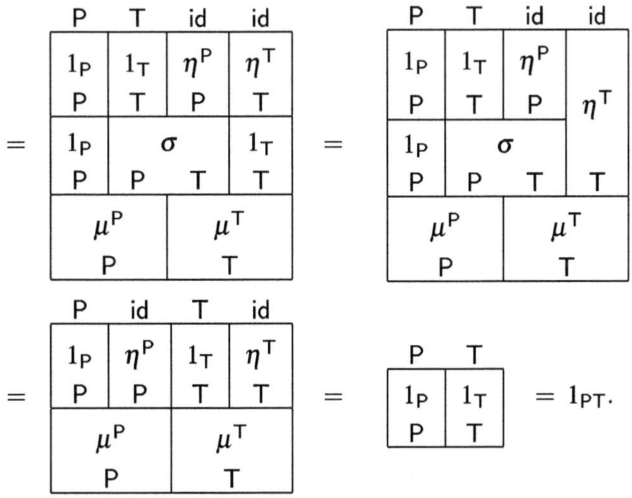

Secondly, we establish the associativity axiom. As a first step we give a graphical representation of the assertion of Lemma 1.2.13:

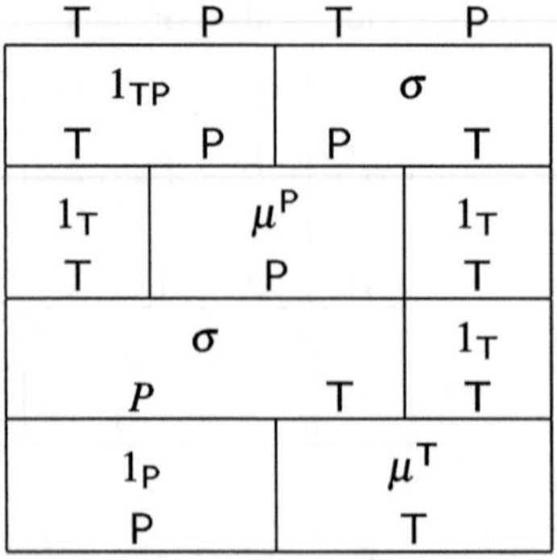

$=$

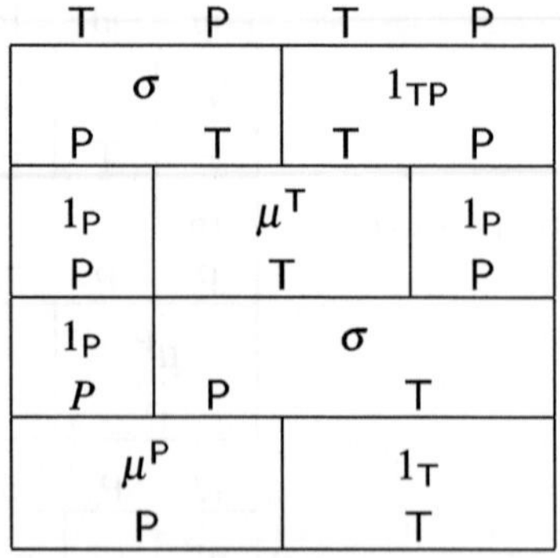

.

Then we can proceed as follows.

$\mu\circ(1_{PT}\star\mu) =$

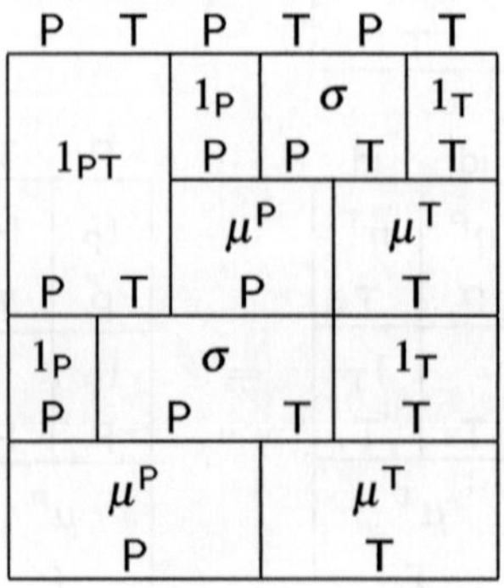

$=$

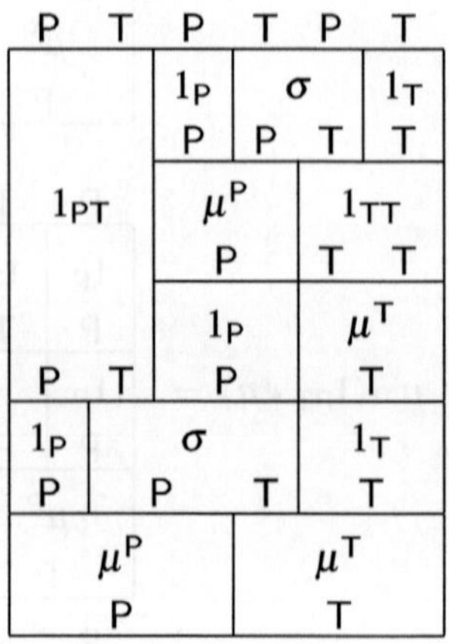

$=$

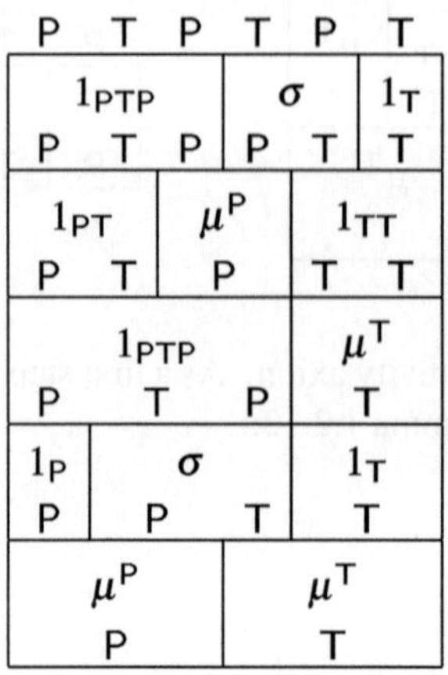

$=$

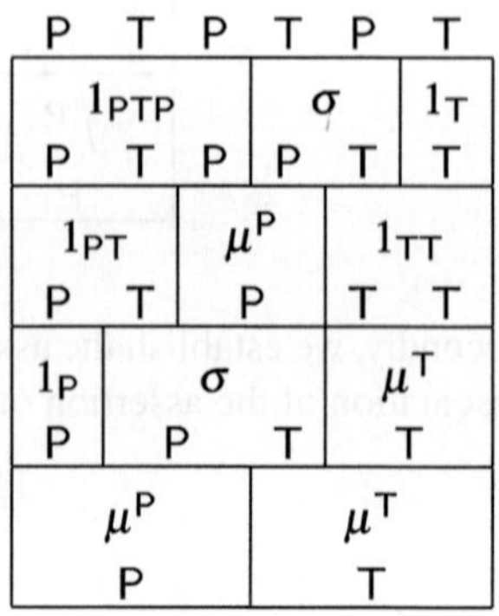

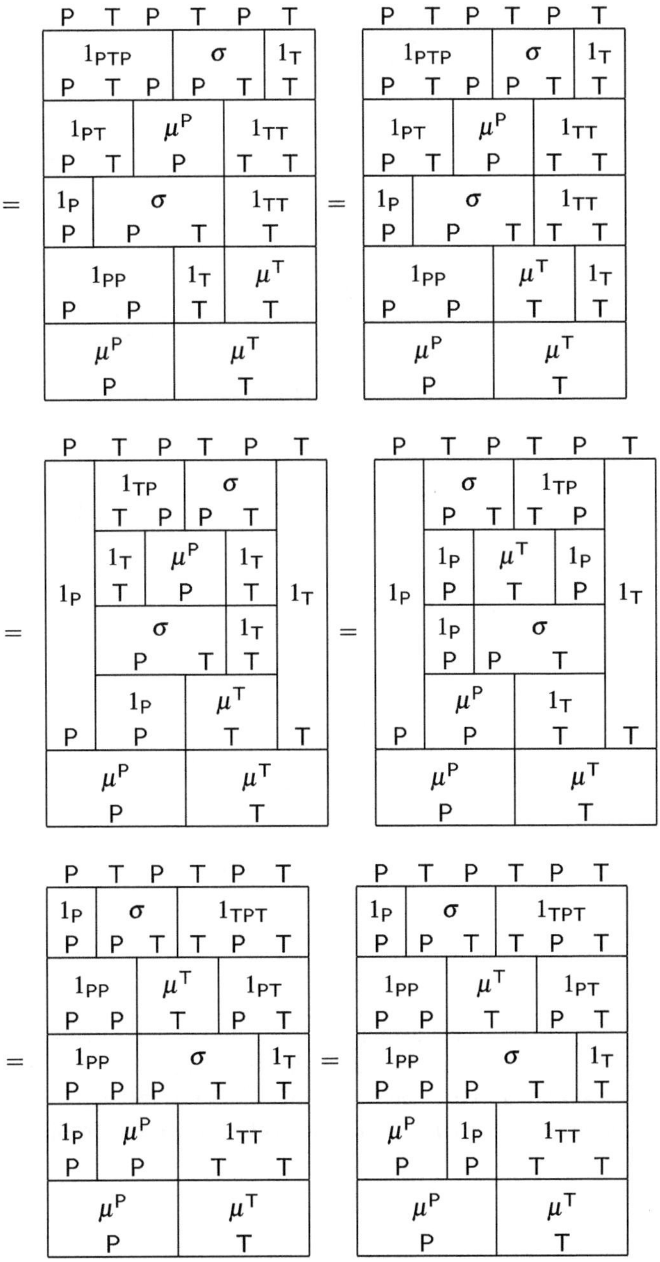

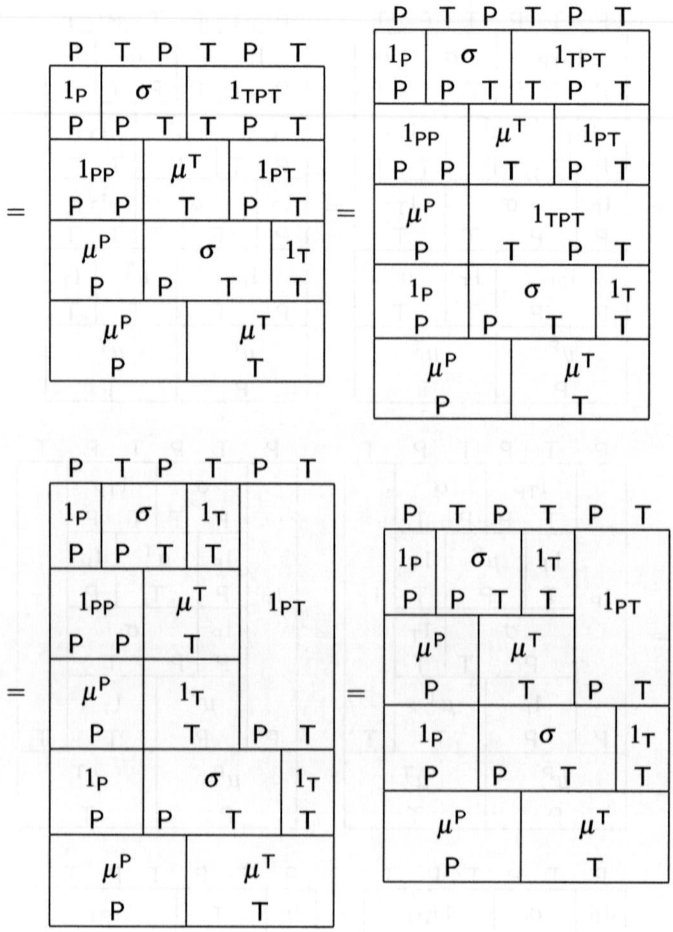

$= \mu \circ (\mu \star 1_{PT}).$

Comment. In order to prove the associativity axiom of composite monads we have already seen in Sect. 1.2 that Lemma 1.2.13 plays a strategic rôle. The graphical representation of this proof as presented above underlines this fact and shows explicitly that an application of Lemma 1.2.13 permits us to reverse the proof strategy after block 8.

References

1. Adámek, J.: Free algebras and automata realizations in the language of categories. Commentat. Math. Univ. Carol. **15**(4), 589–602 (1974)
2. Adámek, J., Herrlich, H., Strecker, G.E.: Abstract and Concrete Categories. Wiley, New York (1990)
3. Adámek, J., Trnková, V.: Automata and Algebras in Categories. Mathematics and Its Applications (East European Series), vol. 37. Kluwer, London (1990)
4. Arbib, M.A., Manes, E.G.: Adjoint machines, state behavior machines, and duality. J. Pure Appl. Algebra **6**(3), 313–344 (1975)
5. Arveson, W.: An Invitation to C^*-Algebra. Graduate Texts in Mathematics, vol. 39. Springer, New York, Heidelberg, Berlin (1976)
6. Banaschewski, B., Nelson, E.: Tensor products and bimorphisms. Can. Math. Bull. **19**, 385–402 (1976)
7. Bandelt, H.-J.: The tensor product of continuous lattices. Math. Z. **172**(1), 89–96 (1980)
8. Barr, M.: $*$-Autonomous Categories. Lecture Notes in Mathematics, vol. 752. Springer, Berlin, Heidelberg (1979)
9. Barr, M.: $*$-Autonomous categories revisited. J. Pure Appl. Algebra **111**, 1–20 (1996)
10. Beck, J.: Distributive laws. In: Eckmann, B. (ed.) Seminar on Triples and Categorical Homology Theory. Lecture Notes in Mathematics, vol. 80, pp. 119–140. Springer, Berlin, Heidelberg (1969) (Reprint in Theory Appl. Categ. **18**, 95-112 (2008))
11. Belluce, L.P.: Semi-simple and complete MV-algebras. Algebra Univ. **29**(1), 1–9 (1992)
12. Bénabou, J.: Catégories avec multiplication. C. R. Acad. Sci., Paris, Sér. I, Math. **256**, 1887–1890 (1963)
13. Birkhoff, G.: Rings of sets. Duke Math. J. **3**(3), 443–454 (1937)
14. Birkhoff, G.: Lattice Theory. Colloquium Publications, vol. XXV, 3rd edn. American Mathematical Society, Providence, Rhode Island (1995). eighth printing
15. Borceux, F.: Handbook of Categorical Algebra 1. Basic Category Theory. Encyclopedia of Mathematics and Its Applications, vol. 50. Cambridge University Press, Cambridge (1994)
16. Borceux, F.: Handbook of Categorical Algebra 2. Categories and Structures. Encyclopedia of Mathematics and Its Applications, vol. 51. Cambridge University Press, Cambridge (1994)
17. Budach, L., Hoehnke, H.-J.: Über eine einheitliche Begründung der Automatentheorie. Seminarbericht 1. Teil, Humboldt-Universität, Berlin (1969/70)
18. Budach, L., Hoehnke, H.-J.: Automaten und Funktoren. Akademie, Berlin (1975)
19. Burris, S., Sankappanavar, H.P.: A Course in Universal Algebra. Graduate Texts in Mathematics, vol. 78. Springer, New York (1981)

P. Eklund et al., *Semigroups in Complete Lattices*, Developments
in Mathematics 54, https://doi.org/10.1007/978-3-319-78948-4

20. Chang, C.C.: Algebraic analysis of many valued logics. Trans. Am. Math. Soc. **88**, 467–490 (1958)
21. Cignoli, R.L., d'Ottaviano, M., Mundici, D.: Algebraic Foundations of Many-Valued Reasoning. Trends in Logic, vol. 7. Kluwer, Boston (2000)
22. Dedekind, R.: XI Supplement. In: Dirichlet, P.G.L., Vorlesungen über Zahlentheorie. 4th edn., Vieweg, Braunschweig (1894)
23. Dijkstra, E.W.: A Discipline of Programming. Prentice-Hall Series in Automatic Computation. Prentice Hall Inc, Englewood Cliffs, N.J. (1976)
24. Dilworth, R.P.: Non-commutative residuated lattices. Trans. Am. Math. Soc. **46**(3), 426–444 (1939)
25. Ehresmann, C.: Catégorie des foncteurs types. Rev. Unión Math. Argent. **20**, 194–209 (1960)
26. Eilenberg, S., Moore, J.C.: Adjoint functors and triples. Ill. J. Math. **9**(3), 381–398 (1965)
27. Eklund, P., Galán, M.A., Medina, J., Ojeda Aciego, M., Valverde, A.: A graphical approach to monad composition. Electron. Notes Theor. Comput. Sci. **40**, 145–160 (2001)
28. Eklund, P., Höhle, U., Kortelainen, J.: A survey on the categorical term construction with applications. Fuzzy Sets and Syst. **298**, 128–157 (2016)
29. Ehrig, H., Kiermeier, K.-D., Kreowski, H.-J., Kühnel, W.: Universal Theory of Automata. A Categorical Approach. Teubner, Stuttgart (1974)
30. Ehrig, H., Pfender, M.: Kategorien und Automaten. De Gruyter, Berlin, New York (1972)
31. Erné, M., Koslowski, J., Melton, A., Strecker, G.E.: A primer on Galois connections. In: Proceedings in the 1991 Summer Conference on General Topology and Applications in Honor of Mary Ellen Rudin and Her Work, Ann. N. Y. Acad. Sci. **704**(1), 103–125 (1993)
32. Erné, M., Picado, J.: Tensor products and relation quantales. Algebra Univ. **78**, 461–487 (2017)
33. Fraser, G.A.: The semilattice tensor product of distributive lattices. Trans. Am. Math. Soc. **217**, 183–194 (1976)
34. Fuchs, L.: Teilweise geordnete algebraische Strukturen. Studia Mathematica, vol. 19. Vandenhoeck und Ruprecht, Göttingen (1966)
35. Galatos, N., Jipsen, P., Kowalski, T., Ono, H.: Residuated Lattices: An Algebraic Glimpse at Substructural Logics. Studies in Logic, vol. 151. Elsevier, Amsterdam, Boston (2007)
36. Gierz, G., Hofmann, K.H., Keimel, K., Lawson, J.D., Mislove, M., Scott, D.S.: A Compendium of Continuous Lattices. Springer, Berlin, Heidelberg, New York (1980)
37. Gierz, G., Hofmann, K.H., Keimel, K., Lawson, J.D., Mislove, M., Scott, D.S.: Continuous Lattices and Domains. Encyclopedia of Mathematics and Its Applications, vol. 93. Cambridge University Press, Cambridge (2003)
38. Godement, R.: Topologie Algébrique et Théorie des Faisceaux. Hermann, Paris (1958)
39. Goguen, J.A.: L-fuzzy sets. J. Math. Anal. Appl. **18**, 145–174 (1967)
40. Goguen, J.A.: Discrete-time machines in closed monoidal categories I. Quarterly Report No. 30, Institute for Computer Research, University of Chicago (1971); condensed. Bull. Am. Math. Soc. **78**, 777–783 (1972)
41. Goguen, J.A.: Minimal realization of machines in closed categories. Bull. Am. Math. Soc. **78**(5), 777–783 (1972)
42. Goguen, J.A.: Discrete-time machines in closed monoidal categories I. J. Comput. Syst. Sci. **10**(1), 1–43 (1975)
43. Gutiérrez García, J., Höhle, U., Kubiak, T.: Tensor products of complete lattices and their application in constructing quantales. Fuzzy Sets and Syst. **313**, 43–60 (2017)
44. Gutiérrez García, J., Mardones-Pérez, I., de Prada Vicente, M.A., Zhang, D.: Fuzzy Galois connections categorically. Math. Log. Q. **56**(2), 131–147 (2010)
45. Hall, B.C.: Lie Groups, Lie Algebra, and Representations. An Elementary Introduction. Graduate Texts in Mathematics, vol. 222, 2nd edn. Springer, New York (2015)
46. Hay, L.S.: Axiomatization of infinite-valued predicate calculus. J. Symb. Log. **28**(1), 77–86 (1963)
47. Herrlich, H., Strecker, G.E.: Category Theory. Allyn and Bacon, Boston (1973)

48. Höhle, U.: Commutative, residuated ℓ-monoids. In: Höhle, U., Klement, E.P. (eds.) Non-Classical Logics and Their Applications to Fuzzy Subsets. Theory and Decision Library Series B: Mathematical and Statistical Methods, vol. 32, pp. 53–106. Kluwer, Boston (1995)
49. Höhle, U.: Many Valued Topology and Its Applications. Kluwer, Boston (2001)
50. Höhle, U.: Categorical foundations of topology with applications to quantaloid enriched topological spaces. Fuzzy Sets Syst. **256**, 166–210 (2014)
51. Höhle, U.: Topological representation of right-sided and idempotent quantales. Semigroup Forum **90**(3), 648–659 (2015)
52. Höhle, U.: Prime elements of non-integral quantales and their applications. Order **32**(3), 329–346 (2015)
53. Höhle, U.: Fuzzy sets as two-sorted algebras. Fuzzy Sets and Syst. **281**, 155–167 (2015)
54. Höhle, U., Kubiak, T.: Many valued topology and lower semicontinuity. Semigroup Forum **75**(1), 1–17 (2007)
55. Höhle, U., Kubiak, T.: A non-commutative and non-idempotent theory of quantale sets. Fuzzy Sets Syst. **166**, 1–43 (2011)
56. Hopcroft, J.E., Motwani, R., Ullman, J.D.: Introduction to Automata Theory, Languages, and Computation, 3rd edn. Addison-Wesley, Reading, Massachusetts (2006)
57. Huber, P.J.: Homotopy theory in general categories. Math. Ann. **144**, 361–385 (1961)
58. Johnstone, P.T.: Stone Spaces. Cambridge Studies in Advanced Mathematics, vol. 3. Cambridge University Press, Cambridge (1982)
59. Joyal, A., Tierney, M.: An Extension of the Galois Theory of Grothendieck. Memoirs of the American Mathematical Society, vol. 51 No. 309. American Mathematical Society, Providence, Rhode Island (1984)
60. Kadison, R.V., Ringrose, J.R.: Fundamentals of the Theory of Operator Algebras. Volume I Elementary Theory. Graduate Studies in Mathematics, vol. 15. American Mathematical Society, Providence, Rhode Island (1997)
61. Kadison, R.V., Ringrose, J.R.: Fundamentals of the Theory of Operator Algebras. Volume II Advanced Theory. Graduate Studies in Mathematics, vol. 16. American Mathematical Society, Providence, Rhode Island (1997)
62. Kan, D.M.: Adjoint functors. Trans. Am. Math. Soc. **87**(2), 294–329 (1958)
63. Kelly, G.M.: Basic Concepts of Enriched Category Theory. London Mathematical Society Lecture Notes Series, vol. 64. Cambridge University Press, Cambridge (1982)
64. Kleene, S.C.: Representation of events in nerve nets and finite automata. Project RAND Research Memorandum RM-704, 101 pages, 15 Dec 1951
65. Kleene, S.C.: Representation of events in nerve nets and finite automata. In: Shannon, C.E., McCarthy, J. (eds.) Automata Studies, pp. 3–42. Princeton University Press, Princeton, N.J. (1956)
66. Kleisli, H.: Every standard construction is induced by a pair of adjoint functors. Proc. Am. Math. Soc. **16**, 544–546 (1965)
67. Klement, E.P., Mesiar, R., Pap, E.: Triangular Norms. Trends in Logic, vol. 8. Kluwer, Boston (2000)
68. Krull, W.: Axiomatische Begründung der allgemeinen Idealtheorie. Sitzungsber. Phys.-med. Sozietät zu Erlangen **56**, 47–63 (1924)
69. Krull, W.: Zur Theorie der zweiseitigen Ideale in nichtkommutativen Bereichen. Math. Z. **28**, 481–503 (1928)
70. Kruml, D., Resende, P.: On quantales that classify C^*-algebras. Cah. Topologie Géom. Différ. Catégoriques **XLV**(4), 287–296 (2004)
71. Lai, H., Zhang, D.: Complete and directed complete Ω-categories. Theor. Comput. Sci. **388**(1–3), 1–25 (2007)
72. Mac Lane, S.: Natural associativity and commutativity. Rice Univ. Stud. **49**, 28–46 (1963)
73. Mac Lane, S.: Categories for the Working Mathematician. Graduate Texts in Mathematics, vol. 5, 2nd edn. Springer, Berlin, Heidelberg, New York (1998)
74. Manes, E.G.: Algebraic Theories. Graduate Texts in Mathematics, vol. 26. Springer, Berlin, Heidelberg, New York (1976)

75. Marmolejo, F., Rosebrugh, R.D., Wood, R.J.: A basic distributive law. J. Pure Appl. Algebra **168**(2–3), 209–226 (2002)
76. McCulloch, W.S., Pitts, W.: A logical calculus of the ideas imminent in nervous activity. Bull. Math. Biophys. **5**(4), 115–133 (1943)
77. Meise, R., Vogt, D.: Introduction to Functional Analysis. Oxford Graduate Texts in Mathematics, vol. 2. Oxford University Press, Oxford, New York (1997)
78. Mikkelsen, C.J.: Lattice Theoretic and Logical Aspects of Elementary Topoi. Matematisk Institut, Aarhus Universitet 1976 (Various Publications Series No. 25)
79. Mostert, P.S., Shields, A.L.: On the structure of semigroups on compact manifolds with boundary. Ann. Math. 2nd series **65**(1), 117–143 (1957)
80. Mowat, D.G.: A Galois problem for mappings. Bull. Am. Math. Soc. **74**, 1095–1097 (1968)
81. Mowat, D.G.: A Galois problem for mappings. Dissertation, University of Waterloo, Canada (1968)
82. Mulvey, C.J.: &. Rend. Circ. Mat. Palermo **2**(Suppl. 12), 99–104 (1986)
83. Mulvey, C.J., Pelletier, J.W.: A quantisation of the calculus of relations. In: Seely, R.A.G. (ed.) CMS Conference Proceedings, vol. 13, pp. 345–360. American Mathematical Society, Providence, Rhode Island (1992)
84. Mulvey, C.J., Pelletier, J.W.: On the quantisation of points. J. Pure Appl. Algebra **159**(2–3), 231–295 (2001)
85. Nelson, E.: Galois connections as left adjoint maps. Commentat. Math. Univ. Carol. **17**(3), 523–541 (1976)
86. Nkuimi-Jugnia, C.: Amalgamation property and epimorphisms in the category of modules over a quantale. Séminaire 303. In: Séminaires de mathématique (Prépublications), Département de mathématique. Université Catholique de Louvain (2000)
87. Ore, O.: Galois connections. Trans. Am. Math. Soc. **55**, 493–513 (1944)
88. Ovchinnikov, S.V.: Representation of transitive relations. In: Skala, H.J., et al. (eds.) Aspects of Vagueness. Theory and Decision Library: An International Series in the Philosophy and Methodology of the Social and Behavioral Sciences, vol. 39, pp. 105–118. Reidel, Dordrecht (1984)
89. Paalman-de Miranda, A.B.: Topological Semigroups. Math. Centrum, Amsterdam (1964)
90. Paseka, J.: A note on nuclei of quantale modules. Cah. Topologie Géom. Différ. Catégoriques XLIII **43**(1), 19–34 (2002)
91. Paseka, J., Kruml, D.: Algebraic and categorical aspects of quantales. In: Hazewinkel, M. (ed.) Handbook of Algebra, vol. 5, pp. 323–362 (2008)
92. Paseka, J., Rosický, J.: Quantales. In: Coecke, B., Moore, D., Wilce, A. (eds.) Current Research in Operational Quantum Logic: Algebras, Categories and Languages. Fundamental Theories in Physics, vol. 111, pp. 245–262. Kluwer, Boston (2000)
93. Pelletier, J.W., Rosický, J.: Simple involutive quantales. J. Algebra **195**, 367–386 (1997)
94. Picado, J.: The quantale of Galois connections. Algebra Univ. **52**(4), 527–540 (2004)
95. Pu, Q., Zhang, D.: Preordered sets valued in a GL-monoid. Fuzzy Sets Syst. **187**, 1–32 (2012)
96. Raney, G.N.: A subdirect-union representation for completely distributive complete lattices. Proc. Am. Math. Soc. **4**(4), 518–522 (1953)
97. Resende, P.: Sup-lattice 2-forms and quantales. J. Algebra **276**, 143–167 (2004)
98. Resende, P.: Étale groupoids and their quantales. Adv. Math. **208**(1), 147–209 (2007)
99. Rodabaugh, S.E.: Applications of localic separation axioms, compactness axioms, representations and compactifications to poslat topological spaces. Fuzzy Sets Syst. **73**, 53–87 (1995)
100. Rose, A., Rosser, J.B.: Fragments of many-valued statement calculi. Trans. Am. Math. Soc. **87**(1), 1–53 (1958)
101. Rosenthal, K.I.: Quantales and Their Applications. Pitman Research Notes in Mathematics, vol. 234. Longman Scientific & Technical, Harlow (1990)
102. Rosický, J.: Multiplicative lattices and frames. Acta Math. Hung. **49**(3–4), 391–395 (1987)
103. Rosický, J.: Characterizing spatial quantales. Algebra Univ. **34**(2), 175–178 (1995)
104. Schweizer, B., Sklar, A.: Probabilistic Metric Spaces. North-Holland, Amsterdam (1983)
105. Schubert, H.: Categories. Springer, Berlin, Heidelberg (1972)

106. Shmuely, Z.: The structure of Galois connections. Pac. J. Math. **54**(2), 209–225 (1974)
107. Stubbe, I.: Categorical structures enriched in a quantaloid: tensored and cotensored categories. Theory Appl. Categ. **16**(14), 283–306 (2006)
108. Stubbe, I.: The double power monad is the composite of the power monad. Fuzzy Sets Syst. **313**, 25–42 (2017)
109. Ward, M., Dilworth, R.P.: Residuated lattices. Trans. Am. Math. Soc. **45**, 335–354 (1939)
110. Wigner, D.: Two notes on frames. J. Aust. Math. Soc. Ser. A **28**(3), 257–268 (1979)
111. Zadeh, L.A.: Fuzzy sets. Inf. Control **8**(3), 338–353 (1965)
112. Zadeh, L.A.: Similarity relations and fuzzy orderings. Inf. Sci. **3**(2), 177–200 (1971)
113. Zadeh, L.A.: The concept of a linguistic variable and its application to approximate reasoning I, II, III. Inf. Sci. **8**(3), 199–249; **8**(4), 301–357; **9**(1), 43–80 (1975)

Index

P. Eklund et al., *Semigroups in Complete Lattices*, Developments in Mathematics 54, https://doi.org/10.1007/978-3-319-78948-4

MIX
Papier aus verantwortungsvollen Quellen
Paper from responsible sources
FSC® C105338

If you have any concerns about our products,
you can contact us on
ProductSafety@springernature.com

In case Publisher is established outside the EU,
the EU authorized representative is:
Springer Nature Customer Service Center GmbH
Europaplatz 3, 69115 Heidelberg, Germany

Printed by Libri Plureos GmbH
in Hamburg, Germany